W9-BFQ-077

PROBABILITY, RANDOM PROCESSES, AND ESTIMATION THEORY FOR ENGINEERS

Henry Stark
John W. Woods

PROBABILITY, RANDOM PROCESSES, AND ESTIMATION THEORY FOR ENGINEERS

PRENTICE-HALL, Englewood Cliffs, New Jersey 07632

Library of Congress Cataloging-in-Publication Data

Stark, Henry. (date)
Probability, random processes, and estimation theory for engineers.

Includes bibliographies and index.
1. Probabilities. 2. Stochastic processes.
3. Estimation theory. I. Woods, John W. (John William).
II. Title.
TA340.S7 1986 519.2 85-16882
ISBN 0-13-711706-X

Editorial/production supervision and
interior design: Marlys Lehmann
Cover design: Ben Santora
Manufacturing buyer: Rhett Conklin

Printed in the United States of America

10 9 8 7 6 5 4

ISBN 0-13-711706-X 025

Prentice-Hall International (UK) Limited, *London*
Prentice-Hall of Australia Pty. Limited, *Sydney*
Prentice-Hall Canada Inc., *Toronto*
Prentice-Hall Hispanoamericana, S.A., *Mexico*
Prentice-Hall of India Private Limited, *New Delhi*
Prentice-Hall of Japan, Inc., *Tokyo*
Prentice-Hall of Southeast Asia Pte. Ltd., *Singapore*
Editora Prentice-Hall do Brasil, Ltda., *Rio de Janeiro*
Whitehall Books Limited, *Wellington, New Zealand*

To
Alice, Larry, and Rich
and
Harriet, Anne, and Chris

CONTENTS

PREFACE

This book grew out of course notes used by us to teach two one-semester courses on probability and random processes at Rensselaer Polytechnic Institute (RPI). The probability course at RPI is required of all students in the Computer and Systems Engineering Program and is a highly recommended elective for many others. Most of the students taking the course are engineering or science majors in their junior year. Seniors and many first-year graduate students take the course for credit as well. The emphasis is on introducing fundamental principles and developing skills to solve problems.

The random processes course is typically taken by first-year graduate students and is designed to give students the needed background to take more advanced courses in communications, signal processing, controls, robotics, large-scale systems, and physics-related phenomena. As can be seen, the material goes considerably beyond elementary input-output relations for linear shift-invariant systems. Our experience has been that the present level of the course also gives the student the mathematical background in stochastic processes to pursue M.S.- or Ph.D.-level research in these fields.

In writing this book, we took an "integrated" view of probability and random processes. We felt that probability and random variables were so closely linked to estimation and decision theory that we made a strong attempt at connecting these subjects. Furthermore, we both shared a strong feeling that describing sophisticated applications of the theory of probability and random processes would be, from a pedagological viewpoint, beneficial for students. Hence, we have included applications to pattern recognition, linear systems, parameter estimation, controls, and communication theory. We also felt that in the age of the computer, the theory of random discrete-time (space) sequences should be given as much weight as continuous-time (space) waveforms. For this reason we devoted an entire chapter to this theory.

In reviewing our own experience, we found that certain closely related topics were often not covered in first courses because the material was deemed too advanced and was then not covered in later courses because the material was considered too basic. Such is often the case with covariance matrices, maximization of quadratic forms, least-squares estimation, and still others. Consequently, we included two chapters dealing with these topics.

The normal use of this book would be as follows: For a first course in probability at, say, the junior or senior year, a reasonable goal is to cover Chapters 1 through 3 with a little material from Chapter 4. Starred sections are often not covered. Homework assignments range from five to ten problems per week. For a first-year graduate course on probability in which the students have had some prior exposure to the subject, Chapters 1 to 3 could be covered in much less time, leaving time for covering Chapters 4 and 5. The material in Chapters 4 and 5 is essential for the statistical-pattern recognition course at RPI. The Gauss-Markov theorem for estimating unknown parameters from measurements corrupted by noise is discussed at length as is the problem of finding the best one-dimensional subspace for separating two classes of statistical objects with well-defined means and covariances.

Chapters 6 through 10 along with Section 5.6 provide the material for a first course in random processes. Beginning with random sequences (Chapter 6), the remaining chapters cover, respectively: random processes; mean-square calculus; stationary processes and sequences; and advanced estimation theory. In general, there is more material than can be covered in a one-semester course. Unless the class is very mature mathematically, Kalman and Wiener filtering are omitted as is Martingale theory upon a first reading.

When course time is reduced as, for example, it might be in schools using the quarter system, it is important that Chapters 6, 7, and 9 be covered, essentially in that order, before teaching material from Chapters 8 and 10. Ideally, it would be preferable to cover Chapter 8 before Chapter 9. But, as the latter covers the basic input-output relations for linear systems excited by random signals, we suggest that it be taught out of normal sequence if the instructor feels there is a danger of running out of time.

We acknowledge a tremendous debt of gratitude to our teachers and students. Thanks are due to the administration of Rensselaer Polytechnic Institute, which graciously recognized the creation of such a book as a scholarly activity. The excellent typing skills and cheerful demeanor of Priscilla

Magilligan were crucial to this effort. One of the authors, Henry Stark, is also very grateful to Peggy and Mark Curchack for providing a warm and comfortable home environment while he worked on the book while on leave at the University of Pennsylvania.

Finally, a project like this can be completed only with the cooperation of the authors' spouses. To Alice and Harriet, we extend our gratitude.

HENRY STARK

JOHN W. WOODS

Troy, New York

1

INTRODUCTION TO PROBABILITY

1.1 INTRODUCTION: WHY STUDY PROBABILITY?

One of the most frequent questions posed by beginning students of probability is: "Is anything truly random and if so how does one differentiate between the truly random and that which, because of a lack of information, is treated as random but really isn't?" First, regarding the question of truly random phenomena: "Do such things exist?" A theologian might state the case as follows: "We cannot claim to know the Creator's mind, and we cannot predict His actions because He operates on a scale too large to be perceived by man. Hence there are many things we shall never be able to predict no matter how refined our measurements."

At the other extreme from the cosmic scale is what happens at the atomic level. Our friends the physicists speak of such things as the *probability* of an atomic system being in a certain state. The uncertainty principle says that, try as we might, there is a limit to the accuracy with which the position and momentum can be simultaneously ascribed to a particle. Both quantities are fuzzy and indeterminate.

Many, including some of our most famous physicists, believe in an essential randomness of nature. Eugen Merzbacher in his well-known textbook on quantum mechanics [1–1] writes:

> The probability doctrine of quantum mechanics asserts that the indetermination, of which we have just given an example, is a property inherent in nature and not merely a profession of our temporary ignorance from which we expect to be relieved by a future better and more complete theory. The conventional interpretation thus denies the possibility of an ideal theory which would

> encompass the present quantum mechanics but would be free of its supposed defects, the most notorious "imperfection" of quantum mechanics being the abandonment of strict classical determinism.

But the issue of determinism versus inherent indeterminism need never even be considered when discussing the validity of the probabilistic approach. The fact remains that there is, quite literally, a nearly uncountable number of situations where we cannot make any categorical deterministic assertion regarding a phenomenon because we cannot measure all the contributing elements. Take, for example, predicting the value of the current $i(t)$ produced by a thermally excited resistor R: Conceivably, we might accurately predict $i(t)$ at some instant t in the future if we could keep track, say, of the 10^{23} or so excited electrons moving in each other's magnetic fields and setting up local field pulses that eventually all contribute to producing $i(t)$. Such a calculation is quite inconceivable, however, and therefore we use a probabilistic model rather than Maxwell's equations to deal with resistor noise. Similar arguments can be made for predicting weather, the outcome of a coin toss, the time to failure of a computer, and many other situations.

Thus to conclude: Regardless of which position one takes, that is, determinism versus indeterminism, we are forced to use probabilistic models in the real world because we do not know, cannot calculate, or cannot measure all the forces contributing to an effect. The forces may be too complicated, too numerous, or too faint.

Probability is a mathematical model to help us study physical systems in an *average sense.* Thus we cannot use probability in any meaningful sense to answer questions such as: "What is the probability that a comet will strike the earth tomorrow?" or "What is the probability that there is life on other planets?"†

R. A. Fisher and R. van Mises, in the first third of the twentieth century, were largely responsible for developing the groundwork of modern probability theory. The modern axiomatic treatment upon which this book is based is largely the result of the work by Andrei N. Kolmogorov [1–2].

1.2 THE DIFFERENT KINDS OF PROBABILITY

There are essentially four kinds of probability. We briefly discuss them here.

A. Probability as Intuition

This kind of probability deals with judgments based on intuition. Thus "She will probably marry him," and "He probably drove too fast," are in this category. A mathematical theory dealing with intuitive probability was developed by B. O. Koopman [1–3]. However, we shall not discuss this subject in this book.

† Nevertheless, certain evangelists deal with this question rather fearlessly, and even a popular astronomer has come up with a figure for this probability. However, whatever probability system these people use, it is not the system that we shall discuss in this book.

B. Probability as the Ratio of Favorable to Total Outcomes (Classical Theory)

In this approach, which is not experimental, the probability of an event is computed *a priori*† by counting the number of ways N_E that E can occur and forming the ratio N_E/N where N is the number of all possible outcomes, that is, the number of all alternatives to E plus N_E. An important notion here is that all outcomes are equally likely. Since equally likely is really a way of saying equally probable, the reasoning is somewhat circular. Suppose we throw a pair of unbiased dice and ask what is the probability of getting a seven? We partition the outcome into 36 equally likely outcomes as shown in Table 1.2–1 where each entry is the sum of the numbers on the two dice.

TABLE 1.2–1 Outcomes of Throwing Two Dice

		1st die					
		1	2	3	4	5	6
2nd die	1	2	3	4	5	6	7
	2	3	4	5	6	7	8
	3	4	5	6	7	8	9
	4	5	6	7	8	9	10
	5	6	7	8	9	10	11
	6	7	8	9	10	11	12

The total number of outcomes is 36 if we keep the dice distinct. The number of ways of getting a seven is $N_7 = 6$. Hence

$$P[\text{getting a seven}] = \tfrac{6}{36} = \tfrac{1}{6}.$$

Example 1.2–1: Throw a fair coin twice (note that since no physical experimentation is involved, there is no problem in postulating an ideal "fair coin"). The possible outcomes are HH, HT, TH, TT. The probability of getting at least one tail T is computed as follows: With E denoting the event of getting at least one tail, the event E is the set of outcomes

$$E = \{HT, TH, TT\}.$$

Thus E occurs whenever the outcome is HT or TH or TT. The number of elements in E is $N_E = 3$; the number of all outcomes, N, is four. Hence

$$P[\text{at least one } T] = \frac{N_E}{N} = \tfrac{3}{4}.$$

The classical theory suffers from at least two significant problems: (1) It cannot deal with outcomes that are not equally likely; and (2) it cannot handle uncountably infinite outcomes without ambiguity (see the example by Athanasios Papoulis [1–4]). Nevertheless, in those problems where it is impractical to actually determine the outcome probabilities by experimentation and where, because of symmetry considerations, one can indeed argue equally likely outcomes the classical theory is useful.

† *A priori* means relating to reasoning from self-evident propositions or presupposed by experience. *A posteriori* means relating to reasoning from observed facts.

Historically, the classical approach was the predecessor of Richard Von Mises' [1–5] relative frequency approach developed in the 1930s.

C. Probability as a Measure of Frequency of Occurrence

The relative-frequency approach to defining the probability of an event E is to perform an experiment n times. The number of times that E appears is denoted by n_E. Then it is tempting to define the probability of E occurring by

$$P[E] = \lim_{n\to\infty} \frac{n_E}{n}. \tag{1.2–1}$$

Quite clearly since $n_E \le n$, we must have $0 \le P[E] \le 1$. One difficulty with this approach is that we can never perform the experiment an infinite number of times so that we can only estimate $P[E]$ from a finite number of trials. Secondly, we *postulate* that n_E/n approaches a limit as n goes to infinity. But consider flipping a fair coin 1000 times. The likelihood of getting exactly 500 heads is very small; in fact, if we flipped the coin 10,000 times, the likelihood of getting exactly 5000 heads is even smaller. As $n \to \infty$, the event of observing exactly $n/2$ heads becomes vanishingly small. Yet our intuition demands that $P[\text{head}] = \frac{1}{2}$ for a fair coin. Suppose we choose a $\delta > 0$; then we shall find experimentally that if the coin is truly fair, the number of times that

$$\left| \frac{n_E}{n} - \frac{1}{2} \right| > \delta \tag{1.2–2}$$

as n becomes large, becomes very small. Thus although it is very unlikely that at any stage of this experiment, especially when n is large, n_E/n is exactly $\frac{1}{2}$, this ratio will nevertheless hover around $\frac{1}{2}$, and the number of times it will make significant excursion away from the vicinity of $\frac{1}{2}$ according to Equation 1.2–2 becomes very small indeed.

Despite these problems with the frequency definition of probability, the relative-frequency concept is essential in applying the probability theory to the physical world.

D. Probability Based on an Axiomatic Theory

This is the approach followed in most modern textbooks on the subject. To develop it we must introduce certain ideas, especially those of a random experiment, a sample description space, and an event. Briefly stated, a random experiment is simply an experiment in which the outcomes are nondeterministic, that is, probabilistic. Hence the word *random* in *random experiment.* The *sample description space* is the set of all outcomes of the experiment. An *event* is a subset of the sample description space that satisfies certain constraints. In general, however, almost any subset of the sample description space is an event.

These notions are refined in the next two sections.

1.3 SETS, FIELDS, AND EVENTS

A set is a collection of objects, either concrete or abstract. As example of a concrete set is the set of all New York residents whose height equals or exceeds 6 feet. A subset of a set is a collection that is contained within the larger set. Thus the set of all New York City residents whose height is between 6 and $6\frac{1}{2}$ feet is a subset of the previous set. In the study of probability we are particularly interested in the set of all outcomes of a random experiment and subsets of this set. We sometimes denote the experiment by the symbol $\mathscr{H}$ and the set of all outcomes by Ω. The set Ω is called the *sample space* or *sample description space* of the random experiment $\mathscr{H}$. Subsets of Ω are called *events*. Since any set is a subset of itself, Ω is itself an event. In particular Ω is called the *certain event*. Thus Ω is used to denote two objects: the set of all elementary outcomes of a random event and the certain event. We shall see that using the same symbol for both objects is entirely consistent. A little later we shall be somewhat more precise as to our definition of events.

Examples of Sample Spaces

Example 1.3–1: The experiment consists of flipping a coin once. Then $\Omega = \{H, T\}$ where H is a head and T is a tail.

Example 1.3–2: The experiment consists of flipping a coin twice. Then $\Omega = \{HH, HT, TH, TT\}$. A typical subset of Ω is $\{HH, HT, TH\}$; it is the event of getting at least one head in two flips.

Example 1.3–3: The experiment consists of choosing a person at random and counting the hairs on his or her head. Then

$$\Omega = \{0, 1, 2, \ldots, 10^7\}$$

that is, the set of all nonnegative integers up to 10^7, it being assumed that no human head has more than 10^7 hairs.

Example 1.3–4: The experiment consists of determining the age, to the nearest year, of each member of a married couple chosen at random. Then with x denoting the age of the man and y denoting the age of the woman, Ω is described by

$$\Omega = \{\text{2-tuples } (x, y)\text{: } x \text{ any integer in 10–200;}$$
$$y \text{ any integer in 10–200}\}.$$

Note that in Example 1.3–4 we have assumed that no human lives beyond 200 years and that no married person is ever less than ten years old. Similarly, in Example 1.3-1, we assumed that the coin never lands on edge. If the latter is a possible outcome, it must be included in Ω in order for it to denote the set of *all* outcomes as well as the certain event.

Example 1.3–5: The experiment consists of observing the angle of deflection of a nuclear particle in an elastic collision. Then

$$\Omega = \{\theta : -\pi \leq \theta \leq \pi\}.$$

An example of a subset of Ω is

$$E = \left\{\frac{\pi}{4} \leq \theta \leq \frac{\pi}{2}\right\} \subset \Omega.$$

Example 1.3–6: The experiment consists of measuring the instantaneous power, P, across a thermally agitated resistor. Then

$$\Omega = \{P : P \geq 0\}.$$

Since power cannot be negative, we leave out negative values of P in Ω. A subset of Ω is $E = \{P > 10^{-3} \text{ watts}\}$.

Note that in Examples 1.3–5 and 1.3–6, the number of elements in Ω is uncountably infinite. Therefore there are an uncountably infinite number of subsets. When, as in Example 1.3–4, the number of outcomes is finite, the number of distinct subsets is also finite, and each represents an event. Thus if $\Omega = \{\zeta_1, \ldots, \zeta_N\}$, the number of subsets of Ω is 2^N. We can see this by noting that each descriptor ζ_i, $i = 1, \ldots, N$ either is or is not present in any given subset of Ω. This gives rise to 2^N distinct events, one of which being the one in which none of the ζ_i appear. This event, involving none of the $\zeta \in \Omega$, is called the *impossible* event and is denoted by ϕ. It is the event that cannot happen and is only included for completeness.

Set Algebra. The *union* (sum) of two sets E and F, written $E \cup F$ or $E + F$ is the set of all elements that are in at least one of the sets E or F. Thus with $E = \{1, 2, 3, 4\}$ and $F = \{1, 3, 4, 5, 6\}$†

$$E \cup F = \{1, 2, 3, 4, 5, 6\}.$$

If E is a subset of F, we indicate this by writing $E \subset F$. Clearly for $E \subset F$ $E \cup F = F$. We indicate that ζ is an element of Ω or "belongs" to Ω by writing $\zeta \in \Omega$. This we can write

$$E \cup F = \{\zeta : \zeta \in E \text{ or } \zeta \in F \text{ or } \zeta \text{ lies in both}\}. \tag{1.3–1}$$

The *intersection* or set product of two sets E and F, written $E \cap F$ or EF is the set of elements common to both E and F. Thus in the preceding example

$$EF = \{1, 3, 4\}.$$

Formally, $EF \triangleq \{\zeta : \zeta \in E \textit{ and } \zeta \in F\}$. The *complement* of a set E, written E^c is the set of *all elements not in E*. From this it follows that if Ω is the sample description space or, more generally, the universal set, then

$$E \cup E^c = \Omega. \tag{1.3–2}$$

Also $EE^c = \phi$. The *difference* of two sets or, more appropriately, the *reduction* of E by F, written $E - F$ is the set of elements in E that are not in F. It should be clear that

$$E - F = EF^c$$

$$F - E = FE^c$$

† The order of the elements in a set is not important.

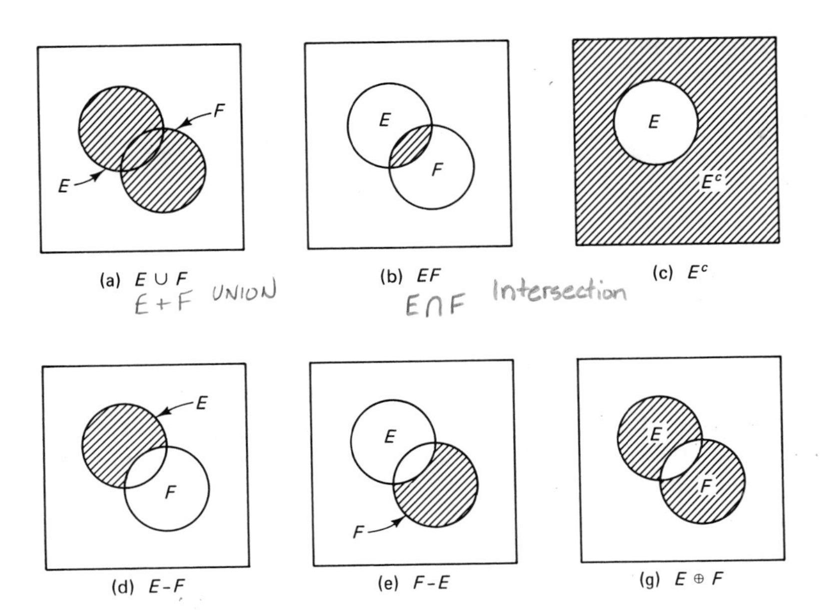

(a) $E \cup F$ (b) EF (c) E^c

(d) $E-F$ (e) $F-E$ (g) $E \oplus F$

Figure 1.3–1

and, in general $E - F \neq F - E$. The *exclusive-or* of two sets, written $E \oplus F$ is the set of all elements in E *or* F but not both. It is readily shown that†

$$E \oplus F = (E - F) \cup (F - E). \tag{1.3–3}$$

The operation of unions, intersections, and so forth, can be symbolically represented by Venn diagrams, which are useful as aids in reasoning and in establishing probability relations. The various set operations $E \cup F$, EF, E^c, $E - F$, $F - E$, $E \oplus F$ are shown in Figure 1.3–1 in hatch lines.

Two sets E, F are said to be disjoint if $EF = \phi$; that is, they have no elements in common. Given any set E, an *n-partition* of E consists of a sequence of sets E_i, $i = 1, \ldots, n$ such that $E_i \subset E$, $\bigcup_{i=1}^{n} E_i = E$ and $E_iE_j = \phi$ all $i \neq j$. Thus given two sets E, F a 2-partition of F is

$$F = FE \cup FE^c. \tag{1.3–4}$$

It is easy to demonstrate, using Venn diagrams, the following results:

$$(E \cup F)^c = E^cF^c \tag{1.3–5}$$

$$(EF)^c = E^c \cup F^c \tag{1.3–6}$$

† Equation 1.3–3 shows why $\cup$ is preferable, at least initially, to + to indicate union. The beginning student might—in error—write $(E - F) + (F - E) = E - F + F - E = 0$, which is meaningless.

and, by induction, given sets $E_1, \ldots, E_n$:

$$\left[\bigcup_{i=1}^{n} E_i\right]^c = \bigcap_{i=1}^{n} E_i^c \tag{1.3-7}$$

$$\left[\bigcap_{i=1}^{n} E_i\right]^c = \bigcup_{i=1}^{n} E_i^c. \tag{1.3-8}$$

These relations are known as De Morgan's laws after the English mathematician Augustus De Morgan (1806–1871).

Two sets E and F are said to be equal if every element in E is in F and vice versa. Equivalently,

$$E = F \quad \text{if} \quad E \subset F, \qquad F \subset E. \tag{1.3-9}$$

Sigma Fields. Consider a universal set Ω and a collection of subsets of Ω. Let $E, F, \ldots$ denote subsets in this collection. This collection of subsets forms a *field* $\mathscr{M}$† if

1. $\phi \in \mathscr{M}, \Omega \in \mathscr{M}$
2. If $E \in \mathscr{M}$ and $F \in \mathscr{M}$, then $E \cup F \in \mathscr{M}$, and $EF \in \mathscr{M}$.‡
3. If $E \in \mathscr{M}$ then $E^c \in \mathscr{M}$.

A *sigma* (σ) *field*¶ $\mathscr{F}$ is a field that is closed under any countable set of unions, intersections, and combinations. Thus if $E_1, \ldots, E_n, \ldots$ belong to $\mathscr{F}$ so do

$$\bigcup_{i=1}^{\infty} E_i \quad \text{and} \quad \bigcap_{i=1}^{\infty} E_i,$$

where these are defined as

$$\bigcup_{i=1}^{\infty} E_i \triangleq \{\text{the set of all elements in } \textit{at least one } E_i\}$$

and

$$\bigcap_{i=1}^{\infty} E_i \triangleq \{\text{the set of all elements in } \textit{every } E_i\}.$$

Events. Consider an experiment $\mathscr{H}$ with sample description space Ω. If Ω has a countable number of elements, then every subset of Ω may be assigned a probability in a way consistent with the axioms given in the next section. Then the class of all subsets make up a σ-field and each subset is an *event*. Thus we speak of the σ-field $\mathscr{F}$ of events. However, when Ω is not countable, that is, when $\Omega = R$ = the real line, then not every subset of Ω can be assigned a probability that will be consistent with the axiomatic theory. Only those subsets to which a probability can be assigned consistent with the axioms will be called events. The collection of those subsets is smaller than the collection of all possible subsets that one can define on Ω.

† Also sometimes called an algebra.

‡ From this it follows that if $E_1, \ldots, E_n$ belongs to $\mathscr{M}$, so do $\bigcup_{i=1}^{n} E_i \in \mathscr{M}$ and $\bigcap_{i=1}^{n} E_i \in \mathscr{M}$.

¶ Also sometimes called a σ-algebra.

This smaller collection forms a σ-field. On the real line R the σ-field is sometimes called the Borel field of events and, as a practical matter, includes all subsets of engineering and scientific interest.†

At this stage of our development, we have two of the three objects required for the axiomatic theory of probability, namely, a sample description space Ω and a σ-field $\mathscr{F}$ of events defined on Ω. We still need a probability measure P. The three objects $(\Omega, \mathscr{F}, P)$ form a triplet called a *probability space* $\mathscr{P}$.

1.4 AXIOMATIC DEFINITION OF PROBABILITY

Probability is a set function $P[\cdot]$ that assigns to every event $E \in \mathscr{F}$ a number $P[E]$ called the probability of E such that

(1) $P[E] \geq 0$ (1.4–1)

(2) $P[\Omega] = 1$ (1.4–2)

(3) $P[E \cup F] = P[E] + P[F]$ if $EF = \phi$. (1.4–3)

These axioms are sufficient‡ to establish the following basic results all but one of which we leave as exercises for the reader:

(4) $P[\phi] = 0$ (1.4–4)

(5) $P[EF^c] = P[E] - P[EF]$ where $E \in \mathscr{F}$, $F \in \mathscr{F}$ (1.4–5)

(6) $P[E] = 1 - P[E^c]$ (1.4–6)

(7) $P[E \cup F] = P[E] + P[F] - P[EF]$. (1.4–7)

From Equation 1.4–3 we can establish by mathematical induction that

$$P\left[\bigcup_{i=1}^{n} E_i\right] = \sum_{i=1}^{n} P[E_i] \quad \text{if} \quad E_iE_j = \phi \quad \text{all} \quad i \neq j.$$

From this result and Equation 1.4–7 we establish the general result that $P\left[\bigcup_{i=1}^{n} E_i\right] \leq \sum_{i=1}^{n} P[E_i]$. This result is sometimes known as the *union bound.*

Example 1.4–1. We wish to prove result (7). First we decompose the event $E \cup F$ into three disjoint events as follows

$$E \cup F = EF^c \cup E^cF \cup EF.$$

† For two-dimensional Euclidean sample spaces, the Borel field of events would be subsets of $R \times R$; for three-dimensional sample spaces it would be subsets of $R \times R \times R$.

‡ A fourth axiom: $P\left[\bigcup_{i=1}^{\infty} E_i\right] = \sum_{i=1}^{\infty} P[E_i]$ if $E_iE_j = \phi$ all $i \neq j$ must be included to enable one to deal rigorously with limits and countable unions. This axiom is of no concern to us here but will be in Chapter 6.

By axiom (3)

$$\begin{aligned} P[E \cup F] &= P[EF^c \cup E^cF] + P[EF] \\ &= P[EF^c] + P[E^cF] + P[EF], \quad \text{by axiom (3) again} \\ &= P[E] - P[EF] + P[F] - P[EF] + P[EF] \\ &= P[E] + P[F] - P[EF]. \end{aligned} \tag{1.4–8}$$

In obtaining line three from line two, we used result (5).

Example 1.4–2: The experiment consists of throwing a coin once. Hence

$$\Omega = \{H, T\}.$$

The σ-field of events consists of the following sets: $\{H\}$, $\{T\}$, Ω, ϕ. With the coin assumed fair we have†

$$P[\{H\}] = P[\{T\}] = \tfrac{1}{2}, \qquad P[\Omega] = 1, \qquad P[\phi] = 0.$$

Example 1.4–3: The experiment consists of throwing a die once. The outcome is the number of dots n_i appearing on the upface of the die. The set Ω is given by $\Omega = \{1, 2, 3, 4, 5, 6\}$. The σ-field of events consists of 2^6 elements. Some are

$$\phi, \Omega, \{1\}, \{1, 2\}, \{1, 2, 3\}, \{1, 4, 6\}, \{1, 2, 4, 5\},$$

and so forth. We assign

$$P[\{n_i\}] = \tfrac{1}{6} \qquad i = 1, \ldots, 6.$$

All probabilities can now be computed from the basic axioms and the assumed probabilities for the elementary events. Thus with $A = \{1\}$ and $B = \{2, 3\}$ we obtain $P[A] = \frac{1}{6}$. Also $P[A \cup B] = P[A] + P[B]$, since $AB = \phi$. Furthermore, $P[B] = P[\{2\}] + P[\{3\}] = \frac{2}{6}$ so that

$$P[A \cup B] = \tfrac{1}{6} + \tfrac{2}{6} = \tfrac{1}{2}.$$

Example 1.4–4: The experiment consists of picking at random a numbered ball from 12 balls numbered 1 to 12 from an urn.

$$\Omega = \{1, \ldots, 12\}.$$

$$\text{Let} \quad A\ddagger = \{1, \ldots, 6\} \qquad B = \{3, \ldots, 9\}$$

$$A \cup B = \{1, \ldots, 9\}, \qquad AB = \{3, 4, 5, 6\}, \qquad AB^c = \{1, 2\}$$

$$B^c = \{1, 2, 10, 11, 12\}, \qquad A^c = \{7, \ldots, 12\}, \qquad A^cB^c = \{10, 11, 12\}$$

$$(AB)^c = \{1, 2, 7, 8, 9, 10, 11, 12\}.$$

Hence

$$\begin{aligned} P[A] &= P[\{1\}] + P[\{2\}] + \ldots + P[\{6\}] \\ P[B] &= P[\{3\}] + \ldots + P[\{9\}] \\ P[AB] &= P[\{3\}] + \ldots + P[\{6\}]. \end{aligned}$$

If $P[\{1\}] = \ldots = P[\{12\}] = \frac{1}{12}$, then $P[A] = \frac{1}{2}$, $P[B] = \frac{7}{12}$, $P[AB] = \frac{4}{12}$, and so forth.

† To be precise we should distinguish between *outcomes* ζ and *elementary events* $\{\zeta\}$. The outcomes ζ are elements of Ω. The elementary events $\{\zeta\}$ are subsets of Ω. Probability is a set function; it assigns a number to every event. Thus we write $P[\{\zeta\}]$ rather than $P[\zeta]$ and more generally $P[\{\zeta_1, \zeta_2, \ldots, \zeta_n\}]$ rather than $P[\zeta_1, \zeta_2, \ldots, \zeta_n]$. However, we shall frequently dispense with this added notational complication, especially when writing joint probabilities.

‡ Thus A occurs whenever any of the numbers 1 through 6 is an outcome. AB occurs whenever any of the 3 through 6 occurs, and so forth, for the other events.

We point out that a theory of probability could be developed from a different set of axioms [1–6]. However, whatever axioms are used and whatever theory is developed, for it to be useful in solving problems in the physical world, it must model our empirical concept of probability as a relative frequency and the consequences that follow from it.

1.5 JOINT, CONDITIONAL, AND TOTAL PROBABILITIES; INDEPENDENCE

Assume that we perform the following experiment: We are in a certain U.S. city and wish to collect weather data about it. In particular we are interested in three events, call them A, B, C, where

A is the event that on any particular day, the temperature equals or exceeds 10°C;

B is the event that on any particular day, the amount of precipitation equals or exceeds five millimeters;

C is the event that on any particular day A and B both occur, that is, $C \triangleq AB$.

Since C is an event $P[C]$ is a probability that satisfies the axioms. But $P[C] = P[AB]$; we call $P[AB]$ the *joint probability of the events A and B.* This notion can obviously be extended to more than two events, that is, $P[EFG]$ is the joint probability of events E, F, and G.† Now let n_i denote the number of days on which event i occurred. Over a thousand-day period ($n = 1000$), the following observations are made: $n_A = 811$, $n_B = 306$, $n_{AB} = 283$. By the relative frequency interpretation of probability

$$P[A] \simeq \frac{n_A}{n} = \frac{811}{1000} = 0.811$$

$$P[B] \simeq \frac{n_B}{n} = 0.306$$

$$P[AB] \simeq \frac{n_{AB}}{n} = 0.283.$$

Consider now the ratio n_{AB}/n_A. This is the relative frequency with which event AB occurs when event A occurs. Put into words, it is the fraction of time that the amount of precipitation equals or exceeds five millimeters on those days *when the temperature equals or exceed* 10°C. Thus we are dealing with the frequency of an event, given that or *conditioned upon the fact that another event has occurred.* Note that

$$\frac{n_{AB}}{n_A} = \frac{n_{AB}/n}{n_A/n} \simeq \frac{P[AB]}{P[A]}. \tag{1.5–1}$$

† E, F, G are any three events defined on the same probability space.

This empirical concept suggests that we introduce in our theory a conditional probability measure $P[B \mid A]$ defined by

$$P[B \mid A] \triangleq \frac{P[AB]}{P[A]}, \qquad P[A] > 0 \tag{1.5–2}$$

and described in words as the probability that B occurs given that A has occurred, and similarly,

$$P[A \mid B] \triangleq \frac{P[AB]}{P[B]}, \qquad P[B] > 0. \tag{1.5–3}$$

Definitions 1.5–2 and 1.5–3 can be used to compute the joint probability of AB. We illustrate this with an example.

Example 1.5–1: In a binary communication system (Figure 1.5–1), that is, one in which a two-symbol alphabet is used to communicate, the two symbols are "zero" and "one." In this system let Y stand for the received symbol and X for the transmitted symbol. The sample description space Ω for this experiment is $\Omega = \{(X, Y): X = 0 \text{ or } 1,\ Y = 0 \text{ or } 1\} = \{(0, 0), (0, 1), (1, 0), (1, 1)\}$. The event $\{X\}$ is given by $\{X\} = \{(X, 0)\} \cup \{(X, 1)\}$. The probability functions are $P[\{(X, Y) = (i, j)\}] = P[\{X = i\}]P[\{Y = j|X = i\}]$ $i, j = 0, 1$. We shall dispense with the curly-bracket notation when writing joint probabilities. Thus $P[\{(X, Y) = (i, j)\}]$ will be written $P[X = i, Y = j]$. The reader should bear in mind, however, that we are speaking of the event $\{X = i\} \cap \{Y = j\}$ and not of the event $\{X = i\} \cup \{Y = j\}$.

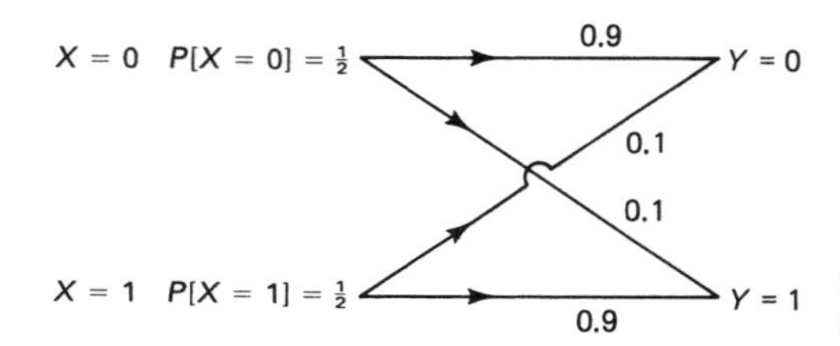

Figure 1.5–1 A binary communication system.

Because of noise a transmitted zero sometimes gets decoded as a received one and vice versa. From measurements it is known that

$$P[Y = 1 \mid X = 1] = 0.9 \qquad P[Y = 1 \mid X = 0] = 0.1$$
$$P[Y = 0 \mid X = 1] = 0.1 \qquad P[Y = 0 \mid X = 0] = 0.9$$

and by design† $P[X = 0] = P[X = 1] = 0.5$. The various joint probabilities are then

$$P[X = 0, Y = 0] = P[Y = 0 \mid X = 0]P[X = 0] = 0.45$$
$$P[X = 0, Y = 1] = P[Y = 1 \mid X = 0]P[X = 0] = 0.05$$
$$P[X = 1, Y = 0] = P[Y = 0 \mid X = 1]P[X = 1] = 0.05$$
$$P[X = 1, Y = 1] = P[Y = 1 \mid X = 1]P[X = 1] = 0.45.$$

The introduction of conditional probabilities raises the intriguing question of whether conditional probabilities satisfy the axioms in Equations 1.4–1 to

† It is good practice to design a code in which the zeros and ones appear at the same rate.

1.4–3. In other words, given any two events E, F such that $EF = \phi$ and a third event A all belonging to the σ-field of events $\mathscr{F}$ in the probability space $(\Omega, \mathscr{F}, P)$ does

$$P[E \mid A] \geq 0?$$
$$P[\Omega \mid A] = 1?$$
$$P[E \cup F \mid A] = P[E \mid A] + P[F \mid A] \quad \text{for } EF = \phi?$$

The answer is yes. We leave the details as an exercise to the reader.

The next problem illustrates the use of joint and conditional probabilities.

Example 1.5–2:† Assume that a beauty contest is being judged by the following rules: (1) there are N contestants not seen by the judges before the contest, and (2) the contestants are individually presented to the judges in a random sequence. Only one contestant appears before the judges at any one time. (3) The judges must decide on the spot whether the contestant appearing before them is the most beautiful. If they decide in the affirmative the contest is over but the risk is that a still more beautiful contestant is in the group as yet not displayed. In that case the judges would have made the wrong decision. On the other hand, if they pass over the candidate, the contestant is disqualified from further consideration even if it turns out that all subsequent contestants are less beautiful. What is the probability of correctly choosing the most beautiful contestant? What is a good strategy to follow?

Solution: To make the problem somewhat more quantitative assume that all the virtues of each contestant are summarized into a single "beauty" number. Thus the most beautiful contestant is associated with the highest number, the least beautiful has the lowest number. The numbers, unseen by the judges, are placed in a bag and the numbers are drawn individually from the bag. Imagine that the draws are ordered along a line as shown in Figure 1.5–2. Thus the first draw is number 1, the second is 2, and so forth. At each draw, a number appears; is it the largest of all N numbers?

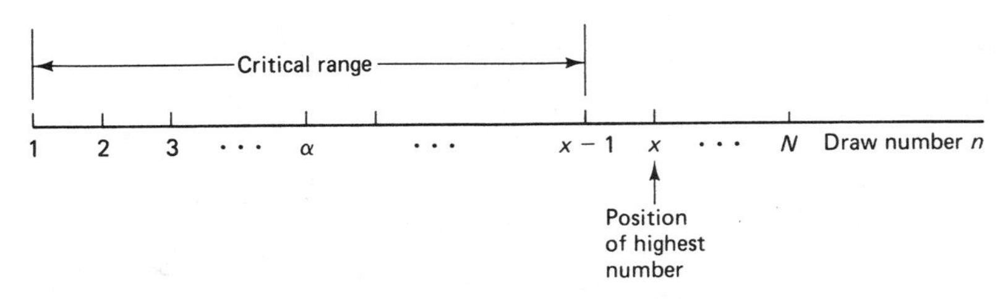

Figure 1.5–2 The numbers along the axis represent the draws, *not* the number actually drawn from the bag.

Assume that the following "wait-and-see" strategy is adopted: We pass over the first α draws (that is, we reject the first α contestants) but record the highest number (that is, the most beautiful contestant) observed within this group of α. Call this number ξ.

† Thanks are due to Jerry Tiemann for valuable discussions regarding this problem.

Now we continue drawing numbers (that is, call for more contestants to appear) but *don't reject them out of hand.* The first draw (contestant) after the α passed-over draws that yields a number exceeding ξ is taken to be the winner. Now given that the highest draw actually occurs at x, two events must occur for the correct decision to occur: (A) $x > \alpha$; and (B) the highest number in the so-called critical range $\{1, \ldots, x-1\}$ must occur in $\{1, \ldots, \alpha\}$. Why? Assume that it occurs in $\{\alpha+1, \ldots, x-1\}$; then by the adopted strategy, we shall choose it, which will be in error since it is assumed that the highest number occurs at x. On the other hand, if the highest number in $\{1, \ldots, x-1\}$ occurs in $\{1, \ldots, \alpha\}$ it will be passed over, which is fine.

Define the four events:

$A \triangleq \{x > \alpha\}$

$B \triangleq \{\text{highest number in } \{1, \ldots, x-1\} \text{ is in } \{1, \ldots, \alpha\}\}$

$C \triangleq \{x \text{ is position of the highest number}\}$

$D \triangleq \{\text{the event of correctly choosing the highest number}\}$.

Then

$$P[D] = P[ABC] = P[AB \mid C]P[C] = \frac{1}{N}\sum_{x=\alpha+1}^{N}\frac{\alpha}{x-1},$$

where (1) $P[C] = N^{-1}$; (2) the probability of the highest number in the $\{1, \ldots, x-1\}$ draws actually being in the $\{1, \ldots, \alpha\}$ draws is just the "length," that is, the number of contestants, α of the set $\{1, \ldots, \alpha\}$ divided by the "length" $x-1$ of the set $\{1, \ldots, x-1\}$; and (3) the summation starting at $x = \alpha + 1$ ensures that $x > \alpha$.

By Euler's summation formula† for large N

$$P[D] = \frac{1}{N}\sum_{x=\alpha+1}^{N}\frac{\alpha}{x-1} \simeq \frac{\alpha}{N}\int_{\alpha}^{N}\frac{dx}{x} = -\frac{\alpha}{N}\ln\frac{\alpha}{N}.$$

The best choice of α, say, α_0 is found by differentiation. Thus by setting

$$\frac{dP[D]}{d\alpha} = 0$$

We find that

$$\alpha_0 = \frac{N}{e}$$

and the maximum probability, P_0, of a correct decision is

$$P_0 \simeq 0.37 \quad (N \text{ large}).$$

Thus we let approximately the first third of the contestants pass by before beginning to judge the contestants in earnest.

Unconditional Probability. In many problems in engineering and science we would like to compute the unconditional probability, $P[B]$, of an event B in terms of the sum of weighted conditional probabilities. Such a computation is easily realized through the following theorem.

† See, for example, V. Mangolis, *Handbook of Series for Scientists and Engineers.* New York: Academic Press, 1965, p. 5.

Theorem. Let $A_1, A_2, \ldots, A_n$ be n mutually exclusive events such that $\bigcup_{i=1}^{n} A_i = \Omega$ (the A_i's are *exhaustive*). Let B be any event defined over the probability space of the A_i's. Then, with $P[A_i] \neq 0$ all i,

$$P[B] = P[B \mid A_1]P[A_1] + \ldots + P[B \mid A_n]P[A_n]. \tag{1.5–4}$$

Sometimes $P[B]$ is called the "average" probability of B because the expression on the right is reminiscent of the operation of averaging.

Proof. We have $A_iA_j = \phi$ $i \neq j$ and $\bigcup_{i=1}^{n} A_i = \Omega$. Also $B\Omega = B = B\bigcup_{i=1}^{n} A_i = \bigcup_{i=1}^{n} BA_i$. But by definition of the intersection operation, $BA_i \subset A_i$; hence $(BA_i)(BA_j) = \phi$ for all $i \neq j$. Thus from axiom 3 (generalized to n events):

$$P[B] = P\left[\bigcup_{i=1}^{n} BA_i\right] = P[BA_1] + P[BA_2] + \ldots + P[BA_n]$$
$$= P[B \mid A_1]P[A_1] + \ldots + P[B \mid A_n]P[A_n]. \tag{1.5–5}$$

The last line follows from Equation 1.5–2.

Example 1.5–3: For the binary communication system shown in Figure 1.5–1 compute $P[Y = 0]$ and $P[Y = 1]$.

Solution: We use Equation (1.5–5) as follows

$$P[Y = 0] = P[Y = 0 \mid X = 0]P[X = 0] + P[Y = 0 \mid X = 1]P[X = 1]$$
$$= (0.9)(0.5) + (0.1)(0.5)$$
$$= 0.5.$$

We can compute $P[Y = 1]$ in a similar fashion or by noting that $\{Y = 0\} \cup \{Y = 1\} = \Omega$ and $\{Y = 0\}\{Y = 1\} = \phi$. Hence $P[Y = 0] + P[Y = 1] = 1$ or

$$P[Y = 1] = 1 - P[Y = 0] = 0.5.$$

Independence. This is an extremely important concept and full justice to all its implications cannot be done until we discuss random variables in Chapter 2.

Definitions. (A) Two events $A \in \mathscr{F}, B \in \mathscr{F}$ with $P[A] > 0$, $P[B] > 0$ are said to be independent if and only if (iff)

$$P[AB] = P[A]P[B]. \tag{1.5–6}$$

Since, in general, $P[AB] = P[B \mid A]P[A] = P[A \mid B]P[B]$ it follows that for independent events

$$P[A \mid B] = P[A] \tag{1.5–7a}$$
$$P[B \mid A] = P[B]. \tag{1.5–7b}$$

Thus the definition satisfies our intuition: If A and B are independent, the outcome B should have no effect on the probability of A and vice versa.

(B) Three events A, B, C defined on $\mathscr{P}$ and having nonzero probabilities are said to be independent if

$$P[ABC] = P[A]P[B]P[C] \tag{1.5–8a}$$

$$P[AB] = P[A]P[B] \tag{1.5–8b}$$

$$P[AC] = P[A]P[C] \tag{1.5–8c}$$

$$P[BC] = P[B]P[C]. \tag{1.5–8d}$$

This is an extension of (A) above and suggests the pattern for the definition of n independent events $A_1, \ldots, A_n$. Note that it is *not enough* to have just $P[ABC] = P[A]P[B]P[C]$. Pairwise independence must also be shown.

(C) Let A_i, $i = 1, \ldots, n$ be n events defined on $\mathscr{P}$. The $\{A_i\}$ are said to be independent if

$$\begin{aligned} P[A_iA_j] &= P[A_i]P[A_j] \\ P[A_iA_jA_k] &= P[A_i]P[A_j]P[A_k] \\ &\vdots \\ P[A_1, \ldots, A_n] &= P[A_1]P[A_2]\ldots P[A_n]. \end{aligned}$$

for all combination of indices such that $1 \leq i < j < k < \ldots \leq n$.

Example 1.5–4: We are given three events A, B, C for which $P[A] = \frac{1}{6}$, $P[B] = \frac{1}{5}$, $P[C] = \frac{1}{4}$ and $P[ABC] = \frac{1}{120}$. Are A, B, and C independent?

Solution: We are not given enough information to make this determination. Clearly $P[ABC] = P[A]P[B]P[C]$. But what about $P[AB] \stackrel{?}{=} P[A]P[B]$, and so forth? Since $P[AB]$, $P[BC]$ and $P[AC]$ are not given, independence cannot be determined.

Suppose $P[AB] = \frac{1}{30}$, $P[AC] = \frac{1}{24}$, $P[BC] = \frac{1}{10}$. Are A, B, and C independent? No, because $P[BC] \neq P[B]P[C]$.

1.6 BAYES' THEOREM AND APPLICATIONS

The previous results enable us now to write a fairly simple formula known as Bayes' Theorem.† Despite its simplicity, this formula is widely used in biometrics, epidemiology, and communication theory.

Bayes' Theorem. Let A_i, $i = 1, \ldots, n$ be a set of disjoint and exhaustive events defined on $\mathscr{P}$. Then $\bigcup_{i=1}^{n} A_i = \Omega$, $A_iA_j = \phi$ $i \neq j$. With B any event defined on $\mathscr{P}$ with $P[B] > 0$ and $P[A_i] \neq 0$ all i

$$P[A_j \mid B] = \frac{P[B \mid A_j]P[A_j]}{\sum_{i=1}^{n} P[B \mid A_i]P[A_i]}. \tag{1.6-1}$$

Proof. The denominator is simply $P[B]$ by Equation 1.5–5 and the numerator is simply $P[A_jB]$. Thus Bayes' Theorem is merely an application of the definition of conditional probability.

† Named after Thomas Bayes, English mathematician/philosopher (1702–1761).

Remark. In practice the terms in Equation 1.6–1 are given various names: $P[A_j \mid B]$ is known as *a posteriori* probability of A_j given B; $P[B \mid A_j]$ is called *a priori* probability of B given A_j; and $P[A_j]$ is the *causal* or *a priori* probability of A_j. In general *a priori* probabilities are estimated from past measurements or presupposed by experience while *a posteriori* probabilities are measured or computed from observations.

Example 1.6–1. In a communication system a zero or one is transmitted with $P[X = 0] = P_0$, $P[X = 1] = 1 - P_0 \triangleq P_1$ respectively. Due to noise in the channel, a zero can be received as a one with probability β and a one can be received as a zero also with probability β. A one is observed. What is the probability that a one was transmitted?

Solution: The structure of the channel is shown in Figure 1.6–1.
We write

$$P[X = 1 \mid Y = 1] = \frac{P[X = 1, Y = 1]}{P[Y = 1]}$$

$$= \frac{P[Y = 1 \mid X = 1]P[X = 1]}{P[Y = 1 \mid X = 1]P[X = 1] + P[Y = 1 \mid X = 0]P[X = 0]}$$

$$= \frac{P_1(1-\beta)}{P_1(1-\beta) + P_0\beta}.$$

If $P_0 = P_1 = \frac{1}{2}$, the *a posteriori* probability $P[X = 1 \mid Y = 1]$ depends on β as shown in Figure 1.6–2.

The channel is said to be noiseless if $\beta = 1$ or $\beta = 0$.

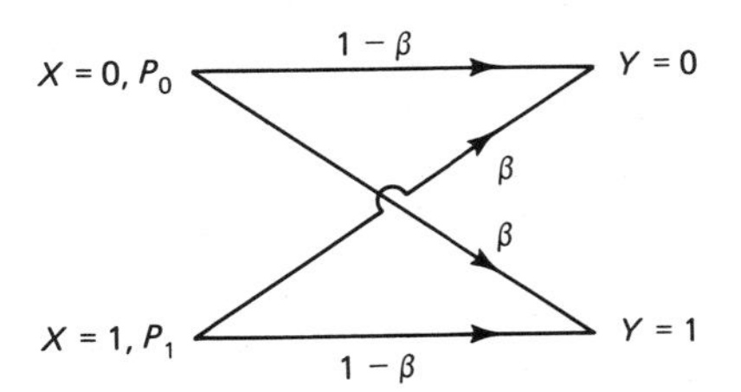

Figure 1.6–1

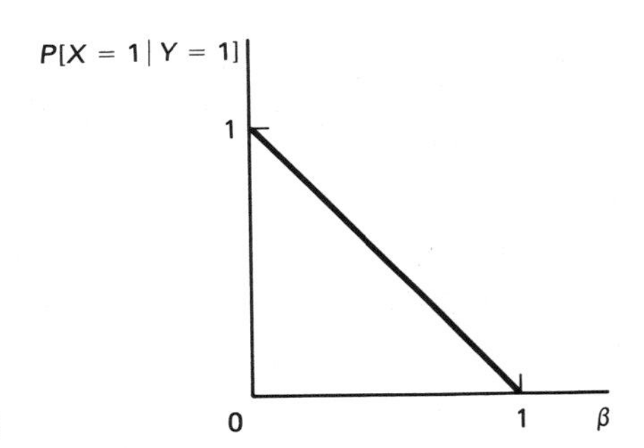

Figure 1.6–2

Example 1.6–2: (Parzen [1–7, p. 119].) Suppose there exists a (fictitious) test for cancer with the following properties. Let

A = event that the test states that tested person has cancer.
B = event that person has cancer.
A^c = event that test states person is free from cancer.
B^c = event that person is free from cancer.

It is known that $P[A \mid B] = P[A^c \mid B^c] = 0.95$ and $P[B] = 0.005$. Is the test a good test?

Solution: To answer this question we should like to know the likelihood that a person actually has cancer if the test so states, that is, $P[B \mid A]$. Hence

$$P[B \mid A] = \frac{P[B]P[A \mid B]}{P[A \mid B]P[B] + P[A \mid B^c]P[B^c]} = \frac{(0.005)(0.95)}{(0.95)(0.005) + (0.05)(0.995)}$$
$$= 0.087.$$

Hence in only 8.7% of the cases where the tests are positive will the person actually have cancer. This test has a very high false-alarm rate and in this sense cannot be regarded as a good test. The fact that, initially, the test seems like a good test is not surprising given that $P[A \mid B] = P[A^c \mid B^c] = 0.95$. However, when the scarcity of cancer in the general public is considered, the test is found to be vacuous.

1.7 COMBINATORICS†

Before proceeding with our study of basic probability we introduce a number of counting formulas of importance in probability. Some of the results presented here will have immediate application in Section 1.8; others will be useful later.

A *population of size n* will be taken to mean a collection of n elements without regard to order. Two populations are considered different if one contains at least one element not contained in the other. A *subpopulation* of size r from a population of size n is a subset of r elements taken from the original population. Likewise, two subpopulations are considered different if one has at least one element different from the other.

Next, consider a population of n elements $a_1, a_2, \ldots, a_n$. Any ordered arrangement $a_{k_1}, a_{k_2}, \ldots, a_{k_r}$ of r symbols is called an *ordered sample* of size r.

Consider now the generic urn containing n distinguishable numbered balls. Balls are removed one by one. How many different ordered samples of size r can be formed? There are two cases:

(*i*) *Sampling with replacement.* Here after each ball is removed, its number is recorded and it is returned to the urn. Thus for the first sample there are n choices, for the second there are again n choices, and so on. Thus we are led to the following result: For a population of n elements,

† This material closely follows that of William Feller [1–8].

there are n^r different ordered samples of size r that can be formed with replacement.

(*ii*) *Sampling without replacement.* After each ball is removed, it is not available anymore for subsequent samples. Thus n balls are available for the first sample, $n-1$ for the second and so forth. Thus we are now led to the result: For a population of n elements there are

$$(n)_r \triangleq n(n-1)(n-2)\ldots(n-r+1)$$
$$= \frac{n!}{(n-r)!} \tag{1.7–1}$$

different ordered samples of size r that can be formed without replacement.†

The number of subpopulations of size r in a population of size n. A basic problem that often occurs in probability is the following: How many groups; that is, subpopulations of size r can be formed from a population of size n? For example, consider 6 balls numbered 1 to 6. How many groups of size 2 can be formed? The following table shows that there are 15 groups of size 2 that can be formed:

12	23	34	45	56
13	24	35	46	
14	25	36		
15	26			
16				

Note that this is different from the number of ordered samples that can be formed without replacement. These are $(6 \cdot 5 = 30)$:

12	21	31	41	51	61
13	23	32	42	52	62
14	24	34	43	53	63
15	25	35	45	54	64
16	26	36	46	56	65

Also it is different from the number of samples that can be formed with replacement $(6^2 = 36)$:

11	21	31	41	51	61
12	22	32	42	52	62
13	23	33	43	53	63
14	24	34	44	54	64
15	25	35	45	55	65
16	26	36	46	56	66

† Different samples will often contain the same subpopulation but with a different *ordering*. For this reason we sometimes speak of $(n)_r$ ordered samples that can be formed without replacement.

A general formula for the number of subpopulations, C_r^n, of size r in a population of size n can be computed as follows: Consider an urn with n distinguishable balls. We already know that the number of ordered samples of size r that can be formed is $(n)_r$. Now consider a specific subpopulation of size r. For this subpopulation there are $r!$ arrangements and therefore $r!$ different ordered samples. Thus for C_r^n subpopulations there must be $C_r^n \cdot r!$ different ordered samples of size r. Hence

$$C_r^n \cdot r! = (n)_r$$

or

$$C_r^n = \frac{(n)_r}{r!} = \frac{n!}{(n-r)!r!} \triangleq \binom{n}{r}. \tag{1.7–2}$$

Equation 1.7–2 is an important result, and we shall apply it in the next section. The symbol

$$C_r^n \triangleq \binom{n}{r}$$

is called a binomial coefficient. Clearly

$$\binom{n}{r} = \frac{n!}{r!(n-r)!} = \frac{n!}{(n-r)!r!} \triangleq \binom{n}{n-r} = C_{n-r}^n. \tag{1.7–3}$$

We already know from Section 1.3 that the total number of subsets of a set of size n is 2^n. The number of subsets of size r is $\binom{n}{r}$. Hence we obtain that

$$\sum_{r=0}^{n} \binom{n}{r} = 2^n.$$

An extension of the binomial formula is given in the following theorem:

Theorem. Let $r_1, r_2, \ldots, r_k$ be integers such that

$$r_1 + r_2 + \ldots + r_k = n. \qquad r_i \geq 0 \qquad i = 1, \ldots k.$$

The number of ways in which a population of n elements can be divided into k ordered parts (partitioned into k subpopulations) of which the first contains r_1 elements, the second r_2 elements, and so forth, is

$$\frac{n!}{r_1!r_2!\ldots r_k!}. \tag{1.7-4}$$

This coefficient is called the *multinomial* coefficient. As an example consider partitioning four balls, labeled 1, 2, 3, 4 into two subpopulations. By Equation 1.7–4 we obtain for the number of partitions: $4!/2!2! = 6$. The 6 partitions are

1. $\{1, 2\}$ $\{3, 4\}$
2. $\{1, 3\}$ $\{2, 4\}$
3. $\{1, 4\}$ $\{2, 3\}$
4. $\{2, 3\}$ $\{1, 4\}$
5. $\{2, 4\}$ $\{1, 3\}$
6. $\{3, 4\}$ $\{1, 2\}$.

In physics, one speaks of the state of atomic particles and the probability of those states. Some particles obey Bose-Einstein statistics while others obey Fermi-Dirac statistics [1–7, p. 71] [1–8, p. 38]. Without going through the details, we can associate the particles with indistinguishable balls and the states with cells. Consider now the following fundamental problem: Given r balls and n cells with r_1 balls in the first cell, r_2 in the second, and so forth, so that

$$r_1 + r_2 + \ldots + r_n = r \qquad r_i \geq 0 \qquad i = 1, \ldots, n.$$

How many distinguishable distributions can be formed from r balls and n cells? Before answering this question we define a distribution by an n-tuple $(r_1, \ldots, r_n)$ where the r_i are called occupancy numbers. Two distributions are then said to be indistinguishable *if their corresponding ordered n-tuples are identical.* The key result is the following: The number of distinguishable distributions is

$$\binom{n + r - 1}{r} = \binom{n + r - 1}{n - 1} \tag{1.7–5}$$

and the number of distinguishable distributions in which no cell remains empty is

$$\binom{r - 1}{n - 1}. \tag{1.7–6}$$

The proof of Equation 1.7–5 is given by William Feller [1–8, p. 38] using a very clever artifice. This artifice consists of representing the n cells by the space between $n + 1$ bars and the balls by stars. Thus

$$| * * | | | * | * * | | * * * * * |$$

represents $n = 7$, $r = 10$ and occupancy numbers 2, 0, 0, 1, 2, 0, 5. Thus n cells require $n + 1$ bars, but since the first and last symbols must be bars, only $n - 1$ bars and r stars can appear in any order. Thus we are asking for the number of subpopulations of size r in a population of size $n - 1 + r$. The result is, by Equation 1.7–2,

$$\binom{n + r - 1}{r} = \binom{n + r - 1}{n - 1}.$$

We leave the proof of Equation 1.7–6 as an exercise to the reader.

1.8 BERNOULLI TRIALS

Consider the very simple experiment consisting of a single trial with a binary outcome: A success $\{\zeta_1 = s\}$ with probability p or a failure $\{\zeta_2 = f\}$ with probability $q = 1 - p$. Thus $P[\{s\}] = p$, $P[\{f\}] = q$ and the sample description space is $\Omega = \{s, f\}$. The σ-field of events $\mathscr{F}$ is ϕ, Ω, $\{s\}$, $\{f\}$.

Suppose we do the experiment twice. The new sample description space Ω_2, written $\Omega_2 = \Omega \times \Omega$ is the set of all ordered 2-tuples

$$\Omega_2 = \{ss, sf, fs, ff\}.$$

$\mathscr{F}$ contains $2^4 = 16$ events. Some are ϕ, Ω, $\{ss\}$, $\{ss, ff\}$, and so forth.

The product $\Omega \times \Omega$ is called the *cartesian product*. If we do n independent trials, the sample space is

$$\Omega_n = \underbrace{\Omega \times \Omega \times \ldots \times \Omega}_{n \text{ times}}$$

and contains 2^n elementary outcomes, each of which is an ordered n-tuple. Thus

$$\Omega_n = \{a_1, \ldots, a_M\} \quad \text{where} \quad M = 2^n$$

and $a_i = z_{i_1} \ldots z_{i_n}$ where $z_{i_j} = s$ or f. Since each outcome z_{i_j} is independent of any other outcome, the joint probability $P[z_{i_1} \ldots z_{i_n}] = P[z_{i_1}]P[z_{i_2}] \ldots P[z_{i_n}]$. Thus the probability of a given ordered set of k successes and $n - k$ failures is simply $p^k q^{n-k}$. For example, suppose we throw a coin 3 times with $p = P[\{H\}]$ and $q = P[\{T\}]$. The probability of the event $\{HTH\}$ is $pqp = p^2 q$. The probability of the event $\{THH\}$ is also $p^2 q$. The different events leading to two heads and one tail are listed here:

$$E_1 = \{HHT\}$$
$$E_2 = \{HTH\}$$
$$E_3 = \{THH\}.$$

If F denotes the event of getting two heads and one tail without regard to order, then $F = E_1 \cup E_2 \cup E_3$. Since $E_i E_j = \phi$ for $i \neq j$ we obtain $P[F] = P[E_1] + P[E_2] + P[E_3] = 3p^2 q$.

Let us now generalize the previous result by considering an experiment consisting of n Bernoulli trials. The sample description space Ω_n contains $M = 2^n$ outcomes $a_1, a_2, \ldots, a_M$, where each a_i is a string of n symbols, and each symbol represents a success s or a failure f. Consider the event $A_k \triangleq \{k$ successes in n trials$\}$ and let the primed outcomes, that is a_i', denote strings with k successes and $n - k$ failures. Then, with K denoting the number of ordered arrangements involving k successes and $n - k$ failures, we write

$$A_k = \bigcup_{i=1}^{K} \{a_i'\}.$$

To determine how large K is, we use an artifice similar to that used in proving Equation 1.7–5. Let bars represent failures and stars represent successes. Then, as an example,

$$| * * | * * | | *$$

represents five successes in nine tries in the order $f s s f s s f f s$. How many such arrangements are there? The solution is given by Equation 1.7–5 with $r = k$ and $(n - 1) + r$ replaced by $(n - k) + k = n$. (Note that there is no restriction that the first and last symbols must be bars.) Thus

$$K = \binom{n}{k}$$

and, since the $\{a_i\}$ are disjoint, that is, $\{a_i\} \cap \{a_j\} = \phi$ for $i \neq j$, we obtain

$$P[A_k] = P\left[\bigcup_{i=1}^{K} \{a'_i\}\right] = \sum_{i=1}^{K} P[\{a'_i\}].$$

Finally, since $P[\{a'_i\}] = p^k q^{n-k}$ regardless of the ordering of the s and f's, we obtain

$$P[A_k] = \binom{n}{k} p^k q^{n-k} \tag{1.8–1}$$

$$\triangleq b(k; n, p).$$

The symbol $b(k; n, p)$ denotes the binomial law defined in Equation 1.8–1, which is the probability of getting k successes in n *independent* tries with individual Bernoulli trial success probability p.

The binomial coefficient

$$C_k^n = \binom{n}{k}$$

was introduced in the previous section and is the number of subpopulations of size k that can be formed from a population of size n. In the preceding example the population has size 3 (three tries) and the subpopulation has size 2 (two heads), and we are interested in getting two heads in three tries with order being irrelevant. Thus the correct result is $C_2^3 = 3$. Note that had we asked for the probability of getting two heads on the first two tosses followed by a tail, that is, $P[E_1]$, we would *not have used the coefficient* C_2^3 because *there is only one way that this event can happen.*

Example 1.8–1: Suppose $n = 4$; that is, there are four balls numbered 1 to 4 in the urn. The number of distinguishable, ordered samples of size 2 that can be drawn without replacement is 12, that is, {1, 2}; {1, 3}; {1, 4}; {2, 1}; {2, 3}; {2, 4}; {3, 1}; {3, 2}; {3, 4}; {4, 1}; {4, 2}; {4, 3}. The number of distinguishable unordered sets is 6, that is,

{1, 2} {2, 1}	{1, 3} {3, 1}	{1, 4} {4, 1}	{2, 4} {4, 2}	{3, 4} {4, 3}	{2, 3} {3, 2}

From Equation 1.7–2 we obtain this result directly; that is ($n = 4$, $k = 2$)

$$\binom{n}{k} = \frac{4!}{2!2!} = 6.$$

Example 1.8–2: 10 independent, binary pulses per second arrive at a receiver. The error probability (that is, a zero received as a one or vice versa) is 0.001. What is the probability of at least one error/second?

$$P[\text{at least one error/sec}] = 1 - P[\text{no errors/sec}]$$

$$= 1 - \binom{10}{0}(0.001)^0(0.999)^{10} = 1 - (0.999)^{10} \simeq 0.01.$$

Observation. Note that

$$\sum_{k=0}^{n} b(k; n, p) = 1. \quad \text{Why?}$$

Example 1.8–3: An odd number of people want to play a game that requires two teams made up of even numbers of players. To decide who shall be left out to act as umpire, each of the N persons tosses a fair coin with the following stipulation: If there is one person whose outcome (be it heads or tails) is different from the rest of the group, that person will be the umpire. Assume that there are 11 players. What is the probability that a player will be "odd-man out"; that is, will be the umpire on the first play?

Solution: Let $E \triangleq \{10H, 1T\}$, where $10H$ means $H, H, \ldots, H$ ten times, and $F \triangleq \{10T, 1H\}$. Then $EF = \phi$ and

$$\begin{aligned} P[E \cup F] &= P[E] + P[F] \\ &= \binom{11}{10}\left(\frac{1}{2}\right)^{10}\left(\frac{1}{2}\right) + \binom{11}{1}\left(\frac{1}{2}\right)\left(\frac{1}{2}\right)^{10} \\ &\simeq 0.01074. \end{aligned}$$

Example 1.8-4: In Example 1.8-3 derive a formula for the probability that the odd-man out will occur for the first time on the nth play. Hint: Consider each play as an independent Bernoulli trial with success if an odd-man out occurs and failure otherwise.

Solution: Let E be the event of odd-man out for first time on nth play. Let F be the event of no odd-man out in $n - 1$ plays and let G be the event of an odd-man out on the nth play. Then

$$E = FG.$$

Since it is completely reasonable to assume F and G are independent events, we can write

$$P[E] = P[F]P[G]$$

$$P[F] = \binom{n-1}{0}(0.0107)^0\ (0.9893)^{n-1} = (0.9893)^{n-1}$$

$$P[G] = 0.0107.$$

Thus $P[E] = (0.0107)(0.893) = (0.0107)(0.9893)^{n-1}$.

Further Discussion of the Binomial Law. We write down some self-evident formulas for further use. The probability $B(k; n, p)$ of k or fewer *successes in n tries* is given by

$$B(k; n, p) = \sum_{i=0}^{k} b(i; n, p) = \sum_{i=0}^{k} \binom{n}{i} p^i q^{n-i}. \qquad (1.8\text{–}2)$$

The probability of at least k successes in n tries is

$$\sum_{i=k}^{n} b(i; n, p) = 1 - B(k - 1; n, p).$$

The probability of more than k successes but no more than j successes is

$$\sum_{k+1}^{j} b(i; n, p).$$

We illustrate this with an example.

Example 1.8–5: Five missiles are fired against an aircraft carrier in the ocean. It takes at least two direct hits to sink the carrier. All five missiles are on the correct trajectory but must get through the "point defense" guns of the carrier. It is known that the point defense guns can destroy a missile with probability $p = 0.9$. What is the probability that the carrier will still be afloat after the encounter?

Solution: Let E be the event that the carrier is still afloat and let F be the probability of a missile getting through the point-defense guns. Then

$$P[F] = 0.1$$

and

$$\begin{aligned} P[E] &= 1 - P[E^c] \\ &= 1 - \sum_{i=2}^{5} \binom{5}{i} (0.1)^i (0.9)^{5-i} \simeq 0.92. \end{aligned}$$

1.9 ASYMPTOTIC BEHAVIOR OF THE BINOMIAL LAW: THE POISSON LAW

Suppose that in the binomial function $b(k; n, p)$ $n \gg 1$, $p \ll 1$, but np remains constant, say $np = a$. Recall that $q = 1 - p$. Hence

$$\binom{n}{k} p^k (1-p)^{n-k} \simeq \frac{1}{k!} a^k \left(1 - \frac{a}{n}\right)^{n-k},$$

where $n(n-1)\ldots(n-k+1) \simeq n^k$ if n is allowed to become large enough and k is held fixed. Hence in the limit as $n \to \infty$, $p \to 0$, and $k \ll n$, we obtain

$$b(k; n, p) \simeq \frac{1}{k!} a^k \left(1 - \frac{a}{n}\right)^{n-k} \to \frac{a^k}{k!} e^{-a}. \tag{1.9-1}$$

Thus in situations where the binomial law applies with $n \gg 1$, $p \ll 1$ but $np = a$ is a finite constant, we can use the approximation

$$b(k; n, p) \simeq \frac{a^k}{k!} e^{-a}. \tag{1.9–2}$$

Example 1.9–1: A computer contains 10,000 components. Each component fails independently from the others and the yearly failure probability per component is 10^{-4}. What is the probability that the computer will be working at the end of the year? Assume that the computer fails if one or more components fail.

Solution:

$$p = 10^{-4} \qquad n = 10{,}000, \qquad k = 0, \qquad np = 1$$

Hence

$$b(0; 10{,}000, 10^{-4}) = \frac{1^0}{0!}e^{-1} = \frac{1}{e} = 0.368.$$

Example 1.9–2: Suppose that n independent points† are placed at random in an interval $(0, T)$. Let $0 < t_1 < t_2 < T$ and $t_2 - t_1 \triangleq \tau$. Let $\tau/T \ll 1$ and $n \gg 1$. What is the probability of observing exactly k points in τ seconds? (Figure 1.9–1.)

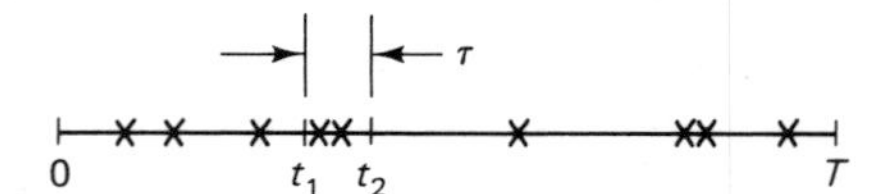

Figure 1.9–1 Points placed at random on a line. Each point is placed with equal likelihood anywhere along the line.

Solution: Consider a single point placed at random in $(0, T)$. The probability of the point appearing in τ is τ/T. Let $p = \tau/T$. Every other point has the same probability of being in τ seconds. Hence the probability of finding k points in τ seconds is the binomial law

$$P[k \text{ points in } \tau \text{ sec.}] = \binom{n}{k} p^k q^{n-k}. \tag{1.9–3}$$

With $n \gg 1$, we use the approximation in Equation 1.9–1 to give

$$b(k; n, p) \simeq \left(\frac{n\tau}{T}\right)^k \frac{e^{-(n\tau/T)}}{k!}, \tag{1.9–4}$$

where n/T can be interpreted as the "average" number of points per unit interval. Equations 1.9–1 and 1.9–4 are examples of the *Poisson probability law.*

The Poisson law with parameter a $(a > 0)$ is defined by‡

$$P[k \text{ events}] = e^{-a}\frac{a^k}{k!}, \tag{1.9–5}$$

where $k = 0, 1, 2, \ldots$. With $a \triangleq \lambda\tau$, where λ is the average number of events per unit time and τ is the length of the interval $(t, t + \tau)$, the probability of k events in τ is

$$P(k; t, t + \tau) = e^{-\lambda\tau}\frac{(\lambda\tau)^k}{k!}. \tag{1.9–6}$$

In Equation 1.9–6 we assumed that λ was independent of t. If λ depends on t, the product $\lambda\tau$ gets replaced by the integral $\int_t^{t+\tau} \lambda(\xi)\, d\xi$, and the probability of k events in the interval $(t, t + \tau)$ is

$$P(k; t, t + \tau) = \exp\left[-\int_t^{t+\tau} \lambda(\xi)\, d\xi\right] \frac{1}{k!}\left[\int_t^{t+\tau} \lambda(\xi)\, d\xi\right]^k. \tag{1.9–7}$$

The Poisson law $P[k \text{ events in } \Delta x]$ or more generally $P[k \text{ events in } (x, x + \Delta x)]$ where x is time, volume, distance, and so forth, and Δx is the interval associated with x is widely used in engineering and sciences. Some typical situations in various fields where the Poisson law is applied are listed on page 27.

† To be more concrete, we can think of the points as the instants (or positions) at which events occur.

‡ The term "event" is more appropriate here than "success."

Physics. In radioactive decay—$P[k$ alpha particles in τ seconds] with λ the average number of emitted alpha particle per second.

Operations research. In planning the size of a switchboard—$P[k$ telephone calls in τ seconds] with λ the average number of calls per second.

Biology. In water pollution monitoring—$P[k$ coliform bacteria in 1000 cubic centimers] with λ the average number of coliform bacteria per cubic centimeter.

Transportation. In planning the size of a highway toll facility – $P[k$ automobiles arriving in τ minutes] with λ the average number of automobiles per minute.

Optics. In designing an optical receiver—$P[k$ photons per second over a surface of area A] with λ the average number of photons-per-second-per-unit area.

Communications. In designing a fiber optical transmitter-receiver link—$P[k$ photoelectrons generated at the receiver in one second] with λ the average number of photoelectrons per second.

The parameter λ is often called the *rate parameter.* Its dimensions are events per unit interval, the interval being time, distance, volume, and so forth. When the form of the Poisson law that we wish to use is as in Equation 1.9–6 or 1.9–7, we speak of the Poisson law with *rate parameter* λ or *rate function* $\lambda(t)$.

Example 1.9–3: A switchboard receives on the average 16 calls per minute. If the switchboard can handle at most 24 calls per minute, what is the probability that in any one minute the switchboard will saturate?

Solution: Saturation occurs if the number of calls in a minute exceeds 24. The probability of this event is:

$$P[\text{saturation}] = \sum_{k=25}^{\infty} [\lambda\tau]^k \frac{e^{-\lambda\tau}}{k!} \tag{1.9–8}$$

$$= \sum_{k=25}^{\infty} [16]^k \frac{e^{-16}}{k!} \simeq 0.017 \simeq 1/60. \tag{1.9–9}$$

Thus only once in every 60 minutes (on the "average") will a caller experience saturation.

The evaluation of the sum in Equation 1.9–9 is a tedious job if done directly. However, there are approximations that can be used, that is, if $\lambda\tau \gg 1$ then we can use the *normal approximation to the Poisson law.* Of course, we have not yet discussed the normal probability law (Chapter 2) but we can still use the approximation

$$\sum_{k=a}^{b} e^{-\lambda} \frac{[\lambda\tau]^k}{k!} \simeq \frac{1}{\sqrt{2\pi}} \int_{l_1}^{l_2} \exp[-\tfrac{1}{2}y^2]\, dy, \tag{1.9–10}$$

where

$$l_2 \triangleq \frac{b - \lambda\tau + 0.5}{\sqrt{\lambda\tau}}$$

and

$$l_1 \triangleq \frac{a - \lambda\tau - 0.5}{\sqrt{\lambda\tau}}.$$

The integral on the right side of Equation 1.9–10 is tabulated (see Table 2.4–1 in Chapter 2). Another useful approximation is

$$\frac{e^{-\lambda\tau}[\lambda\tau]^k}{k!} \simeq \frac{1}{\sqrt{2\pi}} \int_{l_3}^{l_4} \exp[-\tfrac{1}{2}y^2]\, dy, \qquad (1.9\text{–}11)$$

where

$$l_4 \triangleq \frac{k - \lambda\tau + 0.5}{\sqrt{\lambda\tau}}$$

and

$$l_3 \triangleq \frac{k - \lambda\tau - 0.5}{\sqrt{\lambda\tau}}.$$

For example, with $\lambda\tau = 5$, and $k = 5$, the error in using the normal approximation of Equation 1.9–11 is less than 1 percent.

*1.10 DERIVATION OF THE POISSON PROBABILITY LAW†

Given the numerous applications of the Poisson law in the physical and biological sciences, one would think that its origin is of somewhat more noble birth than "merely" as a limiting form of the binomial law. Indeed this is the case, and the Poisson law can be derived once three assumptions are made. Obviously these three assumptions should reasonably mirror the characteristics of the underlying physical process; otherwise our results will be of only academic interest. Fortunately, in many situations these assumptions seem to be quite valid.

In order to be concrete, we shall talk about events‡ taking place in *time* (as opposed to, say, length or volume). The Poisson law is based on the following assumptions:

(1) The probability, $P(1; t, t + \Delta t)$, of a single event occurring in $(t, t + \Delta t)$ is proportional to Δt, that is,

$$P(1; t, t + \Delta t) \simeq \lambda(t)\, \Delta t \qquad \Delta t \to 0. \qquad (1.10\text{–}1)$$

† The material in this section, although fundamental, will not be needed until Chapter 6. In general starred material can be omitted on a first reading.

‡ Examples of events that apply here: the emission of an α-particle, a car arriving at a toll gate, a telephone starting to ring, and the like.

In Equation 1.10–1 $\lambda(t)$ is called the Poisson rate parameter.

(2) The probability of $k(k > 1)$ events in $(t, t + \Delta t)$ goes to zero:

$$P(k; t, t + \Delta t) \simeq 0 \qquad \Delta t \to 0, \qquad k = 2, 3, \ldots. \tag{1.10–2}$$

(3) Events in nonoverlapping time intervals are statistically independent.

We now compute the probability $P(k; t, t + \tau)$ of k events in $(t, t + \tau)$. Consider $P(k; t, t + \tau + \Delta t)$; if Δt is very small then in view of Assumptions (1) and (2) there are only two ways of getting k events in $(t, t + \tau + \Delta t)$:

(1) k in $(t, t + \tau)$ and zero in $(t + \tau, t + \tau + \Delta t)$

(2) $k - 1$ in $(t, t + \tau)$ and one in $(t + \tau, t + \tau + \Delta t)$.

Since events (1) and (2) are disjoint, the probabilities add and we can write

$$\begin{aligned} P(k; t, t + \tau + \Delta t) &= P(k; t, t + \tau)P(0; t + \tau, t + \tau + \Delta t) \\ &\quad + P(k - 1; t, t + \tau)P(1; t + \tau, t + \tau + \Delta t) \\ &= P(k; t, t + \tau)[1 - \lambda(t + \tau)\Delta t] \\ &\quad + P(k - 1; t, t + \tau)\lambda(t + \tau)\Delta t. \end{aligned} \tag{1.10–3}$$

If we rearrange terms, divide by Δt and take limits, we obtain

$$\frac{dP(k; t, t+\tau)}{d\tau} = \lambda(t + \tau)[P(k - 1; t, t + \tau) - P(k; t, t + \tau)]. \tag{1.10–4}$$

Thus we obtain a set of recursive first-order differential equations from which we can solve for $P(k; t, t+\tau)$ $k = 0, 1, \ldots.$ We set $P(-1; t, t + \tau) = 0$, since this is the probability of the impossible event. Also, to shorten our notation, we temporarily write $P(k) \triangleq P(k; t, t + \tau)$; thus the dependences on t and τ are submerged but of course are still there.

k = 0:

$$\frac{dP(0)}{d\tau} = -\lambda(t + \tau)P(0). \tag{1.10–5}$$

This is a simple first-order, homogeneous, differential equation for which the solution is

$$P(0) = C\exp\left[-\int_t^{t+\tau} \lambda(\xi)\, d\xi\right].$$

Since $P(0; t, t) = 1$, $C = 1$ and

$$P(0) = \exp\left[-\int_t^{t+\tau} \lambda(\xi)\, d\xi\right]. \tag{1.10–6}$$

Let us define Z by

$$Z \triangleq \int_t^{t+\tau} \lambda(\xi)\, d\xi. \tag{1.10–7}$$

Then

$$P(0) = e^{-Z}. \tag{1.10–8}$$

k = 1: The differential equation is now

$$\frac{dP(1)}{d\tau} + \lambda(t + \tau)P(1) = \lambda(t + \tau)P(0)$$

$$= \lambda(t + \tau)e^{-Z}. \tag{1.10–9}$$

This elementary first-order, nonhomogeneous equation has a solution that is the sum of the homogeneous and driven solutions. For the homogeneous solution, P_h, we already know from the $k = 0$ case that

$$P_h = C_2 e^{-Z}.$$

For the driven solution P_d we use the method of *variation of parameters* to assume that

$$P_d = v(t + \tau)e^{-Z}, \tag{1.10–10}$$

where $v(t + \tau)$ is to be determined. By substituting Equation 1.10–10 into Equation 1.10–9 we readily find that

$$P_d = Ze^{-Z}. \tag{1.10–11}$$

The complete solution is $P(1) = P_h + P_d$. Since $P(1; t, t) = 0$, we obtain $C_2 = 0$ and thus

$$P(1) = Ze^{Z}. \tag{1.10–12}$$

General case. The DE in the general case is

$$\frac{dP(k)}{d\tau} + \lambda(t + \tau)P(k) = \lambda(t + \tau)P(k - 1)$$

and, proceeding by induction, we find that

$$P(k) = \frac{Z^k}{k!}e^{-Z} \qquad k = 0, 1, \ldots. \tag{1.10–13}$$

Equation 1.10–13 is the key result of this section. Recalling the definition of Z (Equation 1.10–7), we can write Equation 1.10–13 as

$$P(k; t, t + \tau) = \frac{1}{k!}\left[\int_t^{t+\tau} \lambda(\xi)\, d\xi\right]^k \exp\left[-\int_t^{t+\tau} \lambda(\xi)\, d\xi\right]. \tag{1.10–14}$$

Equation 1.10–14 is a very general form of the Poisson law. Fortunately, in a large number of physical situations $\lambda(\xi)$ can be approximated by a constant, say, λ. In order words, the rate parameter (*average* number of events per unit time, or space, or distance, and so forth) doesn't change with time (or space or distance, and so forth). Then Equation 1.10–14 undergoes a dramatic decrease in complexity and appears as

$$P(k; t, t + \tau) = e^{-\lambda\tau}\frac{(\lambda\tau)^k}{k!}$$

$$\triangleq P(k; \tau).$$

The last line merely says that since $P(k; t, t + \tau)$ doesn't depend on t, why not drop t from the argument.

1.11 SUMMARY

In this, the first chapter of the book, we have reviewed some different definitions of probability. We developed the axiomatic theory and showed that for a random experiment three important objects were required: (1) the sample space Ω, the sigma field of events $\mathscr{F}$, and a probability measure P. The triplet $(\Omega, \mathscr{F}, P)$ is called the probability space $\mathscr{P}$. We then briefly discussed some important formulas from combinatorics and applied some of these results to develop the binomial probability law. We saw that the binomial law could, when certain limiting conditions were valid, be approximated by the Poisson law. Finally we considered the Poisson law in its own right and showed how its Poisson probabilities were the solutions of recursive differential equations that followed from three assumptions.

PROBLEMS

1.1. In order for a statement such as "Ralph is probably guilty of theft," to have meaning in the relative frequency approach to probability, what kind of data would one need?

1.2. In a spinning-wheel game, the spinning wheel contains the numbers 1 to 9. The contestant wins if an *even* number shows. What is the probability of a win? What are your assumptions?

1.3. A fair coin is flipped three times. The outcomes on each flip are heads (H) or tails (T). What is the probability of obtaining two heads and one tail?

1.4. An urn contains three balls numbered 1, 2, 3. The experiment consists of drawing a ball at random, recording the number, and replacing the ball before the next ball is drawn. This is called sampling with replacement. What is the probability of drawing the same ball twice in two tries?

1.5. An experiment consists of drawing two balls *without* replacement from an urn containing six balls numbered 1 to 6. Describe the sample description space Ω. What is Ω if the ball is replaced before the second is drawn?

1.6. The experiment consists of measuring the heights of each partner of a randomly chosen married couple. (a) Describe Ω in convenient notation; (b) let E be the event that the man is shorter than the woman. Describe E in convenient notation.

1.7. An urn contains ten balls numbered 1 to 10. Let E be the event of drawing a ball numbered no greater than 5. Let F be the event of drawing a ball numbered greater than 3 but less than 9. Evaluate E^c, F^c, EF, $E \cup F$, EF^c, $E^c \cup F^c$, $EF^c \cup E^cF$, $EF \cup E^cF^c$, $(E \cup F)^c$, and $(EF)^c$. Express these events in words.

1.8. An experiment consists of drawing two balls at random, with replacement from an urn containing five balls numbered 1 to 5. Three students "Dim," "Dense," and "Smart" were asked to compute the probability p that the sum of numbers appearing on the two draws equals 5. Dim computed $p = \frac{2}{15}$, arguing that there are 15 distinguishable unordered pairs and only 2 are favorable, that is, (1, 4) and (2, 3). Dense computed $p = \frac{1}{9}$, arguing that there are 9 distinguishable sums (2 to 10), of which only 1 was favorable. Smart computed $p = \frac{4}{25}$, arguing that there were 25 distinguishable ordered outcomes of which 4 were favorable, that is, (4, 1), (3, 2), (2, 3), and (1, 4). Why is $p = \frac{4}{25}$ the correct answer? Explain what is wrong with the reasoning of Dense and Dim.

1.9. Use the axioms given in Equations 1.4–1 to 1.4–3 to show the following ($E \in \mathscr{F}$, $F \in \mathscr{F}$): (a) $P[\phi] = 0$; (b) $P[EF^c] = P[E] - P[EF]$; (c) $P[E] = 1 - P[E^c]$.

1.10. Use the "exclusive-or" operator in Equation 1.3–3 to show that $P[E \oplus F] = P[EF^c] + P[E^cF]$.

1.11. Show that $P[E \oplus F]$ in Problem 1–10 can be written as $P[E \oplus F] = P[E] + P[F] - 2P[EF]$.

1.12. A fair die is tossed twice (a die is said to be fair if all outcomes 1, . . . , 6 are equally likely). Given that a 3 appears on the first toss, what is the probability of obtaining the sum 7 in two tosses?

1.13. A random-number generator generates integers from 1 to 9 (inclusive). All outcomes are equally likely; each integer is generated independently of any previous integer. Let Σ denote the sum of two consecutively generated integers; that is, $\Sigma = N_1 + N_2$. Given that Σ is odd, what is the conditional probability that Σ is 7? Given that $\Sigma > 10$, what is the conditional probability that at least one of the integers is >7? Given that $N_1 > 8$, what is the conditional probability that Σ will be odd?

1.14. In the ternary communication channel shown in Figure P1.14 a 3 is sent three times more frequently than a 1, and a 2 is sent two times more frequently than a 1. A 1 is observed; what is the conditional probability that a 1 was sent?

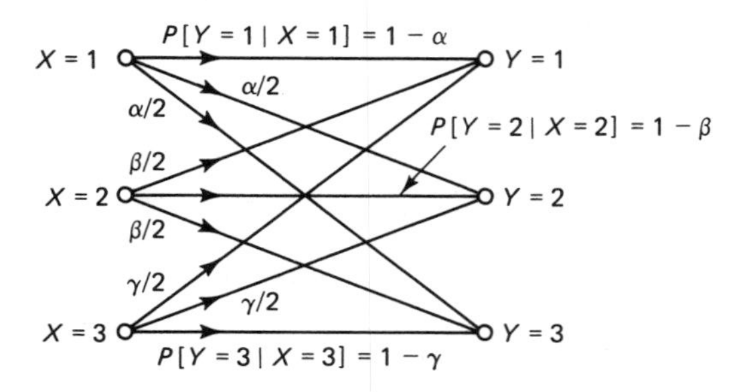

Figure P1.14

1.15. A large class in probability theory is taking a multiple-choice test. For a particular question on the test, the fraction of examinees who know the answer is p; $1 - p$ is the fraction that will guess. The probability of answering a question correctly is unity for an examinee who knows the answer and $1/m$ for a guessee; m is the number of multiple-choice alternatives. Compute the probability that an examinee knew the answer to a question given that he or she has correctly answered it.

1.16. In the beauty-contest problem, Example 1.5–2, what is the probability of picking the most beautiful contestant if we decide *a priori* to choose the *ith* $(1 \leq i \leq N)$?

1.17. In Example 1.5–2, plot the probability of making a correct decision versus α/N, assuming that the "wait-and-see" strategy is adopted. In particular, what is $P[D]$ when $\alpha/N = 0.5$. What does this suggest about the sensitivity of $P[D]$ vis-a-vis α when α is not too far from α_0 and N is large?

1.18. Consider r indistinguishable balls (particles) and n cells (states) where $n > r$. The r balls are placed at random into the n cells (multiple occupancy is possible). What is the probability P that the r balls appear in r preselected cells (one to a cell)?

1.19. Assume that we have r indistinguishable balls and n cells. The cells can at most hold only one ball. As in Problem 1–18 $r < n$. What is the probability P that the r balls appear in r preselected cells?

1.20. Prove that $\sum_{k=0}^{n} b(k; n, p) = 1$, where $b(k; n, p)$ is the binomial law.

1.21. War-game strategists make a living by solving problems of the following type. There are six incoming ballistic missiles (BM's) against which are fired twelve antimissile missiles (AMM's). The AMM's are fired so that two AMM's are directed against each BM. The single-shot-kill probability (SSKP) of an AMM is 0.8. The SSKP is simply the probability that an AMM destroys a BM. Assume that the AMM's don't interfere with each other and that an AMM can, at most, destroy only the BM against which it is fired. Compute the probability that (a) all BM's are destroyed, (b) at least one BM gets through to destroy the target, and (c) exactly one BM gets through.

1.22. Assume in Problem 1.21 that the target was destroyed by the BM's. What is the conditional probability that only one BM got through?

1.23. A smuggler, trying to pass himself off as a glass-bead importer, attempts to smuggle diamonds by mixing diamond beads among glass beads in the proportion of one diamond bead per 1000 glass beads. A harried customs inspector examines a sample of 100 beads. What is the probability that the smuggler will be caught?

1.24. Assume that a faulty receiver produces audible clicks to the great annoyance of the listener. The average number of clicks per second depends on the receiver temperature and is given by $\lambda(\tau) = 1 - e^{-\tau/10}$, where τ is time from turn-on. Derive a formula for the probability of 0, 1, 2, . . . clicks during the first 10 seconds of operation after turn-on. Assume the Poisson law.

1.25. A frequently held lottery sells 100 tickets at $1 per ticket every time it is held. One of the tickets must be a winner. A player has $50 to spend. To maximize the probability of winning at least one lottery, should he buy 50 tickets in one lottery or one ticket in 50 lotteries?

***1.26.** In Problem 1.25, which of the two strategies will lead to a greater expected

* Starred problems can be considered more challenging.

gain for the player? The expected gain if $M(M \leq 50)$ lotteries are played is defined as $\bar{G}_M \triangleq \sum_{i=1}^{M} G_i P(i)$ where G_i is the gain obtained in winning i lotteries and $P(i)$ is the probability of winning i lotteries.

1.27. The circuit shown in Figure P1.27 represents a telephone communication link. Switches α_i $i = 1, \ldots, 6$ are open or closed and operate independently. The probability that a switch is closed is p. Let A_i represent the event that switch i is closed.

(a) In terms of the A_i's write the event that there exists at least one closed path from 1 to 2.

(b) Compute the probability of there being at least one closed path from 1 to 2.

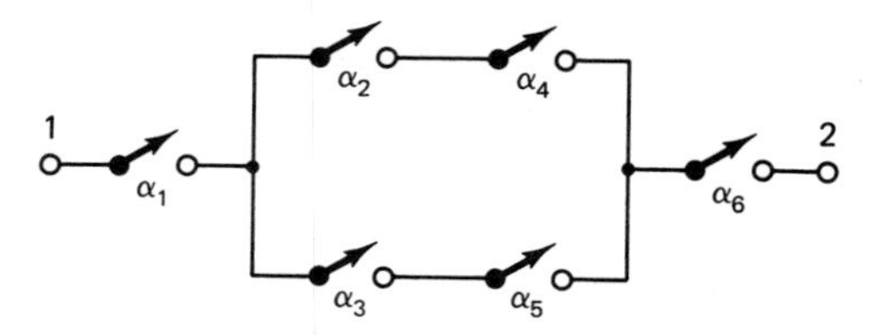

Figure P1.27

***1.28.** (Independence of events in disjoint intervals for Poisson law.) The average number of cars arriving at a tollbooth per minute is λ and the probability of k cars in the interval $(0, T)$ minutes is

$$P(k; 0, T) = e^{-\lambda T}\frac{[\lambda T]^k}{k!}.$$

Consider two disjoint, that is, nonoverlapping intervals, say $(0, t_1)$, and (t_1, T). Then for the Poisson law:

$$\begin{aligned} &P[n_1 \text{ cars in } (0, t_1) \text{ and } n_2 \text{ cars in } (t_1, T)] \\ &\quad = P[n_1 \text{ cars in } (0, t_1)]P[n_2 \text{ cars in } (t_1, T)]. \end{aligned}$$

Thus events in disjoint intervals are independent.

(a) Show that $P[n_1 \text{ cars in } (0, t_1) \mid n_1 + n_2 \text{ cars in } (0, T)]$ is not a function of λ.

(b) In (a) let $T = 2$, $t_1 = 1$ and $n_1 = 5$, $n_2 = 5$. Compute $P[5 \text{ cars in } (0, 1) \mid 10 \text{ cars in } (0, 2)]$.

1.29. An automatic breathing apparatus (B) used in anesthesia fails with probability P_B. A failure means death to the patient unless a monitor system (M) detects the failure and alerts the physician. The monitor system fails with probability P_M. The failure of the systems are independent events. Professor X, an M.D. at a prestigious Eastern medical school, argues that if $P_M > P_B$ installation of M is useless.† Show that Prof. X needs to take a course on probability theory by computing the probability of a patient dying with and without the monitor system in place. Take $P_M = 0.1 = 2P_B$.

† A true story!

1.30. In a particular computer communication network, the host computer broadcasts a packet of data (say L bytes long) to N receivers. The host computer then waits to receive an acknowledgment message from each of the N receivers before proceeding to broadcast the next packet. If the host does not receive all the acknowledgments within a certain time period, it will rebroadcast (retransmit) the same packet. The host computer is then said to be in the "retransmission mode." It will continue retransmitting the packet until all N acknowledgments are received. Then it will proceed to broadcast the next packet.

Let $p \triangleq P$[successful transmission of a single packet to a single receiver along with successful acknowledgment]. Assume that these events are independent for different receivers or separate transmission attempts. Due to random impairments in the transmission media and the variable condition of the receivers (terminals or PCs), we have that $p < 1$.

(a) In a fixed protocol or method of operation, we require that all N of the acknowledgments be received in response to a given transmission attempt for that packet transmission to be declared successful. Let the event $S(m)$ be defined as follows: $S(m) \triangleq$ {a successful transmission of one packet to all N receivers in m or fewer attempts}. Find the probability

$$P(m) \triangleq P[S(m)].$$

(Hint: Consider the complement of the event $S(m)$.)

(b) An improved system operates according to a dynamic protocol as follows. Here we relax the acknowledgment requirement on retransmission attempts, so as to only require acknowledgments from those receivers that have not yet been heard from on previous attempts to transmit the current packet. Let $S_D(m)$ be the same event as in part (a) but using the dynamic protocol. Find the probability

$$P_D(m) \triangleq P[S_D(m)].$$

(Hint: First consider the probabilty of the event $S_D(m)$ for an individual receiver, and then generalize to the N receivers.)

Note: If you try $p = 0.9$ and $N = 5$ you should find that $P(2) < P_D(2)$.

REFERENCES

1–1. E. Merzbacher, *Quantum Mechanics*, New York: John Wiley, 1961.

1–2. A. Kolmogorov, *Foundations of the Theory of Probability*. New York: Chelsea, 1950.

1–3. B. O. Koopman, "The Axioms of Algebra and Intuitive Probability," *Annals of Mathematics* (2), Vol. 41, pp. 269–292.

1–4. A. Papoulis, *Probability, Random Variables, and Stochastic Processes*. New York: McGraw-Hill, 1965, p. 11.

1–5. R. Von Mises, *Wahrscheinlichkeit, Statistic und Wahrheit*. Vienna: Springer-Verlag, 1936.

1–6. W. B. Davenport, Jr., *Probability and Random Processes*. New York: McGraw-Hill, 1970.

1–7. E. Parzen, *Modern Probability Theory and Its Applications*. New York: John Wiley, 1960, p. 119.

1–8. W. Feller, *An Introduction to Probability Theory and Its Applications*, Vol. 1, (2nd Edition), New York: John Wiley, 1950, Chapter 2.

2

RANDOM VARIABLES

2.1 INTRODUCTION

Many random phenomena have outcomes that are sets of real numbers: the voltage $v(t)$, at time t, across a noisy resistor, the arrival time of the next customer at a movie theatre, the number of photons in a light pulse, the brightness level at a particular point on the TV screen, the number of times a light bulb will switch on before failing, the lifetime of a given living person, the number of people on a New York to Chicago train, and so forth. In all these cases the sample description spaces are sets of numbers on the real line.

Even when a sample space Ω is not numerical, we might want to generate a new sample space from Ω that is numerical, that is, converting random speech, color, gray tone, and so forth, to numbers, or converting the physical fitness profile of a person chosen at random into a numerical "fitness" vector consisting of weight, height, blood pressure, heart rate, and so on, or describing the condition of a patient afflicted with, say, black lung disease by a vector whose components are the number and size of lung lesions and the number of lung zones affected.

In science and engineering, we are in almost all instances interested in numerical outcomes, whether the underlying experiment $\mathscr{H}$ is numerical-valued or not. To obtain numerical outcomes, we need a rule or *mapping* from the original sample description space Ω to the real line R. Such a mapping is what a random variable fundamentally is and we discuss it in some detail in the next several sections.

Let us, however, make a remark or two. The concept of a random variable will enable us to replace the original probability space with one in which events are sets of numbers. Thus on the induced probability space of a random variable every event is a subset of R. But is every subset of R always an event? Are there subsets of R that could get us into trouble via violating the axioms of probability? The answer is yes, but fortunately these subsets are not of engineering or scientific importance. We say that they are *nonmeasurable.* Sets of practical importance are of the form $\{x = a\}$, $\{x : a \leq x \leq b\}$, $\{x : a < x \leq b\}$, $\{x : a \leq x < b\}$, $\{x : a < x < b\}$ and their unions and intersections. These five sets are usually abbreviated, respectively, as $[a]$, $[a, b]$ $(a, b]$, $[a, b)$, and (a, b). Intervals that include the end points are said to be *closed*; those that leave out end points are said to be *open*. Intervals can be half-open, and so forth.

The main advantage of dealing with random variables is that we can define certain probability functions that make it both convenient and easy to compute the probabilities of various events. These functions must naturally be consistent with the axiomatic theory. For this reason we must be a little careful in defining events on the real line. Elaboration of the ideas introduced in this section is given next.

2.2 DEFINITION OF A RANDOM VARIABLE

Consider an experiment $\mathscr{H}$ with sample description space Ω. The elements or points of Ω, ζ, are the *random* outcomes of $\mathscr{H}$. If to every ζ, we assign a real number $X(\zeta)$ we establish a correspondence rule between ζ and R, the real line. Such a rule, subject to certain constraints, is called a *random variable.* Thus a random variable $X(\cdot)$ or simply X is not really a variable but a function whose domain is Ω and whose range is some subset of the real line. Being a function, every ζ generates a specific $X(\zeta)$ although for a particular $X(\zeta)$ there may be more than one outcome ζ that produced it. Now consider an event $E_B \subset \Omega$ $(E_B \in \mathscr{F})$.

Through the mapping X, such an event maps into points on the real line (Figure 2.2–1). In particular, the event $\{\zeta : X(\zeta) \leq x\}$, abbreviated

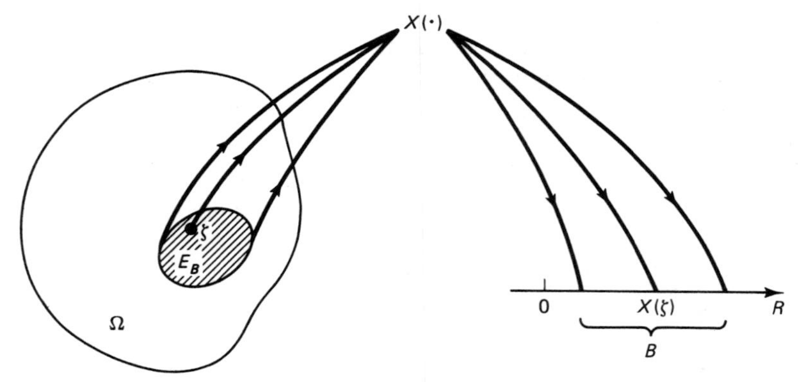

Figure 2.2–1 Symbolic representation of the action of the random variable X.

$\{X \leq x\}$, will denote an event of unique importance, and we should like to assign a probability to it. The probability $P[X \leq x] \triangleq F_X(x)$ is called the *probability distribution function* (PDF) of X: It is shown in more advanced books [2–1], [2–2] that in order for $F_X(x)$ to be consistent with the axiomatic definition of probability, the function X must satisfy the following: For every Borel set of numbers B, the set $\{\zeta : X(\zeta) \in B\}$ must correspond to an event $E_B \in \mathscr{F}$, that is, it must be in the domain of the probability function $P(\cdot)$. Stated somewhat more mathematically, this requirement demands that X can only be a random variable if the *inverse image* under X of all Borel subsets in R, making up the field $\mathscr{B}$† are events. What is an inverse image? Consider an arbitrary Borel set of real numbers B; the set of points E_B in Ω for which $X(\zeta)$ assumes values in B is called the inverse image of the set B under the mapping X. Finally, all sets of engineering interest can be written as countable unions or intersections of events of the form $(-\infty, x]$. The event $\{X \leq x\} \in \mathscr{F}$ gets mapped under X into $(-\infty, x] \in \mathscr{B}$. *Thus if X is a random variable, the set of points $(-\infty, x]$ is an event.*

In many if not most scientific and engineering applications, we are not interested in the actual form of X or the specification of the set Ω. For example, we might conceive of an underlying experiment that consists of heating a resistor and observing the position and velocities of the electrons in the resistor. The set Ω is then the totality of positions and velocities of all N electrons present in the resistor. Let X be the thermal noise current produced by the resistor; clearly $X : \Omega \to R$ although the form of X; that is, the exceedingly complicated equations of quantum electrodynamics that map from electron positions and velocity configurations to current is not specified. What we are really interested in is the behavior of X. Thus although an underlying experiment with sample description space Ω may be implied, it is the real line R and its subsets that will hold our interest and figure in our computations. Under the mapping X we have, in effect, generated a new probability space $(R, \mathscr{B}, P_X)$ where R is the real line, $\mathscr{B}$ is the Borel σ-algebra of all subsets of R generated by countable unions and intersections of sets of the form $(-\infty, x]$ and P_X is a set function assigning a number $P_X[A] \geq 0$ to each set $A \in \mathscr{B}$‡.

In order to assign certain desirable continuity properties to the function $F_X(x)$ at $x = \pm\infty$, we require that the events $\{X = \infty\}$ and $\{X = -\infty\}$ have probability zero. With the latter our specification of a random variable is complete, and we can summarize much of the above discussion in the following definition.

† The σ-algebra of events defined on Ω are denoted by $\mathscr{F}$. The family of Borel subsets of points on R is denoted by $\mathscr{B}$. For definitions see Section 1.3 in Chapter 1.

‡ The extraordinary advantage of dealing with random variables is that a single pointwise function, that is, the distribution function $F_X(x)$ can replace the set function $P_X(\cdot)$ that may be extremely cumbersome to specify, since it must be specified for every event (set) $A \in \mathscr{B}$. See Section 2.3.

Definition. Let $\mathscr{H}$ be an experiment with sample description space Ω. Then the real random variable X is a function whose domain is Ω that satisfies the following: (i) For every Borel set of numbers B, the set $E_B \triangleq \{\zeta \in \Omega, X(\zeta) \in B\}$ is an event and (ii) $P[X = -\infty] = P[X = +\infty] = 0$.

Loosely speaking, when the range of X consists of a countable set of points, X is said to be a discrete random variable; and if the range of X is a continuum, X is said to be continuous. This is a somewhat inadequate definition of discrete and continuous random variables for the simple reason that we often like to take for the range of X the whole real line R. Points in R not actually reached by the transformation X with a nonzero probability are then associated with the impossible event.†

Example 2.2–1: A person, chosen at random in the street, is asked if he or she has a younger brother. If the answer is *no*, the data is encoded as *zero*; if the answer is *yes*, the data is encoded as *one*. The underlying experiment $\mathscr{H}$ has sample description space $\Omega = \{\text{no, yes}\}$, $\mathscr{F} = [\phi, \Omega, \{\text{no}\}, \{\text{yes}\}]$, probabilities $P[\phi] = 0$, $P[\Omega] = 1$, $P[\text{no}] = \frac{3}{4}$ (an assumption), $P[\text{yes}] = \frac{1}{4}$. The associated probabilities for X are $P[\phi] = 0$, $P[X \leq \infty] = P[\Omega] = 1$, $P[X = 0] = P[\text{no}] = \frac{3}{4}$, $P[X = 1] = P[\text{yes}] = \frac{1}{4}$. Take any x_1, x_2 and consider, for example, the probabilities that X lies in sets of the type $[x_1, x_2]$, $[x_1, x_2)$ or $(x_1, x_2]$. Thus

$$P[3 \leq X \leq 4] = P[\phi] = 0$$
$$P[0 \leq X < 1] = P[\text{no}] = \tfrac{3}{4}$$
$$P[0 \leq X \leq 2] = P[\Omega] = 1$$
$$P[0 < X \leq 1] = P[\text{yes}] = \tfrac{1}{4},$$

and so on. Thus every set $\{X = x\}$, $\{x_1 \leq X < x_2\}\{X \leq x_2\}$, and so forth, is related to an event defined on Ω. Hence X is a random variable.

Example 2.2–2: A bus arrives at random in $[0, T]$; let t denote the time of arrival. The sample description space Ω is $\Omega = \{t : t \in [0, T]\}$. A random variable X is defined by

$$X(t) = \begin{cases} 1, & t \in \left[\dfrac{T}{4}, \dfrac{T}{2}\right] \\ 0, & \text{otherwise.} \end{cases}$$

Assume that the arrival time is uniform over $[0, T]$. We can now ask and compute what is $P[X(t) = 1]$ or $P[X(t) = 0]$ or $P[X(t) \leq 5]$.

Example 2.2–3: An urn contains three colored balls. The balls are colored white (W), black (B), and red (R), respectively. The experiment consists of choosing a ball at random from the urn. The sample description space is $\Omega = \{W, B, R\}$. The random variable X is defined by

$$X(\zeta) = \begin{cases} \pi, & \zeta = W \text{ or } B \\ 0, & \zeta = R. \end{cases}$$

† An alternative definition is the following: X is discrete if $F_X(x)$ is a staircase-type function, and X is continuous if $F_X(x)$ is a continuous function. Some random variables cannot be classified as discrete or continuous; they are discussed in Section 2.5.

We can ask and compute the probability $P[X \le x_1]$ where x_1 is any number. Thus $\{X \le 0\} = \{R\}$, $\{2 \le X < 4\} = \{W, B\}$. The computation of the associated probabilities are left as an exercise.

Example 2.2–4: A spinning wheel and pointer game has 50 slots numbered $n = 0, 1, \dots, 49$. The experiment consists of spinning the wheel. Thus $\Omega = \{0, 1, \dots, 49\}$. Because the players are interested only in *even* or *odd* outcomes, the only events that are defined are $\{\phi, \Omega, \text{even}, \text{odd}\}$. Let $X = n$, that is, if n shows up, X assumes that value. Is X a random variable? Note that $\{2 \le X \le 3\}$ cannot be associated with an event. Hence X is not a valid random variable on this probability space.

2.3 PROBABILITY DISTRIBUTION FUNCTION

In Example 2.2–1 the induced event space under X includes $\{0, 1\}$, $\{0\}$, $\{1\}$, ϕ, for which the probabilities are $P[X = 0 \text{ or } 1] = 1$, $P[X = 0] = \frac{3}{4}$, $P[X = 1] = \frac{1}{4}$ and $P[\phi] = 0$. From these probabilities, we can infer any other probabilities such as, for example, $P[X \le 0.5]$. In many cases it is awkward to write down $P[\cdot]$ for every event. For this reason we introduce a pointwise probability function called the probability distribution function (PDF). The PDF† is a function of x, which contains all the information necessary to compute $P[E]$ for any E in the Borel field of events. The PDF, $F_X(x)$ is defined by

$$F_X(x) = P[X \le x] = P_X[(-\infty, x]]. \tag{2.3–1}$$

For the present we shall denote random variables by capital letters, that is, X, Y, Z and the values they can take by lower-case letters x, y, z. The subscript X on $F_X(x)$ associates it with the random variable for which it is the PDF. Thus writing $F_X(y)$ is perfectly consistent notation and refers to $P[X \le y]$. If $F_X(x)$ is discontinuous at a point, say, x_o then $F_X(x_o)$ will be taken to mean the value of the PDF immediately to the right of x_o (continuity from the right property) while $F_X(x_o^-)$ will denote the value immediately to the left.

Properties‡ of $F_X(x)$

(i) $F_X(\infty) = 1$, $F_X(-\infty) = 0$

(ii) $x_1 \le x_2 \rightarrow F_X(x_1) \le F_X(x_2)$, that is, $F_X(x)$ is a nondecreasing function of x.

(iii) $F_X(x)$ is continuous from the right, that is,

$$F_X(x) = \lim_{\varepsilon \to 0} F_X(x + \varepsilon) \qquad \varepsilon > 0.$$

† PDF should not be confused with pdf which will stand for *probability density function*, a function to be introduced shortly.

‡ Properties (i) and (iii) require proof. This is furnished with the help of extended axioms in Chapter 6. Also see Wilbur F. Davenport [2–3, Chapter 4].

Proof of (ii). Consider the event $\{x_1 < X \le x_2\}$ with $x_2 > x_1$. The set $(x_1, x_2]$ is nonempty and $\in \mathscr{B}$. Hence

$$0 \le P[x_1 < X \le x_2] \le 1.$$

But

$$\{X \le x_2\} = \{X \le x_1\} \cup \{x_1 < X \le x_2\}$$

and

$$\{X \le x_1\}\{x_1 < X \le x_2\} = \phi.$$

Hence

$$F_X(x_2) = F_X(x_1) + P[x_1 < X \le x_2]$$

or

$$P[x_1 < X \le x_2] = F_X(x_2) - F_X(x_1) \ge 0 \quad \text{for} \quad x_2 > x_1. \qquad (2.3\text{–}2)$$

√ **Example 2.3–1:** The experiment consists of observing voltage of the parity bit in a word in computer memory. If the bit is *on*, then $X = 1$; if *off* then $X = 0$. Assume that the *off* state has probability q and the *on* state has probability $1 - q$. The sample space has only two points: $\Omega = \{\text{off, on}\}$.

Computation of $F_X(x)$

(i) $x < 0$: The event $\{X \le x\} = \phi$ and $F_X(x) = 0$.

(ii) $0 \le x < 1$: The event $\{X \le x\}$ is equivalent to the event {off} and excludes the event {on} since

$$X(\text{on}) = 1 > x$$
$$X(\text{off}) = 0 \le x.$$

Hence $F_X(x) = q$.

(iii) $x \ge 1$: The event $\{X \le x\} =$ is the certain event since

$$X(\text{on}) = 1 \le x$$
$$X(\text{off}) = 0 \le x.$$

The solution is shown in Figure 2.3–1.

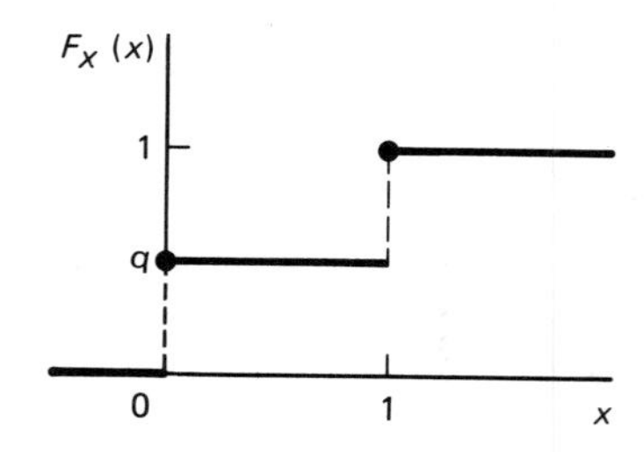

Figure 2.3–1 Probability distribution function associated with the parity bit observation experiment.

√ **Example 2.3–2:** A bus arrives at random in $(0, T]$. Let the random variable X denote the time of arrival. Then clearly $F_X(t) = 0$ for $t \le 0$ and $F_X(T) = 1$ because the former is the probability of the impossible event while the latter is the probability of the certain event. Suppose it is known that the bus is equally or *uniformly* likely to

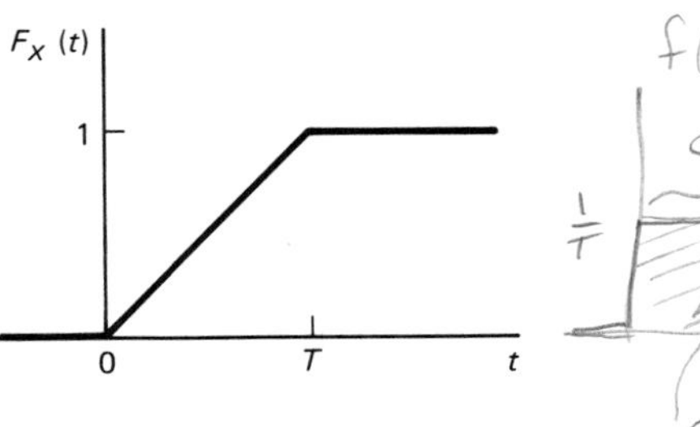

Figure 2.3–2 Probability distribution function of the uniform random variable X of Example 2.3–2.

come at any time within $(0, T]$. Then

$$F_X(t) \begin{cases} 0 & t \le 0 \\ \dfrac{t}{T} & 0 < t \le T \\ 1 & t > T. \end{cases} \tag{2.3–3}$$

Actually Equation 2.3–3 defines "equally likely," not the other way around. The PDF is shown in Figure 2.3–2. In this case we say that X is *uniformly* distributed.

If $F_X(x)$ is a continuous function of x, then

$$F_X(x) = F_X(x^-). \tag{2.3–4}$$

However, if $F_X(x)$ is discontinuous at the point x then, from Equation 2.3.–2

$$\begin{aligned} F_X(x) - F_X(x^-) &= P[x^- < X \le x] \\ &= \lim_{\varepsilon \to 0} P[x - \varepsilon < X \le x] \\ &\triangleq P[X = x]. \end{aligned} \tag{2.3–5}$$

Typically $P[X = x]$ is a discontinuous function of x; it is zero whenever $F_X(x)$ is continuous and nonzero only at discontinuities in $F_X(x)$.

2.4 PROBABILITY DENSITY FUNCTION (pdf)

The pdf, if it exists, is given by

$$f(x) = \frac{dF(x)}{dx}, \tag{2.4–1}$$

where $F(x) \equiv F_X(x)$, since we are dealing with only a single random variable.†

Properties. If $f(x)$ exists then:

(i) $$\int_{-\infty}^{\infty} f(\xi)\, d\xi = F(\infty) - F(-\infty) = 1 \tag{2.4–2}$$

(ii) $$F(x) = \int_{-\infty}^{x} f(\xi)\, d\xi = P[X \le x] \tag{2.4–3}$$

† For the present we shall dispense with the subscript X on $F(\cdot)$ unless we deal with more than one r.v.

(iii) $F(x_2) - F(x_1) = \int_{-\infty}^{x_2} f(\xi)\,d\xi - \int_{-\infty}^{x_1} f(\xi)\,d\xi$

$$= \int_{x_1}^{x_2} f(\xi)\,d\xi = P[x_1 < X \le x_2]. \tag{2.4–4}$$

Interpretation of f(x)

$$P[x < X \le x + \Delta x] = F(x + \Delta x) - F(x).$$

If $F(x)$ is continuous in its first derivative then, for sufficiently small Δx,

$$F(x + \Delta x) - F(x) = \int_{x}^{x+\Delta x} f(\xi)\,d\xi \simeq f(x)\,\Delta x.$$

Hence for small Δx

$$P[x < X \le x + \Delta x] \simeq f(x)\,\Delta x. \tag{2.4–5}$$

Observe that if $f(x)$ exists, then $F(x)$ is continuous and therefore, from Equation 2.3–5, $P(X = x) = 0$.

The univariate normal (Gaussian†) pdf. The pdf is given by

$$f(x) = \frac{1}{\sqrt{2\pi\sigma^2}} e^{-\frac{1}{2}\left[\frac{x-u}{\sigma}\right]^2}. \tag{2.4–6}$$

There are two independent parameters: σ, the standard deviation (σ^2 is called the variance) and μ, the mean. In Chapter 3 we shall discuss what these quantities actually are. When we want to say that a random variable X obeys the normal probability law with mean μ and standard deviation σ, we shall use the symbols $X : N(\mu, \sigma^2)$. The normal pdf is shown in Figure 2.4–1.

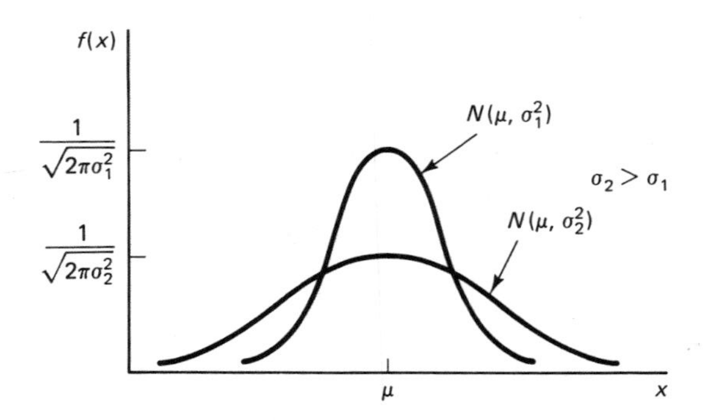

Figure 2.4–1 The normal pdf.

The normal pdf is widely encountered in all branches of science and engineering as well as in social and demographic studies. For example, the IQ of children, the heights of men (or women), the noise voltage produced by a thermally agitated resistor, all are postulated to be approximately normal over a large range of values.

† After the German mathematician/physicist Carl F. Gauss (1777–1855).

Conversion of the Gaussian pdf to the standard normal. Suppose we are given $X:N(\mu, \sigma^2)$ and must evaluate $P[a < X \leq b]$. We have

$$P[a < X \leq b] = \frac{1}{\sqrt{2\pi\sigma^2}} \int_a^b e^{-\frac{1}{2}[\frac{x-u}{\sigma}]^2}\, dx. \tag{2.4-7}$$

With $\beta \triangleq (x - \mu)/\sigma$, $d\beta = (1/\sigma)\, dx$, $b' \triangleq (b - \mu)/\sigma$, $a' \triangleq (a - \mu)/\sigma$, we obtain

$$P[a < X \leq b] = \frac{1}{\sqrt{2\pi}} \int_{a'}^{b'} e^{-\frac{1}{2}x^2}\, dx,$$

$$= \frac{1}{\sqrt{2\pi}} \int_0^{b'} e^{-\frac{1}{2}x^2}\, dx - \frac{1}{\sqrt{2\pi}} \int_0^{a'} e^{-\frac{1}{2}x^2}\, dx.$$

=erf(b')-erf(a')

The function

$$\text{erf}(x) \triangleq \frac{1}{\sqrt{2\pi}} \int_0^x e^{-\frac{1}{2}t^2}\, dt \tag{2.4-8}$$

is sometimes called the error function [erf(x)] although other definitions of erf(x) exist.† The erf(x) is tabulated in Table 2.4–1 and is plotted in Figure 2.4–2.

Hence if $X:N(\mu, \sigma^2)$ then

$$P[a < X \leq b] = \text{erf}\left(\frac{b - \mu}{\sigma}\right) - \text{erf}\left(\frac{a - \mu}{\sigma}\right). \tag{2.4-9}$$

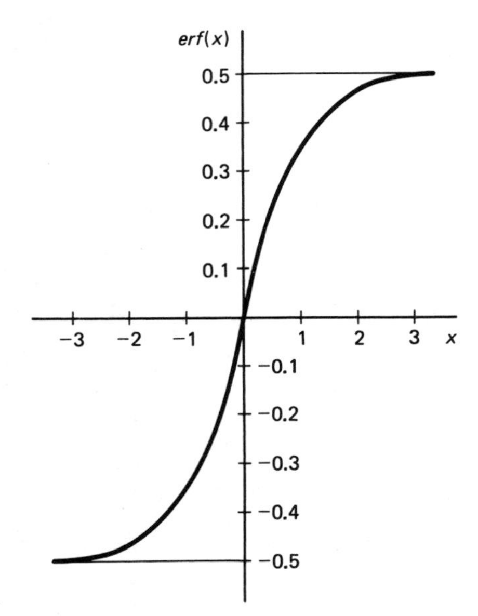

Figure 2.4–2 erf(x) versus x.

† For example, a widely used definition of erf(x) is $\text{erf}(x) \triangleq (2/\sqrt{\pi}) \int_0^x e^{-t^2}\, dt$. If we call this $\text{erf}(x) \triangleq \text{erf}_2(x)$ and we call erf(x) of Equation 2.4–8 $\text{erf}_1(x)$, then $\text{erf}_1(x) = \frac{1}{2}\text{erf}_2(x/\sqrt{2})$.

erf(-x) = -erf(x)

TABLE 2.4–1

$$\text{erf}(x) = \frac{1}{\sqrt{2\pi}} \int_0^x \exp(-\tfrac{1}{2}t^2)\, dt$$

x	erf(x)	x	erf(x)
0.05	0.01994	2.05	0.47981
0.10	0.03983	2.10	0.48213
0.15	0.05962	2.15	0.48421
0.20	0.07926	2.20	0.48609
0.25	0.09871	2.25	0.48777
0.30	0.11791	2.30	0.48927
0.35	0.13683	2.35	0.49060
0.40	0.15542	2.40	0.49179
0.45	0.17364	2.45	0.49285
0.50	0.19146	2.50	0.49378
0.55	0.20884	2.55	0.49460
0.60	0.22575	2.60	0.49533
0.65	0.24215	2.65	0.49596
0.70	0.25803	2.70	0.49652
0.75	0.27337	2.75	0.49701
0.80	0.28814	2.80	0.49743
0.85	0.30233	2.85	0.49780
0.90	0.31594	2.90	0.49812
0.95	0.32894	2.95	0.49840
1.00	0.34134	3.00	0.49864
1.05	0.35314	3.05	0.49884
1.10	0.36433	3.10	0.49902
1.15	0.37492	3.15	0.49917
1.20	0.38492	3.20	0.49930
1.25	0.39434	3.25	0.49941
1.30	0.40319	3.30	0.49951
1.35	0.41149	3.35	0.49958
1.40	0.41924	3.40	0.49965
1.45	0.42646	3.45	0.49971
1.50	0.43319	3.50	0.49976
1.55	0.43942	3.55	0.49980
1.60	0.44519	3.60	0.49983
1.65	0.45052	3.65	0.49986
1.70	0.45543	3.70	0.49988
1.75	0.45993	3.75	0.49990
1.80	0.46406	3.80	0.49992
1.85	0.46783	3.85	0.49993
1.90	0.47127	3.90	0.49994
1.95	0.47440	3.95	0.49995
2.00	0.47724	4.00	0.49996

Example 2.4–1. Suppose we choose a resistor with resistance R from a batch of resistors with parameters: $\mu = 1000$ ohms with $\sigma = 200$ ohms. What is the probability that R will have a value between 900 and 1100 ohms?

Solution: Assuming that $R : N[1000, (200)^2]$ we compute from Equation 2.4–9

$$P[900 < R \leq 1100] = \text{erf}(0.5) - \text{erf}(-0.5).$$

But $\text{erf}(-x) = -\text{erf}(x)$ (deduced from Equation 2.4–8). Hence

$$P[900 < R \leq 1100] = 0.38.$$

Using Figure 2.4–3 as an aid in our reasoning, we readily deduce the

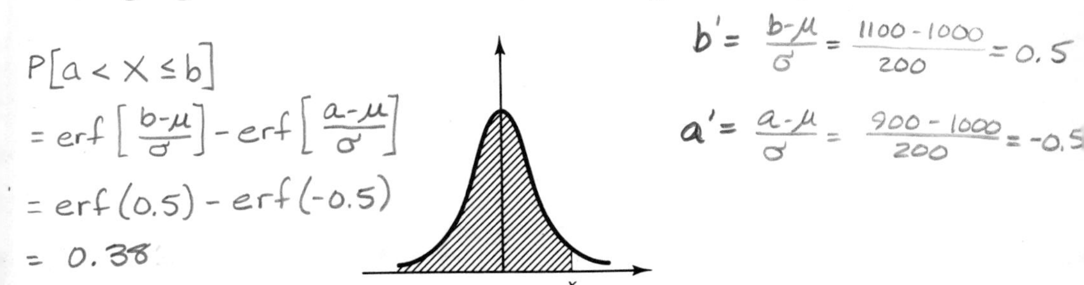

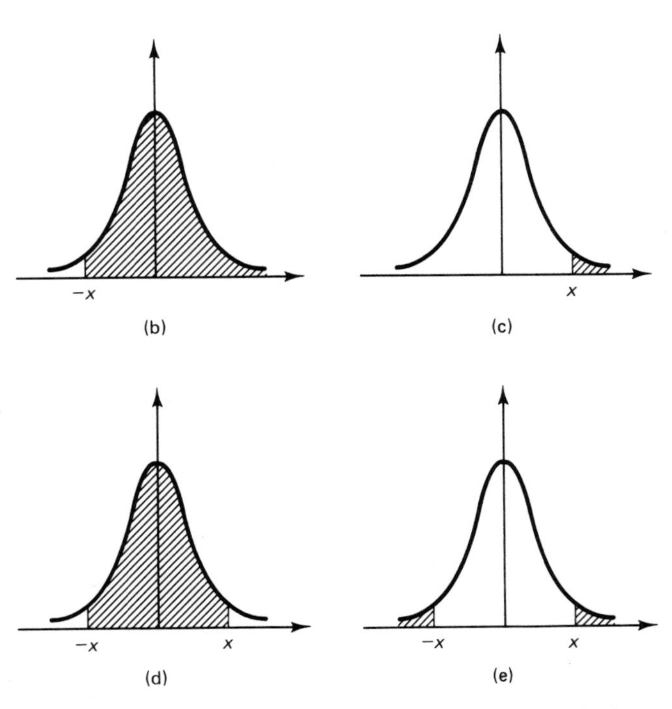

Figure 2.4–3 The areas of the shaded region under curves are: (a) $P[X \leq x]$; (b) $P(X > -x)$; (c) $P(X > x)$; (d) $P(-x < X \leq x)$; and (e) $P(|X| > x)$.

following for $X: N(0, 1)$. Assume $x > 0$ then

$$P[X \le x] = \tfrac{1}{2} + \text{erf}(x) \tag{2.4–10a}$$

$$P[X > -x] = \tfrac{1}{2} + \text{erf}(x) \tag{2.4–10b}$$

$$P[X > x] = \tfrac{1}{2} - \text{erf}(x) \tag{2.4–10c}$$

$$P[-x < X \le x] = 2\,\text{erf}(x) \tag{2.4–10d}$$

$$P[|X| > x] = 1 - 2\,\text{erf}(x). \tag{2.4–10e}$$

Note that since $\text{erf}(-x) = -\text{erf}(x)$, the first three formulas remain valid for $x < 0$ also.

Three Other Common-Density Functions

1. Rayleigh ($\sigma > 0$):

$$f(x) = \frac{x}{\sigma^2} e^{-x^2/2\sigma^2} u(x). \tag{2.4–11}$$

The function $u(x)$ is the unit step, that is, $u(x) = 1$, $x \ge 0$, $u(x) = 0$, $x < 0$. Thus $f(x) = 0$ for $x < 0$. Examples of where the Rayleigh pdf shows up are in rocket-landing errors, random fluctuations in the envelope of certain waveforms, and radial distribution of misses around the bull's-eye at a rifle range.

2. Exponential ($\mu > 0$):

$$f(x) = \frac{1}{\mu} e^{-x/\mu} u(x). \tag{2.4–12}$$

The exponential law occurs, for example, in waiting-time problems, lifetime of machinery, and in describing the intensity variations of incoherent light.

3. Uniform ($b > a$)

$$\begin{aligned} f(x) &= \frac{1}{b - a} \qquad && a < x < b \\ &= 0 && \text{otherwise.} \end{aligned} \tag{2.4–13}$$

The uniform pdf is used in communication theory, in queueing models, and in situations where we have no *a priori* knowledge favoring the distribution of outcomes except for the end points; that is, we don't know when a business call will come but it must come, say, between 9 A.M. and 5 P.M.

The three pdf's are shown in Figure 2.4–4.

There are many other pdf's of importance in engineering and science, and we shall encounter them as we continue our study of probability. They all, however, share the properties that

$$f(x) \ge 0 \tag{2.4–14}$$

$$\int_{-\infty}^{\infty} f(x)\,dx = 1. \tag{2.4–15}$$

When $F(x)$ is not continuous, its finite derivative does not exist and—in the

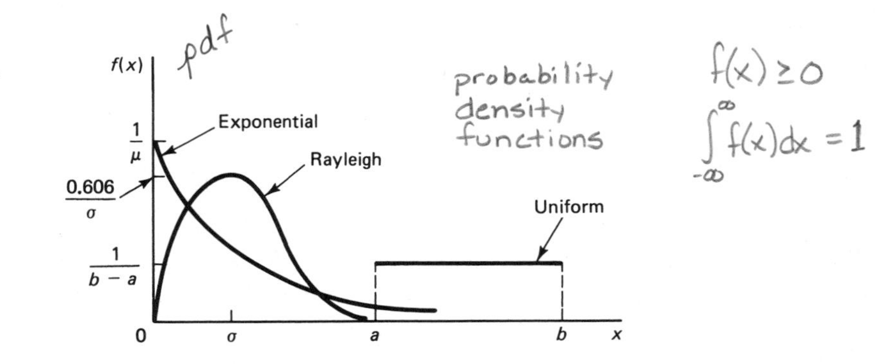

Figure 2.4–4 The Rayleigh, exponential, and uniform pdf's.

classical sense—the pdf doesn't exist. The question of when X has a pdf depends on the classification of X. We consider this next.

2.5 CONTINUOUS, DISCRETE, AND MIXED RANDOM VARIABLES

If $F(x)$ is continuous for every x and its derivative exists everywhere except at a countable set of points, then we say that X is a *continuous random variable* (r.v.). At points x where $F'(x)$ exists, the pdf is $f(x) = F'(x)$. At points where $F'(x)$ doesn't exist, we can assign any positive number to $f(x)$; $f(x)$ will then be defined for every x, and we are free to use the following important formulas:

$$F(x) = \int_{-\infty}^{x} f(\xi)\, d\xi \tag{2.5–1}$$

$$P[x_1 < X \leq x_2] = \int_{x_1}^{x_2} f(\xi)\, d\xi, \tag{2.5–2}$$

and

$$P[B] = \int_{\text{all } \xi:\xi\in B} f(\xi)\, d\xi, \tag{2.5–3}$$

where, in Equation 2.5–3, $B \in \mathscr{B}$, that is, B is an event. Equation 2.5–3 follows from the fact that for a continuous random variable, events can be written as a union of disjoint intervals in R. Thus, for example, let $B = \{\xi : \xi \in \bigcup_{i=1}^{n} I_i,\ I_iI_j = \phi \text{ for } i \neq j\}$, where $I_i = (a_i, b_i]$. Then, clearly

$$\begin{aligned} P[B] &= \int_{a_1}^{b_1} f(\xi)\, d\xi + \int_{a_2}^{b_2} f(\xi)\, d\xi + \ldots + \int_{a_n}^{b_n} f(\xi)\, d\xi \\ &= \int_{\xi:\xi\in B} f(\xi)\, d\xi. \end{aligned} \tag{2.5–4}$$

A discrete *random variable* has a staircase type of distribution function (Figure 2.5–1).

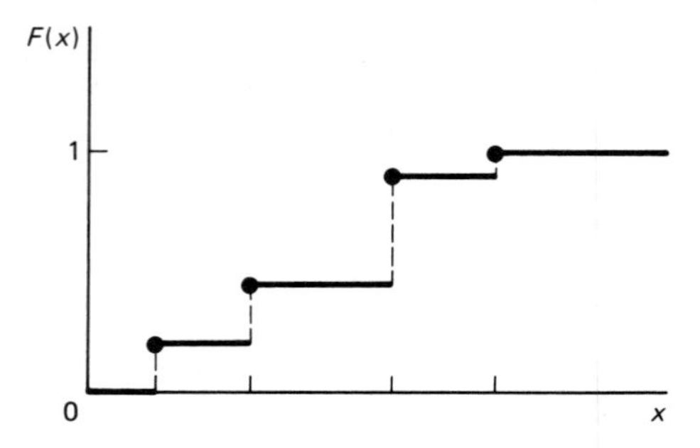

Figure 2.5–1 The probability distribution function for a discrete random variable.

A probability measure for discrete r.v. is the *probability mass*† *function* (PMF). It is defined by

$$P_X(x_i) \triangleq P[X = x_i] = F(x_i) - F(x_i^-),\ddagger \tag{2.5–5}$$

where x_i^- is the point taken on the left of the jump at x_i. The probability mass function is used when there are at most a countable set of outcomes of the random experiment. Indeed $P_X(x_i)$ lends itself to the following frequency interpretation: Perform an experiment n times and let n_i be the number of tries that x_i appears as an outcome. Then, for n large,

$$P_X(x_i) \simeq \frac{n_i}{n}. \tag{2.5–6}$$

Because the PMF is so closely related to the frequency notion of probability, it is sometimes called the *frequency function*.

Since for a discrete r.v. $F(x)$ is not continuous, $f(x)$, strictly speaking, does not exist. Nevertheless, with the introduction of Dirac delta functions,¶ we shall be able to assign pdf's to discrete r.v.'s as well. The PDF for a discrete r.v. is given by

$$F(x) \triangleq P(X \leq x) = \sum_{\text{all } x_i \leq x} P_X(x_i), \tag{2.5–7}$$

more generally, for any event B when X is discrete:

$$P[B] = \sum_{\text{all } x_i \in B} P_X(x_i). \tag{2.5–8}$$

Examples of Probability Mass Functions

1. Bernoulli ($p > 0$, $q > 0$, $p + q = 1$):

$$\begin{aligned} P_X(0) &= p \qquad & P_X(1) &= q \\ P_X(x) &= 0, \qquad & x &\neq 0, 1. \end{aligned} \tag{2.5–9}$$

† Like mass, probability is nonnegative and conserved. Hence the term "mass" in probability *mass* function.

‡ The subscript on the PMF is added to remind us that it is the PMF of a r.v. X. This reminder is important in such equations as (2.5–10) and (2.5–11) where even the argument x, which would normally serve to remind us, is missing.

¶ Also called impulses or impulse functions. Named after the English physicist Paul A. M. Dirac (1902–1984).

The Bernoulli law applies in those situations where the outcome is one of two possible states, that is, whether a particular bit in a digital sequence is "one" or "zero." A r.v. that has the Bernoulli PMF is said to be a Bernoulli r.v.

2. Binomial ($n = 1, 2, \ldots; 0 < p < 1$):

$$P_X(k) = \binom{n}{k} p^k q^{n-k}, \qquad k = 0, 1, 2, \ldots n$$

$$= 0, \qquad \text{otherwise.} \tag{2.5–10}$$

The binomial law applies in games of chance, military defense strategies, failure analysis, and many other situations. A binomial r.v. has a PMF as in Equation 2.5–10.

3. Poisson ($a > 0$):

X is discrete

$$P_X(k) = e^{-a}\frac{a^k}{k!} \qquad k = 0, 1, 2, \ldots$$

$$= 0, \quad \text{otherwise.} \tag{2.5–11}$$

The Poisson law is widely used in every branch of science and engineering (see Section 1.9). A r.v. whose PMF is given by Equation 2.5–11 is said to be a Poisson r.v.

Sometimes a r.v. is neither purely discrete nor purely continuous. We call such a r.v. a *mixed r.v.* The PDF of a mixed r.v. is shown in Figure 2.5–2. Thus $F(x)$ is discontinuous but not a staircase type-function.

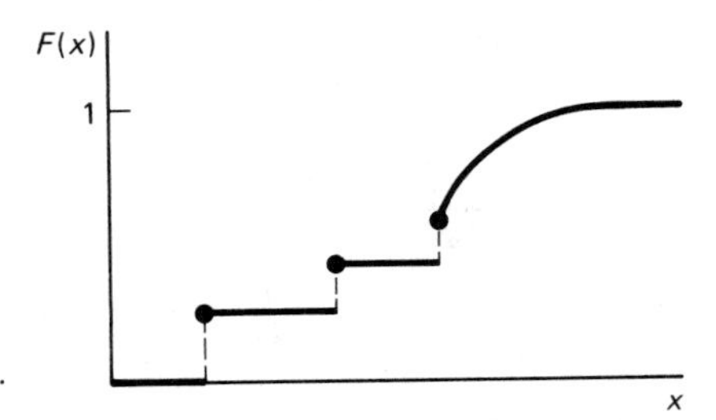

Figure 2.5–2 The PDF of a mixed r.v.

The distinction between continuous and discrete r.v.'s is somewhat artificial. Continuous and discrete r.v.'s are often regarded as different objects even though the only real difference between them is that for the former the PDF is continuous while for the latter it is not. By introducing Dirac delta functions we can, to a large extent, treat them in the same fashion and compute probabilities for both continuous and discrete r.v.'s by using pdf's. For the reader's convenience we furnish a brief review of delta functions.

2.6 THE DIRAC DELTA FUNCTION

The delta function $\delta(x)$ is often "defined" as a function that is zero everywhere except at $x = 0$ where it is infinite such that

$$\int_{-\infty}^{\infty} \delta(x)\,dx = 1. \tag{2.6–1}$$

Another definition is to regard $\delta(x)$ as the limit of one of several pulses; for instance†

$$\delta(x) = \lim_{a\to\infty} a\,\text{rect}(ax) \tag{2.6–2}$$

or‡

$$\delta(x) = \lim_{a\to\infty} ae^{-\pi a^2 x^2}. \tag{2.6–3}$$

These are shown in Figure 2.6–1. The function a rect(ax) in Equation 2.6–2 has discontinuous derivatives whereas $a \exp(-\pi a^2 x^2)$ in Equation 2.6–3 has continuous derivatives.

The exact shape of these functions is immaterial. Their important features are (1) unit area and (2) rapid decrease to zero for $x \neq 0$.

Still another definition is to call any object a delta function if for any function $f(\cdot)$ continuous at x it satisfies the integral equation¶

$$\int_{-\infty}^{\infty} f(y)\delta(y - x)\,dy = f(x). \tag{2.6–4}$$

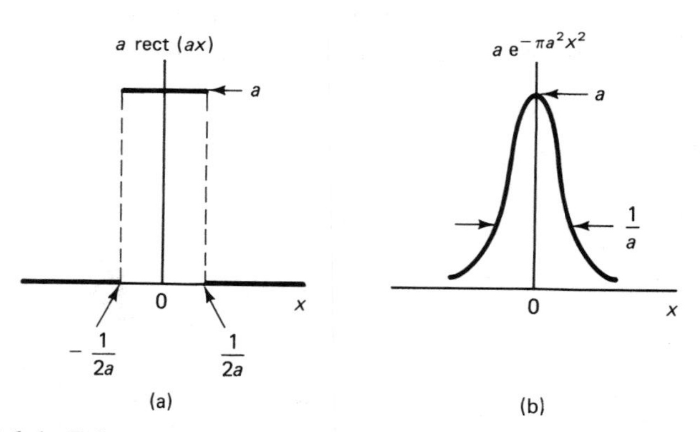

Figure 2.6–1 Pulses approximating a δ-function at $x = 0$ for large a: (a) rectangular; (b) Gaussian.

† rect$(x/b) \triangleq 1$ for $|x| < b/2$ and zero otherwise. This function is called the *rectangular* function.

‡ To show that $\int_{-\infty}^{\infty} \delta(x)\,dx = 1$, show that $a \exp(-\pi a^2 x^2) \triangleq g(x)$ integrates to unity. Do this by considering $(\int g(x)\,dx)^2$ and using polar coordinates.

¶ A word of caution is in order here. Since $\delta(x)$ is zero everywhere except at a single point, its integral (in the Riemann sense) is not defined. Hence Equation 2.6–4 is essentially symbolic, i.e., it implies a limiting operation such as is given in Equations 2.6–2 or 2.6–3. The same comment applies to Equation 2.6–1.

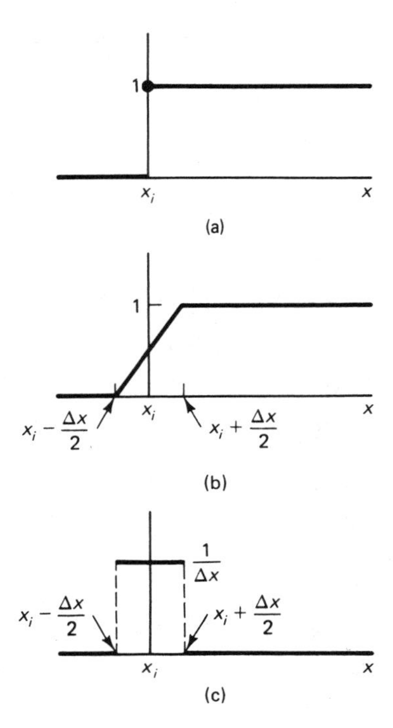

Figure 2.6–2 (a) unit step $u(x - x_i)$; (b) approximation to unit step; (c) derivative of function in (b).

This definition can, of course, be related to the previous one, since either of the pulses in Equations 2.6–2 or 2.6–3 when substituted for $\delta(x)$ in Equation 2.6–4 will essentially furnish the same result when a is large. This follows because the integrand is significantly nonzero only for $x \simeq y$. The integral can, therefore, be approximately evaluated by replacing $f(y)$ by $f(x)$ and moving it outside the integral. Then, since both pulses have unit-area, the result follows. Note that $\delta(x) = \delta(-x)$.

Consider now the unit step $u(x - x_i)$, which is discontinuous at $x = x_i$ with $u(0) \triangleq 1$ (Figure 2.6–2a). The discontinuity can be viewed as the limit of the function shown in Figure 2.6–2b. The derivative is shown in Figure 2.6–2c.

The derivative of the function shown in Figure 2.6–2b is given by

$$\left.\frac{dF}{dx}\right|_{x_i} \triangleq \frac{dF(x_i)}{dx_i} = \lim_{\Delta x \to 0} \frac{1}{\Delta x_i} \operatorname{rect}\left[\frac{x - x_i}{\Delta x}\right]$$
$$= \delta(x - x_i). \tag{2.6–5}$$

Equation 2.6–5 results from Equations 2.6–1 and 2.6–2. Thus, formally, the derivative at a step discontinuity is a delta function with weight† proportional to the height of the jump. It is not uncommon to call $\delta(x - x_i)$ the delta function at "x_i."

† It is also called the *strength* of the delta function and—in sketching the function—the height of the arrow is drawn proportional to its strength.

Returning now to Equation 2.5–7, which can be written as

$$F(x) = \sum_i P_X(x_i)u(x - x_i)$$

and using the result of Equation 2.6–5 enables us to write for a discrete r.v.:

$$f(x) = \frac{dF(x)}{dx} = \sum_i P_X(x_i)\delta(x - x_i), \tag{2.6–6}$$

where we recall that $P_X(x) \triangleq F(x_i) - F(x_i^-)$ and the unit step assures that the summation is over all i such that $x_i \leq x$.

Example 2.6–1: Let X be a discrete r.v. with distribution function as shown in Figure 2.6–3a. The pdf of X is

$$f(x) = \frac{dF}{dx} = 0.2\delta(x) + 0.6\delta(x - 1) + 0.2\delta(x - 3)$$

and is shown in Figure 2.6–3b. To compute probabilities from the pdf for a discrete r.v. great care must be used in choosing the interval of integration. Thus

$$F(x) = \int_{-\infty}^{x^+} f(\xi)\, d\xi,$$

which includes the delta function at x if there is one there.

Similarly $P[x_1 < X \leq x_2]$ involves the interval

$$\underset{x_1}{(}\!\!\longrightarrow\!\!\underset{x_2}{]}$$

and includes the impulse at x_2 (if there is one there) but excludes what happens at x_1. On the other hand $P[x_1 \leq X < x_2]$ involves the interval

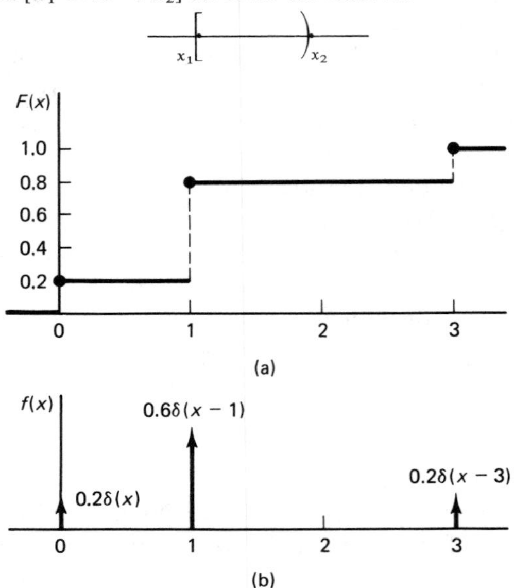

Figure 2.6–3 (a) PDF of a discrete r.v. X; (b) pdf of X using delta functions.

and therefore

$$P(x_1 \le X < x_2) = \int_{x_1^-}^{x_2^-} f(\xi)\, d\xi.$$

Applied to the foregoing example, these formulas give

$$P[X \le 1.5] = F(1.5) = 0.8$$
$$P[1 < X \le 3] = 0.2$$
$$P[1 \le X < 3] = 0.6.$$

9/17

✓ **Example 2.6–2:** The pdf associated with the Poisson law with parameter a

$$f(x) = e^{-a} \sum_{k=0}^{\infty} \frac{a^k}{k!} \delta(x - k).$$

✓ **Example 2.6–3:** The pdf associated with the binomial law $b(k; n, p)$

$$f(x) = \sum_{k=0}^{n} \binom{n}{k} p^k q^{n-k} \delta(x - k).$$

Example 2.6–4: The pdf of a mixed r.v. is shown in Figure 2.6–4. (1) What is the constant K? (2) Compute $P[X \le 5]$, $P[5 \le X < 10]$. (3) Draw the distribution function.

Solution: (1) Since

$$\int_{-\infty}^{\infty} f(\xi)\, d\xi = 1$$

we obtain $10K + 0.25 + 0.25 = 1 \Rightarrow K = 0.05$.

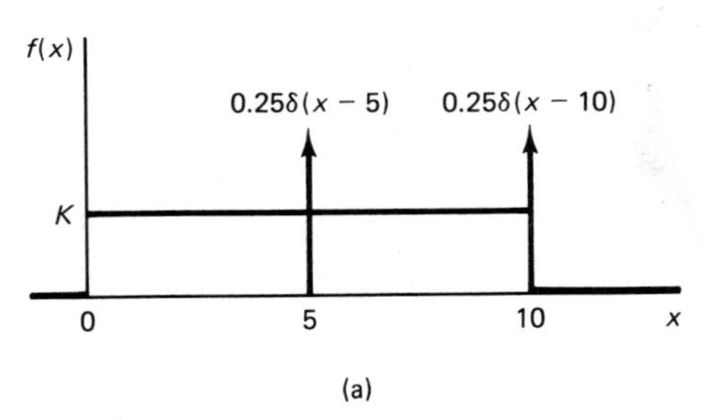

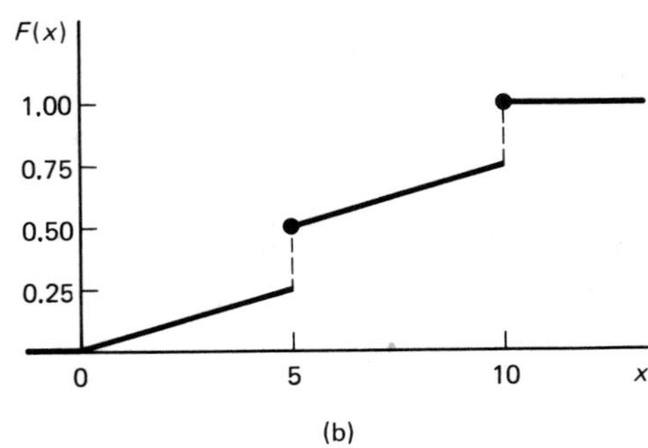

Figure 2.6–4 (a) pdf of a mixed r.v. for Example 2.6–4; (b) computed PDF.

(2) Since $P[X \leq 5] = P[X < 5] + P[X = 5]$ the impulse at $x = 5$ must be included. Hence

$$P[X \leq 5] = \int_0^{5^+} [0.05 + 0.25\delta(\xi - 5)]\, d\xi$$
$$= 0.5.$$

To compute $P(5 \leq X < 10)$, we leave out the impulse at $x = 10$ but include the impulse at $x = 5$. Thus

$$P[5 \leq X < 10] = \int_{5^-}^{10^-} [0.05 + 0.25\delta(\xi - 5)]\, d\xi$$
$$= 0.5.$$

2.7 CONDITIONAL AND JOINT DISTRIBUTIONS AND DENSITIES

Consider the event C consisting of all outcomes $\zeta \in \Omega$ such that $X(\zeta) \leq x$ and $\zeta \in B \subset \Omega$ where B is another event. The event C is then the set product of the two events $\{\zeta : X(\zeta) \leq x\}$ and $\{\zeta : \zeta \in B\}$. We define the *conditional distribution function of X given the event B* as

$$\checkmark \quad F(x \mid B) \triangleq \frac{P[C]}{P[B]} = \frac{P[X \leq x, B]}{P[B]}, \tag{2.7–1}$$

where $P[X \leq x, B]$ is the probability of the joint event $\{X \leq x\} \cap B$ and $P[B] \neq 0$. If $x = \infty$, the event $\{X \leq \infty\}$ is the certain event Ω and since $\Omega \cap B = B$, $F(\infty \mid B) = 1$. Similarly, if $x = -\infty$, $\{X \leq -\infty\} = \phi$ and since $\Omega \cap \phi = \phi$, $F(-\infty \mid B) = 0$. Continuing in this fashion, it is not difficult to show that $F(x \mid B)$ has all the properties of an ordinary distribution, that is, $x_1 \leq x_2 \rightarrow F(x_1 \mid B) \leq F(x_2 \mid B)$.

For example, consider the event $\{X \leq x_2, B\}$ and write (assuming $x_2 \geq x_1$)

$$\{X \leq x_2, B\} = \{X \leq x_1, B\} \cup \{x_1 < X \leq x_2, B\}.$$

Since the two events on the right are disjoint, their probabilities add and we obtain

$$P[X \leq x_2, B] = P[X \leq x_1, B] + P[x_1 < X \leq x_2, B]$$

or

$$P[X \leq x_2 \mid B]P[B] = P[X \leq x_1 \mid B]P[B] + P[x_1 < X \leq x_2 \mid B]P[B].$$

Thus when $P[B] \neq 0$, we obtain after rearranging terms

$$P[x_1 < X \leq x_2 \mid B] = P[X \leq x_2 \mid B] - P[X \leq x_1 \mid B]$$
$$\triangleq F(x_2 \mid B) - F(x_1 \mid B). \tag{2.7–2}$$

Generally the event B will be expressed on the probability space $(R, \mathscr{B}, P_X)$ rather than the original space $(\Omega, \mathscr{F}, P)$. The conditional pdf is simply

$$\checkmark \quad f(x \mid B) \triangleq \frac{dF(x \mid B)}{dx}. \tag{2.7–3}$$

Following are some examples.

Example 2.7–1: Let $B \triangleq \{X \le 10\}$. We wish to compute $F(x \mid B)$.

(i) for $x \ge 10$, the event $\{X \le 10\}$ is a subset of the event $\{X \le x\}$. Hence $P[X \le 10, X \le x] = P[X \le 10]$ and use of Equation 2.7–1 gives

$$F(x \mid B) = \frac{P[X \le x, X \le 10]}{P[X \le 10]} = 1$$

(ii) for $x \le 10$, the event $\{X \le x\}$ is a subset of the event $\{X \le 10\}$. Hence $P[X \le 10, X \le x] = P[X \le x]$ and

$$F(x \mid B) = \frac{P[X \le x]}{P[X \le 10]}.$$

The result is shown in Figure 2.7–1. We leave as an exercise to the reader to compute $F(x \mid B)$ when $B = \{b < X \le a\}$.

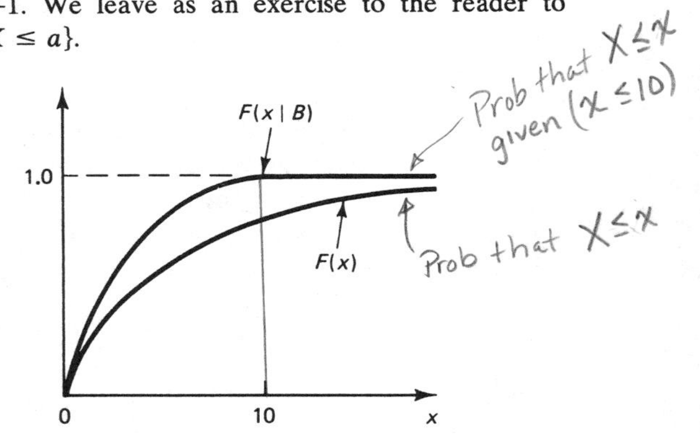

Figure 2.7–1 Conditional and unconditional PDF of X.

Example 2.7–2: Let X be a Poisson r.v. with parameter a. We wish to compute the conditional PMF and pdf of X given $B \triangleq \{X \text{ (is) even}\} = \{X = 0, 2, 4, \ldots\}$. First observe that $P[X \text{ even}]$ is given by

$$P[X = 0, 2, \ldots] = \sum_{k=0,2,\ldots}^{\infty} e^{-a} \frac{a^k}{k!}$$

while for X odd, we have

$$P[X = 1, 3, \ldots] = \sum_{k=1,3,\ldots}^{\infty} e^{-a} \frac{a^k}{k!}.$$

From these relations, we obtain

$$\sum_{k \text{ even}} e^{-a} \frac{a^k}{k!} - \sum_{k \text{ odd}} e^{-a} \frac{a^k}{k!} = \sum_{k=0}^{\infty} e^{-a} \frac{a^k}{k!} (-)^k$$
$$= e^{-2a}$$

and

$$\sum_{k \text{ even}} e^{-a} \frac{a^k}{k!} + \sum_{k \text{ odd}} e^{-a} \frac{a^k}{k!} = 1.$$

Hence $P[X = 0, 2, \ldots] = \frac{1}{2}(1 + e^{-2a})$. Using the definition of a conditional PMF, we obtain

$$P_X(k \mid X \text{ is even}) = \frac{P[X = k, X \text{ is even}]}{P[X \text{ even}]}.$$

If k is even, then $\{X = k\}$ is a subset of $\{X \text{ even}\}$. If k is odd, $\{X = k\} \cap \{X \text{ even}\} = \phi$. Hence

$$P_X(k \mid X \text{ is even}) = \begin{cases} \dfrac{2e^{-a}a^k}{k!\,(1 + e^{-2a})}, & k \text{ even} \\ 0, & k \text{ odd.} \end{cases}$$

The pdf is

$$f(x \mid X \text{ is even}) = \sum_{k=0,2,\ldots}^{\infty} \frac{2e^{-a}a^k}{k!\,(1 + e^{-2a})}\delta(x - k)$$

and the conditional PDF is

$$F(x \mid X \text{ is even}) = \sum_{\text{all } k \le x} P_X(k \mid X \text{ even}).$$

Let us next derive some important formulas involving conditional PDF's and pdf's.

The Distribution Function Written as a Weighted Sum of Conditional Distribution Functions. Equation 1.5–4 in Chapter 1 gave the probability of the event B in terms of n mutually exclusive and exhaustive events $\{A_i\}$ $i = 1, \ldots, n$ defined on the same probability space as B. With $B \triangleq \{X \le x\}$, we immediately obtain from Equation 1.5–4:

$$F(x) = \sum_{i=1}^{n} F(x \mid A_i)P[A_i]. \tag{2.7–4}$$

Equation 2.7–4 describes $F(x)$ as a weighted sum of conditional distribution functions. One way to view Equation 2.7–4 is an "average" over all the conditional PDF's.† Since we haven't yet made concrete the notion of average (this will be done in Chapter 3) we ask only that the reader recall the nomenclature since it is in use in the technical literature.

Example 2.7–3: In the automated manufacturing of computer memory chips, company Z produces one defective chip for every five good chips. The defective chips (DC) have a time of failure X that obeys the PDF

$$F(x \mid DC) = (1 - e^{-x/2})u(x) \quad (x \text{ in months})$$

while the time of failure for the good chips (GC) obeys the PDF

$$F(x \mid GC) = (1 - e^{-x/10})u(x) \quad (x \text{ in months}).$$

The chips are visually indistinguishable. A chip is purchased. What is the probability that the chip will fail before 6 months of use?

Solution: The unconditional PDF for the chip is, from Equation 2.7–4,

$$F(x) = F(x \mid DC)P[DC] + F(x \mid GC)P[GC],$$

where $P[DC]$ and $P[GC]$ are the probabilities of selecting a defective and good chip respectively. From the given data $P[DC] = 1/6$ and $P[GC] = 5/6$. Thus

$$F(6) = [1 - e^{-3}]\tfrac{1}{6} + [1 - e^{-0.6}]\tfrac{5}{6}$$
$$0.158 + 0.376 = 0.534.$$

† For this reason, when $F(x)$ is written as in Equation 2.7–4, it is sometimes called the *average* distribution function.

Bayes' Formula for Probability Density Functions. Consider the events B and $\{X = x\}$ defined on the same probability space. Then from the definition of conditional probability, it seems reasonable to write

$$P[B \mid X = x] = \frac{P[B, X = x]}{P[X = x]}. \tag{2.7–5}$$

The problem with Equation 2.7–5 is that if X is a continuous r.v. then $P[X = x] = 0$. Hence Equation 2.7–5 is undefined. Nevertheless, we can compute $P[B \mid X = x]$ by taking appropriate limits of probabilities involving the event $\{x < X \leq x + \Delta x\}$. Thus consider the expression

$$P[B \mid x < X \leq x + \Delta x] = \frac{P[x < X \leq x + \Delta x \mid B]P[B]}{P[x < X \leq x + \Delta x]}.$$

If we (i) divide numerator and denominator of the expression on the right by Δx, (ii) use the fact that $P[x < X \leq x + \Delta x \mid B] = F(x + \Delta x \mid B) - F(x \mid B)$, and (iii) take the limit as $\Delta x \rightarrow 0$, we obtain

$$P[B \mid X = x] = \lim_{\Delta x \rightarrow 0} P[B \mid x < X \leq x + \Delta x]$$

$$= \frac{f(x \mid B)P[B]}{f(x)}, \qquad f(x) \neq 0. \tag{2.7–6}$$

The quantity on the left is sometimes called the *a posteriori* probability (or *a posteriori* density) of B given $X = x$. Multiplying both sides of Equation 2.7–6 by $f(x)$ and integrating enables us to obtain the important result

$$P[B] = \int_{-\infty}^{\infty} P[B \mid X = x]f(x)\, dx. \tag{2.7–7}$$

In line with the terminology used in this Section, $P[B]$ is sometimes called the *average probability* of B, the usage being suggested by the form of Equation 2.7–7.

Poisson Transform. An important specific example of Equation 2.7–7 is the so-called *Poisson transform* in which B is the event that a random variable Y takes on integer values $k = 0, 1, \ldots$, that is, $B \triangleq \{Y = k\}$ and X is the Poisson parameter, treated here as a random variable with pdf $f_X(x)$. The ordinary Poisson law

$$P[Y = k] = e^{-a}\frac{a^k}{k!}, \tag{2.7–8}$$

where a is the average number of events in a given interval (time, distance, volume, and so forth), treats the parameter a as a constant. But in many situations the underlying phenomenon that determines a is itself random and a must be viewed as a random outcome, that is, the outcome of a random experiment. Thus there are two elements of randomness:† the

† For this reason this situation is sometimes called doubly random or more commonly *doubly stochastic*, stochastic being derived from the Greek word "stochastiko" which means "skillful in aiming at."

random value of a and the random outcome $\{Y = k\}$. When a is random it seems appropriate to replace it by the notation of a random variable, say X. Thus for any given outcome $\{X = x\}$ $P(Y = k \mid X = x)$ is Poisson; but the *unconditional* probability of the event $\{Y = k\}$ is not necessarily Poisson. From Equation 2.7–7 we obtain for the unconditional PMF of Y:

$$\checkmark \quad P_Y(k) = \int_{-\infty}^{\infty} \frac{x^k}{k!} e^{-x} f_X(x)\, dx. \tag{2.7–9}$$

Equation 2.7–9 is known as the Poisson transform and can be used to obtain $f_X(x)$ if $P_Y(k)$ is obtained by experimentation. The mechanism by which $f_X(x)$ is obtained from $P_Y(k)$ is the *inverse Poisson transform*. The derivation of the latter is as follows. Let

$$F(\omega) \triangleq \frac{1}{2\pi} \int_0^{\infty} e^{j\omega x} e^{-x} f_X(x)\, dx \tag{2.7–10}$$

that is, the inverse Fourier transform of $e^{-x} f_X(x)$. Since

$$e^{j\omega x} = \sum_{k=0}^{\infty} [j\omega x]^k / k! \tag{2.7–11}$$

we obtain

$$\checkmark \quad F(\omega) = \frac{1}{2\pi} \sum_{k=0}^{\infty} (j\omega)^k \int_0^{\infty} \frac{x^k}{k!} e^{-x} f_X(x)\, dx$$

$$= \frac{1}{2\pi} \sum_{k=0}^{\infty} (j\omega)^k P_Y(k) \text{ from Equation 2.7–9.} \tag{2.7–12a}$$

Thus $F(\omega)$ *is known if* $P_Y(k)$ *is known.* Taking the forward Fourier transforms of $F(\omega)$ yield

$$e^{-x} f_X(x) = \int_{-\infty}^{\infty} F(\omega) e^{-j\omega x}\, d\omega$$

or

$$\checkmark \quad f_X(x) = e^x \int_{-\infty}^{\infty} F(\omega) e^{-j\omega x}\, d\omega. \tag{2.7–12b}$$

Equation 2.7–12 is the inverse relation we have been seeking. Thus to summarize: If we know $P_Y(k)$ we can compute $F(\omega)$. Knowing $F(\omega)$ enables us to obtain $f_X(x)$ by a Fourier transform. We illustrate the Poisson transform with an important application from optical communication theory.

Example 2.7–4: In an optical communication system, light from the transmitter strikes a photodetector, which generates a photocurrent consisting of valence electrons having become conduction electrons (Figure 2.7–2).

It is known from physics that if the transmitter uses coherent laser light of constant intensity the Poisson parameter X has pdf

$$f_X(x) = \delta(x - x_0) \qquad x_o > 0 \tag{2.7–13}$$

where x_o, except for a constant, is the laser intensity. On the other hand, if the

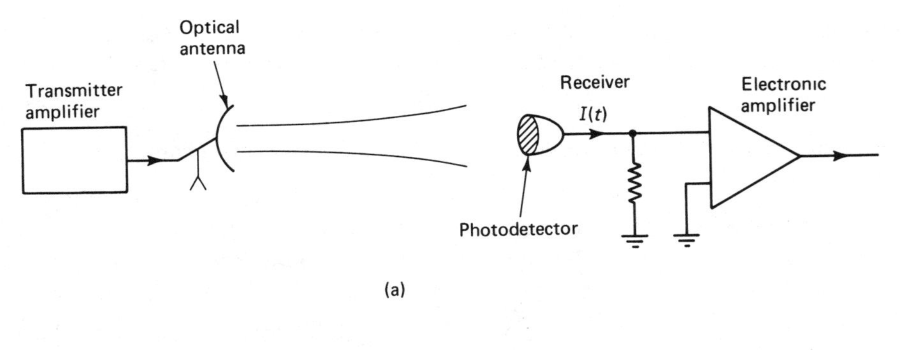

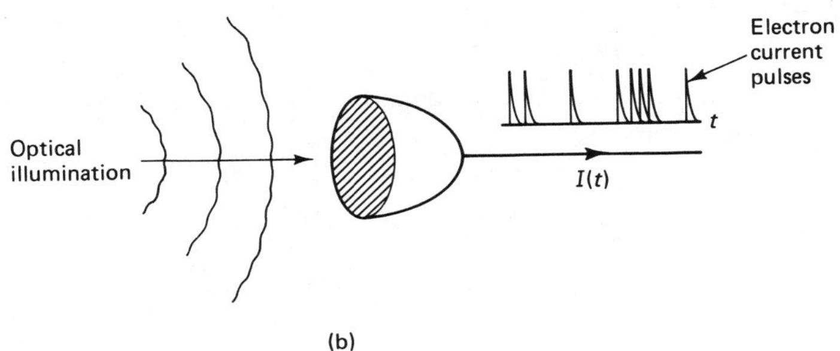

Figure 2.7–2 (a) Optical communication system; (b) output current from photodetector.

transmitter uses thermal illumination, then the Poisson parameter X obeys the exponential law:

$$f_X(x) = \frac{1}{\mu} e^{-(1/\mu)x} u(x), \tag{2.7–14}$$

where $\mu > 0$ is a parameter called the mean value of X. Compute the PMF for the electron-count variable Y.

Solution: For coherent laser illumination we obtain from Equation 2.7–9

$$P_Y(k) = \int_0^\infty \frac{x^k}{k!} e^{-x}\, \delta(x - x_0)\, dx \tag{2.7–15}$$

$$= \frac{x_o^k}{k!} e^{-x_0}. \tag{2.7–16}$$

Thus for coherent laser illumination, the photoelectrons *obey* the Poisson law. For thermal illumination, we obtain†

$$P_Y(k) = \int_0^\infty \frac{x^k}{k!} e^{-x} \frac{1}{\mu} e^{-x/\mu}\, dx$$

$$= \frac{\mu^k}{[1 + \mu]^{k+1}}. \tag{2.7–17}$$

† The evaluation of the integral, while not difficult, is left as an exercise.

This PMF law is known as the geometric distribution and is sometimes called *Bose-Einstein statistics* [2–4]. It obeys the interesting recurrence relation

$$P_Y(k+1) = \frac{\mu}{1+\mu} P_Y(k) \tag{2.7–18}$$

Depending on which illumination applies, the statistics of the photocurrents are widely dissimilar.

Joint Distributions and Densities. Suppose we are given two random variables X, Y defined on the same underlying probability space $(\Omega, \mathcal{F}, P)$. The event $\{X \le x, Y \le y\} \triangleq \{X \le x\} \cap \{Y \le y\}$ †consists of all outcomes $\zeta \in \Omega$ such that $X(\zeta) \le x$ *and* $Y(\zeta) \le y$. The *joint distribution function* of X and Y is defined by

$$F_{XY}(x, y) = P[X \le x, Y \le y]. \tag{2.7–19}$$

Since $\{X \le \infty\}$ and $\{Y \le \infty\}$ are certain events, and for any event B, $B \cap \Omega = B$, we obtain

$$\{X \le x, Y \le \infty\} = \{X \le x\} \tag{2.7–20a}$$

$$\{X \le \infty, Y \le y\} = \{Y \le y\}, \tag{2.7–20b}$$

so that

$$F_{XY}(x, \infty) = F_X(x) \tag{2.7–21a}$$

$$F_{XY}(\infty, y) = F_Y(y). \tag{2.7–21b}$$

The joint pdf, if it exists is given by

$$f_{XY}(x, y) = \frac{\partial^2}{\partial x\, \partial y}[F_{XY}(x, y)]. \tag{2.7–22}$$

By twice integrating Equation 2.7–22 we obtain

$$F_{XY}(x, y) = \int_{-\infty}^{x} d\xi \int_{-\infty}^{y} d\eta f_{XY}(\xi, \eta). \tag{2.7–23}$$

The functions $F_X(x)$ and $F_Y(y)$ are called *marginal* distributions if they are derived from a joint distribution as in Equations 2.7–21. Thus

$$F_X(x) = F_{XY}(x, \infty) = \int_{-\infty}^{x} d\xi \int_{-\infty}^{\infty} dy\, f(\xi, y) \tag{2.7–24}$$

$$F_Y(y) = F_{XY}(\infty, y) = \int_{-\infty}^{y} d\eta \int_{-\infty}^{\infty} dx\, f(x, \eta). \tag{2.7–25}$$

Since the marginal densities are given by

$$f_X(x) = \frac{dF_X(x)}{dx}$$

$$f_Y(y) = \frac{dF_Y(y)}{dy}$$

† Events defined by constraints on random variables e.g. $\{X \le x, Y \le y\}$ are taken to mean that all constraints must be satisfied simultaneously. Thus $\{X \le x, Y \le y\}$ is the *intersection* of $\{X \le x\}$ and $\{Y \le y\}$ not their union.

we obtain by differentiating Equations 2.7–24 and 2.7–25:

$$f_X(x) = \int_{-\infty}^{\infty} f_{XY}(x, y)\, dy \tag{2.7–26}$$

$$f_Y(y) = \int_{-\infty}^{\infty} f_{XY}(x, y)\, dx. \tag{2.7–27}$$

For discrete random variables we obtain similar results. Given the joint probability mass function $P_{XY}(x_i, y_k)$ for all x_i, y_k, we compute the marginal probability mass functions from

$$P_X(x_i) = \sum_{\text{all } y_k} P_{XY}(x_i, y_k) \tag{2.7–28}$$

$$P_Y(y_k) = \sum_{\text{all } x_i} P_{XY}(x_i, y_k). \tag{2.7–29}$$

Independent Random Variables. Two r.v.'s X and Y are said to be independent if the events $\{X \leq x\}$ and $\{Y \leq y\}$ are independent for every combination of x, y. In Section 1.5 two events A and B were said to be independent if $P[AB] = P[A]P[B]$. Taking $AB \triangleq \{X \leq x\} \cap \{Y \leq y\}$, where $A \triangleq \{X \leq x\}$, $B = \{Y \leq y\}$, and recalling that $F_X(x) \triangleq P[X \leq x]$, and so forth, for $F_Y(y)$, it then follows immediately that we can write

$$F_{XY}(x, y) = F_X(x)F_Y(y) \tag{2.7–30}$$

for every x, y if and only if X and Y are independent. Also

$$f_{XY}(x, y) = \frac{\partial^2 F_{XY}(x, y)}{\partial x\, \partial y} \tag{2.7–31}$$

$$= \frac{\partial F_X(x)}{\partial x} \cdot \frac{\partial F_Y(y)}{\partial y}$$

$$= f_X(x)f_Y(y). \tag{2.7–32}$$

From the definition of conditional probability we obtain for independent X, Y:

$$F_X(x \mid Y \leq y) = \frac{F_{XY}(x, y)}{F_Y(y)}$$

$$= F_X(x), \tag{2.7–33}$$

and so forth, for $F_Y(y \mid X \leq x)$. From these results it follows (by differentiation) that for independent events the conditional pdf's are equal to the marginal pdf's, that is,

$$f_X(x \mid Y \leq y) = f_X(x) \tag{2.7–34}$$

$$f_Y(y \mid X \leq x) = f_Y(y). \tag{2.7–35}$$

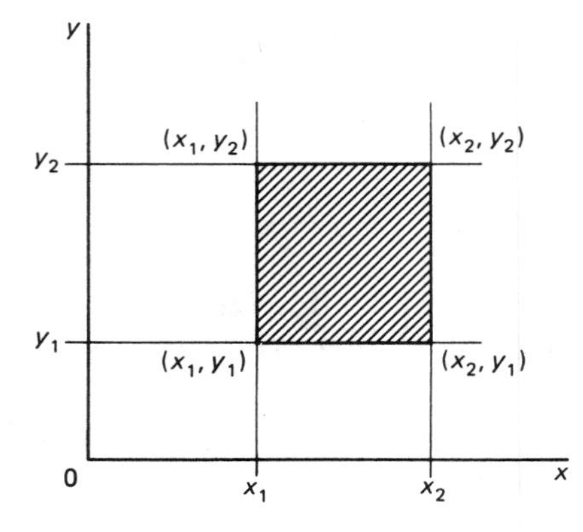

Figure 2.7–3 Point set for the event $\{x_1 < X \le x_2\} \cap \{y_1 < Y \le y_2\}$.

We show next that the events $\{x_1 < X \le x_2\}$ and $\{y_1 < Y \le y_2\}$ are independent if X and Y are independent. The point set associated with the event $\{x_1 < X \le x_2\} \cap \{y_1 < Y \le y_2\}$ is shown in Figure 2.7–3.

Thus, assuming continuous r.v.'s,

$$P[x_1 < X \le x_2, y_1 < Y \le y_2] = \int_{y_1}^{y_2} \int_{x_1}^{x_2} f_{XY}(x, y)\, dx\, dy \tag{2.7–36}$$

$$= F_{XY}(x_2, y_2) - F_{XY}(x_1, y_2) - F_{XY}(x_2, y_1) + F_{XY}(x_1, y_1), \tag{2.7–37}$$

where in the previous line $F_{XY}(x_1, y_1)$ had to be added back because it was subtracted twice. Equation 2.7–37 holds generally. If X and Y are independent $F_{XY}(x, y) = F_X(x)F_Y(y)$. Using this result and a little algebra, we obtain

$$P[x_1 < X \le x_2, y_1 < Y \le y_2] = [F_X(x_2) - F_X(x_1)][F_Y(y_2) - F_Y(y_1)]. \tag{2.7–38}$$

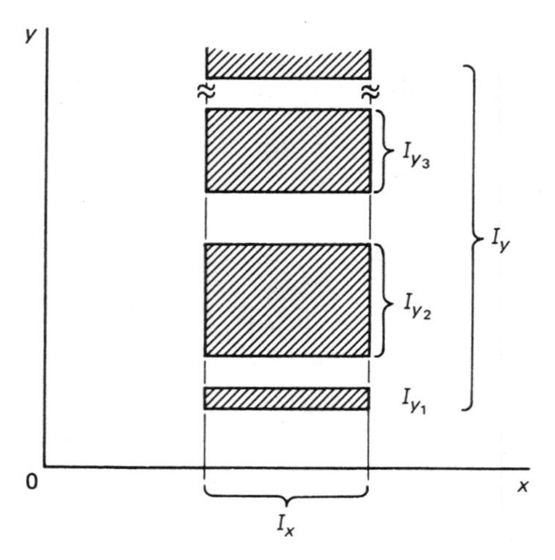

Figure 2.7–4 Generalization of independence over rectangular regions. If X and Y are independent then the events $\{X \in I_x\}$ and $\{Y \in I_y\}$ are independent.

But

$$F_X(x_2) - F_X(x_1) = P[x_1 < X \leq x_2]$$

and

$$F_Y(y_2) - F_Y(y_1) = P[y_1 < Y \leq y_2].$$

Hence

$$P[x_1 < X \leq x_2, y_1 < Y \leq y_2] = P[x_1 < X \leq x_2]P[y_1 < Y \leq y_2]. \tag{2.7–39}$$

The result of Equation 2.7–39 can be generalized to certain multiple rectangular regions in the xy plane. For example, $P[X \in I_x, Y \in I_y] = P[X \in I_x]P[Y \in I_y]$ for the point sets I_x and I_y shown in Figure 2.7–4. The demonstration of this result is left as an exercise.

Example 2.7–5:

$$f_{XY}(x, y) = \frac{1}{2\pi\sigma^2} e^{-(1/2\sigma^2)(x^2+y^2)}$$

$$= \frac{1}{\sqrt{2\pi\sigma^2}} e^{-\frac{1}{2}(x^2/\sigma^2)} \frac{1}{\sqrt{2\pi\sigma^2}} e^{-\frac{1}{2}(y^2/\sigma^2)}. \tag{2.7–40}$$

Hence X and Y are independent r.v.'s.

Joint Densities Involving Nonindependent r.v.'s. Lest the reader thinks that all joint PDF's or pdf's factor, we next consider a case involving nonindependent random variables.

Example 2.7–6:

$$\text{Let } f_{XY}(x, y) = A(x + y) \qquad 0 < x \leq 1, \qquad 0 < y \leq 1$$
$$= 0, \quad \text{otherwise.}$$

(i) What is A? We know that

$$\int_{-\infty}^{\infty} \int_{-\infty}^{\infty} f_{XY}(x, y)\, dxdy = 1.$$

Hence

$$A \int_0^1 dy \int_0^1 xdx + A \int_0^1 dx \int_0^1 ydy = 1 \Rightarrow A = 1.$$

(ii) What are the marginal pdf's?

$$f_X(x) = \int_{-\infty}^{\infty} f_{XY}(x, y)\, dy = \int_0^1 (x + y)\, dy = (xy + y^2/2)\Big|_0^1$$

$$= \begin{cases} x + \frac{1}{2} & 0 < x \leq 1. \\ 0, & \text{otherwise.} \end{cases}$$

Similarly,

$$f_Y(y) = \int_{-\infty}^{\infty} f_{XY}(x, y)\, dx$$

$$= \begin{cases} y + \frac{1}{2} & 0 < y \leq 1 \\ 0, & \text{otherwise.} \end{cases}$$

(iii) What is $F_{XY}(x, y)$? $F_{XY}(x, y) \triangleq P[X \le x, Y \le y]$ so we must integrate over the infinite rectangle with vertices (x, y), $(x, -\infty)$, $(-\infty, -\infty)$, and $(-\infty, y)$. However, only where this rectangle actually overlaps with the region over which $f_{XY}(x, y) \neq 0$, will be there be a contribution to the integral

$$F_{XY}(x, y) = \int_{-\infty}^{x} dx' \int_{-\infty}^{y} dy' f_{XY}(x', y').$$

(A) $x \ge 1$, $y \ge 1$ (Figure 2.7–5a)

$$F_{XY}(x, y) = \int_0^1 \int_0^1 dx'\, dy' f_{XY}(x', y') = 1.$$

(B) $0 < x \le 1$, $y \ge 1$ (Figure 2.7–5b)

$$\begin{aligned} F_{XY}(x, y) &= \int_{y'=0}^{1} dy' \int_{x'=0}^{x} dx'(x' + y') \\ &= \int_{y'=0}^{1} dy' \int_{x'=0}^{x} x'\, dx + \int_{y'=0}^{1} dy'\, y' \int_{x'=0}^{x} dx' \\ &= \frac{x}{2}(x + 1). \end{aligned}$$

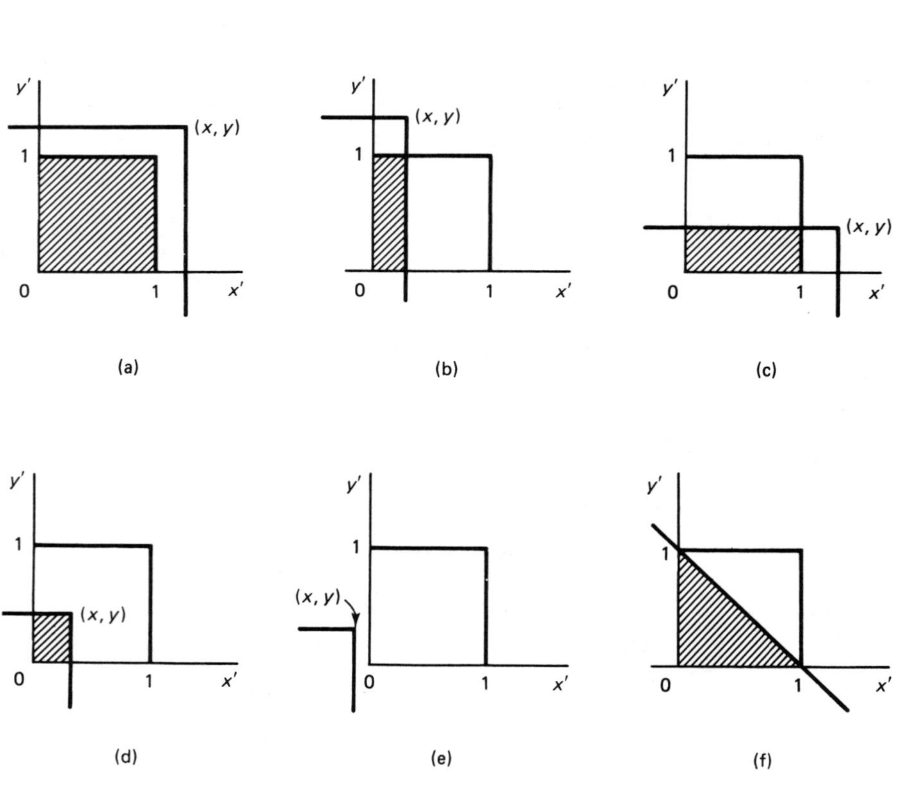

Figure 2.7–5 Shaded region is intersection of supp(f_{XY}) with point set associated with event.

(C) $0 < y \leq 1,\ x \geq 1$ (Figure 2.7–5c)

$$F_{XY}(x, y) = \int_{y'=0}^{y} dy' \int_{x'=0}^{1} dx'(x' + y') = \frac{y}{2}(y + 1).$$

(D) $0 < x \leq 1,\ 0 < y \leq 1$ (Figure 2.7–5d)

$$F_{XY}(x, y) = \int_{y'=0}^{y} dy' \int_{x'=0}^{x} dx'(x' + y') = \frac{yx}{2}(x + y).$$

(E) $x < 0$, any y; or $y < 0$, any x. (Figure 2.7–5e)

$$F_{XY}(x, y) = 0.$$

(F) Compute $P[X + Y \leq 1]$. The point set is the half-space separated by the line $x + y = 1$ or $y = 1 - x$. However, only where this half-space intersects the region over which $f_{XY}(x, y) \neq 0$, will there be a contribution to the integral

$$P[X + Y \leq 1] = \iint_{x'+y' \leq 1} f_{XY}(x', y')\, dx'\, dy'.$$

(See Figure 2.7–5f). Hence

$$\begin{aligned} P[X + Y \leq 1] &= \int_{x'=0}^{1} \int_{y'=0}^{1-x'} (x' + y')\, dx'\, dy' \\ &= \int_{x'=0}^{1} x'(1 - x')\, dx' + \int_{x'=0}^{1} \frac{(1 - x')^2}{2}\, dx' \\ &= \tfrac{1}{3}. \end{aligned}$$

In the previous example we dealt with a pdf that was not factorable. Another example of a joint pdf that is not factorable is

$$f_{XY}(x, y) = \frac{1}{2\pi\sigma^2\sqrt{1 - \rho^2}} \exp\left(\frac{-1}{2\sigma^2(1 - \rho^2)}(x^2 + y^2 - 2\rho xy)\right) \tag{2.7–41}$$

when $\rho \neq 0$. Thus X and Y are not independent. As we shall see in Chapter 3, Equation 2.7–41 is a special case of the *jointly Gaussian probability law* of two r.v.'s. We defer a fuller discussion of this important pdf until we discuss the meaning of the parameter ρ. This we do in Chapter 3.

Conditional Densities. We shall now derive a useful formula for conditional densities involving two r.v. The formula is based on the definition of conditional probability given in Equation 1.5–2. From Equation 2.7–37 we obtain

$$\begin{aligned} &P[x < X \leq x + \Delta x,\ y < Y \leq y + \Delta y] \\ &\quad = F_{XY}(x + \Delta x, y + \Delta y) - F_{XY}(x, y + \Delta y) - F_{XY}(x + \Delta x, y) + F_{XY}(x, y). \end{aligned}$$

Now dividing both sides by $\Delta x\, \Delta y$ and taking limits enables us to write that

$$\lim_{\substack{\Delta x \to 0 \\ \Delta y \to 0}} \frac{P[x < X \leq x + \Delta x, y < Y \leq y + \Delta y]}{\Delta x\, \Delta y} = \frac{\partial^2 F_{XY}}{\partial x\, \partial y} \triangleq f_{XY}(x, y).$$

Hence for $\Delta x, \Delta y$ small

$$P[x < X \leq x + \Delta x, y < Y \leq y + \Delta y] \simeq f_{XY}(x, y)\,\Delta x\,\Delta y, \tag{2.7–42}$$

which is the two-dimensional equivalent of Equation 2.4–5. Now consider $P[y < Y \leq y + \Delta y \mid x < X \leq x + \Delta x]$

$$= \frac{P[x < X \leq x + \Delta x, y < Y \leq y + \Delta y]}{P[x < X \leq x + \Delta x]}$$

$$\simeq \frac{f_{XY}(x, y)\,\Delta x\,\Delta y}{f_X(x)\,\Delta x}.$$

But the quantity on the left is merely

$$F_{Y|B}(y + \Delta y \mid x < X \leq x + \Delta x) - F_{Y|B}(y \mid x < X \leq x + \Delta x),$$

where $B \triangleq \{x < X \leq x + \Delta x\}$. Hence

$$\lim_{\substack{\Delta x \to 0 \\ \Delta y \to 0}} \frac{F_{Y|B}(y + \Delta y \mid x < X \leq x + \Delta x) - F_{Y|B}(y \mid x < X \leq x + \Delta x)}{\Delta y}$$

$$= \frac{f_{XY}(x, y)}{f_X(x)}$$

$$= \frac{\partial F_{Y|X}(y \mid X = x)}{\partial y}$$

$$= f_{Y|X}(y \mid x) \tag{2.7–43}$$

by Equation 2.7–3. The notation $f_{Y|X}(y \mid x)$ reminds us that it is the conditional pdf of Y given the event $\{X = x\}$. We thus obtain the important formula

$$f_{Y|X}(y \mid x) = \frac{f_{XY}(x, y)}{f_X(x)}, \qquad f_X(x) \neq 0. \tag{2.7–44}$$

If we use Equation 2.7–44 in Equation 2.7–27 we obtain the useful formula

$$f_Y(y) = \int_{-\infty}^{\infty} f_{Y|X}(y \mid x) f_X(x)\,dx. \tag{2.7–45}$$

Also

$$f_{X|Y}(x \mid y) = \frac{f_{XY}(x, y)}{f_Y(y)}, \qquad f_Y(y) \neq 0. \tag{2.7–46}$$

The quantity $f_{X|Y}(x \mid y)$ is called the conditional pdf of X given the event $\{Y = y\}$. From Equations 2.7–44 and 2.7–45 it follows that

$$f_{X|Y}(x \mid y) = \frac{f_{Y|X}(y \mid x) f_X(x)}{f_Y(y)}. \tag{2.7–47}$$

We illustrate with an example.

Example 2.7–7: (Laser coherence.) Suppose we observe the light field $U(t)$ being emitted from a laser. Laser light is said to be *temporally coherent*, which means that the light at any two times t_1 and t_2 is statistically dependent if $t_2 - t_1$ is not too large [2–5]. Let $X \triangleq U(t_1)$, $Y \triangleq U(t_2)$ and $t_2 > t_1$. Suppose X and Y are jointly Gaussian as in Equation 2.7–41 with $\sigma^2 = 1$. For $\rho \neq 0$, they are dependent. Still it is easy to show from Equations 2.7–26 and 2.7–27 that the marginal pdf's $f_X(x)$ and $f_Y(y)$ are given by

$$f_X(x) = \frac{1}{\sqrt{2\pi}} e^{-\frac{1}{2}x^2}$$

$$f_Y(y) = \frac{1}{\sqrt{2\pi}} e^{-\frac{1}{2}y^2}$$

and that both are centered about zero. Now suppose that we measure the light at t_1, that is, X and find that $X = x$. Is the pdf of Y still centered at zero, that is, is the average† value of Y still zero?

Solution: We wish to compute $f_{Y|X}(y \mid x)$. From Equation 2.7–44 we obtain

$$f_{Y|X}(y \mid x) = \frac{1}{\sqrt{2\pi(1 - \rho^2)}} \exp\left(-\frac{(y - \rho x)^2}{2(1 - \rho^2)}\right).$$

Thus when $X = x$, the pdf of Y is centered at $y = \rho x$ and not zero as previously. If $\rho x > 0$, Y is more likely to take on positive values and if $\rho x < 0$, Y is more likely to take on negative values. This is in contrast to what happens when X is not observed: The most likely value of Y is then zero!

2.8 A FUNCTION OF A RANDOM VARIABLE (FRV)

The subject we shall discuss in this and the next few sections is one of the most important and interesting in the theory of probability. Before turning to formal definitions we shall introduce the subject via some simple examples.

Example 2.8–1: As is well known from electric-circuit theory the current I flowing through a resistor R (Figure 2.8–1) dissipates an amount of power W given by

$$W \triangleq W(I) = I^2R. \tag{2.8–1}$$

Equation 2.8–1 is an explicit rule that generates for every value of I a number $W(I)$. This rule or correspondence between I and $W(I)$ is called a function and denoted by $W(\cdot)$ or merely W or sometimes even $W(I)$—although the latter notation

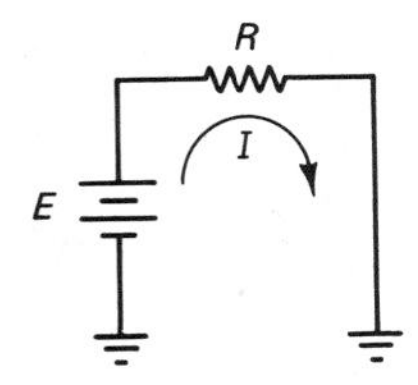

Figure 2.8–1 Ohmic power dissipation in a resistor.

† A concept to be fully developed in Chapter 3.

confuses the difference between the rule and the actual number. Clearly if I were a random variable, the rule $W = I^2R$ generates a new random variable W whose probability distribution function might be quite different from that of I.† Indeed, this alludes to the heart of the problem: Given a rule $g(\cdot)$, and a random variable X with pdf $f_X(x)$, what is the pdf $f_Y(y)$ of the random variable $Y = g(X)$?

The computation of $f_Y(y)$, $F_Y(y)$ or the PMF of Y, that is, $P_Y(y_i)$ can be very simple or quite complex. We illustrate such a computation with a second example, one that comes from communication theory.

Example 2.8–2: A two-level waveform is made multivalued because of the effect of additive Gaussian noise (Figure 2.8–2). A decoder samples the analog waveform $x(t)$ at t_0 and decodes according to the following rule:

Input to Decoder x	Output of Decoder y
If $x(t_0)$:	Then y is assigned:
$\geq \frac{1}{2}$	1
$< \frac{1}{2}$	0

What is the PMF or pdf of Y?

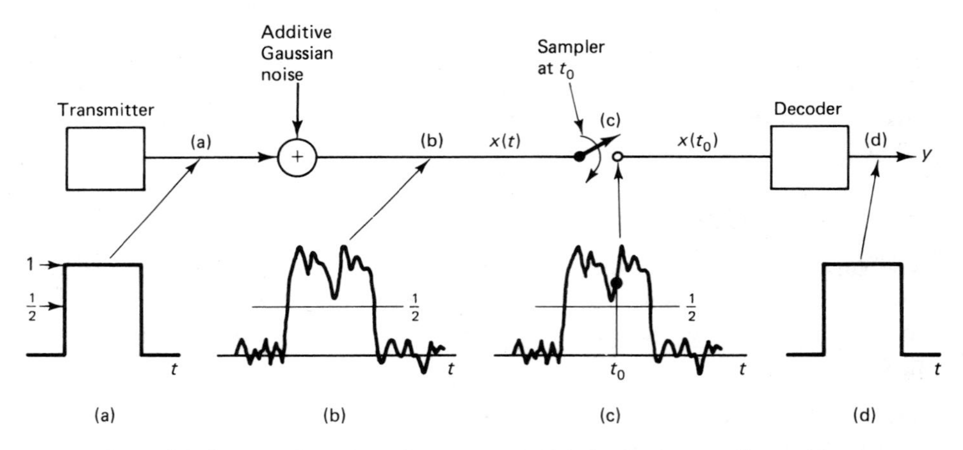

Figure 2.8–2 Decoding of a noise-corrupted digital pulse by sampling and hard clipping.

Solution: Clearly with Y (an r.v.) denoting the output of the decoder, we can write the following events

$$\{Y = 0\} = \{X < 0.5\} \tag{2.8–2a}$$

$$\{Y = 1\} = \{X \geq 0.5\}, \tag{2.8–2b}$$

where $X \triangleq x(t_0)$. Hence if we assume $X : N(1, 1)$, we obtain the following

$$P[Y = 0] = P[X < 0.5] = \frac{1}{\sqrt{2\pi}} \int_{-\infty}^{0.5} e^{-\frac{1}{2}(x-1)^2}\, dx$$

$$= 0.31. \tag{2.8–3}$$

† This is assuming that the composite function $I^2(\zeta)R$ satisfies the required properties of a r.v. (see Section 2.2).

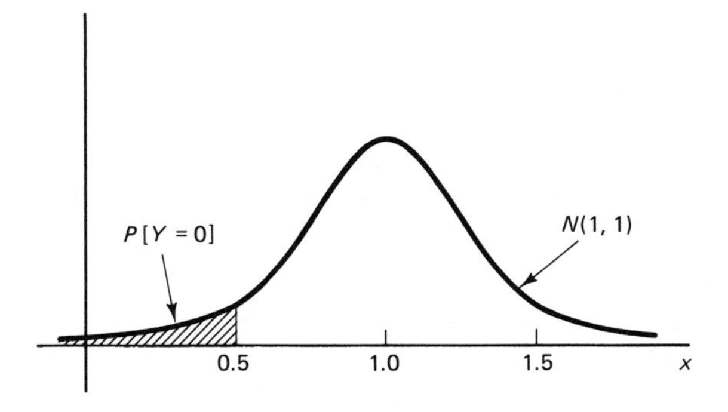

Figure 2.8–3 The area associated with $P[Y=0]$ in Example 2.8–2.

In arriving at Equation 2.8–3 from the previous line we use the normalization procedure explained in Section 2.4 and the fact that for $X:N(0, 1)$ and any $x < 0$ $F_X(x) = \frac{1}{2} - \text{erf}(|x|)$. The area under the normal curve $N(1, 1)$ associated with $P[Y = 0]$ is shown in Figure 2.8–3.

In a similar fashion we compute $P[Y = 1] = 0.69$. Hence the PMF of Y is

$$\begin{aligned} P_Y(0) &= 0.31 \\ P_Y(1) &= 0.69 \\ P_Y(y) &= 0, \quad \text{for} \quad y \neq 0, 1. \end{aligned} \tag{2.8–4}$$

Using delta functions, we can write for the pdf of Y:

$$f_Y(y) = 0.31\,\delta(y) + 0.69\,\delta(y - 1). \tag{2.8–5}$$

Unfortunately not all function-of-a-random-variable (FRV) problems are this easy to evaluate. To gain a deeper insight into the FRV problem we take a closer look at the underlying concept of FRV. The gain in insight will be useful for us in a practical sense when we discuss random processes beginning in Chapter 6.

Functions of a Random Variable (Several Views). There are several different but essentially equivalent views of a FRV. We will now present them. The differences between them are mainly ones of *emphasis.*

Assume as always an underlying probability space $(\Omega, \mathscr{F}, P)$ and a random variable X defined on it. Recall that X is a rule that assigns to every $\zeta \in \Omega$ a number $X(\zeta)$. X transforms the σ-field of events $\mathscr{F}$ into the Borel σ-field $\mathscr{B}$ of sets of numbers on the real line. If R_X denotes the subset of the real line actually reached by X as ζ roams over Ω, then we can regard X as an ordinary function with domain Ω and range R_X. Now, additionally, consider a measurable real function $g(x)$ of the real variable x.

First view $(\mathbf{Y:\Omega \to R_Y})$**.** For every $\zeta \in \Omega$, we generate a number $g[X(\zeta)] \triangleq Y(\zeta)$. The rule Y, which generates the numbers $\{Y(\zeta)\}$ for random outcomes $\{\zeta \in \Omega\}$, is a r.v. with domain Ω and range $R_Y \subset R$.

Finally for every Borel set of real numbers B_Y, the set $\{\zeta : Y(\zeta) \in B_Y\}$ is an event. In particular the event $\{\zeta : Y(\zeta) \leq y\}$ is equal to the event $\{\zeta : g[X(\zeta)] \leq y\}$.

In this view, the stress is on Y as a mapping from Ω to R_Y. The intermediate role of X is suppressed.

Second view (input/output systems view). For every value of $X(\zeta)$ in the range R_X we generate a new number $Y = g(X)$ whose range is R_Y. The rule Y whose domain is R_X and range is R_Y is a function of the random variable X. In this view the stress is on viewing Y as a mapping from one set of real numbers to another. A model for this view is to regard X as the input to a system with transmittance $g(\cdot)$.† For such a system, an input x gets transformed to an output $y = g(x)$ and an input function X gets transformed to an output function $Y = g(X)$. (See Figure 2.8–4.)

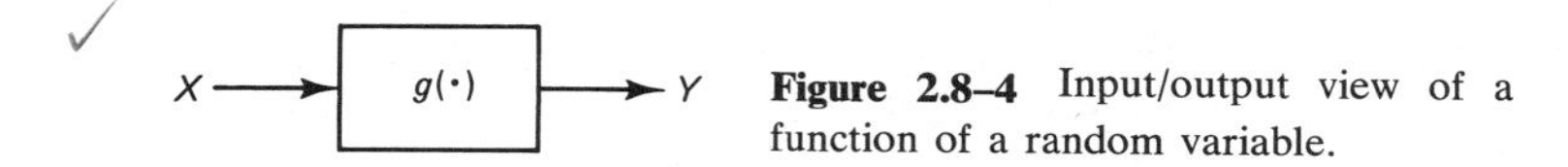

Figure 2.8–4 Input/output view of a function of a random variable.

The input/output viewpoint is the one we stress, partly because it is particularly useful in dealing with random processes where the input consists of waveforms or sequence of random variables. The central problem in computations involving FRV's is: given $g(x)$ and $F_X(x)$ find the point set C_y such that the following events are equal

$$\{\zeta : Y(\zeta) \leq y\} = \{\zeta : g[X(\zeta)] \leq y\}$$
$$= \{\zeta : X(\zeta) \in C_y\}. \tag{2.8–6a}$$

In general we shall submerge the role of ζ and write $\{Y \leq y\} = \{X \in C_y\}$, and so forth. For C_y so determined it follows that

$$P[Y \leq y] = P[X \in C_y] \tag{2.8–6b}$$

since the underlying event is the same. If C_y is empty, then the probability of $\{Y \leq y\}$ is zero.

In dealing with the input/output model it is generally convenient to omit any references to an abstract underlying experiment $\mathscr{H}$ and deal, instead, directly with the r.v.'s X and Y. In this approach the underlying experiments are the observations on X, events are Borel subsets of the real line R, and the set function $P[\cdot]$ is replaced by the distribution function $F_X(\cdot)$. Then Y is a mapping (an r.v.) whose domain is the range, R_X, of X and whose range is a subset, R_Y, of R. The functional properties of X are ignored in favor of viewing X as a mechanism that gives rise to numerically valued random phenomena. In this view the domain of X is irrelevant. (See Problem 2.32.)

† g can be any *measurable* function, i.e., if R_Y is the range of Y then the inverse image of every subset in R_Y generated by countable unions and intersections of sets of the form $\{Y \leq y\}$ is an event.

Third view (composition). In this view Y is regarded as a composition of the functions g and X.

Definition. Let g and X be real functions with domains D_g and D_X respectively. If $D_X = \Omega$ and $D_g = R_X$ (the range of X) then the composition† $Y \triangleq g \circ X$ is said to be a function of the random variable X. In this view Y is regarded as a mapping $\Omega \rightarrow R_X \rightarrow R_Y$.

For additional discussion on the various views of a function of a random variable, see Wilbur Davenport [2–3, p. 174].

2.9 SOLVING PROBLEMS OF THE TYPE $Y = g(X)$

We shall now demonstrate how to solve problems of the type $Y = g(X)$. Eventually we shall develop a formula that will enable us to solve problems of this type very rapidly. However, use of the formula at too early a stage of the development will tend to mask the underlying principles that we shall need later to deal with more difficult problems.

Example 2.9–1: Let X be a uniform r.v. in (0, 1) and let $Y = 2X + 3$. Then we need to find the point set C_y in Equation 2.8–6 to compute $F_Y(y)$. Clearly

$$\{Y \leq y\} = \{2X + 3 \leq y\} = \{X \leq \tfrac{1}{2}(y - 3)\}.$$

Hence C_y is the interval $(-\infty, \frac{1}{2}(y - 3)]$ and

$$F_Y(y) = F_X\left(\frac{y - 3}{2}\right).$$

The pdf of Y is

$$f_Y(y) = \frac{d}{dy}[F_Y(y)] = \frac{d}{dy}\left[F_X\left(\frac{y - 3}{2}\right)\right]$$
$$= \tfrac{1}{2}f_X\left(\frac{y - 3}{2}\right).$$

The solution is shown in Figure 2.9–1.

Generalization. Let $Y = aX + b$ with X a continuous r.v. with pdf $f_X(x)$. Then for $a > 0$:

$$\{Y \leq y\} = \{aX + b \leq y\} = \left\{X \leq \frac{y - b}{a}\right\}.$$

From the definition of PDF:

$$F_Y(y) = F_X\left(\frac{y - b}{a}\right), \tag{2.9–1}$$

and

$$f_Y(y) = \frac{1}{a} f_X\left(\frac{y - b}{a}\right). \tag{2.9–2}$$

† $Y = g \circ X$ is the function $g(X(\cdot))$.

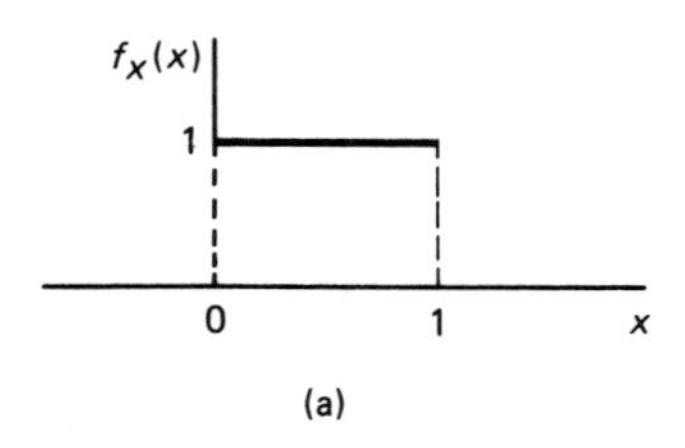

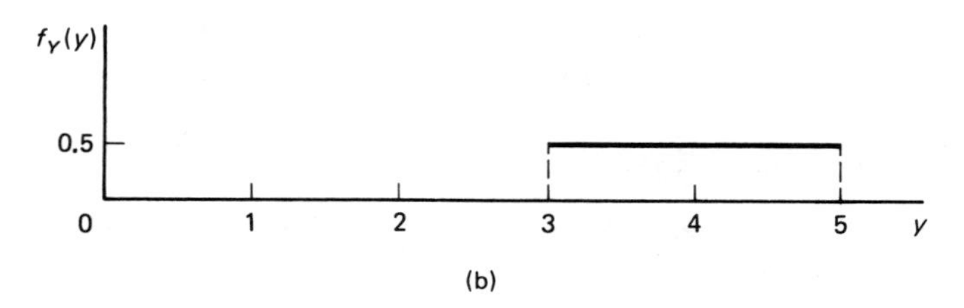

Figure 2.9–1 (a) Original pdf of X; (b) the pdf of $Y = 2X + 3$.

For $a < 0$, the following events are equal†

$$\{Y \leq y\} = \{aX + b \leq y\} = \left\{X \geq \frac{y - b}{a}\right\}.$$

Since the events $\left\{X < \frac{y - b}{a}\right\}$ and $\left\{X \geq \frac{y - b}{a}\right\}$ are disjoint and their union is the certain event, we obtain from Axiom 3

$$P\left[X < \frac{y - b}{a}\right] + P\left[X \geq \frac{y - b}{a}\right] = 1.$$

Finally for a continuous r.v.

$$P\left[X < \frac{y - b}{a}\right] = P\left[X \leq \frac{y - b}{a}\right]$$

and

$$P\left[X \geq \frac{y - b}{a}\right] = P\left[X > \frac{y - b}{a}\right].$$

Thus for $a < 0$

$$F_Y(y) = 1 - F_X\left(\frac{y - b}{a}\right), \tag{2.9–3}$$

and

$$f_Y(y) = \frac{1}{|a|} f_X\left(\frac{y - b}{a}\right), \qquad a \neq 0. \tag{2.9–4}$$

We leave the computation of $F_Y(y)$ when X is not necessarily continuous as an exercise for the reader.

† Recall that we mean the event $\{\zeta : Y(\zeta) \leq y\} = \left\{\zeta : X(\zeta) \geq \frac{y - b}{a}\right\}$.

Example 2.9–2: (Square-law detector.) Let X be a r.v. with continuous PDF $F_X(x)$ and let $Y = X^2$. Then

$$\{Y \leq y\} = \{X^2 \leq y\} = \{-\sqrt{y} \leq X \leq \sqrt{y}\}$$
$$= \{-\sqrt{y} < X \leq \sqrt{y}\} \cup \{X = -\sqrt{y}\}.$$

The probability of the union of disjoint events is the sum of their probabilities. Using the result of Equation 2.4–4 we obtain

$$F_Y(y) = F_X(\sqrt{y}) - F_X(-\sqrt{y}) + P[X = -\sqrt{y}]. \quad (2.9\text{–}6)$$

If X is continuous $P[X = -\sqrt{y}] = 0$. Then for $y > 0$

$$f_Y(y) = \frac{d}{dy}[F_Y(y)]$$
$$= \frac{1}{2\sqrt{y}} f_X(\sqrt{y}) + \frac{1}{2\sqrt{y}} f_X(-\sqrt{y}). \quad (2.9\text{–}7)$$

For $y < 0$, $f_Y(y) = 0$. How do we know this? Recall from Equation 2.8–6 that if C_y is empty, then $P[Y \leq y] = 0$ and hence $f_Y(y) = 0$. For $y < 0$, there are no values of the r.v. X on the real line that satisfy

$$\{-\sqrt{y} \leq X \leq \sqrt{y}\}.$$

Hence $f_Y(y) = 0$ for $y < 0$. If $X:N(0, 1)$ then from Equation 2.9–7:

$$f_Y(y) = \frac{1}{\sqrt{2\pi y}} e^{-\frac{1}{2}y}, \quad y > 0 \quad (2.9\text{–}8)$$
$$= 0, \quad y < 0.$$

Example 2.9–4: Let X be a Bernoulli r.v. with $P[X = 0] = p$ and $P[X = 1] = q$. Then

$$f_X(x) = p\,\delta(x) + q\,\delta(x - 1)$$
$$F_X(x) = pu(x) + q\,u(x - 1),$$

where $u(x)$ is the unit step.

Let $Y \triangleq X - 1$. Then (Figure 2.9–2)

$$F_Y(y) = P[X - 1 \leq y]$$
$$= P[X \leq y + 1]$$
$$= F_X(y + 1)$$
$$= pu(y + 1) + q\,u(y).$$

The pdf is

$$f_Y(y) = \frac{d}{dy}[F_Y(y)] = p\,\delta(y + 1) + q\,\delta(y).$$

Example 2.9–5: (Transformation of PDF's.) Let X have a continuous PDF $F_X(x)$, which is a strict monotone increasing function† of x. Let Y be a r.v. formed from X by the transformation

$$Y = F_X(X). \quad (2.9\text{–}9)$$

To compute $F_Y(y)$, we proceed as usual:

$$\{Y \leq y\} = \{F_X(X) \leq y\}$$
$$= \{X \leq F_X^{-1}(y)\},$$

† In other words $x_2 > x_1$ implies $F_X(x_2) > F_X(x_1)$.

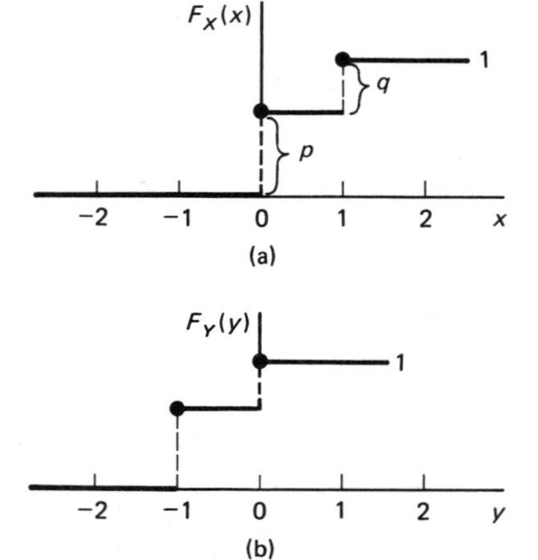

Figure 2.9–2 (a) PDF of X; (b) PDF of $Y = X - 1$.

where F_X^{-1} is the inverse function of $F_X(\cdot)$. Hence

$$0 \leq F_Y(y) = F_X(F_X^{-1}(y)) \leq 1$$

or

$$F_Y(y) = \begin{cases} 1, & y > 1 \\ y, & 0 \leq y \leq 1 \\ 0, & \text{otherwise.} \end{cases} \tag{2.9–10}$$

Equation 2.9–10 results from the facts that (1) for $y \geq 1$, the event $\{F_X(X) \leq y\} = \Omega$ and for $y < 0$, the event $\{F_X(X) \leq y\} = \phi$; and (2) from the definition of an inverse function $F^{-1}F(y) = y$. Thus the result says that whatever probability law X obeys, $Y \triangleq F_X(X)$ will be a *uniform r.v.* Conversely, given a uniform r.v. Y, the transformation $X \triangleq F_X^{-1}(Y)$ will generate a r.v. with PDF $F_X(x)$ (Figure 2.9–3). This technique is sometimes used in *simulation* to generate r.v.'s with specified distributions from a uniform r.v.

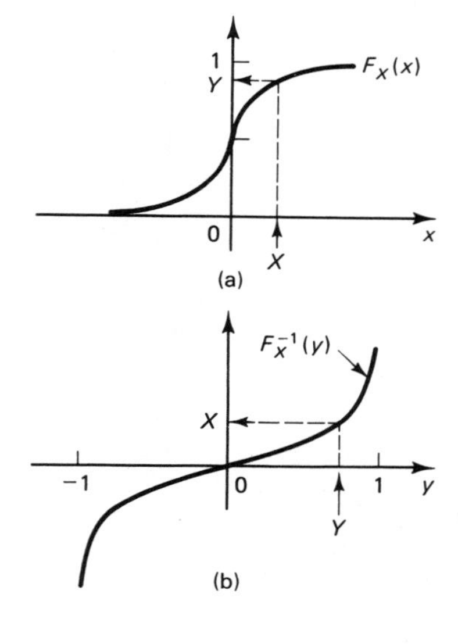

Figure 2.9–3 Generating an r.v. with PDF $F_X(x)$ from a uniform r.v. (a) creating a uniform r.v. Y from an r.v. X with PDF $F_X(x)$; (b) creating a r.v. with PDF $F_X(x)$ from a uniform r.v. Y.

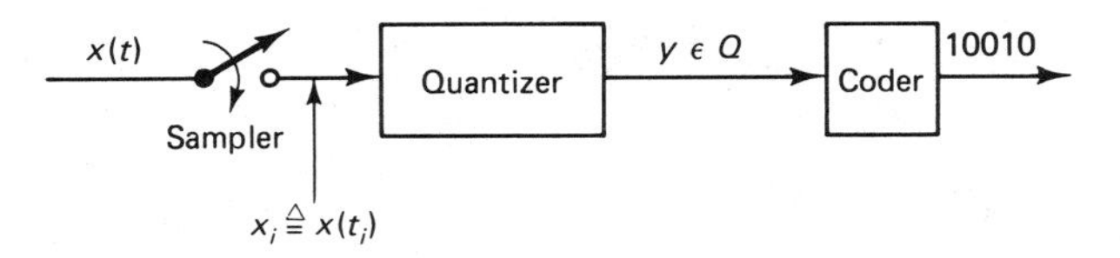

Figure 2.9–4 An analog to digital converter.

Example 2.9–6: (Quantizing.) In analog-to-digital conversion, an analog waveform is sampled, quantized, and coded (Figure 2.9–4).

A quantizer is a function that assigns to each sample x_i, $i = 1, 2, \ldots$ a value from a set $Q \triangleq \{y_{-N}, \ldots, y_0, \ldots, y_N\}$ of $2N + 1$ predetermined values [2–6]. Thus an uncountably infinite set of values (the analog input x_i) is reduced to a finite set (the digital output y_j). Note that a practical quantizer is also a limiter, that is, for x greater than some x_0 or less than some x_1, $y = y_N$ or y_{-N} respectively.

A common quantizer is the uniform quantizer, which is a staircase function of uniform step size a, that is,

$$g(x) = ia \qquad (i-1)a < x \le ia \qquad i \text{ an integer} \tag{2.9–11}$$

Thus the quantizer assigns to each x the closest value of ia above the continuous sample value and is shown by the staircase function in Figure 2.9–5.

If X is a r.v. denoting the sampled value of the input and Y denotes the quantizer output, then with $a = 1$:

$$\begin{aligned} P[Y = i] &= P[i - 1 < X \le i] \\ &= F_X(i) - F_X(i-1). \end{aligned}$$

Also

$$\begin{aligned} F_Y(y) &= \sum_i P[Y = i]u(y - i) \\ &= \sum_i [F_X(i) - F_X(i-1)]u(y - i). \end{aligned} \tag{2.9–12}$$

If limiting is ignored, then when $y = n$ (an integer) $F_Y(n) = F_X(n)$, otherwise $F_Y(y) < F_X(y)$ (Figure 2.9–6)

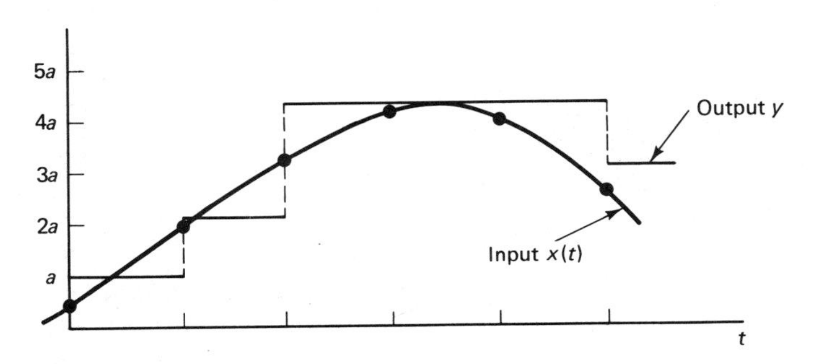

Figure 2.9–5 Quantizer output (staircase function) versus input (continuous line).

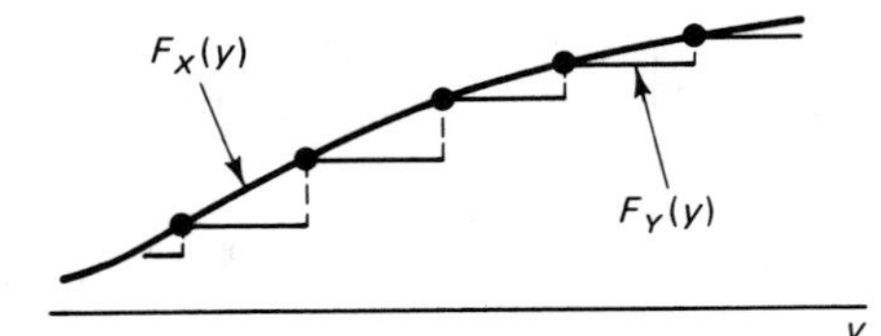

Figure 2.9–6 $F_X(y)$ versus $F_Y(y)$.

Example 2.9–7: (Sine-wave.) A classic problem is to determine the pdf of $Y = \sin X$ where X is a uniform r.v. with pdf

$$f_X(x) = \begin{cases} \dfrac{1}{2\pi}, & -\pi < x \leq \pi \\ 0, & \text{otherwise.} \end{cases} \tag{2.9–13}$$

From Figure 2.9–7 we see that for $0 \leq y \leq 1$, the event $\{Y \leq y\}$ satisfies

$$\begin{aligned} \{Y \leq y\} &= \{\sin X \leq y\} \\ &= \{\pi - \sin^{-1} y < X \leq \pi\} \cup \{-\pi < X \leq \sin^{-1} y\}. \end{aligned}$$

OR

Since the two events on the last line are disjoint, we obtain

$$F_Y(y) = F_X(\pi) - F_X(\pi - \sin^{-1} y) + F_X(\sin^{-1} y) - F_X(-\pi). \tag{2.9–14}$$

$$f_Y(y) = 0 + \frac{1}{\sqrt{1-y^2}} f_X(\pi - \sin^{-1} y) + \frac{1}{\sqrt{1-y^2}} f_X(\sin^{-1} y) - 0$$

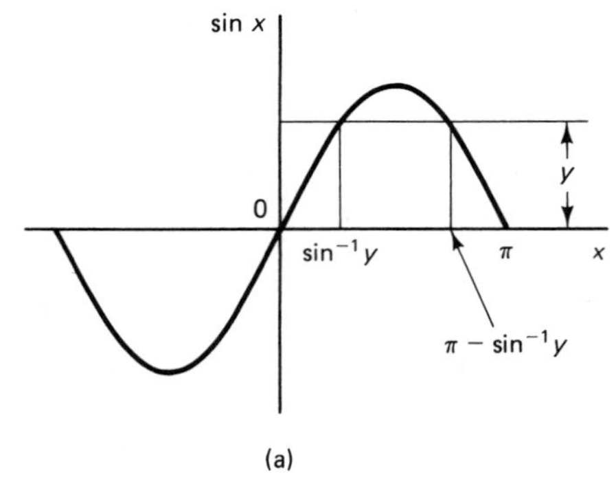

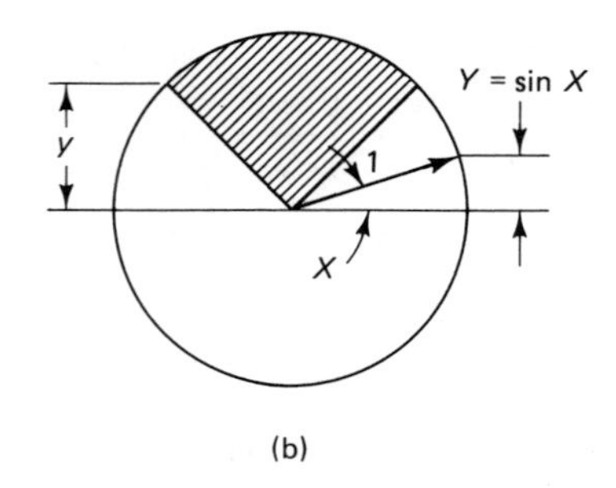

Figure 2.9–7 (a) The function $y = \sin x$; (b) the event $\{\sin X \leq y\}$ is viewed as a unit vector that can rotate to any angle in the clear zone only.

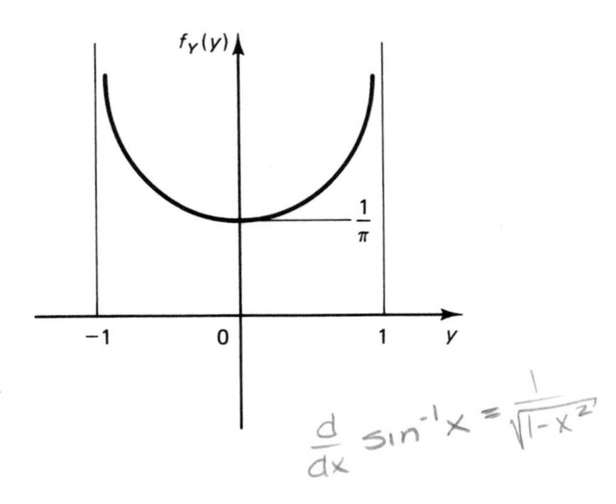

Figure 2.9–8 The probability density function of $Y = \sin X$.

Hence

$$f_Y(y) = \frac{dF_Y(y)}{dy} = f_X(\pi - \sin^{-1} y)\frac{1}{\sqrt{1-y^2}}$$

$$+ f_X(\sin^{-1} y)\frac{1}{\sqrt{1-y^2}} \tag{2.9–15}$$

$$= \frac{1}{\pi}\frac{1}{\sqrt{1-y^2}} \qquad 0 \le y < 1. \tag{2.9–16}$$

If this calculation is repeated for $-1 < y \le 0$ and for $|y| > 1$, we obtain the complete solution (Figure 2.9–8).

$$f_Y(y) = \begin{cases} \dfrac{1}{\pi} \dfrac{1}{\sqrt{1-y^2}}, & |y| < 1 \\ 0, & \text{otherwise.} \end{cases} \tag{2.9–17}$$

The details are left as an exercise.

We shall now derive a formula that will enable us, by "turning the crank," to solve many problems of the type $Y = g(X)$. For some problems, however, the preceding indirect approach may be easier. For lack of a better term we shall call this new approach the *direct* method.

General Formula for Determining the pdf of $Y = g(X)$. We are given the continuous r.v. X with pdf $f_X(x)$ and the differentiable function $g(x)$ of the real variable x. What is the pdf of $Y \triangleq g(X)$?

Solution: The event $\{y < Y \le y + dy\}$ can be written as a union of disjoint elementary events $\{E_i\}$ on the probability space of X. If the equation $y = g(x)$ has n real roots† $x_1, \ldots, x_n$, then the disjoint events have the form $E_i = \{x_i - |dx_i| < X < x_i\}$ if $g'(x_i)$ is negative or $E_i = \{x_i < X < x_i + |dx_i|\}$ if $g'(x_i)$ is positive.‡ (See Figure 2.9–9.) In either case, it follows from the definition of the pdf

† By roots we mean the set of points x_i such that $y - g(x_i) = 0$ $i = 1, \ldots, n$.

‡ The prime indicates derivatives with respect to x.

that $P[E_i] = f(x_i)\,|dx_i|$. Hence

$$P[y < Y \le y + dy] = f_Y(y)\,|dy|$$
$$= \sum_{i=1}^{n} f_X(x_i)\,|dx_i| \qquad (2.9\text{–}18)$$

or, equivalently, if we divide through by $|dy|$

$$f_Y(y) = \sum_{i=1}^{n} f_X(x_i) \left|\frac{dx_i}{dy}\right| = \sum_{i=1}^{n} f_X(x_i) \left|\frac{dy}{dx_i}\right|^{-1}.$$

At the roots of $y = g(x)$, $dy/dx_i = g'(x_i)$, and we obtain the important formula

$$f_Y(y) = \sum_{i=1}^{n} f_X(x_i)/|g'(x_i)| \qquad x_i = x_i(y), \qquad g'(x_i) \neq 0. \qquad (2.9\text{–}19)$$

Equation 2.9–19 is a fundamental equation that is very useful in solving problems where the transmittance $g(x)$ has several roots. If, for a given y, the equation $y - g(x)$ has no real roots then $f_Y(y) = 0$.† Figure 2.9–9 illustrates the case when $n = 2$.

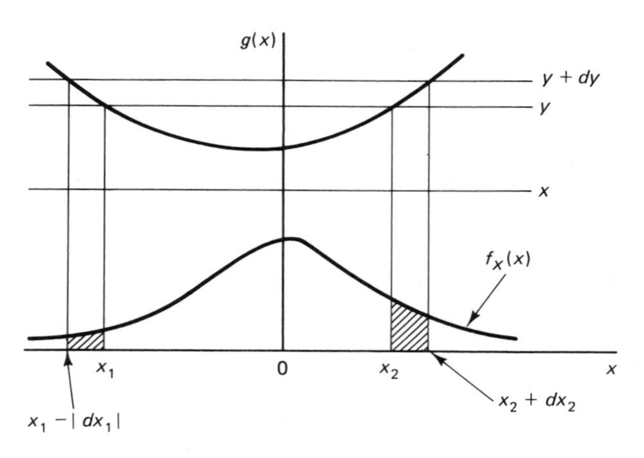

Figure 2.9–9 The event $\{y < Y \le y + dy\}$ is the union of two disjoint events on the probability space of X.

Example 2.9–8: To illustrate the use of Equation 2.9–19 we solve Example 2.9–7 by the formula. Thus we seek the pdf of $Y = \sin X$ with $f_X(x) = 1/2\pi$ for $-\pi < x \le \pi$. The function $g(x)$ is $g(x) = \sin x$. The roots of $y - \sin x = 0$ for $y > 0$ are $x_1 = \sin^{-1} y$, $x_2 = \pi - \sin^{-1} y$. Also

$$\left.\frac{dg}{dx}\right|_{x_1} = \cos x\big|_{x_1=\sin^{-1} y}$$
$$= \cos(\sin^{-1} y)$$
$$= \sqrt{1 - y^2} \qquad = -\left.\frac{dg}{dx}\right|_{x_2}.$$

† The r.v. X, being real, cannot take on values that are imaginary with nonzero probability.

Finally, $f_X(\sin^{-1} y) = f_X(\pi - \sin^{-1} y) = 1/2\pi$. Using these results in Equation 2.9–19 enables us to write

$$f_Y(y) = \frac{1}{\pi}\frac{1}{\sqrt{1 - y^2}} \qquad 0 \le y < 1,$$

which is the same result as in Equation 2.9–16. The solution for all y is as given in Equation 2.9–17.

Example 2.9–10: (Nonlinear devices.) A number of nonlinear zero-memory devices can be modeled by a transmittance function $g(x) = x^n$. Let $Y = X^n$. The pdf of Y depends on whether n is even or odd. We solve the case of n odd, leaving n even as an exercise. For n odd and $y > 0$, the only real root to $y = x^n = 0$ is $x = y^{1/n}$. Also

$$\frac{dg}{dx} = nx^{n-1} = ny^{(n-1)/n}.$$

For $y < 0$, the only real root is $x = -|y|^{1/n}$. Also

$$\frac{dg}{dx} = n\,|y|^{(n-1)/n}.$$

Hence

$$f_Y(y) = \frac{1}{n} y^{(1-n)/n} \cdot f_X(y^{1/n}) \quad , \quad y \ge 0 \tag{2.9–20}$$

$$= \frac{1}{n} |y|^{(n-1)/n} \cdot f_X(-|y|^{1/n}) \, , \; y < 0.$$

In problems in which $g(x)$ assumes a constant value, say $g(x) = c$, over some interval Equation 2.9–19 cannot be used to compute $f_Y(y)$ because $g'(x) = 0$ over the interval. (See Problem 2.17.)

2.10 SOLVING PROBLEMS OF THE TYPE $Z = g(X, Y)$

In many problems in science and engineering a random variable Z is functionally related to two (or more) random variables X, Y. Some examples are:

1. The signal Z at the input of an amplifier consists of a signal X to which is added independent random noise Y. Thus $Z = X + Y$. If X is also a r.v., what is the pdf of Z? (See Figure 2.10–1.)

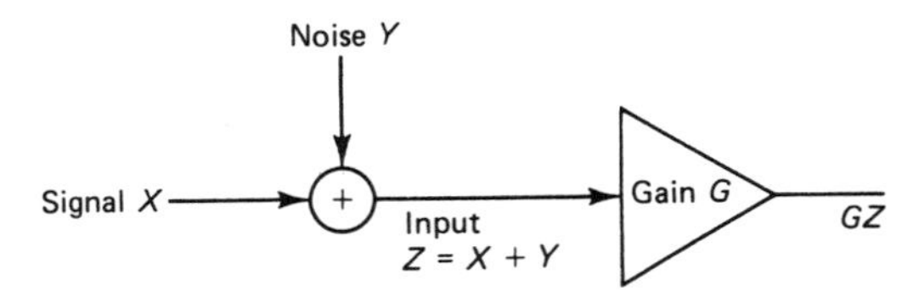

Figure 2.10–1 The signal plus independent additive noise problem.

2. A two-engine airplane is capable of flight as long as at least one of its two engines is working. If the time-to-failures of the starboard and port engines are X and Y respectively, then clearly the time-to-crash of the airplane is $Z \triangleq \max(X, Y)$. What is the pdf of Z?

3. Many signal processing systems use electronic multipliers to multiply two signals together (modulators, demodulators, correlators, and so forth). If X is the signal on one input and Y is the signal on the other input, what is the pdf of the output $Z \triangleq XY$?

4. In the famous "random walk" problem that applies to a number of important physical problems, a particle undergoes random independent displacements X and Y in the x and y directions respectively. What is the pdf of the total displacement $Z \triangleq [X^2 + Y^2]^{1/2}$? (See Figure 2.10–2.)

Problems of the type $Z = g(X, Y)$ are not fundamentally different from the type of problem we discussed in Section 2.9. Recall that for $Y = g(X)$ the basic problem was to find the point set C_y such that the events $\{\zeta : Y(\zeta) \leq y\}$ and $\{\zeta : X(\zeta) \in C_y\}$ were equal. Essentially, the same problem occurs here as well: Find the point set C_z in the (x, y) plane such that the events $\{\zeta : Z(\zeta) \leq z\}$ and $\{\zeta : X(\zeta), Y(\zeta) \in C_z\}$ are equal, this being indicated in our usual shorthand notation by

$$\{Z \leq z\} = \{(X, Y) \in C_z\} \tag{2.10–1}$$

and

$$F_Z(z) = \iint_{(x,y)\in C_z} f_{XY}(x, y)\, dx\, dy. \tag{2.10–2}$$

The point set C_z is determined from $g(x, y) \leq z$. Clearly in problems of the type $Z = g(X, Y)$ we deal with joint densities or distributions and double integrals (or summations) instead of single ones. Thus, in general, the computation of $f_Z(z)$ is more complicated than the computation of $f_Y(y)$ in $Y = g(X)$. However, we have access to two great labor-saving devices, which we shall learn about later: (1) We can solve many $Z = g(X, Y)$-type problems by a "turn-the-crank" type formula through the use of *auxiliary* variables (Section 2.11); and (2) we can solve problems of the type $Z = X + Y$ through the use of *characteristic functions* (Chapter 3). However,

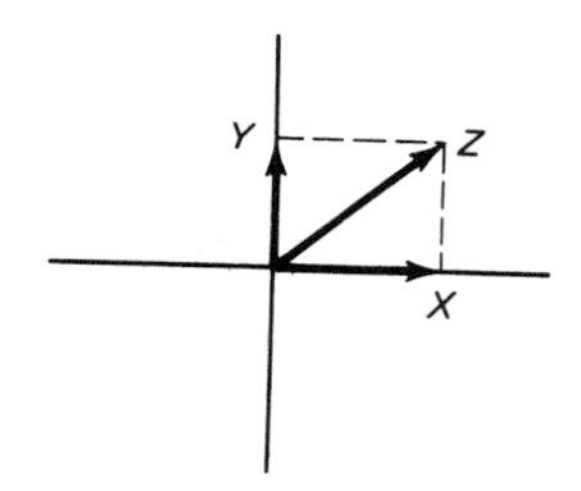

Figure 2.10–2 Diagram relating to the random walk problem.

use of these shortcut methods at this stage would obscure the underlying principles.

Let us now solve the problems mentioned earlier from first principles.

Example 2.10–1: (Multiplier.) To find C_z in Equation 2.10–2 for the PDF of $Z = XY$, we need to determine the region where $g(x, y) \triangleq xy \leq z$. This region is shown in Figure 2.10–3 for $z > 0$.

Thus, reasoning from the diagram, we compute

$$F_Z(z) = \int_0^{\infty} dy \int_{-\infty}^{z/y} f_{XY}(x, y)\, dx + \int_{-\infty}^{0} dy \int_{z/y}^{\infty} f_{XY}(x, y)\, dx, \quad \text{for } z \geq 0. \tag{2.10–3}$$

Now define the indefinite integral $G_{XY}(x, y)$ by

$$G_{XY}(x, y) \triangleq \int f_{XY}(x, y)\, dx. \tag{2.10–4}$$

Then

$$F_Z(z) = \int_0^{\infty} [G_{XY}(z/y, y) - G_{XY}(-\infty, y)]\, dy + \int_{-\infty}^{0} [G_{XY}(\infty, y) - G_{XY}(z/y, y)]\, dy$$

and

$$f_Z(z) = \frac{dF_Z(z)}{dz} = \int_{-\infty}^{\infty} \frac{1}{|y|} f_{XY}(z/y, y)\, dy. \tag{2.10–5}$$

We leave it as an exercise to show that Equation 2.10–5 is valid for $z < 0$ as well.

As a special case, assume X and Y are independent, identically distributed random variables (i.i.d. r.v.) with

$$f_X(x) = f_Y(x) \triangleq \frac{\alpha/\pi}{\alpha^2 + x^2}. \tag{2.10–6}$$

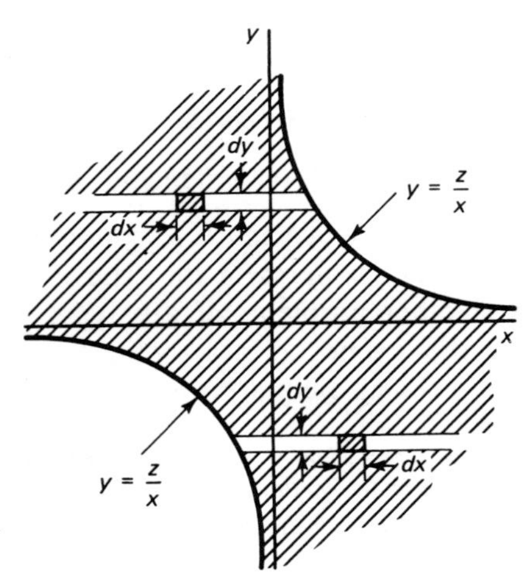

Figure 2.10–3 The region $xy \leq z$ for $z > 0$.

This is known as the Cauchy† probability law. Because of independence

$$f_{XY}(x, y) = f_X(x) f_Y(y)$$

and because of the evenness of the integrand in Equation 2.10–5, we obtain,‡ after a change of variable,

$$\begin{aligned} f_Z(z) &= \left(\frac{\alpha}{\pi}\right)^2 \int_0^\infty \frac{1}{z^2 + \alpha^2 x} \cdot \frac{1}{\alpha^2 + x}\, dx \\ &= \left(\frac{\alpha}{\pi}\right)^2 \frac{1}{z^2 - \alpha^4} \ln \frac{z^2}{\alpha^4}. \end{aligned} \tag{2.10–7}$$

See Figure 2.10–4.

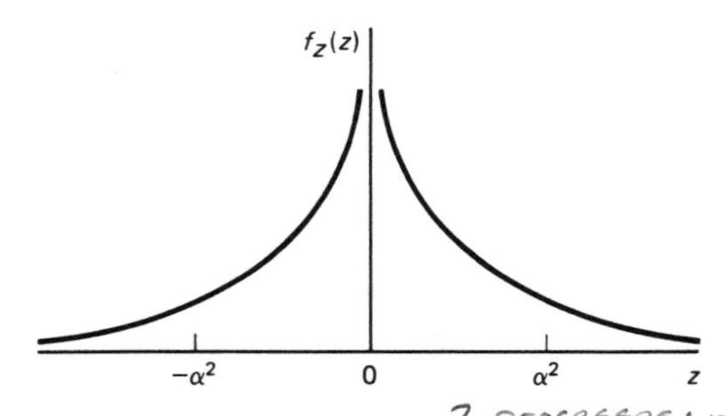

Figure 2.10–4 The pdf $f_Z(z)$ of $Z = XY$ when X and Y are i.i.d. r.v. and Cauchy.

Example 2.10–2: (Parallel operation.) We wish to compute the pdf of $Z = \max(X, Y)$ if X and Y are independent r.v.'s. Then

$$F_Z(z) = P[\max(X, Y) \leq z].$$

But the event $\{\max(X, Y) \leq z\}$ is equal to $\{X \leq z, Y \leq z\}$. Hence

$$P[Z \leq z] = P[X \leq z, Y \leq z] = F_X(z) F_Y(z) \tag{2.10–8}$$

and

$$f_Z(z) = f_Y(z) F_X(z) + f_X(z) F_Y(z) \tag{2.10–9}$$

Again as a special case, let $f_X(x) = f_Y(x)$ be the uniform $[0, 1]$ pdf. Then

$$f_Z(z) = 2z[u(z) - u(z - 1)]. \tag{2.10–10}$$

(See Figure 2.10–5.)

The sum of two independent random variables. The situation modeled by $Z = X + Y$ $\left(\text{and its extension } Z = \sum_{i=1}^{n} X_i\right)$ occurs so frequently in engineering and science that the computation of $f_Z(z)$ is perhaps the most important of all problems of the type $Z = g(X, Y)$.

As in other problems of this type, we must find the set of points C_z such that the event $\{Z \leq z\}$ that, by definition, is equal to the event $\{X + Y \leq z\}$, is also equal to $\{(X, Y) \in C_z\}$. The set of points C_z is the set

† Auguste Louis Cauchy (1789–1857). French mathematician who wrote copiously on astronomy, optics, hydrodynamics, function theory, etc.

‡ See B. O. Peirce and R. M. Foster: *A Short Table of Integrals*, 4th ed. Boston: Ginn & Company, 1956, p. 8.

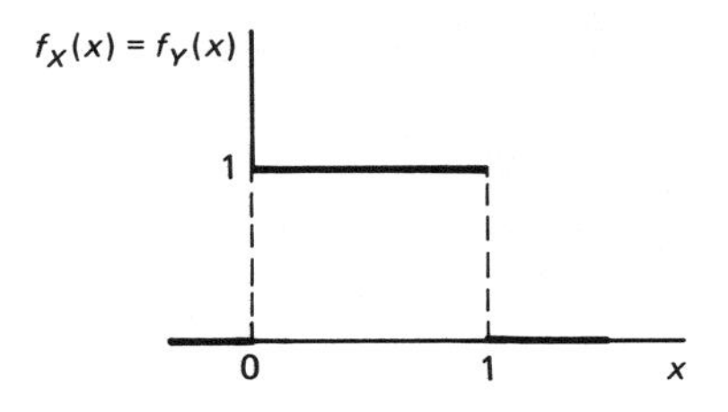

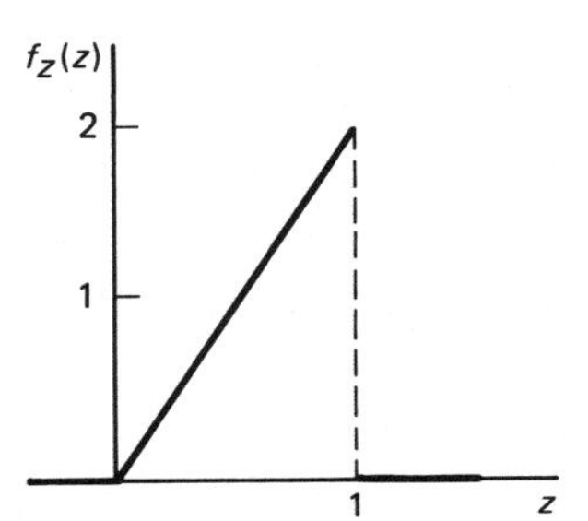

Figure 2.10–5 The pdf of $Z = \max(X, Y)$ for X, Y i.i.d. and uniform in $[0, 1]$.

of all points such that $g(x, y) \triangleq x + y \leq z$ and therefore represents the shaded region to the left of the line in Figure 2.10–6; any point in the shaded region satisfies $x + y \leq z$.

Using Equation 2.10–2 we obtain

$$\begin{aligned} F_Z(z) &= \iint\limits_{x+y\leq z} f_{XY}(x, y)\, dx\, dy \\ &= \int_{-\infty}^{\infty} dy \int_{-\infty}^{z-y} f_{XY}(x, y)\, dx \\ &= \int_{-\infty}^{\infty} [G_{XY}(z - y, y) - G_{XY}(-\infty, y)]\, dy, \qquad (2.10\text{–}11) \end{aligned}$$

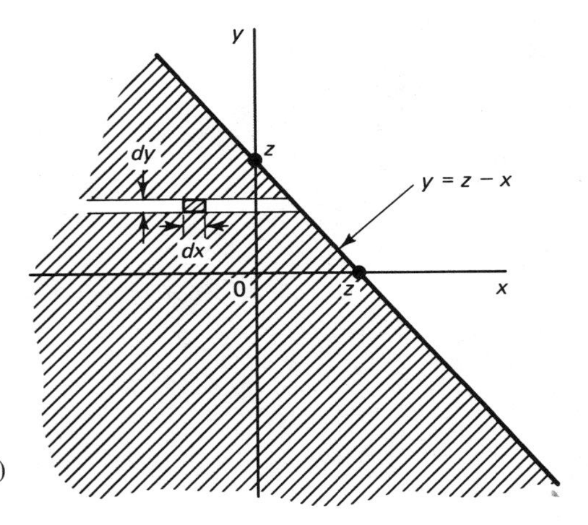

Figure 2.10–6 The region C_z (shaded) for computing the pdf of $Z \triangleq X + Y$.

where $G_{XY}(x, y)$ is the indefinite integral

$$G_{XY}(x, y) \triangleq \int f_{XY}(x, y)\, dx. \tag{2.10-12}$$

The pdf is obtained by differentiation $F_Z(z)$. Thus

$$f_Z(z) = \frac{dF_Z(z)}{dz} = \int_{-\infty}^{\infty} \frac{d}{dz}[G_{XY}(z - y, y)]\, dy$$

$$= \int_{-\infty}^{\infty} f_{XY}(z - y, y)\, dy. \tag{2.10-13}$$

Equation 2.10–13 is an important result (compare with Equation 2.10–5 for $Z = XY$). In many instances X and Y are independent r.v.'s so that $f_{XY}(x, y) = f_X(x)f_Y(y)$. Then Equation 2.10–13 takes the special form

$$f_Z(z) = \int_{-\infty}^{\infty} f_X(z - y)f_Y(y)\, dy \tag{2.10-14}$$

which is known as the *convolution integral* or, more specifically, the *convolution of f_X with f_Y*.† It is a simple matter to prove that Equation 2.10–14 can be rewritten as

$$f_Z(z) = \int_{-\infty}^{\infty} f_X(x)f_Y(z - x)\, dx. \tag{2.10-15}$$

Equations 2.10–14 and 2.10–15 can easily be extended to computing the pdf of $Z = aX + bY$. To be specific, let $a > 0$, $b > 0$. Then the region $g(x, y) \triangleq ax + by \le z$ is to the left of the line $y = z/b - ax/b$ (Figure 2.10–7).

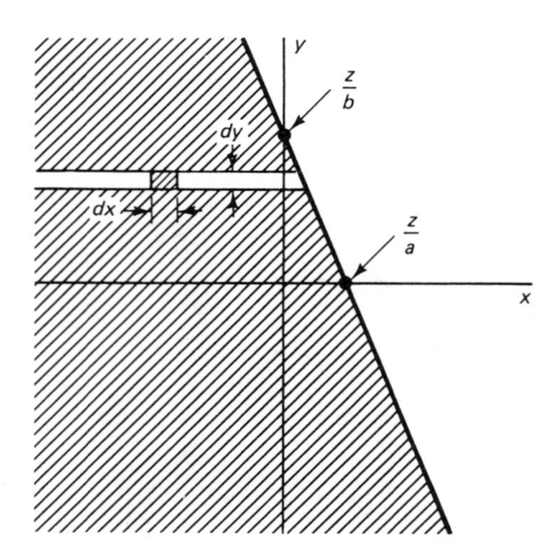

Figure 2.10–7 The region of integration for computing the pdf of $Z = aX + bY$ shown for $a > 0$, $b > 0$.

† A very common notation for the convolution integral as in Equation 2.10–14 is $f_Z = f_X * f_Y$.

Hence

$$F_Z(z) = \iint_{g(x,y)\le z} f_{XY}(x, y)\, dx\, dy$$

$$= \int_{-\infty}^{\infty} dy \int_{-\infty}^{z/a - by/a} f(x, y)\, dx. \tag{2.10–16}$$

As usual, to obtain $f_Z(z)$ we differentiate with respect to z; this furnishes

$Z = aX + bY$ ✓

$$f_Z(z) = \frac{1}{a}\int_{-\infty}^{\infty} f_X\left(\frac{z}{a} - \frac{b}{a}y\right) f_Y(y)\, dy, \tag{2.10–17}$$

where we assumed that X and Y are independent r.v.'s. Equivalently, we can compute $f_Z(z)$ by writing

$$V \triangleq aX$$

$$W \triangleq bY$$

$$Z \triangleq V + W.$$

Then again, assuming $a > 0$, $b > 0$ and X, Y independent, we obtain from Equation 2.10–14

$$f_Z(z) = \int_{-\infty}^{\infty} f_V(z - w) f_W(w)\, dw,$$

where from Equation 2.9–2

$$f_V(v) = \frac{1}{a} f_X\left(\frac{v}{a}\right),$$

and

$$f_W(w) = \frac{1}{b} f_Y\left(\frac{w}{b}\right).$$

Thus

✓

$$f_Z(z) = \frac{1}{ab}\int_{-\infty}^{\infty} f_X\left(\frac{z - w}{a}\right) f_Y\left(\frac{w}{b}\right) dw. \tag{2.10–18}$$

Although Equation 2.10–18 doesn't "look" like Equation 2.10–17, in fact it is identical to it. We need only make the change of variable $y \triangleq w/b$ in Equation 2.10–18 to obtain Equation 2.10–17.

Example 2.10–3: Let X and Y be independent r.v.'s with $f_X(x) = e^{-x}u(x)$ and $f_Y(y) = \frac{1}{2}[u(y + 1) - u(y - 1)]$ and let $Z \triangleq X + Y$. What is the pdf of Z?

Solution: A big help in solving convolution-type problems is to keep track of what is going on graphically. Thus in Figure 2.10–8(a) is shown $f_X(y)$ and $f_Y(y)$; in Figure 2.10–8(b) is shown $f_X(z - y)$. Note that $f_X(z - y)$ is the *reverse and shifted* image of $f_X(y)$. How do we know that the point at the leading edge of the reverse/shifted image is $y = z$? Consider

$$f_X(z - y) = e^{-(z-y)}u(z - y).$$

But $u(z - y) = 0$ for $y > z$. Therefore the reverse/shifted function is nonzero for $(-\infty, z]$.

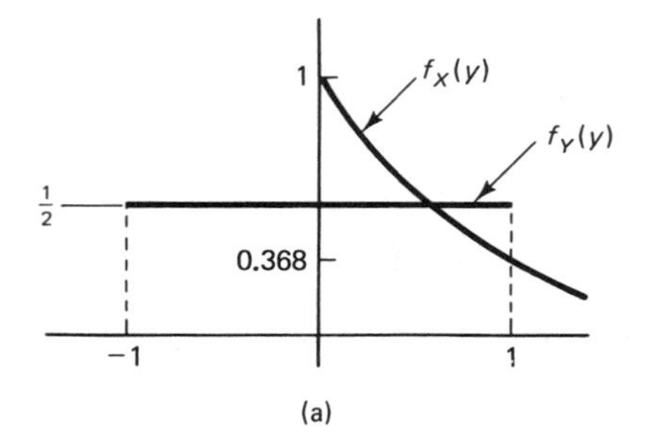

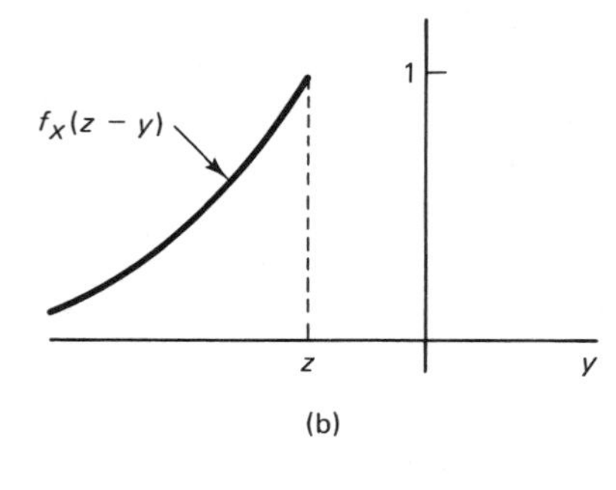

Figure 2.10–8 (a) The pdf's $f_X(y)$, $f_Y(y)$; (b) the reverse/shifted pdf $f_X(z-y)$.

Since f_X and f_Y are discontinuous functions, we do not expect $f_Z(z)$ to be described by the same expression for all values of z. This means that we must do a careful step-by-step evaluation of Equation 2.10–14 for different regions of z-values.

(*a*) *Region 1.* $z < -1$. For $z < -1$ the situation is as shown in Figure 2.10–9(a). Since there is no overlap, Equations 2.10–14 yields zero. Thus $f_Z(z) = 0$ for $z < -1$.

(*b*) *Region 2.* $-1 \leq z < 1$. In this region the situation is as in Figure 2.10–9(b). Thus Equation 2.10–14 yields

$$\begin{aligned} f_Z(z) &= \frac{1}{2}\int_{-1}^{z} e^{-(z-y)}\,dy \\ &= \tfrac{1}{2}[1 - e^{-(z+1)}]. \end{aligned} \tag{2.10–19}$$

(*c*) *Region 3.* $z \geq 1$. In this region the situation is as in Figure 2.10–9(c). From Equation 2.10–14 we obtain

$$\begin{aligned} f_Z(z) &= \frac{1}{2}\int_{-1}^{1} e^{-(z-y)}\,dy \\ &= \tfrac{1}{2}[e^{-(z-1)} - e^{-(z+1)}]. \end{aligned} \tag{2.10–20}$$

Before collecting these results to form a graph we make one final important observation: Since no delta functions were involved in the computation, $f_Z(z)$ must be a *continuous* function of z. Hence, as a check on the solution, the $f_Z(z)$ values at the boundaries of the regions must match. For example, at the junction $z = 1$ between region 2 and region 3:

$$\tfrac{1}{2}[1 - e^{-(z+1)}]_{z=1} \stackrel{?}{=} \tfrac{1}{2}[e^{-(z-1)} - e^{-(z+1)}]_{z=1}. \tag{2.10–21}$$

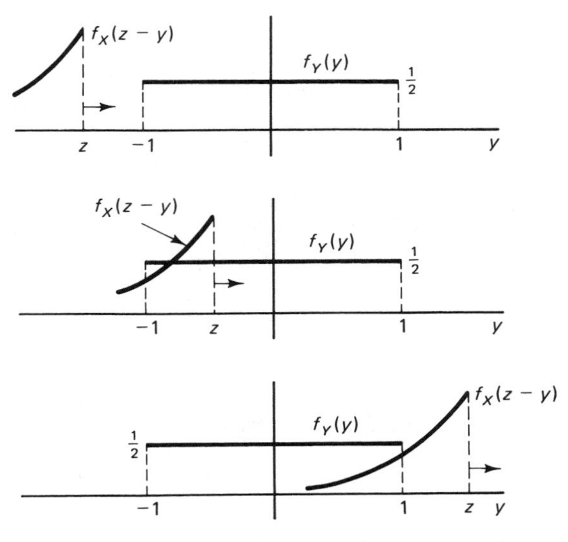

Figure 2.10–9 Relative positions of $f_X(z-y)$ and $f_Y(y)$ for (a) $z<1$; (b) $-1\leq z<1$; (c) $z>1$.

Obviously the right and left sides of Equation 2.10–21 agree so we have some confidence in our solution (Figure 2.10–10).

Example 2.10–4: (Square-law detector.) Let X and Y be independent, uniform r.v.'s in $(-1, 1)$. Compute the pdf of $V \triangleq (X + Y)^2$.

Solution: We solve this problem in two steps. First, we compute the pdf of $Z \triangleq X + Y$; then we compute the pdf of $V = Z^2$. With the notation

$$\text{rect}\left(\frac{x}{D}\right) \triangleq u\left(x + \frac{D}{2}\right) - u\left(x - \frac{D}{2}\right)$$

we obtain

$$f_X(x) = \tfrac{1}{2}\text{rect}\left(\frac{x}{2}\right)$$

$$f_Y(y) = \tfrac{1}{2}\text{rect}\left(\frac{y}{2}\right)$$

$$f_X(z - y) = \tfrac{1}{2}\text{rect}\left(\frac{z - y}{2}\right).$$

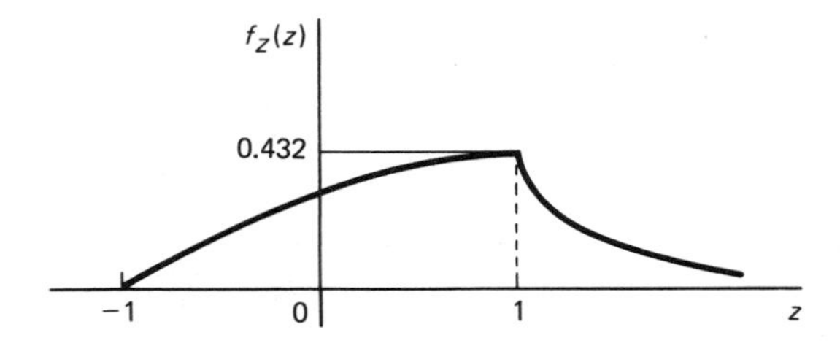

Figure 2.10–10 The pdf $f_Z(z)$ from Example 2.10–3.

From Equation 2.10–14 we get

$$f_Z(z) = \frac{1}{4}\int_{-\infty}^{\infty} \text{rect}\left(\frac{y}{2}\right)\text{rect}\left(\frac{z-y}{2}\right) dy. \tag{2.10–22}$$

The evaluation of Equation 2.10–22 is best done by keeping track graphically of where the "moving," that is, z-dependent function rect$((z - y)/2)$ is centered vis-a-vis the "stationary," that is, z-independent function rect$(y/2)$. The term *moving* is used because as z is varied, the function $f_X((z - y)/2)$ has the appearance of moving past $f_Y(y)$. The situation for four different values of z is shown in Figure 2.10–11.

The evaluation of $f_Z(z)$ for the four distinct regions is as follows:

(a) $z < -2$. In this region, there is no overlap so

$$f_Z(z) = 0.$$

(b) $-2 \le z < 0$. In this region there is overlap in the interval $(-1, z + 1)$ so

$$f_Z(z) = \frac{1}{4}\int_{-1}^{z+1} dy = \tfrac{1}{4}(z + 2).$$

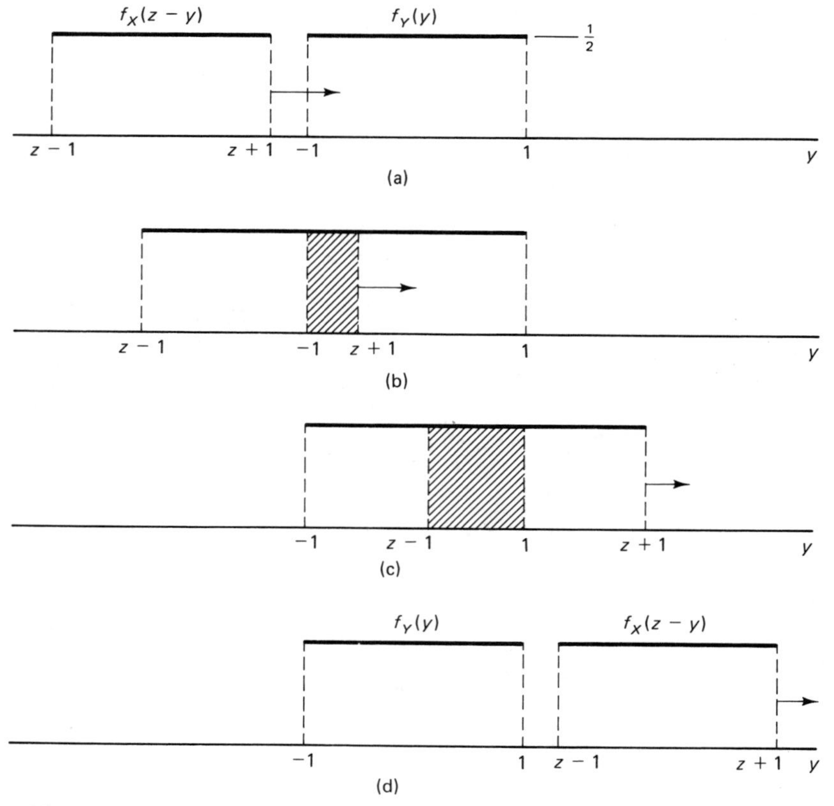

Figure 2.10–11 Four distinct regions in the convolution of two uniform densities: (a) $z < -2$; (b) $-2 \le z < 0$; (c) $0 \le z < 2$; (d) $z \ge 2$.

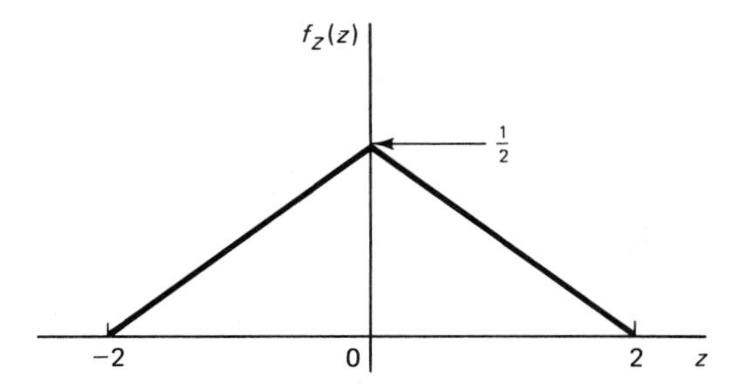

Figure 2.10–12 The pdf of $Z = X + Y$ for X, Y i.i.d. r.v.'s uniform in $(-1, 1)$.

(c) $0 \leq z < 2$. In this region there is overlap in the interval $(z - 1, 1)$ so

$$f_Z(z) = \frac{1}{4}\int_{z-1}^{1} dy = \tfrac{1}{4}(2 - z).$$

(d) $2 \leq z$. In this region there is no overlap so

$$f_Z(z) = 0.$$

If we put all these results together, we obtain

$$f_Z(z) = \tfrac{1}{4}(2 - |z|)\mathrm{rect}\left(\frac{z}{4}\right), \qquad (2.10\text{–}23)$$

which is graphed in Figure 2.10–12.

To complete the solution to this problem, we still need the pdf of $V = Z^2$. We compute $f_V(v)$ using Equation 2.9–19 with $g(z) = z^2$. For $v > 0$, the equation $v - z^2 = 0$ has two real roots, that is, $z_1 = \sqrt{v}$, $z_2 = -\sqrt{v}$; for $v < 0$ there are no real roots. Hence, using Equation 2.10–23 in

$$f_V(v) = \sum_{i=1}^{2} f_Z(z_i)/2\,|z_i|$$

yields

$$f_V(v) = \begin{cases} \dfrac{1}{4}\left(\dfrac{2}{\sqrt{v}} - 1\right) & 0 < v \leq 4 \\ 0, & \text{otherwise} \end{cases} \qquad (2.10\text{–}24)$$

which is shown in Figure 2.10–13.

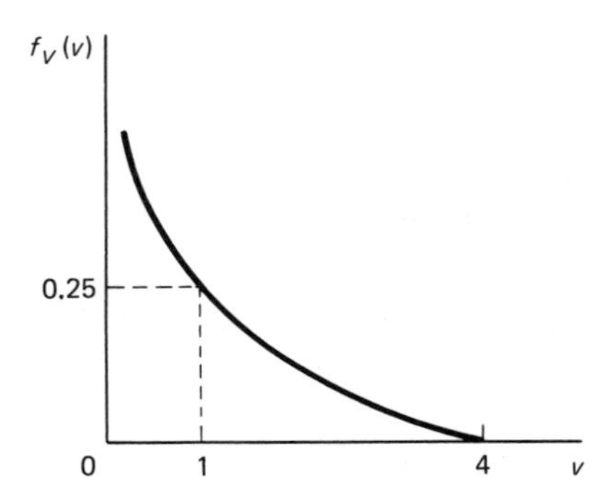

Figure 2.10–13 The pdf of V in Example 2.10–4.

The pdf of the sum of discrete random variables is also computed by convolution, albeit a discrete convolution. For instance, let X and Y be two r.v.'s that take on values $x_1, \ldots, x_k, \ldots$ and $y_1, \ldots, y_j, \ldots$ respectively. Then $Z \triangleq X + Y$ is obviously discrete as well and its PMF is given by

$$P_Z(z_n) = \sum_{x_k + y_j = z_n} P[X = x_k, Y = y_j]. \tag{2.10–25}$$

If X and Y are independent Equation 2.10–25 becomes

$$\begin{aligned} P_Z(z_n) &= \sum_{x_k + y_j = z_n} P_X(x_k)P_Y(y_j) \\ &= \sum_{x_k} P_X(x_k)P_Y(z_n - x_k), \end{aligned} \tag{2.10–26a}$$

If the z_n's and x_k's are equally spaced† then Equation 2.10–26a is recognized as a discrete convolution, in which case it can be written as

$$✓ \quad P_Z(n) = \sum_k P_X(k)P_Y(n - k). \tag{2.10–26b}$$

An illustration of the use of Equation 2.10–26b is given below.

Exercise 2.10–5: (Sum of Poisson random variables.) Let X and Y be two independent Poisson r.v.'s with PMF's $P_X(k) = \frac{1}{k!}e^{-a}a^k$ and $P_Y(i) = \frac{1}{i!}e^{-b}b^i$, where a and b are the Poisson parameters for X and Y respectively. Let $Z \triangleq X + Y$. Then $P_Z(n)$ is given by

$$\begin{aligned} P_Z(n) &= \sum_{k=0}^{n} P_X(k)P_Y(n - k) \\ &= \sum_{k=0}^{n} \frac{1}{k!}\frac{1}{(n - k)!} e^{-(a+b)}a^k b^{n-k}. \end{aligned} \tag{2.10–27}$$

Recall the binomial theorem:

$$\sum_{k=0}^{n} \binom{n}{k} a^k b^{n-k} = (a + b)^n. \tag{2.10–28}$$

Then

$$\begin{aligned} P_Z(n) &= \frac{1}{n!}e^{-(a+b)} \sum_{k=0}^{n} \binom{n}{k} a^k b^{n-k} \\ &= \frac{(a + b)^n}{n!} e^{-(a+b)}, \qquad n \geq 0, \end{aligned} \tag{2.10–29}$$

which is the Poisson law with parameter $a + b$. Thus we obtain the important result that the sum of two independent Poisson r.v.'s with parameters a, b is a Poisson r.v. with parameter $(a + b)$.

We mentioned earlier in Section 2.9 that although the formula in Equation 2.9–19 (and its extensions to be discussed in Section 2.11) is very handy for

† For example let $z_n = n\Delta$, $x_k = k\Delta$, Δ a constant.

solving problems of this type, the indirect approach is sometimes easier. We illustrate with the following Example.

Example 2.10–6: Let X and Y be i.i.d. r.v.'s with $X = N(0, \sigma^2)$. What is the pdf of $Z \triangleq X^2 + Y^2$?

Solution: We begin with the fundamental result given in Equation 2.10–2:

$$F_Z(z) = \iint_{(x,y)\in C_z} f_{XY}(x, y)\, dx\, dy \quad \text{for} \quad z \geq 0$$

$$= \frac{1}{2\pi\sigma^2} \iint_{x^2+y^2\leq z} e^{-(1/2\sigma^2)(x^2+y^2)}\, dx\, dy. \tag{2.10–30}$$

The region C_z consists of the shaded region in Figure 2.10–14.

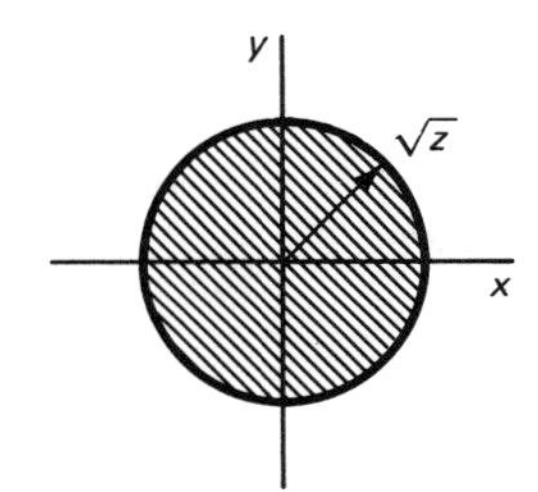

Figure 2.10–14 The region C_z for the event $\{X^2 + Y^2 \leq z\}$ for $z \geq 0$.

Equation 2.10–30 is easily evaluated using polar coordinates. Let

$$x = r\cos\theta \qquad y = r\sin\theta$$

$$dxdy \rightarrow rdrd\theta$$

then $x^2 + y^2 \leq z \rightarrow r \leq \sqrt{z}$ and Equation 2.10–30 is transformed into

$$F_Z(z) = \frac{1}{2\pi\sigma^2}\int_0^{2\pi} d\theta \int_0^{\sqrt{z}} r\exp\left[-\frac{1}{2\sigma^2}r^2\right] dr$$

$$= [1 - e^{z/2\sigma^2}]u(z) \tag{2.10–31}$$

and

$$f_Z(z) = \frac{dF_Z(z)}{dz} = \frac{1}{2\sigma^2} e^{-z/2\sigma^2} u(z). \tag{2.10–32}$$

Thus $Z = X^2 + Y^2$ is an *exponential* r.v. if X and Y are i.i.d. zero-mean Gaussian.

Example 2.10–7: If the previous example is modified to finding the pdf of $Z \triangleq (X^2 + Y^2)^{1/2}$, a radically different pdf results. Again we use Equation 2.10–2 except that now C_z consists of the shaded region in Figure 2.10–15.

Thus

$$F_Z(z) = \frac{1}{2\pi\sigma^2}\int_0^{2\pi} d\theta \int_0^{z} r\exp\left[-\frac{1}{2\sigma^2}r^2\right] dr$$

$$= (1 - e^{-z^2/2\sigma^2})u(z). \tag{2.10–33}$$

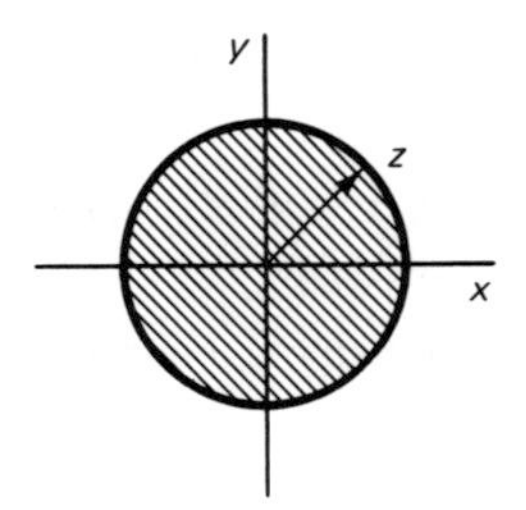

Figure 2.10–15 The region C_z for the event $\{(X^2+Y^2)^{1/2} \leq z\}$.

and

$$f_Z(z) = \frac{z}{\sigma^2} e^{-z^2/2\sigma^2} u(z) \tag{2.10–34}$$

which is the Rayleigh density function. It is also known as the χ ("chi") distribution with two degrees of freedom. The exponential and Rayleigh pdf's are compared in Figure 2.10–16.

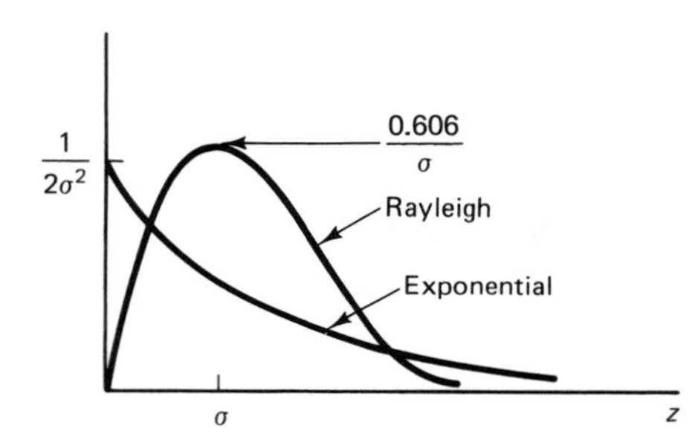

Figure 2.10–16 Rayleigh and exponential pdf's.

Stephen O. Rice [2–7], who in the 1940s did pioneering work in the analysis of electrical noise, showed that narrow-band noise signals at center frequency ω can be represented by the wave

$$Z(t) = X \cos \omega t + Y \sin \omega t \tag{2.10–35}$$

where t is time, ω is the radian frequency in radians per second and, at any particular instant t, X and Y are i.i.d. r.v.'s as $N(0, \sigma^2)$. The envelope $Z \triangleq (X^2 + Y^2)^{1/2}$ has, therefore, a Rayleigh distribution with parameter σ.

The next example generalizes the results of Example 2.10–7 and is a result of considerable importance in communication theory.†

***Example †2.10–8:** (The Rician density.) S. O. Rice considered a version of the following problem: Let $X : N(P, \sigma^2)$ and $Y : N(0, \sigma^2)$ be independent Gaussian r.v.'s. What is the pdf of $Z = (X^2 + Y^2)^{1/2}$? Note that with the parameter $P = 0$, we obtain the solution of Example 2.10–7.

We write

$$F_Z(z) = \begin{cases} \dfrac{1}{2\pi\sigma^2} \displaystyle\iint_{(x^2+y^2)^{1/2} \leq z} \exp\left[-\frac{1}{2}\left(\left[\frac{x-P}{\sigma}\right]^2 + \left(\frac{y}{\sigma}\right)^2\right)\right] dx\, dy, & z > 0 \\ 0, \quad z < 0. & \end{cases} \tag{2.10–36}$$

† Starred examples are somewhat more involved and can be omitted on a first reading.

The usual Cartesian-to-polar transformation $x = r\cos\theta$, $y = r\sin\theta$, $r = x^2 + y^2$, $\theta \tan^{-1}(y/x)$ yields

$$F_Z(z) = \frac{\exp\left[-\frac{1}{2}\left(\frac{P}{\sigma}\right)^2\right]}{2\pi\sigma^2}\int_0^z dr\, r\, e^{-\frac{1}{2}(r/\sigma)^2}\int_0^{2\pi} e^{rP\cos\theta/\sigma^2}\, d\theta \cdot u(z). \qquad (2.10\text{–}37)$$

The function

$$I_o(x) \triangleq \frac{1}{2\pi}\int_0^{2\pi} e^{x\cos\theta}\, d\theta$$

is called the *zero-order modified Bessel function of the first kind* and is monotonically increasing like e^x. With this notation, the cumbersome Equation 2.10–37 can be rewritten as

$$F_Z(z) = \frac{\exp\left[-\frac{1}{2}\left(\frac{P}{\sigma}\right)^2\right]}{\sigma^2}\int_0^z r I_o\left(\frac{rP}{\sigma^2}\right) e^{-\frac{1}{2}[(z-P)/\sigma]^2}\, dr \cdot u(z), \qquad (2.10\text{–}38)$$

where $u(z)$ ensures that the above is valid for $z > 0$. To obtain $f_Z(z)$ we differentiate with respect to z. This produces

$$f_Z(z) = \frac{z}{\sigma^2}\exp\left[-\frac{1}{2}\left(\frac{P^2 + z^2}{\sigma^2}\right)\right] I_o\left(\frac{zP}{\sigma^2}\right) \cdot u(z). \qquad (2.10\text{–}39)$$

The pdf given in Equation 2.10–39 is called the *Rician*† probability density function. Since $I_o(0) = 1$, we obtain the Rayleigh law when $P = 0$. When $zP \gg \sigma^2$, that is, the argument of $I_o(\cdot)$ is large, we use the approximation

$$I_o(x) = \frac{e^x}{(2\pi x)^{1/2}}$$

to obtain

$$f_Z(z) = \frac{1}{\sqrt{2\pi\sigma^2}}\left(\frac{z}{P}\right)^{1/2} e^{-\frac{1}{2}[(z-P)/\sigma]^2}$$

which is almost Gaussian [except for the factor $(z/P)^{1/2}$]. This is the pdf of the *envelope* of the sum of a strong sine wave and weak narrow-band Gaussian noise, a situation that occurs often in electrical communications.

2.11 SOLVING PROBLEMS OF THE TYPE $V = g(X, Y)$, $W = h(X, Y)$

The problem of two functions of two random variables is essentially an extension of the earlier cases except that the algebra is somewhat more involved.

Fundamental problem. We are given two r.v. X, Y with joint pdf $f_{XY}(x, y)$ and two differentiable functions $g(x, y)$ and $h(x, y)$. Two new random variables are constructed according to $V = g(X, Y)$, $W = h(X, Y)$. How do we compute the joint PDF $F_{VW}(v, w)$ (or joint pdf $f_{VW}(v, w)$) of V and W?

† Sometimes called the Rice-Nakagami pdf in recognition of the work of Nakagami around the time of World War II.

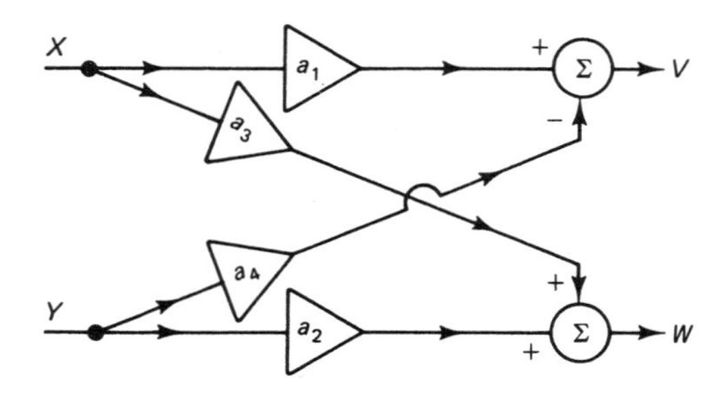

Figure 2.11–1 A two-variable-to-two-variable matrixer.

Illustrations. 1. The circuit shown in Figure 2.11–1 occurs in communication systems such as in the generation of stereo baseband systems [2–6]. The $\{a_i\}$ are gains. When $a_1 = a_2 = \cos\theta$ and $a_3 = a_4 = \sin\theta$ the circuit is known as *θ-rotational transformer.* In another application if X and Y are used to represent, for example, the left and right pick-up signals in stereo broadcasting, then V and W represent the difference and sum signals if all the a_i's are set to unity. The sum and difference signals are then used to generate the signal to be transmitted. Suppose for the moment that there are no source signals and that X and Y therefore represent only Gaussian noise. What is the pdf of V and W?

2. The error in the landing location from a prescribed point of a spacecraft is denoted by (X, Y) in Cartesian coordinates. We wish to specify the error in polar coordinates $V \triangleq (X^2 + Y^2)^{1/2}$, $W = \tan^{-1}(Y/X)$. Given the joint pdf $f_{XY}(x, y)$ of landing error coordinates in Cartesian coordinates, how do we compute the pdf of the landing error in polar coordinates?

The solution to the problem at hand is, as before, to find a point set C_{vw} such that the two events $\{V \le v,\ W \le w\}$ and $\{(X, Y) \in C_{vw}\}$ are equal.† Thus the fundamental relation is

$$P[V \le v, W \le w] \triangleq F_{VW}(v, w) = \iint_{(x,y)\in C_{vw}} f_{XY}(x, y)\, dx\, dy. \tag{2.11–1}$$

The region C_{vw} is given by the points x, y that satisfy

$$C_{vw} = \{(x, y) : g(x, y) \le v,\ h(x, y) \le w\}. \tag{2.11–2}$$

We illustrate the application of Equation 2.11–1 with an example.

Example 2.11–1: We are given $V \triangleq X + Y$ and $W \triangleq X - Y$ and wish to calculate the pdf $f_{VW}(v, w)$. The point set C_{vw} is described by the combined constraints $g(x, y) \triangleq x + y \le v$ and $h(x, y) \triangleq x - y \le w$; it is shown in Figure 2.11–2 for $v > 0$, $w > 0$.

† In more elaborate notation we would write $\{\zeta : V(\zeta) \le v \text{ and } W(\zeta) \le w\} = \{\zeta : (X(\zeta), Y(\zeta)) \in C_{vw}\}$.

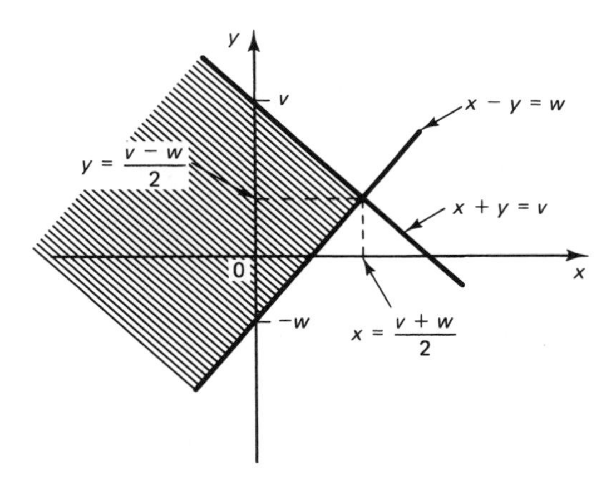

Figure 2.11–2 Point set C_{vw} (shaded region) for Example 2.11–1.

The integration over the shaded region yields

$$F_{VW}(v, w) = \int_{-\infty}^{(v+w)/2} dx \int_{x-w}^{v-x} f_{XY}(x, y)\, dy. \tag{2.11–3}$$

To obtain the joint density $f_{VW}(v, w)$ we use Equation 2.7–22. Hence (after considerable algebra)

$$\begin{aligned} f_{VW}(v, w) &= \frac{\partial^2 F_{VW}(v, w)}{\partial v\, \partial w} \\ &= \tfrac{1}{2} f_{XY}\left(\frac{v + w}{2}, \frac{v - w}{2}\right). \end{aligned} \tag{2.11–4}$$

Thus even this simple problem, involving only linear functions, requires a considerable amount of work and care to obtain a solution. (We leave the details as an exercise—see Problem 2.29.) For this reason, problems of the type discussed in this section and their extensions, that is, $Y_1 = g_1(X_1, \ldots, X_n)$, $Y_2 = g_2(X_1, \ldots, X_n), \ldots, Y_n = g_n(X_1, \ldots, X_n)$, are generally solved by the technique discussed next.

Obtaining f_{VW} Directly from f_{XY}. Instead of attempting to find $f_{VW}(v, w)$ through Equation 2.11–1, we shall instead take a different approach. Consider the elementary event

$$\{v < V \le v + dv, w < W \le w + dw\}$$

and the one-to-one† differentiable functions $v = g(x, y)$, $w = h(x, y)$. The inverse mappings exist and are given by $x = \phi(v, w)$, $y = \psi(v, w)$. Later we shall consider the more general case.

The probability $P[v < V \le v + dv, w < W \le w + dw]$ is the probability that V and W lie in an infinitesimal rectangle of area $dv\, dw$ with vertices at (v, w), $(v + dv, w)$, $(v, w + dw)$ and $(v + dv, w + dw)$. The

† Every point (x, y) maps into a unique (v, w) and vice versa.

image of this square in the x, y coordinate system is (for example, see Marsden and Weinstein [2–8, p. 769]) an infinitesimal parallelogram with vertices at

$$P_1 = (x, y), \qquad P_2 = \left(x + \frac{\partial \phi}{\partial v} dv, y + \frac{\partial \psi}{\partial v} dv\right),$$

$$P_3 = \left(x + \frac{\partial \phi}{\partial w} dw, y + \frac{\partial \psi}{\partial w} dw\right),$$

$$P_4 = \left(x + \frac{\partial \phi}{\partial v} dv + \frac{\partial \phi}{\partial w} dw, y + \frac{\partial \psi}{\partial v} dv + \frac{\partial \psi}{\partial w} dw\right).$$

This mapping is shown in Figure 2.11–3.

With $\mathcal{R}$ denoting the rectangular region shown in Figure 2.11–3(a) and $\mathcal{S}$ denoting the parallelogram in Figure 2.11–3(b) and $A(\mathcal{R})$ and $A(\mathcal{S})$ denoting the areas of $\mathcal{R}$ and $\mathcal{S}$ respectively, we obtain

$$P[v < V \leq v + dv, w < W \leq w + dw]$$

$$= \iint_{\mathcal{R}} f_{VW}(\xi, \eta)\, d\xi\, d\eta \tag{2.11–5}$$

$$= f_{VW}(v, w) A(\mathcal{R}) \tag{2.11–6}$$

$$= \iint_{\mathcal{S}} f_{XY}(\xi, \eta)\, d\xi\, d\eta \tag{2.11–7}$$

$$= f_{XY}(x, y) A(\mathcal{S}). \tag{2.11–8}$$

Equation 2.11–5 follows from the fundamental relation given in Equation 2.11–1; Equation 2.11–6 follows from the interpretation of the pdf given in Equation 2.4–5; Equation 2.11–7 follows by definition of the point set $\mathcal{S}$, that is, $\mathcal{S}$ is the set of points that makes the events $\{(V, W) \in \mathcal{R}\}$ and $\{(X, Y) \in \mathcal{S}\}$ equal and Equation 2.11–8 again follows from the interpretation of pdf.

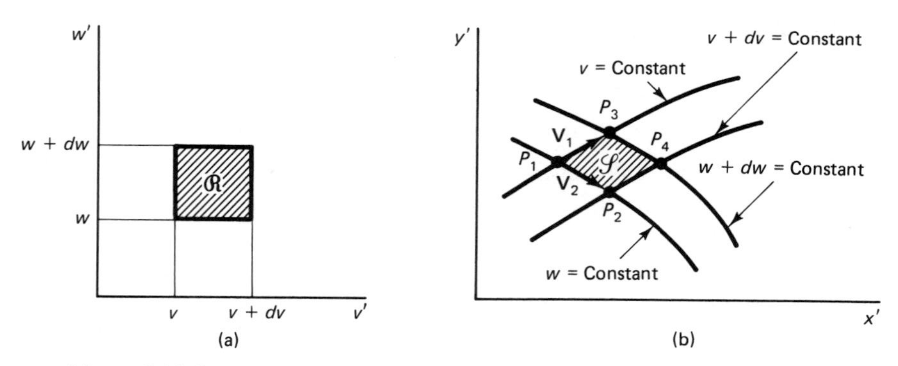

Figure 2.11–3 An infinitesimal rectangle in the v, w system (a) maps into an infinitesimal parallelogram (b) in the x, y system.

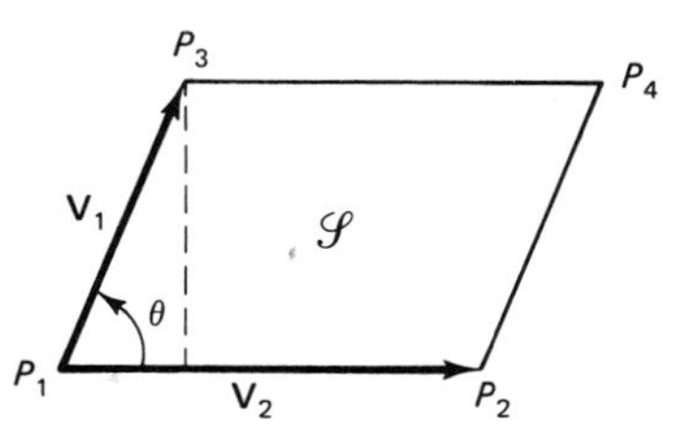

Figure 2.11–4 The area of a parallelogram $\mathcal{S}$ is $|V_1|\,|V_2| \sin\theta$.

Essentially all that remains at this point is to compute $A(\mathcal{S})$. As is well known from vector analysis, the area of a parallelogram $\mathcal{S}$ spanned by the vectors $\mathbf{V}_1$ and $\mathbf{V}_2$ (Figure 2.11–4) is given by

$$A(\mathcal{S}) = |\mathbf{V}_1 \times \mathbf{V}_2| = |\mathbf{V}_1|\,|\mathbf{V}_2| \sin\theta,$$

where $\times$ denotes the cross-product.

Let $\mathbf{i}, \mathbf{j}, \mathbf{k}$ be *unit* vectors in the x, y, and z directions respectively. By the rules of the cross product $\mathbf{i} \times \mathbf{j} = \mathbf{k}$, $\mathbf{j} \times \mathbf{i} = -\mathbf{k}$ and $\mathbf{i} \times \mathbf{i} = \mathbf{j} \times \mathbf{j} = \mathbf{k} \times \mathbf{k} = 0$. Then we can write

$$\mathbf{V}_2 = \frac{\partial \phi}{\partial v}\, dv\, \mathbf{i} + \frac{\partial \psi}{\partial v}\, dv\, \mathbf{j}$$

$$\mathbf{V}_1 = \frac{\partial \phi}{\partial w}\, dw\, \mathbf{i} + \frac{\partial \psi}{\partial w}\, dw\, \mathbf{j}$$

and

$$A(\mathcal{S}) = \left| \frac{\partial \psi}{\partial v} \frac{\partial \phi}{\partial w}\, dv\, dw - \frac{\partial \phi}{\partial v} \frac{\partial \psi}{\partial w}\, dv\, dw \right|. \qquad (2.11\text{–}9)$$

Equation 2.11–9 can be written in determinant form as

$$A(\mathcal{S}) = \text{mag} \begin{vmatrix} \dfrac{\partial \phi}{\partial v} & \dfrac{\partial \phi}{\partial w} \\[2ex] \dfrac{\partial \psi}{\partial v} & \dfrac{\partial \psi}{\partial w} \end{vmatrix} dv\, dw \qquad (2.11\text{–}10)$$

where mag is short for "magnitude of."

A standard notation for the determinant in Equation 2.11–10 is

$$\frac{\partial(\phi, \psi)}{\partial(v, w)}.$$

It is not difficult to show by considering the mapping of a rectangle in the x, y domain to a parallelogram in the v, w domain that

$$\left.\frac{\partial(g, h)}{\partial(x, y)}\right|_{\substack{x = \phi(v,w) \\ y = \psi(v,w)}} = \frac{1}{\dfrac{\partial(\phi, \psi)}{\partial(v, w)}} \qquad (2.11\text{–}11)$$

where

$$J \triangleq \left.\frac{\partial(g, h)}{\partial(x, y)}\right|_{\substack{x = \phi(v,w) \\ y = \psi(v,w)}}$$

is called the Jacobian of the transformation $g(x, y) = v$, $h(x, y) = w$. Equivalently $\tilde{J}$ given by

$$\tilde{J} \triangleq \frac{\partial(\phi, \psi)}{\partial(v, w)} \tag{2.11–12}$$

and is called the Jacobian of the transformation $x = \phi(v, w)$, $y = \psi(v, w)$. From Equations 2.11–6, 2.11–8, 2.11–10, and 2.11–12, we obtain the important results that

$$f_{VW}(v, w) = \frac{f_{XY}(x, y)}{|J|}\Bigg|_{\substack{x=\phi(v,w)\\ y=\psi(v,w)}} \tag{2.11–13a}$$

$$= f_{XY}(x, y)\,|\tilde{J}|\,\Bigg|_{\substack{x=\phi(v,w)\\ y=\psi(v,w)}}. \tag{2.11–13b}$$

Equations 2.11 to 2.13 are important results but hold only when the functions $v = g(x, y)$, $w = h(x, y)$ are one-to-one. Suppose, however, that $(x_1, y_1), (x_2, y_2), \ldots, (x_n, y_n)$ all satisfy $v = g(x_i, y_i)$, $w = h(x_i, y_i)$. Then the inverse image—if we can still call it that—of the region $\mathscr{R}$ is n parallelograms $\mathscr{S}_1, \mathscr{S}_2, \ldots, \mathscr{S}_n$ anchored at the vertices $(x_1, y_1), \ldots, (x_n, y_n)$. More precisely all the parallelograms $\mathscr{S}_1, \mathscr{S}_2, \ldots, \mathscr{S}_n$ map into $\mathscr{R}$ under the transformation $v = g(x, y)$, $w = h(x, y)$. Thus *every* elementary event $\{(X, Y) \in \mathscr{S}_i\}$ $i = 1, \ldots, n$ results in the event $\{(V, W) \in \mathscr{R}\}$, and to properly compute the *total probability* of $\{(V, W) \in \mathscr{R}\}$ we must consider the probability of $\bigcup_{i=1}^{n} \{(X, Y) \in \mathscr{S}_i\}$. Since these elementary events are disjoint we obtain

$$P[\{(V, W) \in \mathscr{R}\}] = P[\{(X, Y) \in \mathscr{S}_1\}] + \ldots + P[\{(X, Y) \in \mathscr{S}_n\}]$$

or

$$f_{VW}(v, w)\, dv\, dw = f_{XY}(x_1, y_1)A(\mathscr{S}_1) + \ldots + f_{XY}(x_n, y_n)A(\mathscr{S}_n). \tag{2.11–14}$$

The $A(\mathscr{S}_i)$ are known to us from Equation 2.11–10. By using the latter and Equation 2.11–11 in Equation 2.11–14 and dividing through by $dv\, dw$ on both sides we can write the following important result for n real roots $(x_1, y_1), \ldots, (x_n, y_n)$:

$$f_{VW}(v, w) = \sum_{i=1}^{n} \frac{f_{XY}(x_i, y_i)}{|J_i|} \tag{2.11–15}$$

$$= \sum_{i=1}^{n} f_{XY}(x_i, y_i)\,|\tilde{J}_i|, \tag{2.11–16}$$

where the subscript i refers to the fact that the Jacobians are evaluated at the ith root. In the remainder of this Section we shall illustrate the application of Equation 2.11–15 or Equation 2.11–16.

Example 2.11–2: Consider the r.v.'s

$$V \triangleq g(X, Y) = \sqrt{X^2 + Y^2} \tag{2.11–17}$$

$$W = h(X, Y) = \begin{cases} \tan^{-1}\left(\dfrac{Y}{X}\right) & X > 0 \\ \tan^{-1}\left(\dfrac{Y}{X}\right) + \pi & X < 0. \end{cases} \tag{2.11–18}$$

The r.v. V is called the magnitude or envelope while W is called the phase. Equation 2.11–18 has been written in this form because we seek a solution for w over a 2π interval while the function $\tan^{-1}(y/x)$ has range $(-\pi/2, \pi/2)$ (i.e., its principle value).

To find the roots of

$$v = \sqrt{x^2 + y^2} \tag{2.11–19a}$$

$$w = \begin{cases} \tan^{-1}\left(\dfrac{y}{x}\right) & x \geq 0 \\ \tan^{-1}\left(\dfrac{y}{x}\right) + \pi & x < 0 \end{cases} \tag{2.11–19b}$$

we observe that for $x > 0$, $\dfrac{-\pi}{2} \leq w \leq \dfrac{\pi}{2}$ and $\cos w \geq 0$. Similarly, for $x < 0$, $\dfrac{\pi}{2} < w < \dfrac{3\pi}{2}$ and $\cos w < 0$. Hence the only solution to Equations 2.11–19 is

$$x = v \cos w \triangleq \phi(v, w)$$
$$y = v \sin w \triangleq \psi(v, w).$$

The Jacobian $\tilde{J}$ is given by

$$\tilde{J} = \frac{\partial(\phi, \psi)}{(v, w)} = \begin{vmatrix} \cos w & -v \sin w \\ \sin w & v \cos w \end{vmatrix} = v.$$

Hence the solution is, from Equation 2.11–16

$$f_{VW}(v, w) = v\, f_{XY}(v \cos w, v \sin w). \tag{2.11–20}$$

Suppose that X and Y are i.i.d. with probability law $N(0, \sigma^2)$, that is,

$$f_{XY}(x, y) = \frac{1}{2\pi\sigma^2} e^{-[(x^2+y^2)/2\sigma^2]}$$

Then from Equation 2.11–20

$$f_{VW}(v, w) = \begin{cases} \left(\dfrac{v}{\sigma^2} e^{-v^2/2\sigma^2}\right) \dfrac{1}{2\pi}, & v > 0, -\dfrac{\pi}{2} \leq w < \dfrac{3\pi}{2} \\ 0, & \text{otherwise} \end{cases} \tag{2.11–21}$$

$$= f_V(v) f_W(w).$$

Thus V and W are independent random variables. The envelope V has a Rayleigh pdf and the phase W is uniform over a 2π interval.

Example 2.11–3: Consider now a modification of the previous problem. Let $V \triangleq \sqrt{X^2 + Y^2}$ and $W \triangleq Y/X$. Then with $g(x, y) = \sqrt{x^2 + y^2}$ and $h(x, y) = y/x$, the equations

$$v - g(x, y) = 0$$
$$w - h(x, y) = 0$$

have two solutions:

$$x_1 = v(1 + w^2)^{-1/2}, \qquad y_1 = wx_1$$

$$x_2 = -v(1 + w^2)^{-1/2}, \qquad y_2 = wx_2$$

for $-\infty < w < \infty$ and $v > 0$, and no real solutions for $v < 0$.

A direct evaluation yields $|J_1| = |J_2| = (1 + w^2)/v$. Hence

$$f_{VW}(v, w) = \frac{v}{1 + w^2}[f_{XY}(x_1, y_1) + f_{XY}(x_2, y_2)].$$

With $f_{XY}(x, y)$ given by

$$f_{XY}(x, y) = \frac{1}{2\pi\sigma^2}\exp[-(x^2 + y^2)/2\sigma^2],$$

we obtain

$$\begin{aligned} f_{VW}(v, w) &= \frac{v}{\sigma^2} e^{-v^2/2\sigma^2} u(v) \cdot \frac{1/\pi}{1 + w^2} \\ &= f_V(v) f_W(v). \end{aligned}$$

Thus the r.v.'s V, W are independent. V is a Rayleigh r.v. as in Example 2.11–2, but W is Cauchy.

Example 2.11–4: Let θ be a prescribed angle and consider the transformation

$$\begin{aligned} V &\triangleq X \cos\theta + Y \sin\theta \\ W &\triangleq X \sin\theta - Y \cos\theta \end{aligned} \tag{2.11–22}$$

with

$$f_{XY}(x, y) = \frac{1}{2\pi\sigma^2} e^{-[(x^2+y^2)/2\sigma^2]}.$$

The only solution to

$$\begin{aligned} v &= x \cos\theta + y \sin\theta \\ w &= x \sin\theta - y \cos\theta \end{aligned}$$

is

$$\begin{aligned} x &= v \cos\theta + w \sin\theta \\ y &= v \sin\theta - w \cos\theta. \end{aligned}$$

The Jacobian $\tilde{J}$ is

$$\begin{vmatrix} \dfrac{\partial x}{\partial v} & \dfrac{\partial x}{\partial w} \\ \dfrac{\partial y}{\partial v} & \dfrac{\partial y}{\partial w} \end{vmatrix} = \begin{vmatrix} \cos\theta & \sin\theta \\ \sin\theta & -\cos\theta \end{vmatrix} = -1.$$

Hence

$$f_{VW}(v, w) = \frac{1}{2\pi\sigma^2} e^{-[(v^2+w^2)/2\sigma^2]}.$$

Thus under the rotational transformation $V = g(X, Y)$, $W = h(X, Y)$ given in Equation 2.11–22, V and W are i.i.d. Gaussian r.v.'s just like X and Y. If X and Y are Gaussian but not independent r.v.'s, it is still possible to find a transformation† so that V, W will be independent Gaussian if the joint pdf of X, Y is Gaussian.

Example 2.11–5: Consider again the problem of solving for the pdf of $Z =$

† Chapter 4.

$\sqrt{X^2+Y^2}$ as in Example 2.10–7. This time we shall use Equation 2.11–16 to compute $f_Z(z)$. First we note that $Z = \sqrt{X^2 + Y^2}$ is one function of two r.v.'s while Equation 2.11–16 applies to two functions of two r.v.'s. To convert from one kind of problem to the other we introduce an *auxiliary r.v.* $W \triangleq X$. The introduction of the r.v. W will enable us to use Equation 2.11–16 directly. Hence

$$Z \triangleq g(X, Y) = \sqrt{X^2 + Y^2}$$
$$W \triangleq h(X, Y) = X.$$

The equations

$$z - g(x, y) = 0$$
$$w - h(x, y) = 0$$

have two real roots, for $|w| < z$, namely

$$x_1 = w \qquad\qquad x_2 = w$$
$$y_1 = \sqrt{z^2 - w^2} \qquad\qquad y_2 = -\sqrt{z^2 - w^2}.$$

At both roots, $|\tilde{J}|$ has the same value:

$$|\tilde{J}_1| = |\tilde{J}_2| = \frac{z}{\sqrt{z^2 - w^2}}.$$

Hence a direct application of Equation 2.11–16 yields

$$f_{ZW}(z, w) = \frac{z}{\sqrt{z^2 - w^2}}[f_{XY}(x_1, y_1) + f_{XY}(x_2, y_2)].$$

Now assume that

$$f_{XY}(x, y) = \frac{1}{2\pi\sigma^2} e^{-[(x^2+y^2)/2\sigma^2]}.$$

Then, since in this case $f_{XY}(x, y) = f_{XY}(x, -y)$, we obtain

$$f_{ZW}(z, w) = \begin{cases} \dfrac{1}{\pi\sigma^2}\dfrac{z}{\sqrt{z^2 - w^2}} e^{-z^2/2\sigma^2}, & z > 0, |w| < z \\ 0, & \text{otherwise.} \end{cases}$$

However, we don't really want $f_{ZW}(z, w)$, but only the marginal pdf $f_Z(z)$. To obtain this we use Equation 2.7–26 with x replaced by z and y replaced by w. This gives

$$f_Z(z) = \int_{-\infty}^{\infty} f_{ZW}(z, w)\, dw$$
$$= \frac{z}{\sigma^2} e^{-z^2/2\sigma^2}\left[\frac{2}{\pi}\int_0^z \frac{dw}{\sqrt{z^2 - w^2}}\right]u(z).$$

But the term in parentheses has value $\pi/2$. Hence we obtain the same result as obtained in Equation 2.10–34, that is, that Z is a Rayleigh r.v., but obtained by a completely different method.

Example 2.11–6: Finally, let us return to the problem considered in Example 2.11–1:

$$V \triangleq X + Y$$
$$W \triangleq X - Y.$$

The only root to

$$v - (x + y) = 0$$
$$w - (x - y) = 0$$

is

$$x = \frac{v + w}{2}$$
$$y = \frac{v - w}{2}$$

and $|\tilde{J}| = \frac{1}{2}$. Hence

$$f_{VW}(v, w) = \tfrac{1}{2} f_{XY}\left(\frac{v + w}{2}, \frac{v - w}{2}\right).$$

We leave it as an exercise to the reader to show that

$$f_V(v) = \int_{-\infty}^{\infty} f_{VW}(\alpha, v - \alpha)\, d\alpha.$$

This important result was derived in Section 2–10 by different means.

2.12 SUMMARY

The material in this chapter is central to the whole book. We began by defining a real random variable as a mapping from the sample description space Ω to the real line R. We then introduced a point function $F_X(x)$ called the *probability distribution function* (PDF), which enabled us to compute the probabilities of events of the type $\{\zeta : \zeta \in \Omega,\ X(\zeta) \le x\}$. The probability density function pdf and probability mass function (PMF) were derived from the PDF, and a number of useful and specific probability laws were discussed. We showed how, by using Dirac delta functions, we could develop a unified theory for both discrete and continuous random variables. We then discussed joint distributions, the Poisson transform and its inverse and the application of these concepts to physical problems.

The second half of the chapter (Sections 2.8 to 2.11) was devoted to functions of random variables. We showed how most problems involving functions of r.v.'s could be computed in at least two ways: (1) the so-called indirect approach and (2) directly through the use of a "turn-the-crank" formula. A number of important problems involving transformations of random variables were worked out including computing the pdf of the sum of two random variables.

PROBLEMS

2.1. The event of k successes in n tries regardless of the order is the binomial law $b(k, n; p)$ (Equation 1.8–1). Let $n = 10$, $p = 0.3$. Define the r.v. (random variable) X by

$$X(k) = \begin{cases} 1 & 0 \le k \le 2 \\ 2 & 2 < k \le 5 \\ 3 & 5 < k \le 8 \\ 4 & 8 < k \le 10 \end{cases}$$

Compute $P[X = j]$ $j = 1, \ldots, 4$. Plot $F_X(x) = P[X \le x]$ for all x.

2.2. In a restaurant known for its unusual service, the time X, in minutes, that a customer has to wait before he captures the attention of a waiter is specified by the following distribution function:

$$\begin{aligned} F_X(x) &= \left(\frac{x}{2}\right)^2 && \text{for } 0 \le x \le 1, \\ &= \frac{x}{4} && \text{for } 1 \le x \le 2, \\ &= \tfrac{1}{2} && \text{for } 2 \le x \le 10, \\ &= \frac{x}{20} && \text{for } 10 \le x \le 20, \\ &= 1 && \text{for } x \ge 20. \end{aligned}$$

(a) Sketch $F_X(x)$. (b) Compute and sketch the pdf $f_X(x)$. Verify that the area under the pdf is indeed unity. (c) What is the probability that the customer will have to wait (1) at least 10 minutes, (2) less than 5 minutes, (3) between 5 and 10 minutes, (4) exactly 1 minute?

2.3. Compute the probabilities of the events $\{X < a\}$, $\{X \le a\}$, $\{a \le X < b\}$, $\{a \le X \le b\}$ $\{a < X \le b\}$, and $\{a < X < b\}$ in terms of $F_X(x)$ and $P[X = x]$ for $x = a, b$.

2.4. In the following pdf's, compute the constant B required for proper normalization:

(a) Cauchy ($\alpha < \infty$, $\beta > 0$):

$$f(x) = \frac{B}{1 + [(x - \alpha)/\beta]^2}, \qquad -\infty < x < \infty.$$

(b) Maxwell ($\alpha > 0$):

$$f(x) = \begin{cases} Bx^2 e^{-x^2/\alpha^2}, & x > 0 \\ 0, & \text{otherwise.} \end{cases}$$

(c) Beta ($b > -1$, $c > -1$):

$$f(x) = \begin{cases} Bx^b(1 - x)^c, & 0 \le x \le 1 \\ 0, & \text{otherwise.} \end{cases}$$

(See formula 6.2.1 on page 258 of [2–9].)

(d) Chi-square ($\sigma > 0$, $n = 1, 2, \ldots$):

$$f(x) = \begin{cases} Bx^{(n/2)-1} \exp(-x/2\sigma^2), & x > 0 \\ 0, & \text{otherwise.} \end{cases}$$

2.5. A noisy resistor produces a voltage $v_n(t)$. At $t = t_1$, the noise level $X \triangleq v_n(t_1)$

is known to be a Gaussian r.v. with pdf

$$f_X(x) = \frac{1}{\sqrt{2\pi\sigma^2}} \exp\left[-\frac{1}{2}\left(\frac{x}{\sigma}\right)^2\right].$$

Compute and plot the probability that $|X| > k\sigma$ for $k = 1, 2, \ldots$.

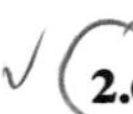

2.6. Compute $F_X(k\sigma)$ for the Rayleigh pdf (Equation 2.4–11) for $k = 0, 1, 2, \ldots$.

2.7. Write the *probability density functions* (using delta functions) for the Bernoulli, binomial, and Poisson PMF's.

2.8. The pdf of a r.v. X is shown in Figure P-2.8. The numbers in parentheses indicate area.

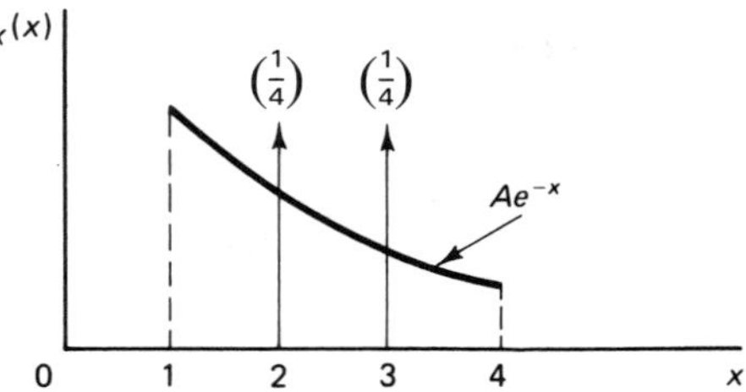

Figure P–2.8

(a) Compute the value of A; (b) sketch the PDF; (c) compute $P[2 \le X < 3]$; (d) compute $P[2 < X \le 3]$; (e) compute $F_X(3)$.

2.9. Consider a binomial r.v. with PMF $b(k; 4, \frac{1}{2})$. Compute $P[X = k \mid X \text{ even}]$ for $k = 0, \ldots, 4$.

2.10. The time-to failure in months, X, of light bulbs produced at two manufacturing plants A and B obey, respectively, the following PDF's

$$F_X(x) = (1 - e^{-x/5})u(x) \text{ for plant } A$$ 0.25

$$F_X(x) = (1 - e^{-x/2})u(x) \text{ for plant } B.$$ 0.75

Plant B produces three times as many bulbs as plant A. The bulbs, indistinguishable to the eye, are intermingled and sold. What is the probability that a bulb purchased at random will burn at least (a) two months; (b) five months; (c) seven months?

2.11. Show that the conditioned distribution of X given the event $A = \{b < X \le a\}$ is

$$F_X(x \mid A) = \begin{cases} 0, & x \le b \\ \dfrac{F_X(x) - F_X(b)}{F_X(a) - F_X(b)}, & b < x \le a \\ 1, & x \ge a. \end{cases}$$

2.12. It has been found that the number of people Y waiting in a queue in the bank on payday obeys the Poisson law as

$$P[Y = k \mid X = x] = e^{-x}\frac{x^k}{k!}$$

given that the normalized serving time of the teller x (that is, the time it takes the teller to deal with a customer) is constant. However, the serving time is more accurately modeled as an r.v. X. For simplicity let X be a uniform r.v.

with

$$f_X(x) = \tfrac{1}{5}[u(x) - u(x - 5)].$$

Then $P[Y = k \mid X = x]$ is still Poisson but $P[Y = k]$ is something else. Compute $P[Y = k]$ for $k = 0, 1, 2, \ldots$. Hint: Use Equation 2.7–9.

2.13. Consider the joint pdf of X and Y

$$f_{XY}(x, y) = \frac{1}{3\pi} e^{-\frac{1}{2}[(x/3)^2+(y/2)^2]} u(x)u(y).$$

Are X and Y independent r.v.? Compute the probability of $\{0 < X \le 3, 0 < Y \le 2\}$.

2.14. Let X have PDF $F_X(x)$ and consider $Y = aX + b$ where $a < 0$. Show that if X is not necessarily continuous, Equation 2.9–3 should be modified to

$$F_Y(y) = 1 - F_X\left(\frac{y - b}{a}\right) + P\left[X = \frac{y - b}{a}\right].$$

2.15. Let Y be a function of the r.v. X as follows

$$Y \triangleq \begin{cases} X, & X \ge 0 \\ X^2, & X \le 0. \end{cases}$$

Compute $f_Y(y)$ in terms of $f_X(x)$. Let $f_X(x)$ be given by

$$f_X(x) = \frac{1}{\sqrt{2\pi}} e^{-\frac{1}{2}x^2}.$$

2.16. Let X have pdf

$$f_X(x) = \alpha e^{-\alpha x} u(x).$$

Compute the pdf of (a) $Y = X^3$; (b) $Y = 2X + 3$.

2.17. (a) Let $X : N(0, 1)$ and let $Y \triangleq g(X)$ where the function $g(\cdot)$ is shown in Figure P-2.17. Use the indirect approach to compute $F_Y(y)$ and $f_Y(y)$ from $f_X(x)$. (b) Compute $f_Y(y)$ from Equation 2.9–19. Why can't Equation 2.9–19 be used to compute $f_Y(y)$ at $y = 0, 1$?

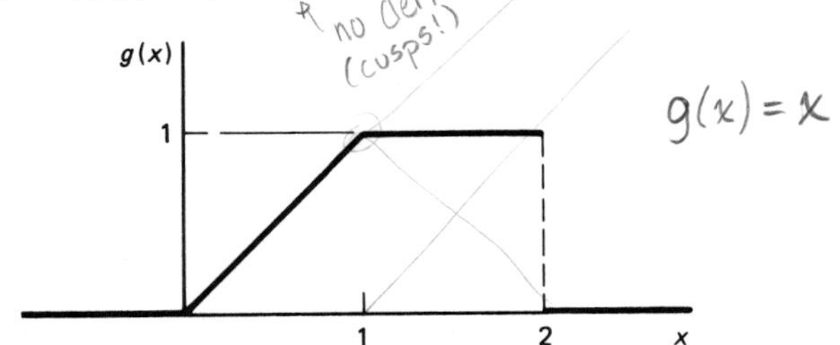

Figure P–2.17

2.18. Compute the pdf of $Y = a/X$ $(a > 0)$. Show that if X is Cauchy with parameter α, Y is Cauchy with parameter a/α.

2.19. Let $Y = \sec X$. Compute $f_Y(y)$ in terms of $f_X(x)$. What is $f_Y(y)$ when $f_X(x)$ is uniform in $(-\pi, \pi]$.

2.20. Let X and Y be independent, identically distributed r.v.'s with

$$f_X(x) = f_Y(x) = \alpha e^{-\alpha x} u(x).$$

Compute the pdf of $Z \triangleq X - Y$.

2.21. In Problem 2.20 compute the pdf of $|Z|$.

2.22. Let X and Y be independent, continuous r.v.'s. Let $Z = \min(X, Y)$. Compute $F_Z(z)$ and $f_Z(z)$. Sketch the result if X and Y are uniform r.v.'s in (0, 1). Repeat for $f_X(x) = f_Y(y) = \alpha \exp[-\alpha x] \cdot u(x)$.

2.23. The length of time Z, an airplane, can fly is given by $Z = \alpha X$ where X is the amount of fuel in its tank and $\alpha > 0$ is a constant of proportionality. Suppose a plane has two independent fuel tanks so that when one gets empty the other switches on automatically. Because of lax maintenance a plane takes off with neither of its fuel tanks checked. Let X be the fuel in the first tank and Y the fuel in the second tank. Let X and Y be modeled as independent, identically distributed r.v.'s with $f_X(x) = f_Y(x) = \frac{1}{b}[u(x) - u(x - b)]$. Compute the pdf of Z, the maximum flying time of the plane. If $b = 100$, say in liters, and α = one hour/10 liters, what is the probability that the plane will fly *at least* five hours?

2.24. Let X and Y be two independent Poisson r.v.'s with

$$P_X(k) = \frac{1}{k!}e^{-2}2^k$$

$$P_Y(j) = \frac{1}{j!}e^{-3}3^j.$$

Compute the PMF of $Z = X + Y$. Hint $\sum_{j=0}^{n} \binom{n}{j} a^j b^{n-j} = (a + b)^n$.

2.25. Let X and Y be independent uniform r.v.'s with $f_X(x) = \frac{1}{2}$, $|x| < 1$ and zero otherwise and $f_Y(y) = \frac{1}{4}$, $|y| < 2$ and zero otherwise. Compute (a) the pdf of $Z \triangleq X + Y$; (b) the pdf of $Z \triangleq 2X - Y$.

2.26. Compute the joint pdf of

$$Z \triangleq X^2 + Y^2$$

$$W \triangleq X$$

when

$$f_{XY}(x, y) = \frac{1}{2\pi\sigma^2} e^{-[(x^2+y^2)/2\sigma^2]}.$$

Compute $f_Z(z)$ from your results.

2.27. Consider the transformation

$$Z = aX + bY$$

$$W = cX + dY$$

Let

$$f_{XY}(x, y) = \frac{1}{2\pi\sigma^2\sqrt{1 - \rho^2}} e^{-Q(x,y)}$$

where

$$Q(x, y) = \frac{1}{2\sigma^2(1 - \rho^2)}[x^2 - 2\rho xy + y^2].$$

What combination of coefficients a, b, c, d will enable Z, W to be independent Gaussian r.v.'s?

2.28. Let

$$f_{XY}(x, y) = \frac{1}{2\pi\sqrt{1-\rho^2}} \exp\left[-\left(\frac{x^2 - 2\rho xy + y^2}{2(1-\rho^2)}\right)\right].$$

Compute the joint pdf $f_{VW}(v, w)$ of

$$V = \tfrac{1}{2}(X^2 + Y^2)$$
$$W = \tfrac{1}{2}(X^2 - Y^2).$$

2.29. Derive Equation 2.11–4 by the indirect method, that is, use Equations 2.11–3 and 2.7–22.

2.30. Consider the transformation

$$Z = X\cos\theta + Y\sin\theta$$
$$W = X\sin\theta - Y\cos\theta.$$

Compute the joint pdf of $f_{ZW}(z, w)$ in terms of $f_{XY}(x, y)$ if

$$f_{XY}(x, y) = \frac{1}{2\pi} e^{-\frac{1}{2}(x^2+y^2)} \quad -\infty < x < \infty, -\infty < y < \infty.$$

2.31. Let $f_{XY}(x, y) = A(x^2 + y^2)$ for $0 \le x \le 1$, $|y| \le 1$ and zero otherwise. Compute the PDF $F_{XY}(x, y)$ for all x, y. Determine the value of A.

2.32. Consider the input–output view mentioned in Section 2.8. Let the underlying experiment be observations on a r.v. X, which is the input to a system that generates an output $Y = g(X)$.

(a) What is the range of Y?
(b) What are reasonable probability spaces for X and Y?
(c) What subset of R consists of the event $\{Y \le y\}$?
(d) What is the inverse image under Y of the event $(-\infty, y]$ if $Y = 2X + 3$?

REFERENCES

2–1. M. Loeve, *Probability Theory*. New York: Van Nostrand, Reinhold, 1962.
2–2. W. Feller, *An Introduction to Probability Theory and Its Applications*, 2 vols. New York: John Wiley, 1950, 1966.
2–3. W. F. Davenport, *Probability and Random Processes: An Introduction for Applied Scientists and Engineers*. New York: McGraw-Hill, 1970.
2–4. B. Saleh, *Photoelectron Statistics*. New York: Springer-Verlag, 1978.
2–5. M. Born and E. Wolf, *Principles of Optics*. New York: Pergamon Press, 1965.
2–6. H. Stark and F. B. Tuteur, *Modern Electrical Communications*. Englewood Cliffs, N.J.: Prentice-Hall, 1979.
2–7. S. O. Rice, "Mathematical Analysis of Random Noise," *Bell System Technical Journal*, Vols. 23, 24, 1944, 1945.
2–8. J. Marsden and A. Weinstein, *Calculus*. Menlo Park, Calif.: Benjamin/Cummings, 1980.
2–9. M. Abramowitz and I. A. Stegun, *Handbook of Mathematical Functions*, Dover, New York, 1965.

ADDITIONAL READING

Cooper, G. R., and C. D. McGillem, *Probabilistic Methods of Signal and System Analysis*. New York: Holt, Rinehart and Winston, 1971.

Helstrom, C. W., *Probability and Stochastic Processes for Engineers*. New York: Macmillan, 1984.

Papoulis, A., *Probability, Random Variables, and Stochastic Processes*. New York: McGraw-Hill, 1965.

Parzen, E., *Modern Probability Theory and Its Applications*. New York: John Wiley, 1960.

3

AVERAGES

3.1 EXPECTED VALUE OF A RANDOM VARIABLE

It is often desirable to summarize certain properties of a random variable (r.v.) and its probability law by a few numbers. Such numbers are furnished to us by the various averages or *expectations* of a r.v.; the term *moments* is often used to describe a broad class of averages, and we shall use it later.

We are all familiar with the notion of the average of a set of numbers, for example, the average class grade for an exam, the average height and weight of children at age five, the expected lifetime of men versus women, and the like. Basically, we compute the average of a set of numbers $x_1, x_2, \ldots, x_N$ as follows:

$$\mu_s = \frac{1}{N} \sum_{i=1}^{N} x_i, \qquad (3.1\text{–}1)$$

where the subscript s is a reminder that μ_s is the average of a set.

Although the average as given in Equation 3.1–1 gives us the "most likely" value or the "center of gravity" of the set, it does not tell us how much the numbers spread or deviate from the average. For example, the sets of numbers $S_1, = \{0.9, 0.98, 0.95, 1.1, 1.02, 1.05\}$ and $S_2 = \{0.2, -3, 1.8, 2, 4, 1\}$ have the same average but the spread of the numbers in S_2 is much greater than that of S_1. An average that summarizes this spread is the *standard deviation of the set*, σ_s, computed from

$$\sigma_s = \left[\frac{1}{N} \sum_{i=1}^{N} (x_i - \mu_s)^2\right]^{1/2}. \qquad (3.1\text{–}2)$$

Equations 3.1–1 and 3.1–2, important as they are, fall far short of disclosing the usefulness of averages. To exploit the full range of applications of averages we must develop a calculus of averages from probability theory.

Consider a probability space $(\Omega, \mathcal{F}, P)$ associated with an experiment $\mathcal{H}$ and a discrete r.v. X. Associated with each outcome ζ_i of $\mathcal{H}$, there is a value $X(\zeta_i) \triangleq x_i$, which the r.v. X takes on. Let $x_1, x_2, \ldots, x_M$ be the M distinct values that X can take. Now assume that $\mathcal{H}$ is repeated N times and let $x^{(k)}$ be the observed outcome at the kth trial. Note that $x^{(k)}$ must assume one of the numbers $x_1, \ldots, x_M$. Suppose that in the N trials x_1 occurs n_1 tries, x_2 occurs n_2 times, and so forth. Then for N large, we can estimate the average value $\bar{X}$ of X from the formula

$$\bar{X} \simeq \frac{1}{N} \sum_{k=1}^{N} x^{(k)} \tag{3.1–3}$$

$$= \frac{1}{N} \sum_{i=1}^{M} n_i x_i = \sum_{i=1}^{M} x_i \left(\frac{n_i}{N}\right)$$

$$\simeq \sum_{i=1}^{M} x_i P[X = x_i]. \tag{3.1–4}$$

Equation 3.1–4, which follows from the frequency definition of probability, leads us to our first definition.

Definition 3.1–1. The *expected* or *average* value of a discrete r.v. X taking on values x_i with PMF $P_X(x_i) \triangleq P[X = x_i]$ $i = 1, 2, \ldots,$ is defined by

$$E[X] \triangleq \sum_i x_i P_X(x_i) \tag{3.1–5}$$

A definition that applies to both continuous and discrete r.v.'s is the following:

Definition 3.1–2. The expected value, if it exists,† of a real r.v. X with pdf $f_X(x)$ is defined by

$$E[X] = \int_{-\infty}^{\infty} x f_X(x)\, dx. \tag{3.1–6}$$

The symbols $E[X]$, $\bar{X}$, μ_X, or simply μ are often used interchangeably for the expected value of X. Consider now a function of a r.v. $Y = g(X)$. The expected value of Y is, from Equation 3.1–6,

$$E[Y] = \int_{-\infty}^{\infty} y f_Y(y)\, dy. \tag{3.1–7}$$

However, Equation 3.1–7 requires computing $f_Y(y)$ from $f_X(x)$. If all we

† The expected value will exist if the integral is absolutely convergent, i.e., if $\int_{-\infty}^{\infty} |x|\, f(x)\, dx < \infty$.

want is $E[Y]$, is there a way to compute it without first computing $f_Y(y)$? The answer follows.

Theorem 3.1–1. The expected value of $Y = g(X)$ can be computed from

$$E[Y] = \int_{-\infty}^{\infty} g(x) f_X(x)\, dx \tag{3.1–8}$$

where $g(\cdot)$ is a measurable (Borel) function.† Equation 3.1–8 is one of the most important results in the theory of probability. Unfortunately, a rigorous proof of Equation 3.1–8 requires some knowledge of Lebesgue integration. We offer instead an informal argument to argue that Equation 3.1–8 is valid; readers interested in more rigorous arguments can consult William Feller [3–1, p. 5] or Wilbur Davenport [3–2, p. 223].

On the Validity of Equation 3.1–8. Recall from Section 2.9 that if $Y = g(X)$ then for any y_j (Figure 3.1–1)

$$\{y_j < Y \leq y_j + \Delta y_j\} = \bigcup_{k=1}^{r_j} \{x_j^{(k)} < X \leq x_j^{(k)} + \Delta x_j^{(k)}\}, \tag{3.1–9}$$

where r_j is the number of real roots of the equation $y_j - g(x) = 0$, that is,

$$y_j = g(x_j^{(1)}) = \ldots = g(x_j^{(r_j)}). \tag{3.1–10}$$

The equal sign in Equation 3.1–9 means that the underlying event is the same for both mappings X and Y. Hence the probabilities of the events on either side of the equal sign are equal. The events on the right side of Equation 3.1–9 are disjoint and therefore the probability of the union is the sum of the probabilities of the individual events. Now partition the y-axis into many fine subintervals $y_1, y_2, \ldots, y_j, \ldots$. Then approximating Equation

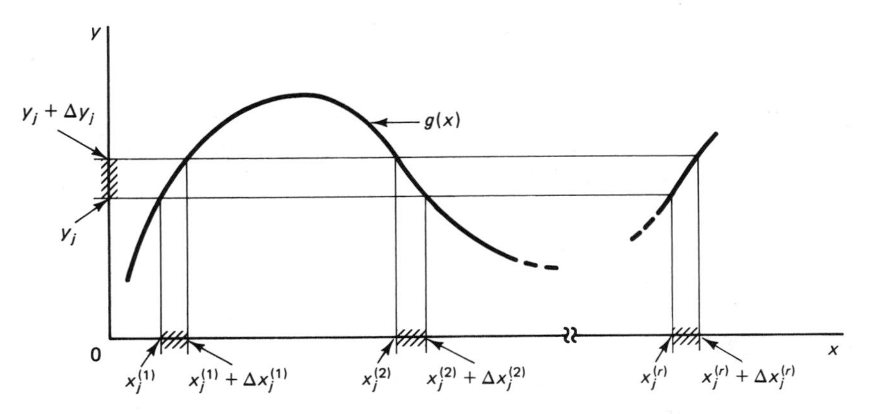

Figure 3.1–1 Equivalence between the events given in Equation 3.1–9.

† See definition of a measurable function in Section 2.8.

3.1–7 with a Riemann† sum and using Equation 2.4–5, we can write

$$E[Y] = \int_{-\infty}^{\infty} y f_Y(y)\, dy$$

$$\simeq \sum_{j=1}^{m} y_j P[y_j < Y \le y_j + \Delta y_j]$$

$$= \sum_{j=1}^{m} \sum_{k=1}^{r_j} g(x_j^{(k)}) P[x_j^{(k)} < X \le x_j^{(k)} + \Delta x_j^{(k)}]. \qquad (3.1\text{–}11)$$

The last line of Equation 3.1–11 is obtained with the help of Equations 3.1–9 and 3.1–10. But the points $x_j^{(k)}$ are distinct, so that the cumbersome double indices j and k can be replaced with a single subscript index, say, i. Then Equation 3.1–11 becomes

$$E[Y] \simeq \sum_{i=1}^{n} g(x_i) P[x_i < X \le x_i + \Delta x_i],$$

and—as Δy, $\Delta x \to 0$—we obtain the exact result that

$$E[Y] = \int_{-\infty}^{\infty} g(x) f_X(x)\, dx. \qquad (3.1\text{–}12)$$

Equation 3.1–12 follows from the Riemann sum approximation and Equation 2.4–5; the x_i have been arranged in increasing order $x_1 < x_2 < x_3 < \ldots$. As Athanasios Papoulis [3–3, p. 141] points out, with the use of Lebesgue integration one can almost immediately and rigorously establish Equation 3.1–12.

In the special case where X is a discrete r.v. then

$$E[Y] = \sum_{i} g(x_i) P_X(x_i). \qquad (3.1\text{–}13)$$

This result follows immediately from Equation 3.1–12, since the pdf of a discrete r.v. involves delta functions that have the property given in Equation 2.6–4.

Example 3.1–1: Let $X:N(\mu, \sigma^2)$. The expected value of X is

$$E[X] = \int_{-\infty}^{\infty} x\left(\frac{1}{\sqrt{2\pi\sigma^2}} \exp\left(-\frac{1}{2}\left(\frac{x-\mu}{\sigma}\right)^2\right)\right) dx.$$

Let $z \triangleq (x - \mu)/\sigma$. Then

$$E[X] = \frac{\sigma}{\sqrt{2\pi}} \int_{-\infty}^{\infty} z e^{-\frac{1}{2}z^2}\, dz + \mu\left(\frac{1}{\sqrt{2\pi}} \int_{-\infty}^{\infty} e^{-\frac{1}{2}z^2}\, dz\right).$$

The first term is zero because the integrand is odd, and the second term is μ because the term in brackets is $P[Z \le \infty]$ for $Z:N(0, 1)$. Hence

$$E[X] = \mu \quad \text{for } X:N(\mu, \sigma^2).$$

Thus the parameter μ in $N(\mu, \sigma^2)$ is rightfully called the mean.

† Bernhard Riemann (1826–1866). German mathematician who made numerous contributions to the theory of integration.

Example 3.1–2: Let X be a Poisson r.v. with parameter a. Then

$$\begin{aligned} E[X] &= \sum_{k=0}^{\infty} k \frac{e^{-a}}{k!} a^k \\ &= a \sum_{k=0}^{\infty} \frac{e^{-a}}{(k-1)!} a^{k-1} \\ &= a \sum_{k=1}^{\infty} \frac{e^{-a}}{(k-1)!} a^{k-1} \\ &= a \sum_{i=0}^{\infty} \frac{e^{-a}}{i!} a^{i} \\ &= a. \end{aligned} \tag{3.1–14}$$

Thus the expected value of a Poisson r.v. with parameter a is a.

Example 3.1–3: Let $Y = X^2$ with $X:N(0, \sigma^2)$. Then using Equation 3.1–12 we write

$$E[Y] = \frac{1}{\sqrt{2\pi\sigma^2}} \int_{-\infty}^{\infty} x^2 \exp\left(-\frac{1}{2}\left(\frac{x}{\sigma}\right)^2\right) dx.$$

With $z \triangleq x/\sigma$ this becomes (for $\sigma > 0$)

$$E[Y] = \sigma^2 \left(\frac{1}{\sqrt{2\pi}} \int_{-\infty}^{\infty} z^2 e^{-\frac{1}{2}z^2} dz\right).$$

The term in brackets can be reduced to

$$\frac{1}{\sqrt{2\pi}} \int_{-\infty}^{\infty} e^{-\frac{1}{2}z^2} dz = 1$$

with the help of integration by parts. Thus for $Y = X^2$ and $X:N(0, \sigma^2)$

$$E[Y] = \sigma^2. \tag{3.1–15}$$

More generally, for $X:N(\mu, \sigma^2)$, $E[X^2] = \mu^2 + \sigma^2$ and $E[(X - \mu)^2] = \sigma^2$. We leave the proof of these important results as exercises for the reader. The parameter σ^2 is called the variance, and μ is the mean. In practice these are often estimated by making many observations on X and using Equation 3.1–1 to estimate the mean and Equation 3.1–2 to estimate σ.

Example 3.1–4: The Cauchy pdf with parameters $\alpha(-\infty < \alpha < \infty)$ and $\beta(\beta > 0)$ is given by

$$f_X(x) = \frac{1}{\pi\beta\left(1 + \left(\frac{x - \alpha}{\beta}\right)^2\right)} \qquad -\infty < x < \infty. \tag{3.1–16}$$

Let X be Cauchy with $\beta = 1$, $\alpha = 0$. Then

$$E[X] = \int_{-\infty}^{\infty} x\left(\frac{1}{\pi(x^2 + 1)}\right) dx$$

is an improper integral and doesn't converge in the ordinary sense. However, if we evaluate the integral in the Cauchy principal value sense, that is,

$$E[X] = \lim_{x_0 \to \infty} \left[\int_{-x_0}^{x_0} x\left(\frac{1}{\pi(x^2 + 1)}\right) dx\right] \tag{3.1–17}$$

then $E[X] = 0$. Note, however, that with $Y \triangleq X^2$, $E[Y]$ doesn't exist in any sense because

$$E[Y] = \int_{-\infty}^{\infty} x^2 \left[\frac{1}{\pi(x^2 + 1)}\right] dx \tag{3.1–18}$$

fails to converge in any sense. Thus the variance of a Cauchy r.v. does not exist.

From the linearity of the expectation operator† we can easily obtain the important result that for any X

$$E\left[\sum_{i=1}^{N} g_i(X)\right] = \sum_{i=1}^{N} E[g_i(X)] \tag{3.1–19}$$

provided that these exist. The demonstration of Equation 3.1–19 is left as an exercise. For two random variables X, Y with joint pdf $f_{XY}(x, y)$, we obtain

$$\begin{aligned} E[X] &= \int_{-\infty}^{\infty} \int_{-\infty}^{\infty} x f_{XY}(x, y)\, dx\, dy \\ &= \int_{-\infty}^{\infty} dx\, x\left[\int_{-\infty}^{\infty} f_{XY}(x, y)\, dy\right]. \end{aligned} \tag{3.1–20}$$

By Equation 2.7–26, the integral in brackets is the marginal pdf $f_X(x)$. Hence Equation 3.1–20 is completely consistent with the definition

$$E[X] \triangleq \int_{-\infty}^{\infty} x f_X(x)\, dx.$$

With the help of marginal densities we can conclude that

$$\begin{aligned} E[X + Y] &= \int_{-\infty}^{\infty} \int_{-\infty}^{\infty} (x + y) f_{XY}(x, y)\, dx\, dy \\ &= \int_{-\infty}^{\infty} dx\, x \int_{-\infty}^{\infty} f_{XY}(x, y)\, dy + \int_{-\infty}^{\infty} dy\, y \int_{-\infty}^{\infty} f_{XY}(x, y)\, dx \\ &= E[X] + E[Y]. \end{aligned} \tag{3.1–21}$$

Equation 3.1–21 can be extended to N random variables $X_1, X_2, \ldots, X_N$. Thus

$$E\left[\sum_{i=1}^{N} X_i\right] = \sum_{i=1}^{N} E[X_i]. \tag{3.1–22}$$

Note that *independence is not required.*

Example 3.1–5: Let X, Y be jointly normal, independent r.v.'s with pdf

$$f_{XY}(x, y) = \frac{1}{2\pi\sigma_1\sigma_2} \exp\left\{-\frac{1}{2}\left[\left(\frac{x - \mu_1}{\sigma_1}\right)^2 + \left(\frac{y - \mu_2}{\sigma_2}\right)^2\right]\right\}.$$

It is clear that X and Y are independent from the pdf alone since $f_{XY}(x, y) =$

† An operator $\mathscr{A}$ is said to be linear if for any two functions g_1 and g_2 in its domain, $\mathscr{A}[a_1 g_1 + a_2 g_2] = a_1 \mathscr{A} g_1 + a_2 \mathscr{A} g_2$, where a_1 and a_2 are arbitrary coefficients.

$f_X(x)f_Y(y)$. The marginal pdf's are obtained using Equations 2.7–26 and 2.7–27:

$$f_X(x) = \frac{1}{\sqrt{2\pi\sigma_1^2}} \exp\left[-\frac{1}{2}\left(\frac{x-\mu_1}{\sigma_1}\right)^2\right]$$

$$f_Y(y) = \frac{1}{\sqrt{2\pi\sigma_2^2}} \exp\left[-\frac{1}{2}\left(\frac{y-\mu_2}{\sigma_2}\right)^2\right].$$

Thus Equation 3.1–21 yields

$$E[X+Y] = \mu_1 + \mu_2.$$

3.2 CONDITIONAL EXPECTATIONS

In many practical situations we want to know the average of a subset of the population: the average of the *passing grades* of an exam; the average lifespan of people who are still alive at age seventy; the average height of fighter pilots (many air forces have both an upper and lower limit on the acceptable height of a pilot); the average blood pressure of long-distance runners, and so forth. Problems of this type fall within the realm of *conditional expectations.*

Definition 3.2–1. The conditional expectation of X given that the event B has occurred is

$$E[X \mid B] \triangleq \int_{-\infty}^{\infty} x f_{X|B}(x \mid B)\, dx. \tag{3.2–1}$$

If X is a discrete then Equation 3.2–1 can be replaced with

$$E[X \mid B] \triangleq \sum_i x_i P_{X|B}(x_i \mid B). \tag{3.2–2}$$

To give the reader a feel for the notion of conditional expectation, consider the following exam scores in a course on Probability Theory: 28, 35, 44, 66, 68, 75, 77, 80, 85, 87, 90, 100, 100. Assume that the passing grade is 65. Then the average score is 71.9; however, the *average passing score* is 82.8. A closely related example is worked out as follows.

Example 3.2–1: Consider a continuous r.v. X and the event $B \triangleq \{X \geq a\}$. From Equations 2.7–1 and 2.7–2 and a little bit of work we obtain

$$F_{X|B}(x \mid X \geq a) = \begin{cases} 0, & x < a \\ \dfrac{F_X(x) - F_X(a)}{1 - F_X(a)}, & x \geq a. \end{cases} \tag{3.2–3}$$

Hence

$$f_{X|B}(x \mid X \geq a) = \begin{cases} 0, & x < a \\ \dfrac{f_X(x)}{1 - F_X(a)}, & x > a \end{cases} \tag{3.2–4}$$

and

$$E[X \mid X \geq a] = \frac{\int_a^\infty x f_X(x)\,dx}{\int_a^\infty f_X(x)\,dx}. \tag{3.2–5}$$

Assume that X is a uniform r.v. in $[0, 100]$. Then

$$E[X] = \frac{1}{100}\int_0^{100} x\,dx = 50,$$

but using Equation 3.2–5

$$E[X \mid X \geq 65] = 82.5.$$

Conditional expectations often occur when dealing with random variables that are related in some way. For example let Y denote the lifetime of a person chosen at random, and let X be a binary r.v. that denotes whether the person smokes or not, that is, $X = 0$ if a nonsmoker, $X = 1$ if a smoker. Then clearly $E[Y \mid X = 0]$ is expected to be larger† than $E[Y \mid X = 1]$. Or let X be the intensity of the incident illumination and let Y be the instantaneous photocurrent generated by a photodetector. Clearly the expected value of Y will be larger for stronger illumination and smaller for weaker illumination. We define some important concepts as follows.

Definition 3.2–2. Let X and Y be discrete r.v.'s with joint PMF $P_{X,Y}(x_i, y_j)$. Then the conditional expectation of Y given $X = x_i$ denoted by $E[Y \mid X = x_i]$ is

$$E[Y \mid X = x_i] \triangleq \sum_j y_j P_{Y|X}(y_j \mid x_i) \tag{3.2–6}$$

Here $P_{Y|X}(y_j \mid x_i)$ is the conditional probability that $\{Y = y_i\}$ occurs given that $\{X = x_i\}$ has occurred and is of course given by $P_{X,Y}(x_i, y_j)/P_X(x_i)$.

We can derive an interesting and useful formula for $E[Y]$ in terms of the conditional expectation of Y given $X = x$. The reasoning is much the same as that which we used in computing the average or total probability of an event in terms of its conditional probabilities (see Equations 1.5–4 or 2.7–4). Thus

$$E[Y] = \sum_j y_j P_Y(y_j) \tag{3.2–7}$$

$$= \sum_j y_j \sum_i P_{X,Y}(x_i, y_j)$$

$$= \sum_i \left[\sum_j y_j P_{Y|X}(y_j \mid x_i)\right] P_X(x_i)$$

$$= \sum_i E[Y \mid X = x_i] P_X(x_i). \tag{3.2–8}$$

† Statistical evidence indicates that each cigarette smoked reduces longevity by about eight minutes. Hence smoking one pack a day for a whole year reduces the expected longevity of the smoker by 40 days!

Equation 3.2–8 is a very neat result and says that we can compute $E[Y]$ by averaging the conditional expectation of Y given X with respect to X.† Thus in the smoking-longevity example discussed earlier, suppose $E[Y \mid X = 0] = 79.2$ years and $E[Y \mid X = 1] = 69.4$ years and $P_X(0) = 0.75$ and $P_X(1) = 0.25$. Then

$$E[Y] = (79.2 \times 0.75) + (69.4 \times 0.25) = 76.75$$

is the expected lifetime of the general population.

A result similar to Equation 3.2–8 holds for the continuous case as well. It is derived using Equation 2.7–44 from Chapter 2, that is,

$$f_{Y|X}(y \mid x) = \frac{f_{XY}(x, y)}{f_X(x)} \qquad f_X(x) \neq 0. \tag{3.2–9}$$

The definition of conditional expectation for a continuous r.v. follows.

Definition 3.2–3. Let X and Y be continuous r.v.'s with joint pdf $f_{XY}(x, y)$. Let the conditional pdf of Y given that $X = x$ be denoted as in Equation 3.2–9. Then the conditional expectation of Y given that $X = x$ is given by

$$E[Y \mid X = x] \triangleq \int_{-\infty}^{\infty} y f_{Y|X}(y \mid x)\, dy. \tag{3.2–10}$$

Since

$$E[Y] = \int_{-\infty}^{\infty} \int_{-\infty}^{\infty} y f_{XY}(x, y)\, dx\, dy \tag{3.2–11}$$

it follows from Equations 3.2–9 and 3.2–10 that

$$E[Y] = \int_{-\infty}^{\infty} dx f_X(x) \int_{-\infty}^{\infty} y f_{Y|X}(y \mid x)\, dy$$

$$= \int_{-\infty}^{\infty} E[Y \mid X = x] f_X(x)\, dx. \tag{3.2–12}$$

Equation 3.2–12 is the continuous r.v. equivalent of Equation 3.2–8. It can be used to good advantage (over the direct method) for computing $E[Y]$. We illustrate this point with an important problem from optical communications.

Example 3.2–2: In the photoelectric detector shown in Figure 3.2–1, the number of photoelectrons Y produced in time τ depends on the (normalized) incident energy X. If X were constant, say $X = x$, Y would be a Poisson r.v. [3–4] with parameter x, but as real light sources—except for gain-stabilized lasers—do not emit constant energy signals, X must be treated as a r.v. In certain situations the pdf of X is

† Notice that this statement implies that the conditional expectation of Y given X is a r.v., We shall elaborate on this important concept shortly. For the moment we assume that X assumes the fixed value x_i (or x).

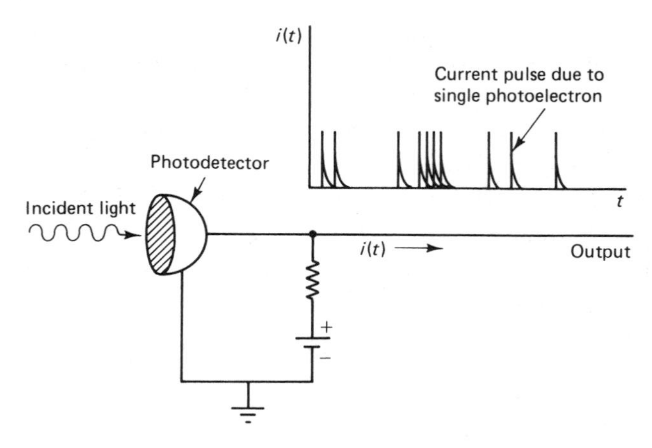

Figure 3.2–1 In a photoelectric detector, incident illumination generates a current consisting of photo-generated electrons.

accurately modeled by

$$f_X(x) = \begin{cases} \dfrac{1}{\mu_X}\exp\left(-\dfrac{x}{\mu_X}\right), & x \geq 0 \\ 0, & x < 0, \end{cases} \tag{3.2–13}$$

where μ_X is a parameter that equals $E[X]$. We shall now compute $E[Y]$ by Equation 3.2–12 and by the direct method.

Solution: Since for $X = x$, Y is Poisson, we can write

$$P[Y = k \mid X = x] = \frac{x^k}{k!}e^{-x} \qquad k = 0, 1, 2, \ldots$$

and, from Example 3.1–2,

$$E[Y \mid X = x] = x.$$

Finally, using Equation 3.2–12 with the appropriate substitutions, that is,

$$E[Y] = \int_0^\infty x\left[\frac{1}{\mu_X}\exp\left(-\frac{x}{\mu_X}\right)\right]dx$$

we easily obtain

$$E[Y] = \mu_X.$$

In contrast to the simplicity with which we obtained this result, consider the direct approach, that is,

$$E[Y] = \sum_{k=0}^{\infty} kP_Y(k). \tag{3.2–14}$$

To compute $P_Y(k)$ we use the Poisson transform (Equation 2.7–9) with $f_X(x)$, as given by Equation 3.2–13. This furnishes (see Equation 2.7–17)

$$P_Y(k) = \frac{\mu_X^k}{(1 + \mu_X)^{k+1}}. \tag{3.2–15}$$

Finally, using Equation 3.2–15 in 3.2–14 yields

$$E[Y] = \sum_{k=0}^{\infty} k \frac{\mu_X^k}{(1 + \mu_X)^{k+1}}.$$

It is known that this series sums to μ_X. Alternatively one can evaluate the sum indirectly using some clever tricks involving derivatives.

Example 3.2–3: Let X and Y be two zero-mean r.v.'s with joint density

$$f_{XY}(x, y) = \frac{1}{2\pi\sigma^2\sqrt{1 - \rho^2}} \exp\left(-\frac{x^2 + y^2 - 2\rho xy}{2\sigma^2(1 - \rho^2)}\right) \quad (3.2\text{–}16)$$

$$|\rho| < 1.$$

We shall soon find out (Section 3.3) that the pdf in Equation 3.2–16 is a special case of the general joint Gaussian law for two r.v.'s. First we see that when $\rho \neq 0$ $f_{XY}(x, y) \neq f_X(x) f_Y(y)$; hence X and Y are not independent when $\rho \neq 0$. When $\rho = 0$ we can indeed write $f_{XY}(x, y) = f_X(x) f_Y(y)$ so that $\rho = 0$ implies independence. For the present, however, our unfamiliarity with the meaning of ρ (ρ is called the normalized *covariance* or *correlation coefficient*) is not important. From Equations 2.7–26 and 2.7–27 it is easy to show that X and Y are zero-mean Gaussian r.v.'s, that is,

$$f_X(x) = f_Y(x) = \frac{1}{\sqrt{2\pi\sigma^2}} e^{-x^2/2\sigma^2}.$$

However, the conditional expectation of Y given $X = x$ is not zero even though Y is a zero-mean r.v.! In fact from Equation 3.2–9:

$$f_{Y|X}(y \mid x) = \frac{1}{\sqrt{2\pi\sigma^2(1 - \rho^2)}} \exp\left(-\frac{(y - \rho x)^2}{2\sigma^2(1 - \rho^2)}\right). \quad (3.2\text{–}17)$$

Hence $f_{Y|X}(y \mid x)$ is Gaussian with mean ρx. Thus†

$$E[Y \mid X = x] = \int_{-\infty}^{\infty} y f_{Y|X}(y \mid x)\, dx$$

$$= \rho x.$$

When ρ is close to unity $E[Y] \simeq x$, which implies that Y tracks X quite closely (exactly if $\rho = 1$), and if we wish to predict Y, say, with Y_p upon observing $X = x$, a good bet is to choose our *predicted value* $Y_p = x$. On the other hand, when $\rho = 0$, observing X doesn't help us to predict Y. Thus we see that in the Gaussian case at least and somewhat more generally, ρ is related to the predictability of one r.v. from observing another. A cautionary note should be sounded, however: The fact that one r.v. doesn't help us to linearly predict another doesn't generally mean that the two r.v.'s are independent.

Conditional Expectation as a Random Variable. Consider, for the sake of being specific, a function $Y = g(X)$ of a discrete r.v. X. Then its expected value is

$$E[Y] = \sum_i g(x_i) P_X(x_i)$$

$$= E[g(X)].$$

† We have already encountered this problem in a different context in Example 2.7–7.

This suggests that we could write Equation 3.2–8 in similar notation, that is,

$$E[Y] = \sum_i E[Y \mid X = x_i]P_X(x_i)$$

$$= E[E[Y \mid X]]. \tag{3.2–18}$$

It is important to note that the object $E[Y \mid X = x_i]$ is a number, as is $g(x_i)$, but the object $E[Y \mid X]$ is a function of the r.v. X *and therefore is itself a r.v.* Given a probability space $\mathscr{P} = (\Omega, \mathscr{F}, P)$ and a r.v. X defined on $\mathscr{P}$, for each outcome $\zeta \in \Omega$ we generate the real number $E[Y \mid X(\zeta)]$. Thus $E[Y \mid X]$ is a r.v. that assumes the value $E[Y \mid X(\zeta)]$ when ζ is the outcome of the underlying experiment. As always, the functional dependence of X on ζ is suppressed, and we specify X rather than the underlying probability space $\mathscr{P}$. The following example illustrates the use of the conditional expectation as a r.v.

Example 3.2–4: Consider a communication system in which the message delay (in milliseconds) is Y and the channel choice is X. Let $X = 1$ for a satellite channel, $X = 2$ for a coaxial cable channel, $X = 3$ for a microwave surface link, and $X = 4$ for a fiber-optical link. A channel is chosen based on availability, which is a random phenomenon. Suppose $P_X(k) = 1/4$ $k = 1, \ldots, 4$. Assume that it is known that $E[Y \mid X = 1] = 500$, $E[Y \mid X = 2] = 300$, $E[Y \mid X = 3] = 200$, and $E[Y \mid X = 4] = 100$. Then the r.v. $g(X) \triangleq E[Y \mid X]$ is defined by

$$g(X) = \begin{cases} 500, & \text{for } X = 1 \quad P_X(1) = \frac{1}{4} \\ 300, & \text{for } X = 2 \quad P_X(2) = \frac{1}{4} \\ 200, & \text{for } X = 3 \quad P_X(3) = \frac{1}{4} \\ 100, & \text{for } X = 4 \quad P_X(4) = \frac{1}{4} \end{cases}$$

and $E[Y] = E[g(X)] = (500 \times \frac{1}{4}) + (300 \times \frac{1}{4}) + (200 \times \frac{1}{4}) + (100 \times \frac{1}{4}) = 275.$

The notion of $E[Y \mid X]$ being a r.v. is equally valid for discrete, continuous, or mixed r.v.'s X. For example, Equation 3.2–12

$$E[Y] = \int_{-\infty}^{\infty} E[Y \mid X = x]f_X(x)\,dx$$

can also be written as $E[Y] = E[E(Y \mid X)]$, where $E[Y \mid X]$ in this case is a function of the continuous r.v. X. The inner expectation is with respect to Y and the outer with respect to X.

The foregoing can be extended to more complex situations. For example, the object $E[Z \mid X, Y]$ is a function of the r.v.'s X and Y and therefore is a function of two r.v.s. For a particular outcome $\zeta \in \Omega$, it assumes the value $E[Z \mid X(\zeta), Y(\zeta)]$. To compute $E[Z]$ we would write $E[Z] = E[E[Z \mid X, Y]]$

$$E[Z] = E[E[Z \mid X, Y]] \tag{3.2–19a}$$

$$= \int_{-\infty}^{\infty}\int_{-\infty}^{\infty}\int_{-\infty}^{\infty} z f_{Z|X,Y}(z \mid x, y) f_{XY}(x, y)\,dx\,dy\,dz. \tag{3.2–19b}$$

In Equation 3.2–19a the inner expectation is with respect to Z and the outer is with respect to X and Y.

We conclude this section by summarizing some properties of conditional expectations that will be useful in Chapters 5 and 10.

✓ **Property (i).** $E[Y] = E[E[Y \mid X]]$. *

Proof. See arguments leading up to Equation 3.2–8 for the discrete case and Equation 3.2–12 for the continuous case. The inner expectation is with respect to Y, the outer with respect to X.

✓ **Property (ii).** If X and Y are independent, then $E[Y \mid X] = E[Y]$. *

Proof.

$$E[Y \mid X = x] = \int_{-\infty}^{\infty} y f_{Y|X}(y \mid x)\, dy.$$

But $f_{XY}(x, y) = f_{Y|X}(y \mid x) f_X(x) = f_Y(y) f_X(x)$ if X and Y are independent. Hence $f_{Y|X}(y \mid x) = f_Y(y)$ and

$$E[Y \mid X = x] = \int_{-\infty}^{\infty} y f_Y(y)\, dy = E[Y]$$

for each x. Thus

$$E[Y \mid X] = \int_{-\infty}^{\infty} y f_Y(y)\, dy = E[Y].$$

An analogous proof holds for the discrete case.

✓ **Property (iii).** $E[Z \mid X] = E[E[Z \mid X, Y] \mid X]$ *

Proof.

$$\begin{aligned} E[Z \mid X = x] &= \int_{-\infty}^{\infty} z f_{Z|X}(z \mid x)\, dz \\ &= \int_{-\infty}^{\infty} \int_{-\infty}^{\infty} z f_{Z|X,Y}(z \mid x, y) f_{Y|X}(y \mid x)\, dz\, dy \\ &= \int_{-\infty}^{\infty} dy f_{Y|X}(y \mid x) \int_{-\infty}^{\infty} z f_{Z|X,Y}(z \mid x, y)\, dz \\ &= E[E[Z \mid X, Y] \mid X = x], \end{aligned}$$

where the inner expectation is with respect to Z and the outer with respect to Y. Since this is true for all x, we have $E[Z \mid X] = E[E[Z \mid X, Y] \mid X]$. The mean $\mu_Y = E[Y]$ is an estimate of the random variable Y. The mean-square error in this estimate is $\varepsilon = E[(Y - \mu_Y)^2]$. In fact this estimate is optimal in that any constant other than μ_Y would lead to an increased ε. Similarly, it is shown in Chapter 5 (Section 5.6) that the conditional mean $E[Y \mid X]$ is the optimal mean-square error estimate of Y given observing X; that is, $g(X) \triangleq E[Y \mid X]$ minimizes $E[(Y - g(X)^2]$ over all Borel functions g.

3.3 MOMENTS

Although the expectation is an important "summary" number for the behavior of a random variable, it is far from adequate in describing the complete behavior of the r.v. Indeed, we saw in Section 3.1 that two sets of numbers could have the same sample mean but the sample deviations could be quite different. Likewise, for two r.v.'s: Their expectations could be the same but their standard deviations could be very different. Summary numbers like $\bar{X}$, σ_X^2, $E[X^2]$ and others are called *moments*. Generally, a r.v. will have many higher-order moments and, under certain conditions (Section 3.5), it is possible to completely describe the behavior of the r.v., that is, reconstruct its pdf from knowledge of all the moments. In the following definitions we shall assume that the moments exist.

Definition 3.3–1. The rth moment of X is defined as

$$\xi_r \triangleq E[X^r] = \int_{-\infty}^{\infty} x^r f_X(x)\,dx, \quad \text{where } r = 0, 1, 2, 3, \ldots. \tag{3.3–1}$$

If X is a discrete r.v., the rth moment can be computed from the PMF as

$$\xi_r \triangleq \sum_i x_i^r P_X(x_i).$$

We note that $\xi_0 = 1$, $\xi_1 = \mu$ (the mean).

Definition 3.3–2. The rth *central moment* of X is defined as

$$m_r \triangleq E[(X - \mu)^r] \quad \text{where } r = 0, 1, 2, 3, \ldots. \tag{3.3–2a}$$

For a discrete r.v. we can compute m_r from

$$m_r \triangleq \sum_i (x_i - \mu)^r P_X(x_i). \tag{3.3–2b}$$

The most frequently used central moment is m_2. It is called the variance and is denoted by σ^2 and also sometimes as Var $[X]$. Note that $m_0 = 1$, $m_1 = 0$, $m_2 = \sigma^2$. A very important formula that connects the variance to $E[X^2]$ and μ is obtained as follows:

$$\sigma^2 = E[[X - \mu]^2] = E[X^2] - E[2\mu X] + E[\mu^2].$$

But for any constant c, $E[cX] = cE[X]$ and $E[c^2] = c^2$. Thus

$$\sigma^2 = E[X^2] - 2\mu E[X] + \mu^2$$

$$\sigma^2 = E[X^2] - \mu^2 \tag{3.3–3}$$

since $E[X] \triangleq \mu$. As has already been done for the mean, in order to save symbology, an overbar $^-$ is often used to denote expectation. Thus $\overline{X^r} \triangleq E[X^r]$, and so forth, for other moments. Using this notation, Equation 3.3–3 appears as

$$\sigma^2 = \overline{X^2} - \mu^2 \tag{3.3–4a}$$

or, equivalently,

$$\overline{X^2} = \sigma^2 + \mu^2. \tag{3.3–4b}$$

Equations 3.3–4 relates to the second central moment m_2 to ξ_2 and μ. We can generalize this result as follows. Observe that

$$(X - \mu)^r = \sum_{i=0}^{r} \binom{r}{i} (-1)^i \mu^i X^{r-i}. \tag{3.3–5a}$$

By taking the expectation of both sides of Equation 3.3–5a and recalling the linearity of the expectation operator we obtain

$$m_r = \sum_{i=0}^{r} \binom{r}{i} (-1)^i \mu^i \xi_{r-i}. \tag{3.3–5b}$$

Example 3.3–1: Let us compute ξ_2 for X, a binomial r.v. By definition

$$P_X(k) = \binom{n}{k} p^k q^{n-k}$$

and

$$\begin{aligned} \xi_2 &= \sum_{k=0}^{n} k^2 \binom{n}{k} p^k q^{n-k} \\ &= p^2 n(n-1) + np \\ &= n^2 p^2 + npq. \end{aligned} \tag{3.3–6}$$

In going from line 1 to line 2, a considerable amount of algebra was used whose duplication we leave as an exercise (Problem 3.13). In going from line 2 to line 3, we rearranged terms and used the fact that $q \triangleq 1 - p$. The expected value of X is

$$\begin{aligned} \xi_1 &= \sum_{k=0}^{n} k \frac{n!}{k!(n-k)!} p^k q^{n-k} \\ &= np = \mu. \end{aligned} \tag{3.3–7}$$

Using this result in Equation 3.3–6 and recalling Equation 3.3–4 allows us to conclude that for a binomial r.v. with PMF $b(k; n, p)$

$$\sigma^2 = npq. \tag{3.3–8}$$

For any given n, maximum variance is obtained when $p = q = 0.5$ (Figure 3.3–1).

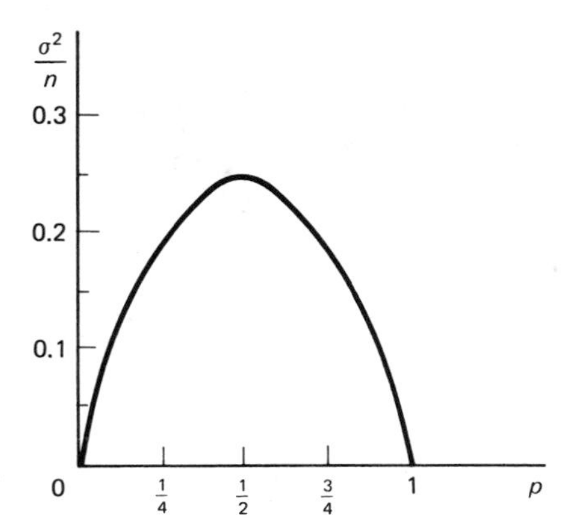

Figure 3.3–1 Variance of a binomial r.v. versus p.

Example 3.3–2: Let us compute m_2 for $X:N(0, \sigma^2)$. Since $\mu = 0$, $m_2 = \xi_2$ and

$$m_2 = \frac{1}{\sqrt{2\pi\sigma^2}} \int_{-\infty}^{\infty} x^2 e^{-\frac{1}{2}(x/\sigma)^2}\, dx.$$

But this integral was already evaluated in Example 3.1–3, where we found $\overline{X^2} = \sigma^2$. Thus the variance of a Gaussian r.v. is indeed the parameter σ^2 regardless of whether X is zero-mean or not.

An interesting and somewhat more difficult example that illustrates a useful application of moments is given in the next example.

Example 3.3–3: The maximum entropy (ME) principle states that if we don't know the pdf $f_X(x)$ of X but would like to estimate it with a function, say $p(x)$, a very good choice is the function $p(x)$ which maximizes the *entropy* defined by [3–5].

$$H[X] \triangleq -\int_{-\infty}^{\infty} p(x) \ln p(x)\, dx \tag{3.3–9}$$

and which satisfies the constraints

$$p(x) \geq 0 \tag{3.3–10a}$$

$$\int_{-\infty}^{\infty} p(x) dx = 1 \tag{3.3–10b}$$

$$\int_{-\infty}^{\infty} x p(x)\, dx = \mu \tag{3.3–10c}$$

$$\int_{-\infty}^{\infty} x^2 p(x) dx = \xi_2, \text{ and so forth.} \tag{3.3–10d}$$

Suppose we know from measurements only that $\int_{-\infty}^{\infty} xp(x)dx = \mu$ and that $x \geq 0$. Thus we wish to find $p(x)$ that maximizes $H[X]$ of Equation 3.3–9 subject to the first three constraints of Equations 3.3–10. According to the method of Lagrange multipliers [3–6], the solution is obtained by maximizing the expression

$$-\int_0^{\infty} p(x) \ln p(x)\, dx + \lambda_1 \int_0^{\infty} p(x)\, dx + \lambda_2 \int_0^{\infty} x p(x)\, dx$$

by differentiation with respect to $p(x)$. The constants λ_1 and λ_2 are Lagrange multipliers and must be determined. After differentiating we obtain

$$\ln p(x) = -(1 + \lambda_1) - \lambda_2 x$$

or

$$p(x) = e^{-(1 + \lambda_1 + \lambda_2 x)}. \tag{3.3–11}$$

When this result is substituted in Equations 3.3–10b and 3.3–10c we find that

$$e^{-(1-\lambda_1)} = \frac{1}{\mu}$$

and

$$\lambda_2 = \frac{1}{\mu}.$$

Hence our ME estimate of $f_X(x)$ is

$$p(x) = \begin{cases} \dfrac{1}{\mu} e^{-x/\mu}, & x \geq 0 \\ = 0, & x < 0. \end{cases} \tag{3.3–12}$$

The problem of obtaining the ME estimate of $f_X(x)$ when both μ and σ^2 are known is left as an exercise (Problem 3.18).

Less useful than ξ_r or m_r are the absolute moments and generalized moments about some arbitrary point, say a, defined by, respectively,

$$E[|X|^r] \triangleq \int_{-\infty}^{\infty} |x|^r f_X(x)\,dx \quad \text{(absolute moment)}$$

$$E[(X-a)^r] \triangleq \int_{-\infty}^{\infty} (x-a)^r f_X(x)\,dx \quad \text{(generalized moment).}$$

Note that if we set $a = \mu$, the generalized moments about a are then the central moments. If $a = 0$, the generalized moments are simply the moments ξ_r.

Joint Moments. Let us now turn to a topic first touched upon in Example 3.2–3. Suppose we are given two random variables X and Y and wish to have a measure of how good a prediction we can make of the value of, say, Y upon observing what value X has. At one extreme if X and Y are independent, observing X tells us nothing about Y. At the other extreme if, say, $Y = aX + b$ then observing the value of X immediately tells us the value of Y. However, in many situations in the real world, two random variables are neither completely independent nor linearly dependent. Given this state of affairs, it then becomes important to have a measure of how much can be said about one random variable from observing another. The quantities called *joint moments* offer us such a measure. Not all joint moments, to be sure, are equally important in this task; especially important are certain *second-order joint moments* (to be defined shortly). However, as we shall see later, in various applications other joint moments are important as well and so we shall deal with the general case below.

Definition 3.3–3. The ijth joint moment of X and Y is given by

$$\xi_{ij} \triangleq E[X^i Y^j] = \int_{-\infty}^{\infty}\int_{-\infty}^{\infty} x^i y^j f_{XY}(x, y)\,dx\,dy. \tag{3.3–13}$$

If X and Y are discrete, we can compute ξ_{ij} from the PMF as

$$\xi_{ij} \triangleq \sum_l \sum_m x_l^i y_m^j P_{X,Y}(x_l, y_m). \tag{3.3–14}$$

Definition 3.3–4. The ijth joint *central moment* of X and Y is defined by

order = i+j

$$m_{ij} \triangleq E[(X-\bar{X})^i (Y-\bar{Y})^j], \tag{3.3–15}$$

where, in the notation introduced earlier, $\bar{X} \triangleq E[X]$, and so forth, for $\bar{Y}$. The order of the moment is $i + j$. Thus all of the following are second-order moments

$$\xi_{02} = E[Y^2] \qquad m_{02} = E[(Y - \bar{Y})^2]$$
$$\xi_{20} = E[X^2] \qquad m_{20} = E[(X - \bar{X})^2]$$
$$\xi_{11} = E[XY] \qquad m_{11} = E[(X - \bar{X})(Y - \bar{Y})]$$
$$= E[XY] - \bar{X}\bar{Y}$$
$$\triangleq \text{Cov}[X, Y].$$

As measures of predictability and in some cases statistical dependence, the most important joint moments are ξ_{11} and m_{11}; they are known as the *correlation* and *covariance* of X and Y respectively. The *correlation coefficient* defined by

$$\rho \triangleq \frac{m_{11}}{\sqrt{m_{20}m_{02}}} \tag{3.3–16}$$

was already introduced in Section 3.2 (Equation 3.2–16). It satisfies $|\rho| \leq 1$.† To show this consider the nonnegative expression

$$E[(\lambda(X - \bar{X}) - (Y - \bar{Y}))^2] \geq 0,$$

where λ is any real constant. To verify that the left side is indeed nonnegative, we merely rewrite it in the form

$$Q(\lambda) \triangleq \int_{-\infty}^{\infty}\int_{-\infty}^{\infty} [\lambda(x - \bar{X}) - (y - \bar{Y})]^2 f_{XY}(x, y)\,dx\,dy \geq 0,$$

where the $\geq$ follows from the fact that the integral of a nonnegative quantity cannot be negative.

The previous equation is a quadratic in λ. Indeed, after expanding we obtain

$$Q(\lambda) = \lambda^2 m_{20} + m_{02} - 2\lambda m_{11} \geq 0.$$

Thus $Q(\lambda)$ can have at most one real root. Hence its discriminant must satisfy

$$\left(\frac{m_{11}}{m_{20}}\right)^2 - \frac{m_{02}}{m_{20}} \leq 0$$

or

$$m_{11}^2 \leq m_{02}m_{20} \tag{3.3–17}$$

whence the condition $|\rho| \leq 1$ follows.

When $m_{11}^2 = m_{02}m_{20}$, that is, $|\rho| = 1$ then it is easy to establish (Problem 3.20) that

$$E\left[\left(\frac{m_{11}}{m_{20}}(X - \bar{X}) - (Y - \bar{Y})\right)^2\right] = 0 \tag{3.3–18}$$

or, equivalently, that

$$\int_{-\infty}^{\infty}\int_{-\infty}^{\infty} \left(\frac{m_{11}}{m_{20}}(x - \bar{X}) - (y - \bar{Y})\right)^2 f_{XY}(x, y)\,dx\,dy = 0. \tag{3.3–19}$$

† Note that it would be more properly termed the *covariance coefficient* or *normalized covariance.*

independence → uncorrelated ; but not vice-versa !!

Since $f_{XY}(x, y)$ is never negative, Equation 3.3–19 implies that the term in brackets is zero everywhere.† Thus we have from Equation 3.3–18 that when $|\rho| = 1$

$$Y = \alpha X + \beta \tag{3.3–20}$$

that is, Y is a linear function of X. The constants α and β are easily related to the moments (Problem 3.20). When $\text{Cov}[X, Y] = 0$ then $\rho = 0$ and X and Y are *said to be uncorrelated.* The following results are important:

(A) If X and Y are uncorrelated then

$$\sigma^2_{X+Y} = \sigma^2_X + \sigma^2_Y,$$

where

$$\sigma^2_{X+Y} \triangleq E[(X + Y)^2] - (E[X + Y])^2.$$

(B) If X and Y are independent, they are uncorrelated. Proof of (A): We leave this as an exercise to the reader; proof of (B): Since $\text{Cov}[X, Y] = E[XY] - E[X]E[Y]$, we must show that $E[XY] = E[X]E[Y]$. But

$$\begin{aligned} E[XY] &= \int_{-\infty}^{\infty}\int_{-\infty}^{\infty} xyf_{XY}(x, y)\,dx\,dy \\ &= \int_{-\infty}^{\infty} xf_X(x)\,dx \int_{-\infty}^{\infty} yf_y(y)\,dy \quad \text{(by independence assumption)} \\ &= E[X]E[Y]. \end{aligned}$$

Example 3.3–4: (Linear prediction.) Suppose we wish to predict the values of an r.v. Y by observing the values of another r.v. X. In particular, the available data (Figure 3.3–2) suggests that a good prediction model for Y is the linear function

$$Y_p \triangleq \alpha X + \beta. \tag{3.3–21}$$

Now although Y may be related to X, the values it takes on may be influenced by other sources that do not affect X. Thus, in general, $|\rho| \neq 1$ and we expect that there will be an error between the predicted value of Y, that is, Y_p and the value that Y actually assumes. Our task becomes then to adjust the coefficients α and β in order to minimize the mean-square error

$$\varepsilon \triangleq E[(Y - Y_p)^2]. \tag{3.3–22}$$

This problem is a simple version of *optimum linear prediction.* In statistics it is called *linear regression.*

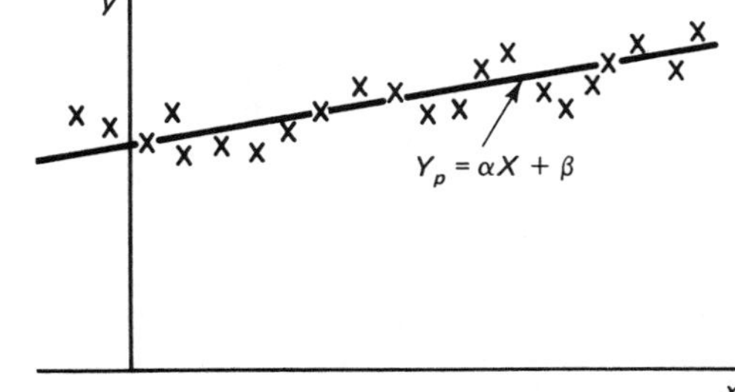

Figure 3.3–2 Pairwise observations on (X, Y) constitute a scatter diagram. The relationship between X and Y is approximated with a straight line.

† Except possibly over a set of points of zero probability. To be more precise we should exchange the word "everywhere" in the text to "almost everywhere," often abbreviated a.e.

Solution: Upon expanding Equation 3.3–22 we obtain

$$\varepsilon = \overline{Y^2} - 2\alpha\overline{XY} - 2\beta\overline{Y} + 2\alpha\beta\overline{X} + \alpha^2\overline{X^2} + \beta^2.$$

To minimize ε with respect to α and β we solve for α and β that satisfy

$$\frac{\partial\varepsilon}{\partial\alpha} = 0 \qquad \frac{\partial\varepsilon}{\partial\beta} = 0 \tag{3.3–23}$$

This yields the best α and β, which we denote by α_0, β_0 in the sense that they minimize ε. A little algebra establishes that

$$\alpha_0 = \frac{\text{Cov}[X, Y]}{\sigma_X^2} = \frac{\rho\sigma_Y}{\sigma_X} \tag{3.3–24a}$$

and

$$\beta_0 = \bar{Y} - \frac{\text{Cov}[X, Y]}{\sigma_X^2}\bar{X}$$

$$= \bar{Y} - \rho\frac{\sigma_Y}{\sigma_X}\bar{X}. \tag{3.3–24b}$$

Thus the best linear predictor is given by

$$Y_p - \bar{Y} = \rho\frac{\sigma_Y}{\sigma_X}(X - \bar{X}) \tag{3.3–25}$$

and passes through the point $(\bar{X}, \bar{Y})$. If we use α_0, β_0 in Equation 3.3–22 we obtain the smallest mean-square error $\varepsilon_{\min}$, which is (Problem 3.21)

$$\varepsilon_{\min} = \sigma_Y^2(1 - \rho^2). \tag{3.3–26}$$

Something rather strange happens when $\rho = 0$. From Equation 3.3–25 we see that for $\rho = 0$ $Y_p = \bar{Y}$ regardless of X! This means that observing X has no bearing on our prediction of Y, and the best predictor is merely $Y_p = \bar{Y}$. We encountered somewhat the same situation in Example 3.2–3. Thus associating the correlation coefficient with ability to predict seems justified in problems involving linear prediction and the joint Gaussian pdf. In some fields, a lack of correlation between two r.v.'s is taken to be *prima facie* evidence that they are unrelated, that is, independent. No doubt this conclusion arises in part from the fact that if two r.v.'s, say, X and Y, are indeed independent, they will be uncorrelated. As stated earlier, the opposite is generally not true. An example follows.

Example 3.3–5: Consider two r.v.'s X and Y with joint PMF $P_{X,Y}(x_i, y_j)$ as shown.

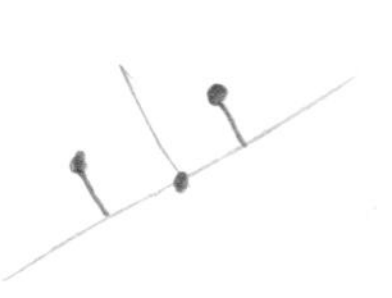

Values of $P_{X,Y}(x_i, y_j)$

	$x_1 = -1$	$x_2 = 0$	$x_3 = +1$
$y_1 = 0$	0	$\frac{1}{3}$	0
$y_2 = 1$	$\frac{1}{3}$	0	$\frac{1}{3}$

X and Y are not independent, since $P_{X,Y}(0, 1) = 0 \neq P_X(0)P_Y(1) = \frac{2}{9}$. Furthermore, $\bar{X} = 0$ so that $\text{Cov}(X, Y) = \overline{XY} - \bar{X}\bar{Y} = \overline{XY}$. We readily compute

$$\overline{XY} = (-1)(1)\tfrac{1}{3} + (1)(1)\tfrac{1}{3} = 0.$$

Hence X and Y are uncorrelated but not independent.

There is an important special case for which $\rho = 0$ always implies independence. We now discuss this case.

Jointly Gaussian Random Variables. We say that two r.v.'s are jointly Gaussian (or jointly normal) if their joint pdf is

$$f_{XY}(x, y) = \frac{1}{2\pi\sigma_X\sigma_Y\sqrt{1-\rho^2}}$$

$$\times \exp\left(\frac{-1}{2(1-\rho^2)}\left\{\left(\frac{x-\bar{X}}{\sigma_X}\right)^2 - 2\rho\frac{(x-\bar{X})(y-\bar{Y})}{\sigma_X\sigma_Y} + \left(\frac{y-\bar{Y}}{\sigma_Y}\right)^2\right\}\right). \quad (3.3\text{–}27)$$

Five parameters are involved: σ_x, σ_Y, $\bar{X}$, $\bar{Y}$, and ρ. If $\rho = 0$ we observe that

$$f_{XY}(x, y) = f_X(x)f_Y(y),$$

where

$$f_X(x) = \frac{1}{\sqrt{2\pi\sigma_X^2}} \exp\left(-\frac{1}{2}\left(\frac{x-\bar{X}}{\sigma_X}\right)^2\right) \quad (3.3\text{–}28)$$

and

$$f_Y(y) = \frac{1}{\sqrt{2\pi\sigma_Y^2}} \exp\left(-\frac{1}{2}\left(\frac{y-\bar{Y}}{\sigma_Y}\right)^2\right). \quad (3.3\text{-}29)$$

Thus two jointly Gaussian r.v.'s that are uncorrelated (that is, $\rho = 0$) are also independent. The marginal densities $f_X(x)$ and $f_Y(y)$ for jointly normal r.v.'s are always normal regardless of what ρ is. However, the converse does not hold; that is, if $f_X(x)$ and $f_Y(y)$ are Gaussian, one cannot conclude that X and Y are jointly Gaussian (Papoulis [3–3, p. 184]).

To illustrate a joint normal distribution consider the following somewhat idealized situation. Let X and Y denote the height of the husband and wife respectively of a married pair picked at random from the population of married people. It is often assumed that X and Y are individually Gaussian although this is obviously only an approximation since heights are bounded from below by zero and from above by physiological constraints. Conventional wisdom has it that in our society tall people prefer tall mates and short people prefer short mates. If this is indeed true then X and Y are *positively correlated* that is, $\rho > 0$. On the other hand, in certain societies it may be fashionable for tall men to marry short women and for tall women to marry short men. Again we can expect X and Y to be correlated albeit negatively this time, that is, $\rho < 0$. Finally, if all marriages are the result of a lottery, we would expect ρ to be zero or very small.

Contours of Constant Density of the Joint Gaussian pdf. It is of interest to determine the locus of points in the xy plane when $f_{XY}(x, y)$ is set constant. Clearly $f_{XY}(x, y)$ will be constant if the exponent is set to a constant, say, c^2:

$$\left(\frac{x-\bar{X}}{\sigma_X}\right)^2 - 2\rho\frac{(x-\bar{X})(y-\bar{Y})}{\sigma_X\sigma_Y} + \left(\frac{y-\bar{Y}}{\sigma_Y}\right)^2 = c^2.$$

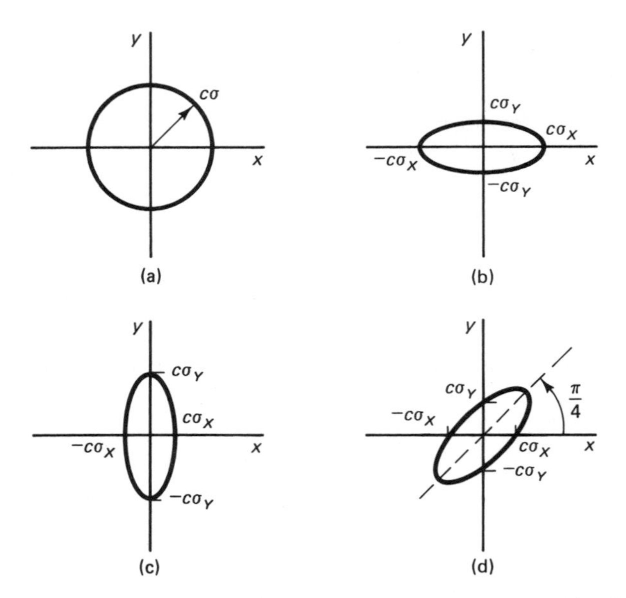

Figure 3.3–3 Contours of constant density for the joint normal ($\bar{X} = \bar{Y} = 0$): (a) $\sigma_X = \sigma_Y$, $\rho = 0$; (b) $\sigma_X > \sigma_Y$, $\rho = 0$; (c) $\sigma_X < \sigma_Y$, $\rho = 0$; (d) $\sigma_X = \sigma_Y$; $\rho \neq 0$.

This is the equation of an ellipse centered at $x = \bar{X}$, $y = \bar{Y}$. For simplicity we set $\bar{X} = \bar{Y} = 0$. When $\rho = 0$, the major and minor diameters of the ellipse are parallel to the x and y axes, a condition we know to associate with independence of X and Y. If $\rho = 0$ and $\sigma_X = \sigma_Y$, the ellipse degenerates into a circle. Several cases are shown in Figure 3.3–3.

3.4 CHEBYSHEV AND SCHWARZ INEQUALITIES

The Chebyshev† inequality furnishes a bound on the probability of how much a r.v. X can deviate from its mean value $\bar{X}$.

Theorem 3.4–1: Let X be an arbitrary r.v. with mean $\bar{X}$ and finite variance σ^2. Then for any $\delta > 0$

$$P[|X - \bar{X}| \geq \delta] \leq \frac{\sigma^2}{\delta^2}. \tag{3.4–1}$$

Proof. Equation 3.4–1 follows directly from the following observation

$$\begin{aligned}\sigma^2 &\triangleq \int_{-\infty}^{\infty} (x - \bar{X})^2 f_X(x)\, dx \geq \int_{|x-\bar{X}|\geq\delta} (x - \bar{X})^2 f_X(x)\, dx \\ &\geq \delta^2 \int_{|x-\bar{X}|\geq\delta} f_X(x)\, dx \\ &= \delta^2 P[|X - \bar{X}| \geq \delta]\end{aligned}$$

† Pafnuti L. Chebyshev (1821–1894), Russian mathematician.

whence Equation 3.4–1 follows. Since

$$\{|X - \bar{X}| \geq \delta\} \cup \{|X - \bar{X}| < \delta\} = \Omega \quad (\Omega \text{ being the certain event})$$

it immediately follows that

$$P[|X - \bar{X}| < \delta] \geq 1 - \frac{\sigma^2}{\delta^2}. \tag{3.4–2}$$

Sometimes it is convenient to express δ in terms of σ, that is, $\delta \triangleq k\sigma$, where k is a constant. Then Equations 3.4–1 and 3.4–2 become, respectively,

$$P[|X - \bar{X}| \geq k\sigma] \leq \frac{1}{k^2} \tag{3.4–3}$$

$$P[|X - \bar{X}| < k\sigma] \geq 1 - \frac{1}{k^2}. \tag{3.4–4}$$

Example 3.4–1: (Deviation from the mean for a normal r.v.) Let $X : N(\bar{X}, \sigma^2)$. How do $P[|X - \bar{X}| < k\sigma]$ and $P[|X - \bar{X}| \geq k\sigma]$ compare with the Chebyshev bound (CB)?

Solution: Using Equations 2.4–10d and 2.4–10e it is easy to show that $P[|X - \bar{X}| < k\sigma] = 2\,\text{erf}(k)$ and $P[|X - \bar{X}| \geq k\sigma] = 1 - 2\,\text{erf}(k)$, where $\text{erf}(k)$ is defined in Equation 2.4–8. Using Table 2.4–1 and Equations 3.4–3 and 3.4–4 we obtain Table 3.4–1.

TABLE 3.4–1

k	$P[\|X - \bar{X}\| < k\sigma]$	CB	$P[\|X - \bar{X}\| > k\sigma]$	CB
0	0	0	1	1
0.5	0.383	0	0.617	1
1.0	0.683	0	0.317	1
1.5	0.866	0.556	0.134	0.444
2.0	0.955	0.750	0.045	0.250
2.5	0.988	0.840	0.012	0.160
3.0	0.997	0.889	0.003	0.111

From Table 3.4–1 we see that the Chebyshev bound is not very good; however, it must be recalled that the bound applies to *any* r.v. X as long as σ^2 exists.

There are a number of extensions of the Chebyshev inequality (see Wilbur Davenport, [3–2, p. 256]). We consider such an extension in what follows.

Random Variables with Positive Values. Consider a r.v. X for which $f_X(x) = 0, x < 0$. Then X is said to take on only positive values.† For such a r.v., the following inequality applies

$$P[X \geq \delta] \leq \frac{E[X]}{\delta}. \tag{3.4–5}$$

† Strictly speaking we should say *nonnegative values*. If X is a continuous r.v. this distinction is not important and the word positive is o.k.

In contrast to the Chebyshev bound, this bound involves only the mean of X.

Proof of Equation 3.4–5.

$$E[X] = \int_0^\infty x f_X(x)\,dx \geq \int_\delta^\infty x f_X(x)\,dx \geq \delta \int_\delta^\infty f_X(x)\,dx$$
$$= \delta P[X \geq \delta]$$

whence Equation 3.4–5 follows. Equation 3.4–5 puts a bound on what fraction of a population can exceed δ.

Example 3.4–2: Assume that in the manufacturing of very low-grade electrical 1000-ohm resistors the average resistance, as determined by measurements, is indeed 1000 ohms but there is a large variation about this value. If all resistors over 1500 ohms are to be discarded, what is the maximum fraction of resistance to meet such a fate?

Solution: With $\bar{X} = 1000$, and $\delta = 1500$, we obtain

$$P[X \geq 1500] \leq \frac{1000}{1500} = 0.67.$$

Thus, if nothing else, the manufacturer has the assurance that the percentage of discarded resistors cannot exceed 67 percent of the total.

The Schwarz Inequality. We have already encountered the probabilistic form of the Schwarz† inequality in Equation 3.3–17 repeated here as

$$\text{Cov}^2(X, Y) \leq E[(X - \bar{X})^2]E[(Y - \bar{Y})^2]$$

with equality if and only if Y is a linear function of X. In later work we shall need another version of the Schwarz inequality that is commonly used in obtaining results in signal processing and stochastic processes. Consider two ordinary, that is, nonrandom functions h and g not necessarily real and define the *norm* of h, if it exists, by

$$\|h\| \triangleq \left(\int_{-\infty}^{\infty} |h(x)|^2\,dx\right)^{1/2}, \tag{3.4–6}$$

and so forth, for the norm of g. The *scalar* or *inner* product of h with g is denoted by (h, g) and is defined by

$$(h, g) \triangleq \int_{-\infty}^{\infty} h(x)g^*(x)\,dx$$
$$= (g, h)^*. \tag{3.4–7}$$

Then another form of the Schwarz inequality is

$$|(h, g)| \leq \|h\|\,\|g\| \tag{3.4–8}$$

with equality if and only if h is proportional to g, that is, $h(x) = c\,g(x)$. The

† H. Amandus Schwarz (1843–1921), German mathematician.

proof of Equation 3.4–8 is easily obtained by considering the norm of $\lambda h(x) + g(x)$ as a function of the real variable λ,

$$\|\lambda h(x) + g(x)\|^2 = |\lambda|^2 \|h\|^2 + \lambda(h, g) + \lambda^*(h, g)^* + \|g\|^2 \geq 0 \tag{3.4–9}$$

with

$$\lambda = -\frac{(h, g)^*}{\|h\|^2} \tag{3.4–10}$$

whence Equation 3.4–8. In the special case where h and g are real functions of real random variables, that is, $h(X)$, $g(X)$, then Equation 3.4–8 still is valid provided that the definitions or norm and inner product are modified as follows

$$\|h\|^2 \triangleq \int_{-\infty}^{\infty} h^2(x) f_X(x)\, dx = E[h^2(X)] \tag{3.4–11}$$

$$(h, g) \triangleq \int_{-\infty}^{\infty} h(x) g(x) f_X(x)\, dx = E[h(X)g(X)] \tag{3.4–12}$$

whence we obtain

$$\checkmark \quad |E[h(X)g(X)]| \leq (E[h^2(X)])^{1/2}(E[g^2(X)]^{1/2}. \tag{3.4–13}$$

Example 3.4–3: (The [weak] law of large numbers [LLN].) Let $X_1, \ldots, X_n$ be independent and identically distributed random variables with mean μ_X and variance σ_X^2. Assume that we don't know the value of μ_X (or σ_X) and thus consider the sample mean

$$\hat{\mu}_n \triangleq \frac{1}{n} \sum_{i=1}^{n} X_i$$

as an estimate of μ_X. We can use the Chebyshev inequality to show that $\hat{\mu}_n$ is asymptotically a perfect estimate of μ_X. First we compute

$$\begin{aligned} E[\hat{\mu}_n] &= \frac{1}{n} \sum_{i=1}^{n} E[X_i] \\ &= \left(\frac{1}{n}\right) n\mu_X \\ &= \mu_X. \end{aligned}$$

Next we compute

$$\begin{aligned} \text{Var}[\hat{\mu}_n] &= \frac{1}{n^2} \text{Var}\left[\sum_{i=1}^{n} X_i\right] \\ &= \left(\frac{1}{n^2}\right) n\sigma_X^2 \\ &= \frac{1}{n} \sigma_X^2. \end{aligned}$$

Thus by the Chebyshev inequality (Equation 3.4–1) we have

$$P[|\hat{\mu}_n - \mu_X| \geq \delta] \leq \sigma_X^2/n\delta^2$$

Clearly for any fixed $\delta > 0$, the right side can be made arbitrarily small by choosing n large enough. Thus

$$\lim_{n \to \infty} P[|\hat{\mu}_n - \mu_X| \geq \delta] = 0$$

for every $\delta > 0$.

The law of large numbers is the theoretical basis for estimating μ_X from measurements. When an experimenter takes the sample mean of n measurements, he is relying on the LLN in order to use the sample mean as an estimate of the unknown theoretical expectation μ_X.

3.5 MOMENT-GENERATING FUNCTIONS

The moment-generating function, if it exists, of a random variable X is defined by†

$$\theta(t) \triangleq E[e^{tX}] \tag{3.5-1}$$

$$= \int_{-\infty}^{\infty} e^{tx} f_X(x)\, dx, \tag{3.5-2}$$

where t is a complex variable.

For discrete r.v.'s, we can define $\theta(t)$ using the PMF as

$$\theta(t) = \sum_i e^{tx_i} P_X(x_i). \tag{3.5-3}$$

From Equation 3.5–2 we see that except for a sign reversal in the exponent, the characteristic function is the two-sided Laplace transform for which there is a known inversion formula. Thus, in general, knowing $\theta(t)$ is equivalent to knowing $f_X(x)$ and vice versa.

The main reasons for introducing $\theta(t)$ are: (1) it enables a convenient computation of the moments of X; (2) it can be used to estimate $f_X(x)$ from experimental measurements of the moments; (3) it can be used to solve problems involving the computation of the sums of random variables; and (4) it is an important analytical instrument that can be used to demonstrate basic results such as *the central limit theorem*.‡

Proceeding formally, if we expand e^{tX} and take expectations, then

$$E[e^{tX}] = E\left[1 + tX + \frac{(tX)^2}{2!} + \ldots + \frac{(tX)^n}{n!} + \ldots\right]$$

$$= 1 + t\mu + \frac{t^2}{2!}\xi_2 + \ldots + \frac{t^n}{n!}\xi_n + \ldots. \tag{3.5-4}$$

Since the moments ξ_i may not exist, for example, none of the moments above the first exist for the Cauchy pdf, $\theta(t)$ may not exist. However, if $\theta(t)$ does

† The terminology is in dispute (see Feller [3–1], p. 411).

‡ To be discussed in Section 3.7.

exist, computing any moment is easily obtained by differentiation. Indeed, if we allow the notation

$$\theta^{(k)}(0) \triangleq \frac{d^k}{dt^k}(\theta(t))\bigg|_{t=0},$$

then

$$\boxed{\xi_k = \theta^{(k)}(0) \qquad k = 0, 1, \ldots .} \tag{3.5–5}$$

Example 3.5–1: Let $X{:}N(\mu, \sigma^2)$. Its moment-generating function is

$$\theta(t) = \frac{1}{\sqrt{2\pi\sigma^2}} \int_{-\infty}^{\infty} \exp\left(-\frac{1}{2}\left(\frac{x-\mu}{\sigma}\right)^2\right) e^{tx}\, dx. \tag{3.5–6}$$

Using the trick known as "completing the square"† in the exponent, we can write Equation 3.5–6 as

$$\theta(t) = \exp[\mu t + \sigma^2 t^2/2]$$
$$\times \frac{1}{\sqrt{2\pi\sigma^2}} \int_{-\infty}^{\infty} \exp\left\{-\frac{1}{2\sigma^2}(x - (\mu + \sigma^2 t))^2\right\} dx.$$

But the factor on the second line is unity since it is the integral of a Gaussian pdf with mean $\mu + \sigma^2 t$ and variance σ^2. Hence

$$\theta(t) = \exp(\mu t + \sigma^2 t^2/2), \tag{3.5–7}$$

from which we obtain

$$\theta^{(1)}(0) = \mu$$
$$\theta^{(2)}(0) = \mu^2 + \sigma^2.$$

Example 3.5–2: Let X be a binomial r.v. with parameters n (number of tries), p (probability of a success per trial), and $q = 1 - p$. Then

$$\begin{aligned}\theta(t) &= \sum_{k=0}^{n} e^{tk}\binom{n}{k}p^k q^{n-k} \\ &= \sum_{k=0}^{n}\binom{n}{k}[e^t p]^k q^{n-k} \\ &= (pe^t + q)^n.\end{aligned} \tag{3.5–8}$$

We obtain

$$\begin{aligned}\theta^{(1)}(0) &\triangleq np = \mu \\ \theta^{(2)}(0) &= \{npe^t(pe^t + q)^{n-1} + n(n-1)p^2e^{2t}(pe^t + q)^{n-2}\}_{t=0} \\ &= npq + \mu^2.\end{aligned} \tag{3.5–9}$$

Hence

$$\text{Var}[X] = npq. \tag{3.5–10}$$

We make the observation that if all the moments exist and are known, then

† For example, $(x - a)^2 + bx = x^2 - 2ax + a^2 + bx = x^2 - 2x[a - b/2] + a^2 = [x - (a - b/2)]^2 + a^2 - (a - b/2)^2 = [x - (a - b/2)]^2 + ab - (b/2)^2$.

$\theta(t)$ is known as well (see Equation 3.5–4). Since $\theta(t)$ is related to $f_X(x)$ through the Laplace transform, we can, in principle at least, determine $f_X(x)$ from its moments. In practice, if X is the r.v. whose pdf is desired and X_i represents our ith observation of X, then we can *estimate* the rth moment of X, ξ_r, from

$$\hat{\Theta}_r = \frac{1}{n} \sum_{i=1}^{n} X_i^r, \tag{3.5–11}$$

where $\hat{\Theta}_r$ is called a *moment estimator* and is a random variable, and n is the number of observations. Even though Θ_r is an r.v., its variance becomes small as n becomes large and for n large enough we can have confidence that $\hat{\Theta}_r$ is reasonably close to ξ_r† (not a r.v.).

The moment-generating function $\theta_{XY}(t_1, t_2)$ of two r.v.'s X and Y is defined by

$$\begin{aligned} \theta_{XY}(t_1, t_2) &\triangleq E[e^{(t_1X+t_2Y)}], \\ &= \int_{-\infty}^{\infty} \int_{-\infty}^{\infty} \exp(t_1x + t_2y) f_{XY}(x, y)\, dx\, dy. \end{aligned} \tag{3.5–12}$$

Proceeding as we did in Equation 3.5–4, we can easily establish with the help of a power series expansion that

$$\theta_{XY}(t_1, t_2) = \sum_{i=0}^{\infty} \sum_{j=0}^{\infty} \frac{t_1^i t_2^j}{i!j!} \xi_{ij}, \tag{3.5–13}$$

where ξ_{ij} is defined in Equation 3.3–13. Using the notation

$$\theta_{XY}^{(l,n)}(0, 0) \triangleq \left. \frac{\partial^{l+n} \theta_{XY}(t_1, t_2)}{\partial t_1^l\, \partial t_2^n} \right|_{t_1=t_2=0}$$

we can easily show from Equation 3.5–12 or 3.5–13 that

$$\xi_{ln} = \theta_{XY}^{(l,n)}(0, 0). \tag{3.5–14}$$

In particular

$$\theta_{XY}^{(1,0)}(0, 0) = \bar{X}, \qquad \theta_{XY}^{(0,1)}(0, 0) = \bar{Y} \tag{3.5–15}$$

$$\theta_{XY}^{(2,0)}(0, 0) = \overline{X^2}, \qquad \theta_{XY}^{(0,2)}(0, 0) = \overline{Y^2} \tag{3.5–16}$$

$$\theta_{XY}^{(1,1)}(0, 0) = \overline{XY} = \text{Cov}[X, Y] + \bar{X}\bar{Y}. \tag{3.5–17}$$

One could similarly define a joint moment-generating function for N random $X_1, \ldots, X_N$ variables by

$$\begin{aligned} \theta(t_1, t_2, \ldots, t_N) &= E\left[\exp \sum_{i=1}^{N} t_i X_i\right] \\ &= \sum_{k_1=0}^{\infty} \sum_{k_2=0}^{\infty} \cdots \sum_{k_N=0}^{\infty} \frac{t_1^{k_1}}{k_1!} \cdots \frac{t_N^{k_N}}{k_N!} \overline{X_1^{k_1} X_2^{k_2} \ldots X_N^{k_N}} \end{aligned} \tag{3.5–18}$$

from which the moments can be computed using straightforward extensions of the earlier cases.

† These notions are explored in greater detail in the first half of Chapter 5.

3.6 CHARACTERISTIC FUNCTIONS

If in Equation 3.5–1 we replace the parameter t by $j\omega$ where $j \triangleq \sqrt{-1}$, we obtain the characteristic function of X defined by

$$\Phi_X(\omega) \triangleq E[e^{j\omega X}]$$

$$= \int_{-\infty}^{\infty} f_X(x) e^{j\omega x}\, dx, \tag{3.6–1}$$

which except for a minus sign difference in the exponent, we recognize as the Fourier transform of $f_X(x)$. For discrete r.v.'s we can define $\Phi_X(\omega)$ in terms of the PMF by

$$\Phi_X(\omega) = \sum_i e^{j\omega x_i} P_X(x_i). \tag{3.6–2}$$

For our purposes, the characteristic function has all the properties of the moment-generating function. The Fourier transform is widely used in statistical communication theory, and since the inversion of Equation 3.6–1 is often easy to achieve, either by direct integration or through the availability of extensive tables of Fourier transforms (for example, Reference 3–7), the characteristic function is widely used to solve problems involving the computation of the sums of independent r.v.'s. We have seen that the pdf of the sum of independent random variables involves the convolution of their pdf's. Thus if $Z = X_1 + \ldots + X_N$, where X_i, $i = 1, \ldots, N$ are independent r.v.'s the pdf of Z is furnished by

$$f_Z(z) = f_{X_1}(z) * f_{X_2}(z) * \ldots * f_{X_N}(z) \tag{3.6–3}$$

that is, the repeated convolution product.

The actual evaluation of Equation 3.6–3 can be very tedious. However, we know from our studies of Fourier transforms that *the Fourier transform of a convolution product is the product of the individual transforms.* We illustrate the use of characteristic functions in the following examples.

Example 3.6–1: Let X_i, $i = 1, \ldots, N$ be a sequence of *independent, identically distributed* (i.i.d.) r.v.'s with $X : N(0, 1)$. Compute the pdf of

$$Z = \sum_{i=1}^{N} X_i.$$

Solution: The pdf of Z is given by Equation 3.6–3. With $\Phi_{X_i}(\omega)$ denoting the characteristic function of X_i we have

$$\Phi_Z(\omega) = \Phi_{X_1}(\omega) \times \ldots \times \Phi_{X_N}(\omega). \tag{3.6–4}$$

However, since the X_i's are i.i.d. $N(0, 1)$ the characteristic function of all the X_i's are the same, and we define $\Phi_X(\omega) \triangleq \Phi_{X_1}(\omega) = \ldots = \Phi_{X_N}(\omega)$. Thus

$$\Phi_X(\omega) = \int_{-\infty}^{\infty} \frac{1}{\sqrt{2\pi}} e^{-\frac{1}{2}x^2} e^{j\omega x}\, dx$$

$$= e^{-\frac{1}{2}\omega^2}, \tag{3.6–5}$$

and

$$\Phi_Z(\omega) = [\Phi_X(\omega)]^n = e^{-\frac{1}{2}n\omega^2}. \tag{3.6–6}$$

From the form of $\Phi_Z(\omega)$ we deduce that $f_Z(z)$ must also be Gaussian. To obtain $f_Z(z)$ we use the Fourier inversion formula:

$$f_Z(z) = \frac{1}{2\pi}\int_{-\infty}^{\infty} \Phi_Z(\omega)e^{-j\omega z}\, d\omega. \tag{3.6–7}$$

Inserting Equation 3.6–6 into Equation 3.6–7 and manipulating terms enables us to obtain

$$f_Z(z) = \frac{1}{\sqrt{2\pi n}} e^{-\frac{1}{2}(x^2/n)}.$$

Hence $f_Z(z)$ is indeed Gaussian. The variance of Z is n, and its mean is zero.

Example 3.6–2: Consider two i.i.d. r.v.'s X and Y with

$$f_X(x) = f_Y(x) = \frac{1}{a}\operatorname{rect}\left(\frac{x}{a}\right).$$

Compute the pdf of $Z = X + Y$ using characteristic functions.

Solution: We can, of course, compute $f_Z(z)$ by convolving $f_X(x)$ and $f_Y(y)$. However, using characteristic functions, we obtain $f_Z(z)$ from

$$f_Z(z) = \frac{1}{2\pi}\int_{-\infty}^{\infty} \Phi_X(\omega)\Phi_Y(\omega)e^{-j\omega z}\, d\omega,$$

where

$$\Phi_X(\omega)\Phi_Y(\omega) = \Phi_Z(\omega).$$

Since the pdf's of X and Y are the same we can write

$$\begin{aligned}\Phi(\omega) &\triangleq \Phi_X(\omega) = \Phi_Y(\omega)\\ &= \frac{1}{a}\int_{-a/2}^{a/2} e^{j\omega x}\, dx\\ &= \frac{\sin(a\omega/2)}{a\omega/2}.\end{aligned}$$

Hence

$$\Phi_Z(\omega) = \left(\frac{\sin(a\omega/2)}{a\omega/2}\right)^2 \tag{3.6–8}$$

and

$$\begin{aligned}f_Z(z) &= \frac{1}{2\pi}\int_{-\infty}^{\infty} \Phi_Z(\omega)e^{-j\omega z}\, d\omega\\ &= \frac{1}{a}\left(1 - \frac{|z|}{a}\right)\operatorname{rect}\left(\frac{z}{2a}\right),\end{aligned} \tag{3.6–9}$$

which is shown in Figure 3.6–1. The easiest way to obtain the result in Equation 3.6–9 is to look up the Fourier transform (or its inverse) of Equation 3.6–8 in a table of elementary Fourier transforms.

As in the case of moment-generating functions, we can compute the moments from the characteristic functions by differentiation, provided that these exist. If we

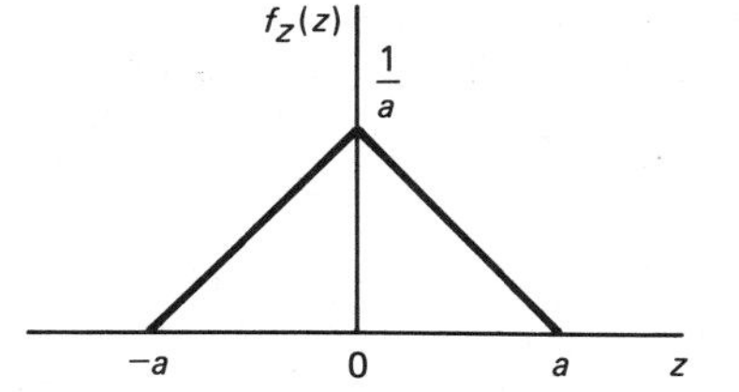

Figure 3.6–1 The pdf of $Z = X + Y$ when X and Y are i.i.d. uniformly distributed in $(-a/2, a/2)$.

expand $\exp(j\omega X)$ into a power series and take the expectation we obtain

$$\begin{aligned}\Phi_X(\omega) &= E[e^{j\omega X}] \\ &= \sum_{n=0}^{\infty} \frac{(j\omega)^n}{n!} \xi_n. \end{aligned} \qquad (3.6\text{–}10)$$

From Equation 3.6–10 it is easily established that

$$\boxed{\xi_n = \frac{1}{j^n} \Phi_X^{(n)}(0)} \qquad (3.6\text{–}11)$$

where we have used the notation

$$\Phi_X^{(n)}(0) \triangleq \left. \frac{d^n}{d\omega^n} \Phi_X(\omega) \right|_{\omega=0}.$$

End 10/3 Start 10/8

Example 3.6–3: Compute the first few moments of $Y = \sin \Theta$ if Θ is uniformly distributed in $[0, 2\pi]$.

Solution: We use the result in Equation 3.1–8; that is, if $Y = g(X)$, then

$$\overline{Y} = \int_{-\infty}^{\infty} y f_Y(y)\, dy = \int_{-\infty}^{\infty} g(x) f_X(x)\, dx.$$

Hence

$$\begin{aligned} E[e^{j\omega Y}] &= \int_{-\infty}^{\infty} e^{j\omega y} f_Y(y)\, dy \\ &= \frac{1}{2\pi} \int_0^{2\pi} e^{j\omega \sin\theta}\, d\theta \\ &= J_0(\omega), \end{aligned}$$

where $J_0(\omega)$ is the Bessel function of the first kind of order zero. A power series expansion of $J_0(\omega)$ gives

$$J_0(\omega) = 1 - \left(\frac{\omega}{2}\right)^2 + \frac{1}{2!2!}\left(\frac{\omega}{2}\right)^4 - \ldots.$$

Hence all the odd-order moments are zero. From Equation 3.6–11 we compute

$$\overline{Y^2} = \xi_2 \triangleq (-1)\Phi_X^{(2)}(0) = \tfrac{1}{2}$$

$$\overline{Y^4} = \xi_4 = (+1)\Phi_X^{(4)}(0) = \tfrac{3}{8}.$$

Example 3.6–4: Let X and Y be i.i.d. binomial r.v.'s with parameters n and p, that is,

$$P_X(k) = P_Y(k) = \binom{n}{k} p^k q^{n-k}.$$

Compute the PMF of $Z = X + Y$.

Solution: Since X and Y take on nonnegative integer values so must Z. We can solve this problem by: (1) convolution of the pdf's, which involves delta functions; (2) *discrete convolution* of the PMF's; and (3) by characteristic functions. The discrete convolution for this case is

$$\begin{aligned} P[Z = k] &= \sum_i P_X(i)P_Y(k - i) \\ &= \sum_i \binom{n}{i} p^i q^{n-i} \binom{n}{k - i} p^{k-i} q^{n-(k-i)} \\ &= p^k q^{2n-k} \sum_i \binom{n}{i}\binom{n}{k - i}, \quad \text{for } k = 0, 1, \ldots, 2n. \end{aligned}$$

The trouble is that we may not immediately recognize the closed form of the sum of products of binomial coefficients. The computation of the PMF of Z by characteristic functions is very simple. First observe that

$$\begin{aligned} \Phi_X(\omega) = \Phi_Y(\omega) &= \sum_{k=0}^{n} e^{j\omega k}\binom{n}{k} p^k q^{n-k} \\ &= (pe^{j\omega} + q)^n. \end{aligned}$$

Thus, by virtue of the independence of X and Y, we obtain

$$\begin{aligned} \Phi_Z(\omega) &= E(\exp j\omega[X + Y]) \\ &= E[\exp(j\omega X)]E[\exp(j\omega Y)] \\ &= (pe^{j\omega} + q)^{2n}. \end{aligned}$$

Thus Z is binomial with parameters $2n$, p, that is,

$$P_Z(k) = \binom{2n}{k} p^k q^{2n-k}, \quad \text{for } k = 0, \ldots, 2n.$$

As a by-product of the compution of $P_Z(k)$ by c.f.'s we obtain the result that

$$\binom{2n}{k} = \sum_{i=0}^{n} \binom{n}{i}\binom{n}{k - i}.$$

An extension of this result is the following: If $X_1, X_2, \ldots, X_N$ are i.i.d. binomial with parameters n, p then $Z = \sum_{i=1}^{N} X_i$ is binomial with parameters Nn, p. Regardless of how large N gets, Z *remains a discrete r.v. with a binomial PMF.*†

Joint Characteristic Functions. As in the case of joint moment-generating functions we can define the joint characteristic function by

$$\Phi_{X_1 \ldots X_N}(\omega_1, \omega_2, \ldots, \omega_N) = E\left[\exp\left(j \sum_{i=1}^{N} \omega_i X_i\right)\right]. \tag{3.6–12}$$

By the Fourier inversion property, the joint pdf is the inverse Fourier

† Recall this statement for future reference in connection with the *Central Limit Theorem.*

transform (with a sign reversal) of $\Phi_{X_1 \ldots X_N}(\omega_1, \ldots, \omega_N)$. Thus

$$f_{X_1 \ldots X_N}(x_1, \ldots, x_N) = \frac{1}{(2\pi)^n} \int_{-\infty}^{\infty} \cdots \int_{-\infty}^{\infty} \Phi_{X_1 \ldots X_N}(\omega_1, \ldots, \omega_N)$$

$$\exp\left(-j\sum_{i=1}^{N} \omega_i x_i\right) d\omega_1\, d\omega_2 \ldots d\omega_N. \tag{3.6–13}$$

We can obtain the moments by differentiation. For instance, with X, Y denoting any two random variables ($N = 2$) we have

$$\xi_{rk} \equiv E[X^r Y^k] = (-j)^{r+k} \Phi_{XY}^{(r,k)}(0, 0), \tag{3.6–14}$$

where

$$\Phi_{XY}^{(r,k)}(0, 0) \triangleq \left. \frac{\partial^{r+k} \Phi_{XY}(\omega_1, \omega_2)}{\partial \omega_1^r\, \partial \omega_2^k} \right|_{\omega_1 = \omega_2 = 0} \tag{3.6–15}$$

Finally for discrete r.v.'s we can define the joint characteristic function in terms of the joint PMF. For instance for two r.v.'s, X and Y, we obtain

$$\Phi_{XY}(\omega_1, \omega_2) \triangleq \sum_i \sum_j e^{j(\omega_1 x_i + \omega_2 y_j)} P_{XY}(x_i, y_j). \tag{3.6–16}$$

The extension to more than two discrete r.v.'s is straightforward, although the notation becomes a little clumsy.

The Central Limit Theorem. It is sometimes said that the sum of a large number of random variables tends toward the normal. Under what conditions is this true? The *Central Limit Theorem* deals with this important point.

Basically the Central Limit Theorem† says that the normalized sum of a large number of mutually independent random variables $X_1, \ldots, X_n$ with zero means and finite variances $\sigma_1^2, \ldots, \sigma_n^2$ tends to the normal probability distribution function provided that the individual variances σ_k^2, $k = 1, \ldots, n$ are small compared to $\sum_{i=1}^{n} \sigma_i^2$. The constraint on the variances are known as the Lindeberg conditions and are discussed in detail by Feller [3–1, p. 262]. We state a general form of the Central Limit Theorem following and furnish a proof for a special case.

Theorem 3.6–1. Let $X_1, \ldots, X_n$ be n mutually independent (scalar) r.v.'s with PDF's $F_1(x_1), F_2(x_2), \ldots, F_n(x_n)$ respectively such that

$$\bar{X}_k = 0, \qquad \text{Var}[X_k] = \sigma_k^2$$

and let

$$s_n^2 \triangleq \sigma_1^2 + \ldots + \sigma_n^2.$$

If for a given $\varepsilon > 0$ and n sufficiently large the σ_k satisfy

$$\sigma_k < \varepsilon s_n \qquad k = 1, \ldots, n$$

† First proved by Abraham De Moivre in 1733 for the special case of Bernoulli r.v.'s.

then the normalized sum

$$Z_n \triangleq (X_1 + \ldots + X_n)/s_n$$

converges to the standard normal PDF, denoted $\mathcal{N}(z)$, that is, $\lim_{n\to\infty} F_{Z_n}(z) = \mathcal{N}(z)$. A discussion of convergence in distribution is given in Chapter 6.

We now prove a special case of the foregoing.

Theorem 3.6–2. Let $X_1, X_2, \ldots, X_n$ be i.i.d. r.v.'s with $\bar{X}_i = 0$, and $\text{Var}[X_i] = 1$, $i = 1, \ldots, n$. Then

$$Z_n \triangleq \frac{1}{\sqrt{n}} \sum_{i=1}^{n} X_i$$

tends to the normal in the sense that its characteristic function Φ_{Z_n} satisfies

$$\lim_{n\to\infty} \Phi_{Z_n}(\omega) = e^{-\frac{1}{2}\omega^2},$$

which is the characteristic function of the $N(0, 1)$ r.v.

Proof. Let $W_i \triangleq X_i/\sqrt{n}$. Also, let $\Phi_{X_i}(\omega)$ and $f_{X_i}(x)$ be the characteristic function and pdf of X respectively. Then

$$\begin{aligned}\Phi_{W_i}(\omega) &\triangleq E[e^{j\omega W_i}]\\ &= E[e^{j(\omega/\sqrt{n})X_i}]\\ &= \Phi_{X_i}\left(\frac{\omega}{\sqrt{n}}\right).\end{aligned}$$

Now expand $\Phi_{X_i}(\omega)$ and $\Phi_{W_i}(\omega)$ into Taylor polynomials of the second degree with remainders, that is,

$$\Phi_{X_i}(\omega) = 1 - \frac{\omega^2}{2} + R_2(\omega) \triangleq \Phi_X(\omega)$$

$$\Phi_{W_i}(\omega) = 1 - \frac{\omega^2}{2n} + \frac{R_2'(\omega)}{n\sqrt{n}} \triangleq \Phi_W(\omega),$$

where

$$\begin{aligned}R_2(\omega) &= \omega^3\Phi_X^{(3)}(\omega_o)/6\\ R_2'(\omega) &= \omega^3\Phi_X^{(3)}(\omega_o/\sqrt{n})/6\\ \omega_o &\in [0, \omega]\end{aligned}$$

and where we used the fact that $\sigma_{X_i}^2 = 1$ and $\bar{X}_i = 0$. The remainder term R_2' will exist if the third derivative of $\Phi_{X_i}(\omega)$ exists for all ω. Since $Z_n = \sum_{i=1}^{n} W_i$, we obtain

$$\Phi_{Z_n}(\omega) = [\Phi_W(\omega)]^n,$$

or

$$\ln \Phi_Z(\omega) = n \ln \Phi_W(\omega).$$

Now recall that for any h such that $|h| < 1$,

$$\ln(1 + h) = h - \frac{h^2}{2} + \frac{h^3}{3} - \ldots.$$

For any *fixed* ω, we can choose an n large enough so that

$$\left|-\frac{\omega^2}{2n} + \frac{R_2'}{n\sqrt{n}}\right| < 1.$$

Assuming this to have been done we can write

$$\begin{aligned} \ln \Phi_{Z_n}(\omega) &= n \ln\left[1 - \frac{\omega^2}{2n} + \frac{R_2'}{n\sqrt{n}}\right] \\ &\simeq n\left[-\frac{\omega^2}{2n} + \frac{R_2'}{n\sqrt{n}} - \left(-\frac{\omega^2}{2n} + \frac{R_2'}{n\sqrt{n}}\right)^2 + \left(-\frac{\omega^2}{2n} + \frac{R_2'}{n\sqrt{n}}\right)^3 + \ldots\right] \\ &= -\frac{\omega^2}{2} + \text{terms involving factors of } n^{-1/2}, n^{-1}, n^{-3/2}, \ldots. \end{aligned}$$

Hence

$$\lim_{n\to\infty} [\ln \Phi_{Z_n}(\omega)] = -\frac{\omega^2}{2}$$

or, equivalently,

$$\lim_{n\to\infty} \Phi_{Z_n}(\omega) = e^{-(\omega^2/2)},$$

which is the characteristic function of the $N(0, 1)$ r.v. Note that to argue that $\lim_{n\to\infty} f_{Z_n}(z)$ is the normal pdf we should have to argue that

$$\begin{aligned} \lim_{n\to\infty} \Phi_{Z_n}(\omega) &\triangleq \lim_{n\to\infty} \left(\int_{-\infty}^{\infty} f_{Z_n}(z) e^{j\omega z}\, dz\right) \\ &\stackrel{?}{=} \int_{-\infty}^{\infty} \left(\lim_{n\to\infty} f_{Z_n}(z)\right) e^{j\omega z}\, dz. \end{aligned}$$

However, the operations of limiting and integrating are not always interchangeable. Hence we cannot say that the pdf of Z_n converges to $N(0, 1)$. Indeed we already know from Example 3.6–4 that the sum of n i.i.d. binomial r.v.'s is binomial regardless of how large n is; moreover, the binomial PMF or pdf is a discontinuous function while the Gaussian is continuous and no matter how large n is, this fact cannot be altered. However, the integrals of the binomial pdf, for large n, behave like integrals of the Gaussian pdf. This is why the distribution function of Z_n tends to a Gaussian distribution function but not necessarily to a Gaussian pdf.

The astute reader will have noticed that in the prior development we showed the normal convergence of the characteristic function but not as yet the normal convergence of the PDF. To prove the latter true we can use a continuity theorem (see Feller [3–1, p. 508]) which states the following: Consider a sequence of r.v.'s Z_i, $i = 1, \ldots, n$ with characteristic functions and PDF's $\Phi_i(\omega)$ and $F_i(z)$, $i = 1, \ldots, n$ respectively; with $\Phi(\omega) \triangleq \lim_{n\to\infty} \Phi_n(\omega)$ and $\Phi(\omega)$ continuous at $\omega = 0$ then $F(z) = \lim_{n\to\infty} F_n(z)$.

Example 3.6–5: (Application of the Central Limit Theorem [CLT].) Let X_i, $i = 1, \ldots, n$ be a sequence of i.i.d. r.v.'s with $E[X_i] = \mu_X$ and $\text{Var}[X_i] = \sigma_X^2$. Let $Y \triangleq \sum_{i=1}^{n} X_i$. We wish to compute $P[a \leq Y \leq b]$ using the CLT. Clearly with $Z \triangleq (Y - E[Y])/\sigma_Y$

$$P[a \leq Y \leq b] = P[a' \leq Z \leq b'],$$

where

$$a' \triangleq \frac{a - E[Y]}{\sigma_Y}$$

$$b' = \frac{b - E[Y]}{\sigma_Y}$$

and

$$\sigma_Y = \sqrt{n}\,\sigma_X.$$

Note that Z is a zero-mean, unity variance r.v. involving the sum of a large number (n assumed large) of i.i.d. r.v.'s. Indeed it is easy to show that

$$Z = \frac{1}{\sqrt{n}} \sum_{i=1}^{n} \left(\frac{X_i - \mu_X}{\sigma_X}\right).$$

Hence

$$P[a' \leq Z \leq b'] \simeq \frac{1}{\sqrt{2\pi}} \int_{a'}^{b'} e^{-\frac{1}{2}z^2}\, dz.$$

Although the CLT might be more appropriately called the "normal convergence theorem" the word *central* in Central Limit Theorem is useful as a reminder that PDF's converge to the normal PDF around the center, that is, around the mean. Although all PDF's converge together at $\pm\infty$, it is in fact in the tails that the CLT frequently gives the poorest estimates of the correct probabilities, if these are small. An illustration of this phenomenon is given in Problem 3.32.

3.7 SUMMARY

In this chapter we discussed the various averages of one or more random variables and the implication of those averages. We began by defining the average or expected value of a random variable X and then showed that the expected value of $Y = g(X)$ could be computed directly from the pdf or PMF of X. We briefly discussed the important notion of conditional expectation and showed how the expected value of an r.v. could be advantageously computed by averaging over its conditional expectation. We then argued that a single summary number such as the average value, $\bar{X}$, of X was insufficient for describing the behavior of X. This led to the introduction of moments, that is, the average of powers of X. We illustrated how moments can be used to estimate pdf's by the maximum entropy principle and introduced the concept of *joint moments.* We showed how the covariance of two random variables could be interpreted as a measure of how well we can predict one r.v. from another using a linear predictor model. By giving a counterexample, we demonstrated that uncorrelatedness does not imply independence of two r.v.'s, the latter being a stronger condition. The joint Gaussian pdf for two r.v.'s was discussed, and it was shown that in the Gaussian case, independence and uncorrelatedness are equivalent. We then introduced the reader to some important bounds known as the Chebyshev and Schwarz inequalities and illustrated how these are used in problems in probability.

The second half of the chapter dealt mostly with moment-generating (m.g.f.) and characteristic functions (c.f.) and the Central Limit Theorem. We showed how the m.g.f. and c.f. are essentially the Laplace and Fourier transforms, respectively, of the pdf of a random variable and how we could compute all the moments, provided that these exist, from either of these functions. Several properties of these important functions were explored. We illustrated how the c.f. could be used to solve problems involving the computation of the pdf's of the sums of random variables.

We ended the chapter by briefly discussing one of the most important results of probability theory: the Central Limit Theorem. The CLT states that under not-overly restrictive circumstances, the probability distribution function of the sum of independent r.v.'s tends toward the normal PDF $\mathcal{N}(z)$. Finally, using c.f.'s, we proved the CLT for a special case.

PROBLEMS

3.1. Compute the average and standard deviation of the following set: 3.02, 5.61, −2.37, 4.94, −6.25, −1.05, −3.25, 5.81, 2.27, 0.54, 6.11, −2.56.

3.2. Compute $E[X]$ when X is a Bernoulli r.v., that is,

$$X = \begin{cases} 1, P_X(1) = p > 0 \\ 0, P_X(0) = 1 - p > 0. \end{cases}$$

3.3. Compute $E[X]$ when X is a binomial r.v., that is,

$$P_X(k) = \binom{n}{k} p^k (1-p)^{n-k} \quad k = 0, \ldots, n, \quad 0 < p < 1.$$

3.4. Let X be a uniform r.v., that is,

$$f_X(x) = \begin{cases} (b-a)^{-1} & 0 < a < x < b \\ 0, & \text{otherwise} \end{cases}$$

Compute $E[X]$.

3.5. In Problem 3.4, let $Y \triangleq X^2$. Compute the pdf of Y and $E[Y]$ by Equation 3.1–7. Then compute $E[Y]$ by Equation 3.1–8.

3.6. Let X be a Poisson r.v. with parameter a. Compute $E[Y]$ when $Y \triangleq X^2 + b$.

3.7. Prove that if $E[X]$ exists and X is a continuous r.v. then $|E[X]| \leq E[|X|]$. Repeat for X discrete.

3.8. Show that if $E[g_i(X)]$ exists for $i = 1, \ldots, N$ then $E\left[\sum_{i=1}^{N} g_i(X)\right] = \sum_{i=1}^{n} E[g_i(X)]$.

3.9. A random sample of 20 households shows the following numbers of children per household: 3, 2, 0, 1, 0, 0, 3, 2, 5, 0, 1, 1, 2, 2, 1, 0, 0, 0, 6, 3. (a) For this set what is the average number of children per household? (b) What is the average number of children in households given that there is at least one child?

3.10. Let $B \triangleq \{a < X \leq b\}$. Derive a general expression for $E[X \mid B]$ if X is a continuous r.v. Let $X: N(0, 1)$ with $B = \{-1 < X \leq 2\}$. Compute $E[X \mid B]$.

3.11. A particular color TV model is manufactured in three different plants, say, A, B, and C of the same company. Because the workers at A, B, and C are not equally experienced, the quality of the sets differs from plant to plant. The pdf's of the time-to-failure, X, in years are

$$f_X(x) = \frac{1}{5}\exp(-x/5)u(x) \text{ for } A$$

$$f_X(x) = \frac{1}{6.5}\exp(-x/6.5)u(x) \text{ for } B$$

$$f_X(x) = \frac{1}{10}\exp(-x/10)u(x) \text{ for } C,$$

where $u(x)$ is the unit step. Plant A produces three times as many sets as B, which produces twice as many as C. The sets are all sent to a central warehouse, intermingled, and shipped to retail stores all around the country. What is the expected lifetime of a set purchased at random?

3.12. A source transmits a signal Θ with pdf

$$f_\Theta(\theta) = \begin{cases} (2\pi)^{-1} & 0 < \theta \leq 2\pi \\ 0, & \text{otherwise.} \end{cases}$$

Because of additive Gaussian noise, the pdf of the received signal Y when $\Theta = \theta$ is

$$f_{Y|\Theta}(y \mid \theta) = \frac{1}{\sqrt{2\pi\sigma^2}}\exp\left[-\frac{1}{2}\left(\frac{y-\theta}{\sigma}\right)^2\right].$$

Compute $E[Y]$.

3.13. Compute the variance of X if X is (a) Bernoulli; (b) binomial; (c) Poisson; (d) Gaussian; (e) Rayleigh.

3.14. Let X and Y be independent r.v.'s, each $N(0, 1)$. Find the mean and variance of $Z \triangleq \sqrt{X^2 + Y^2}$.

3.15. (Papoulis [3–3, p. 151]). Let $Y = h(X)$. We wish to compute approximations to $E[h(X)]$ and $E[h^2(X)]$. Assume that $h(x)$ admits to a power series expansions, that is, all derivatives exist. Assume further that all derivatives above the second are small enough to be omitted. Show that ($\bar{X} = \mu$, $\text{Var}(X) = \sigma^2$)

(a) $E[h(X)] \simeq h(\mu) + h''(\mu)\sigma^2/2$;

(b) $E[h^2(X)] \simeq h^2(\mu) + ([h'(\mu)]^2 + h(\mu)h''(\mu))\sigma^2$.

3.16. Let $f_{XY}(x, y)$ be given by

$$f_{XY}(x, y) = \frac{1}{2\pi\sigma^2\sqrt{1 - \rho^2}} \exp\left(-\frac{x^2 + y^2 - 2\rho xy}{2\sigma^2(1 - \rho^2)}\right),$$

where $|\rho| < 1$. Show that $E[Y] = 0$ but $E[Y \mid X = x] = \rho x$. What does this result say about predicting the value of Y upon observing the value of X?

3.17. Consider a probability space $\mathscr{P} = (\Omega, \mathscr{F}, P)$. Let $\Omega = \{\zeta_1, \ldots, \zeta_5\} = \{-1, -\frac{1}{2}, 0, \frac{1}{2}, 1\}$ with $P[\zeta_i] = \frac{1}{5}$, $i = 1, \ldots, 5$. Define two r.v.'s on $\mathscr{P}$ as follows:

$$X(\zeta) \triangleq \zeta \quad \text{and} \quad Y(\zeta) \triangleq \zeta^2.$$

(a) Show that X and Y are dependent r.v.'s.
(b) Show that X and Y are uncorrelated.

3.18. We wish to estimate the pdf of X with a function $p(x)$ that maximizes the entropy

$$H[X] \triangleq -\int_{-\infty}^{\infty} p(x) \ln p(x)\, dx.$$

It is known from measurements that $E[X] = \mu$ and $\text{Var}[X] = \sigma^2$. Find the maximum entropy estimate of the pdf of X.

3.19. Let $X : N(0, \sigma^2)$. Show that

$$\xi_n \triangleq E[X^n] = 1 \cdot 3 \ldots (n - 1)\sigma^n \qquad n \text{ even}$$

$$\xi_n = 0 \qquad n \text{ odd}.$$

3.20. With $\bar{X} \triangleq E[X]$ and $\bar{Y} \triangleq E[Y]$ show that if $m_{11} = \sqrt{m_{20}m_{02}}$ then

$$E\left[\left(\frac{m_{11}}{m_{20}}(X - \bar{X}) - (Y - \bar{Y})\right)^2\right] = 0.$$

Use this result to show that when $|\rho| = 1$, then Y is a linear function of X, that is, $Y = \alpha X + \beta$. Relate α, β to the moments of X and Y.

3.21. Show that in the optimum linear predictor in Example 3.3–4 the smallest mean-square error is

$$\varepsilon_{\min} = \sigma_Y^2(1 - \rho^2).$$

Explain why $\varepsilon_{\min} = 0$ when $|\rho| = 1$.

3.22. Let $E[X_i] = \mu$, $\text{Var}[X_i] = \sigma^2$. We wish to estimate μ with the *sample mean*

$$\hat{\mu} \triangleq \frac{1}{N} \sum_{i=1}^{N} X_i.$$

Compute the mean and variance of $\hat{\mu}$ assuming the X_i $i = 1, \ldots, N$ are independent.

3.23. In Problem 3.22, how large should N be so that

$$P[|\hat{\mu} - \mu| > 0.1\sigma] \leq 0.01.$$

3.24. Let X have a Cauchy pdf

$$f_X(x) = \frac{\alpha}{\pi(\alpha^2 + x^2)}.$$

Compute the characteristic function (cf) $\Phi_X(\omega)$ of X.

3.25. Let $Y = \frac{1}{N}\sum_{i=1}^{N} X_i$ where the X_i are independent, identically distributed Cauchy r.v.'s with

$$f_{X_i}(x) = \frac{1}{\pi[1 + (x - \mu)^2]} \qquad i = 1, \ldots, N.$$

Show that the pdf of Y is

$$f_Y(x) = \frac{1}{\pi[1 + (x - \mu)^2]},$$

that is, is identical to the pdf of the X_i's and independent of N. Hint: Use the c.f. approach.

3.26. Let X be uniform over $(-a, a)$. Let Y be independent of X and uniform over $([n - 2]a, na)$ $n = 1, 2, \ldots$. Compute the expected value of $Z = X + Y$ for each n. From this result sketch the pdf of Z. What is the only effect of n?

3.27. Consider the recursion known as a *first-order moving average* given by

$$X_n = Z_n - aZ_{n-1} \qquad |a| < 1,$$

where X_n, Z_n, Z_{n-1} are all r.v.'s for $n = \cdot\cdot, -1, 0, 1, \ldots$. Assume $E[Z_n] = 0$ all n; $E[Z_n Z_{n-j}] = 0$ all $n \neq j$; and $E[Z_n^2] = \sigma^2$ all n. Compute $R_n(k) \triangleq E[X_n X_{n-k}]$ for $k = 0, \pm 1, \pm 2, \ldots$.

3.28. Consider the recursion known as a *first-order autoregression*

$$X_n = bX_{n-1} + Z_n \qquad |b| < 1.$$

The following is assumed true: $E[Z_n] = 0$, $E[Z_n^2] = \sigma^2$ all n; $E[Z_n Z_{n-j}] = 0$ all $n \neq j$. Also $E[Z_n X_{n-j}] = 0$ $j = 1, 2, \ldots$. Compute $R_n(k) = E[X_n X_{n-k}]$ for $k = \pm 1, \pm 2, \ldots$. Assume $E[X_n^2] \triangleq K$ independent of n.

3.29. Let $Z \triangleq aX + bY$, $W \triangleq cX + dY$. Compute the joint c.f. of Z and W in terms of the joint c.f. of X and Y.

3.30. Let X and Y be two independent Poisson r.v.'s with

$$P_X(k) = \frac{1}{k!}e^{-2}2^k$$

$$P_Y(k) = \frac{1}{k!}e^{-3}3^k.$$

Compute the PMF of $Z = X + Y$ using moment-generating or characteristic functions.

3.31. Your company manufactures toaster ovens. Let the probability that a toaster oven has a dent or scratch be $p = 0.05$. Assume different ovens get dented or scratched independently. In one week the company makes 2000 of these ovens. What is the approximate probability that in this week more than 109 ovens are dented or scratched?

3.32. Let X_i $i = 1, \ldots, n$ be a sequence of i.i.d. Bernoulli r.v.'s with $P_X(1) = p$ and $P_X(0) = q = 1 - p$. Let the event of a $\{1\}$ be a success and the event of a $\{0\}$ be a failure.

(a) Show that

$$Z_n \triangleq \frac{1}{\sqrt{n}}\sum_{i=1}^{n} W_i,$$

where $W_i \triangleq (X_i - p)/\sqrt{pq}$ is a zero-mean, unity variance r.v., involving the sum of a large number of independent r.v.'s.

(b) For $n = 2000$ and $k = 110, 130, 150$ compute $P[k$ successes in n tries$]$ using: (i) the exact binomial expression; (ii) the Poisson approximation to the binomial; and (iii) the CLT approximations. Do this by writing three BASIC miniprograms. Verify that as the correct probabilities decrease, the error in the CLT approximation increases.

REFERENCES

3–1. W. Feller, *An Introduction to Probability Theory and Its Applications*, Vol. 2, 2nd Ed. New York: John Wiley, 1971.

3–2. W. B. Davenport, Jr., *Probability and Random Processes*. New York: McGraw-Hill, p. 99.

3–3. A. Papoulis, *Probability, Random Variables, and Stochastic Processes*. New York: McGraw-Hill, 1965.

3–4. B. Saleh, *Photoelectron Statistics*. New York: Springer-Verlag, 1978, Chapter 5.

3–5. R. G. Gallagher, *Information Theory and Reliable Communications*. New York: John Wiley, 1968.

3–6. P. M. Morse and H. Feshbach, *Methods of Theoretical Physics*, Part 1. New York: McGraw-Hill, 1953, p. 279.

3–7. G. A. Korn and T. S. Korn, *Mathematical Handbook for Scientists and Engineers*. New York: McGraw-Hill, 1961.

3–8. W. Feller, *An Introduction to Probability Theory and Its Applications*, Vol. 1, 2nd Ed. New York: John Wiley, 1957.

4

VECTOR RANDOM VARIABLES

4.1 JOINT DISTRIBUTION AND DENSITIES

In many practical problems involving random phenomena we make observations that are essentially of a vector nature. We illustrate with three examples.

Example 4.1–1: A seismic waveform $X(t)$ is received at a geophysical recording station and is sampled at the instants $t_1, t_2, \ldots, t_n$. We thus obtain a vector $\mathbf{X} = (X_1, \ldots, X_n)^T$ where $X_i \triangleq X(t_i)$ and T denotes transpose.† For political and military reasons it is important to determine whether the waveform was radiated from an earthquake or an underground explosion. Assume that an expert computer system has available a lot of stored data regarding both earthquakes and underground explosions. The vector $\mathbf{X}$ is compared to the stored data. What is the probability that $X(t)$ is correctly identified?

Example 4.1–2: To evaluate the health of grade-school children, the Health Department of a certain region measures the height, weight, blood pressure, red-blood cell count, white-blood cell count, pulmonary capacity, heart rate, blood-lead level, and vision accuity of each child. The resulting vector $\mathbf{X}$ is taken as a summary of the health of each child. What is the probability that a child chosen at random is healthy?

Example 4.1–3: A computer system equipped with a TV camera is designed to recognize black-lung disease from X-rays. It does this by counting the number of radio-opacities in six lung zones (three in each lung) and estimating the average size of the opacities in each zone. The result is a twelve-component vector $\mathbf{X}$ from which a decision is made. What is the best computer decision?

The three previous problems are illustrative of many problems encountered

† All vectors will be assumed to be column vectors unless otherwise stated.

in engineering and science that involve sets of random variables that are grouped for some purpose. Such sets of random variables are conveniently studied by vector methods. For this reason we treat these grouped random variables as a single object called a *random vector*. As in earlier chapters, capital letters at the lower end of the alphabet will denote random variables; bold capital letters will denote random vectors and matrices and lower-case bold letters are deterministic vectors, that is, the values that random vectors assume.

Let $\mathbf{X}$ be a random vector with probability distribution function (PDF) $F_{\mathbf{X}}(\mathbf{x})$. Then by definition†

$$F_{\mathbf{X}}(\mathbf{x}) \triangleq P[X_1 \le x_1, \ldots, X_n \le x_n]. \tag{4.1–1}$$

By defining $\{\mathbf{X} \le \mathbf{x}\} \triangleq \{X_1 \le x_1, \ldots, X_n \le x_n\}$, we can rewrite Equation 4.1–1 as

$$F_{\mathbf{X}}(\mathbf{x}) \triangleq P[\mathbf{X} \le \mathbf{x}]. \tag{4.1–2}$$

We associate the events $\{\mathbf{X} \le \infty\}$ and $\{\mathbf{X} \le -\infty\}$ with the certain and impossible events respectively. Hence

$$F_{\mathbf{X}}(\infty) = 1 \tag{4.1–3a}$$

$$F_{\mathbf{X}}(-\infty) = 0. \tag{4.1–3b}$$

If the nth-mixed partial of $F_{\mathbf{X}}(\mathbf{x})$ exists we can define a *probability density function* (pdf) as

$$f_{\mathbf{X}}(\mathbf{x}) \triangleq \frac{\partial^n F_{\mathbf{X}}(\mathbf{x})}{\partial x_1 \ldots \partial x_n}. \tag{4.1–4}$$

The reader will observe that these definitions are completely analogous to the scalar definitions given in Chapter 2. We could have defined

$$f_{\mathbf{X}}(\mathbf{x}) \triangleq \lim_{\substack{\Delta x_1 \to 0 \\ \vdots \\ \Delta x_n \to 0}} \frac{P[x_1 < X_1 \le x_1 + \Delta x_1, \ldots, x_n < X_n \le x_n + \Delta x_n]}{\Delta x_1 \ldots \Delta x_n} \tag{4.1–5}$$

and arrived at Equation 4.1–4. For example, for $n = 2$

$$\begin{aligned} P[x_1 < X_1 \le x_1 + \Delta x_1, x_2 < X_2 \le x_2 + \Delta x_2] \\ = F_{\mathbf{X}}(x_1 + \Delta x_1, x_2 + \Delta x_2) - F_{\mathbf{X}}(x_1, x_2 + \Delta x_2) \\ - F_{\mathbf{X}}(x_1 + \Delta x_1, x_2) + F_{\mathbf{X}}(x_1, x_2). \end{aligned}$$

Thus (still for $n = 2$)

$$f_{\mathbf{X}}(\mathbf{x}) = \lim_{\substack{\Delta x_1 \to 0 \\ \Delta x_2 \to 0}} \frac{1}{\Delta x_1 \Delta x_2} (F_{\mathbf{X}}(x_1 + \Delta x_1, x_2 + \Delta x_2) - F_{\mathbf{X}}(x_1 + \Delta x_1, x_2)$$
$$- F_{\mathbf{X}}(x_1, x_2 + \Delta x_2) + F_{\mathbf{X}}(x_1, x_2))$$

† We remind the reader that unlike the notation in Sec. 1.3 the event $\{X_1 \le x_1, \ldots, X_n \le x_n\}$ is an *intersection* of the n events $\}X_i \le x_i\}_i = 1, \ldots$ n (cf. footnote on p. 62).

which is by definition the 2nd mixed partial derivative, and thus

$$f_{\mathbf{X}}(x_1, x_2) = \frac{\partial^2 F_{\mathbf{X}}(x_1, x_2)}{\partial x_1 \, \partial x_2}.$$

From Equation 4.1–5 we make the useful observation that

$$f_{\mathbf{X}}(\mathbf{x}) \, \Delta x_1 \ldots \Delta x_n \simeq P[x_1 < X_1 \leq x_1 + \Delta x_1, \ldots, x_n < X_n \leq x_n + \Delta x_n] \tag{4.1–6}$$

if the increments are small. If we integrate Equation 4.1–4, we obtain the PDF as

$$F_{\mathbf{X}}(\mathbf{x}) = \int_{-\infty}^{x_1} \ldots \int_{-\infty}^{x_n} f_{\mathbf{X}}(\mathbf{x}') \, dx_1' \ldots dx_n',$$

which we can write in compact notation as

$$\checkmark \quad F_{\mathbf{X}}(\mathbf{x}) = \int_{-\infty}^{\mathbf{x}} f_{\mathbf{X}}(\mathbf{x}') \, d\mathbf{x}'.$$

More generally, for any event $B \subset R^N$ (R^N being Euclidean N-space) consisting of the countable union and intersection of parallelepipeds

$$P[B] = \int_{\mathbf{x} \in B} f_X(\mathbf{x}) \, d\mathbf{x}. \tag{4.1–7}$$

(Compare with Equation 2.5–3.) The argument behind the validity of Equation 4.1–7 follows very closely the argument furnished in the one-dimensional case (Section 2–5). Davenport [4–1, p. 149] discusses the validity of Equation 4.1–7 for the case $n = 2$. For $n > 2$ one can proceed by induction.

The conditional distribution function of $\mathbf{X}$ given the event B is defined by

$$\checkmark \quad F_{\mathbf{X}|B}(\mathbf{x} \mid B) \triangleq P[\mathbf{X} \leq \mathbf{x} \mid B]$$
$$= \frac{P[\mathbf{X} \leq \mathbf{x}, B]}{P[B]} \quad (P[B] \neq 0).$$

These and subsequent results closely parallel the one-dimensional case. Consider next the n disjoint and exhaustive events $\{B_i, i = 1, \ldots, n\}$ with $P[B_i] > 0$. Then $\bigcup_{i=1}^{n} B_i = \Omega$ and $B_i B_j = \phi$ all $i \neq j$. We leave it as an exercise to the reader to show that

$$\checkmark \quad F_{\mathbf{X}}(\mathbf{x}) = \sum_{i=1}^{n} F_{\mathbf{X}|B_i}(\mathbf{x} \mid B_i) P[B_i]. \tag{4.1–8}$$

The unconditional PDF on the left is sometimes called a *mixture* distribution function. The conditional pdf of $\mathbf{X}$ given the event B is an nth mixed partial derivative of $F_{\mathbf{X}|B}(\mathbf{x} \mid B)$ if it exists. Thus

$$\checkmark \quad f_{\mathbf{X}|B}(\mathbf{x} \mid B) \triangleq \frac{\partial^n F_{\mathbf{X}|B}(\mathbf{x} \mid B)}{\partial x_1 \ldots \partial x_n}. \tag{4.1–9}$$

It follows from Equations 4.1–8 and 4.1–9 that

$$f_{\mathbf{X}}(\mathbf{x}) = \sum_{i=1}^{n} f_{\mathbf{X}|B}(\mathbf{x} \mid B_i)P[B_i]. \tag{4.1–10}$$

Because $f_{\mathbf{X}}(\mathbf{x})$ is a mixture, that is, a linear combination of conditional pdf's it is sometimes called a *mixture* pdf.†

The joint distribution of two random vectors $\mathbf{X} = (X_1, \ldots, X_n)^T$ and $\mathbf{Y} = (Y_1, \ldots, Y_m)^T$ is

$$F_{\mathbf{XY}}(\mathbf{x}, \mathbf{y}) = P[\mathbf{X} \le \mathbf{x}, \mathbf{Y} \le \mathbf{y}]. \tag{4.1–11a}$$

The *joint density* of $\mathbf{X}$ and $\mathbf{Y}$, if it exists, is given by

$$f_{\mathbf{XY}}(\mathbf{x}, \mathbf{y}) = \frac{\partial^{(n+m)} F_{\mathbf{XY}}(\mathbf{x}, \mathbf{y})}{\partial x_1 \ldots \partial x_n \, \partial y_1 \ldots \partial y_m}. \tag{4.1–11b}$$

The *marginal density* of $\mathbf{X}$ alone, $f_{\mathbf{X}}(\mathbf{x})$, can be obtained from $f_{\mathbf{XY}}(\mathbf{x}, \mathbf{y})$ by integration, that is,

$$f_{\mathbf{X}}(\mathbf{x}) = \int_{-\infty}^{\infty} \cdots \int_{-\infty}^{\infty} f_{\mathbf{XY}}(\mathbf{x}, \mathbf{y}) \, dy_1 \ldots dy_m.$$

Similarly, the marginal pdf of the vector $\mathbf{X}' \triangleq (X_1, \ldots, X_{n-1})^T$ is obtained from the pdf of $\mathbf{X}$ by

$$f_{\mathbf{X}'}(\mathbf{x}') \triangleq \int_{-\infty}^{\infty} f_{\mathbf{X}}(\mathbf{x}) \, dx_n \quad \text{where } \mathbf{x}' \triangleq (x_1, \ldots, x_{n-1})^T. \tag{4.1–12}$$

Obviously, Equation 4.1–12 can be extended to other kinds of marginal pdf's as well by merely integrating over the appropriate variable.

Example 4.1–4: Let $\mathbf{X} = (X_1, X_2, X_3)^T$ denote the position of a particle inside a sphere of radius a centered about the origin. Assume that at the instant of observation, the particle is equally likely to be anywhere in the sphere, that is,

$$f_{\mathbf{X}}(\mathbf{x}) = \begin{cases} \dfrac{3}{4\pi a^3}, & \sqrt{x_1^2 + x_2^2 + x_3^2} < a \\ 0, & \text{otherwise.} \end{cases}$$

Compute the probability that the particle lies within a subsphere of radius $2a/3$ contained within the larger sphere.

Solution: Let E denote the event that the particle lies within the subsphere (centered at the origin for simplicity) and let

$$\mathcal{R} \triangleq \{x_1, x_2, x_3 : \sqrt{x_1^2 + x_2^2 + x_3^2} < 2a/3\}.$$

Then the evaluation of

$$P[E] = \iiint_{\mathcal{R}} f_{\mathbf{X}}(x_1, x_2, x_3) \, dx_1 \, dx_2 \, dx_3$$

† This usage is prevalent in statistical pattern recognition.

is best done using spherical coordinates, that is,

$$P[E] = \frac{3}{4\pi a^3} \int_{r=0}^{2a/3} \int_{\phi=0}^{\pi} \int_{\theta=0}^{2\pi} r^2 \sin\phi \, dr \, d\phi \, d\theta$$

Note that in this simple case the answer can be obtained directly by noting the ratio of volumes i.e., $(2a/3)^3 = 8/27 \simeq 0.3$.

Although a random vector is completely characterized by its distribution or density functions the latter are often hard to come by except for some notable exceptions. By far the two most important exceptions are: (1) when $F_{\mathbf{X}}(\mathbf{x}) = F_{X_1}(x_1) \ldots F_{X_n}(x_n)$, that is, the n components of $\mathbf{X}$ are independent; and (2) when $\mathbf{X}$ obeys the multidimensional Gaussian law. Case (1) is easily handled, since it is essentially an extension of the scalar case. Case (2) will be discussed in Section 4.5. But what to do when neither case (1) nor (2) applies? Estimating multidimensional distributions involving dependent variables is often not practical and even if available might be too complex to be of any real use. Therefore, when we deal with vector r.v.'s we often settle for a less complete but more computable characterization based on moments. For most engineering applications the most important moments are the expectation vector (the first moment) and the covariance matrix (a second moment). These quantities and their use are discussed in the next several sections.

4.2 EXPECTATION VECTORS AND COVARIANCE MATRICES†

Definition 4.2–1. The expectation of the vector $\mathbf{X} = (X_1, \ldots, X_n)^T$ is a vector $\boldsymbol{\mu}$ (or $\bar{\mathbf{X}}$) whose elements $\mu_1, \ldots, \mu_n$ are given by

$$\mu_i \triangleq \int_{-\infty}^{\infty} \cdots \int_{-\infty}^{\infty} x_i f_{\mathbf{X}}(x_1, \ldots, x_n) \, dx_1 \ldots dx_n. \tag{4.2–1}$$

Equivalently with

$$f_{X_i}(x_i) \triangleq \int_{-\infty}^{\infty} \cdots \int_{-\infty}^{\infty} f_{\mathbf{X}}(\mathbf{x}) \, dx_1 \ldots dx_{i-1} \, dx_{i+1} \ldots dx_n$$

the marginal pdf of X_i, we can write

$$\mu_i = \int_{-\infty}^{\infty} x_i f_{X_i}(x_i) \, dx_i \qquad i = 1, \ldots, n.$$

Definition 4.2–2. The covariance matrix $\mathbf{K}$ associated with a real random vector $\mathbf{X}$ is the expected value of the outer vector product $(\mathbf{X} - \boldsymbol{\mu}) \cdot (\mathbf{X} - \boldsymbol{\mu})^T$, that is,

$$\mathbf{K} \triangleq E[(\mathbf{X} - \boldsymbol{\mu})(\mathbf{X} - \boldsymbol{\mu})^T]. \tag{4.2–2}$$

Column vector   Row vector

† This section requires some familiarity with matrix theory.

Define

$$K_{ij} \triangleq E[(X_i - \mu_i)(X_j - \mu_j)]$$
$$= E[(X_j - \mu_j)(X_i - \mu_i)]$$
$$= K_{ji} \quad i, j = 1, \ldots, n. \tag{4.2–3}$$

In particular with $\sigma_i^2 \triangleq K_{ii}$, we can write $\mathbf{K}$ in expanded form as

$$\mathbf{K} = \begin{bmatrix} \sigma_1^2 & \cdots & K_{1n} \\ \vdots & \sigma_i^2 & \vdots \\ K_{n1} & \cdots & \sigma_n^2 \end{bmatrix} \tag{4.2–4}$$

If $\mathbf{X}$ is real, all the elements of $\mathbf{K}$ are real. Also since $K_{ij} = K_{ji}$, covariance matrices fall within the class of matrices called *real symmetric* (r.s.). Such matrices fall within the larger class of Hermitian matrices.† Real symmetric matrices have many interesting properties, several of which we shall discuss in the next section.

The diagonal elements σ_i^2 are the variances associated with the individual r.v.'s X_i $i = 1, \ldots, n$. The covariance matrix $\mathbf{K}$ is closely related to the correlation matrix $\mathbf{R}$ defined by

$$\mathbf{R} \triangleq E[\mathbf{X}\mathbf{X}^T]. \tag{4.2–5}$$

Indeed expanding Equation 4.2–2 yields

$$\mathbf{K} = \mathbf{R} - \boldsymbol{\mu}\boldsymbol{\mu}^T$$

or

$$\mathbf{R} = \mathbf{K} + \boldsymbol{\mu}\boldsymbol{\mu}^T. \tag{4.2–6}$$

The correlation matrix $\mathbf{R}$ is also real symmetric; it is sometimes called the autocorrelation or scatter matrix.

Random vectors are often classified according to whether they are uncorrelated, orthogonal, or independent.

Definition 4.2–3. Consider two real n-dimensional random vectors $\mathbf{X}$ and $\mathbf{Y}$. Then if

(a) $$E[\mathbf{X}\mathbf{Y}^T] = E[\mathbf{X}]E[\mathbf{Y}^T] \tag{4.2–7}$$

$\mathbf{X}$ and $\mathbf{Y}$ are said to be *uncorrelated.* If

(b) $$E[\mathbf{X}^T\mathbf{Y}] \triangleq \sum_{i=1}^{n} E[X_i Y_i] = 0 \tag{4.2–8}$$

$\mathbf{X}$ and $\mathbf{Y}$ are said to be *orthogonal.* Finally, if

(c) $$f_{\mathbf{XY}}(\mathbf{x}, \mathbf{y}) = f_{\mathbf{X}}(\mathbf{x}) f_{\mathbf{Y}}(\mathbf{y}) \tag{4.2–9}$$

$\mathbf{X}$ and $\mathbf{Y}$ are said to be *independent.*

† The class of $n \times n$ matrices for which $K_{ij} = K_{ji}^*$. For a thorough discussion of the properties of such matrices see Reference [4–2]. When $\mathbf{X}$ is complex, the covariance is generally not r.s. but is Hermitian.

We leave it as an exercise to the reader to show that $\mathbf{X}$ can never be uncorrelated from itself unless tra $\mathbf{K} = 0$. Also recalling that the *scalar* or inner product of two n-vectors $\mathbf{u}$ and $\mathbf{v}$ is

$$\mathbf{u}^T\mathbf{v} = \sum_{i=1}^{n} u_i v_i$$

we see that Equation 4.2–8 is the expected inner product of two random vectors. Thus the use of the term *orthogonal* is apt: When random vectors are orthogonal their (expected) inner product is zero. Finally, as in the case of scalar r.v.'s, independence of two random vectors implies their uncorrelatedness, but the converse is generally not true. It is often difficult, in practice, to show that the two random vectors $\mathbf{X}$ and $\mathbf{Y}$ are independent. However, statistical tests exist to determine, with considerable reliability, the extent to which they are correlated.

4.3 PROPERTIES OF COVARIANCE MATRICES

Since covariance matrices are r.s., we study some of the properties of such matrices. Let $\mathbf{M}$ be any $n \times n$ r.s. matrix. The quadratic form associated with $\mathbf{M}$ is the scalar $q(\mathbf{z})$ defined by

$$q(\mathbf{z}) \triangleq \mathbf{z}^T\mathbf{M}\mathbf{z}, \tag{4.3–1}$$

where $\mathbf{z}$ is any column vector. $\mathbf{M}$ is said to be *positive semidefinite* (p.s.d.) if

$$\mathbf{z}^T\mathbf{M}\mathbf{z} \geq 0$$

for all $\mathbf{z}$. If $\mathbf{z}^T\mathbf{M}\mathbf{z} > 0$ for all $\mathbf{z} \neq 0$, $\mathbf{M}$ is said to be *positive definite* (p.d.). A covariance matrix $\mathbf{K}$ is always (at least) p.s.d. since for any vector $\mathbf{z} \triangleq (z_1, \ldots, z_n)^T$

$$\begin{aligned} 0 &\leq E\{[\mathbf{z}^T(\mathbf{X} - \boldsymbol{\mu})]^2\} \\ &= \mathbf{z}^T E[(\mathbf{X} - \boldsymbol{\mu})(\mathbf{X} - \boldsymbol{\mu})^T]\mathbf{z} \\ &= \mathbf{z}^T\mathbf{K}\mathbf{z}, \quad \mathbf{K} \triangleq E[(\mathbf{X} - \boldsymbol{\mu})(\mathbf{X} - \boldsymbol{\mu})^T]. \end{aligned} \tag{4.3–2}$$

We shall show later that when $\mathbf{K}$ is full-rank then $\mathbf{K}$ is p.d.

We now state some definitions and theorems (most without proof) from linear algebra [4–2, Chapter 4] that we shall need for developing important properties of covariance matrices.

Definition 4.3–1. The *eigenvalues* of an $n \times n$ matrix $\mathbf{M}$ are those numbers λ for which the characteristic equation $\mathbf{M}\boldsymbol{\phi} = \lambda\boldsymbol{\phi}$ has a solution $\boldsymbol{\phi} \neq \mathbf{0}$. The column vector $\boldsymbol{\phi} = (\phi_1, \phi_2, \ldots, \phi_n)^T$ is called an *eigenvector.* Eigenvectors are often normalized so that $\boldsymbol{\phi}^T\boldsymbol{\phi} = \|\boldsymbol{\phi}\|^2 = 1$.

Theorem 4.3–1. The number λ is an eigenvalue of the square matrix $\mathbf{M}$ if and only if $\det(\mathbf{M} - \lambda\mathbf{I}) = 0$.†

† *det* is short for determinant and $\mathbf{I}$ is the identity matrix.

Example 4.3–1: Consider the matrix

$$\mathbf{M} = \begin{bmatrix} 4 & 2 \\ 2 & 4 \end{bmatrix}.$$

The eigenvalues are obtained with the help of Theorem 4.3–1, that is,

$$\det\begin{bmatrix} 4-\lambda & 2 \\ 2 & 4-\lambda \end{bmatrix} = (4-\lambda)^2 - 4 = 0,$$

whence

$$\lambda_1 = 6, \qquad \lambda_2 = 2.$$

The (normalized) eigenvector associated with $\lambda_1 = 6$ is obtained from

$$(\mathbf{M} - 6\mathbf{I})\boldsymbol{\phi} = 0,$$

which, written out as a system of equations, yields

$$\left.\begin{aligned} -2\phi_1 + 2\phi_2 = 0 \\ -2\phi_1 + 2\phi_2 = 0 \end{aligned}\right\} \Rightarrow \boldsymbol{\phi}_1 = \frac{1}{\sqrt{2}}(1, 1)^T.$$

The double arrow $\Rightarrow$ means "implies that." The eigenvector associated with $\lambda_2 = 2$, following the same procedure as above, is found from

$$\left.\begin{aligned} 2\phi_1 + 2\phi_2 = 0 \\ 2\phi_1 + 2\phi_2 = 0 \end{aligned}\right\} \Rightarrow \boldsymbol{\phi}_2 = \frac{1}{\sqrt{2}}(1, -1)^T.$$

Not all $n \times n$ matrices have n distinct eigenvalues or n eigenvectors. Sometimes a matrix can have fewer than n distinct eigenvalues but still have n distinct eigenvectors.

Definition 4.3–2. Two $n \times n$ matrices $\mathbf{A}$ and $\mathbf{B}$ are called *similar* if there exists an $n \times n$ matrix $\mathbf{T}$ with $\det \mathbf{T} \neq 0$ such that

$$\mathbf{T}^{-1}\mathbf{A}\mathbf{T} = \mathbf{B}. \tag{4.3–3}$$

Theorem 4.3–2. An $n \times n$ matrix $\mathbf{M}$ is similar to a diagonal matrix if and only if $\mathbf{M}$ has n linearly independent eigenvectors.

Theorem 4.3–3. Let $\mathbf{M}$ be a r.s. matrix with eigenvalues $\lambda_1, \ldots, \lambda_n$. Then $\mathbf{M}$ has n mutually orthogonal unit eigenvectors $\boldsymbol{\phi}_1, \ldots, \boldsymbol{\phi}_n$.

Discussion. Since $\mathbf{M}$ has n mutually orthogonal (and therefore independent) unit eigenvectors, it is similar to some diagonal matrix $\boldsymbol{\Lambda}$ under a suitable transformation $\mathbf{T}$. What is $\boldsymbol{\Lambda}$ and $\mathbf{T}$? The answer is furnished by the following remarkable Theorem.

Theorem 4.3–4. Let $\mathbf{M}$ be a real symmetric matrix with eigenvalues $\lambda_1, \ldots, \lambda_n$. Then $\mathbf{M}$ is similar to the diagonal matrix $\boldsymbol{\Lambda}$ given by

$$\boldsymbol{\Lambda} \triangleq \begin{bmatrix} \lambda_1 & & 0 \\ & \ddots & \\ 0 & & \lambda_n \end{bmatrix}$$

under the transformation

$$\checkmark \quad \mathbf{U}^{-1}\mathbf{M}\mathbf{U} = \mathbf{\Lambda} \tag{4.3–4}$$

where $\mathbf{U}$ is a matrix whose columns are the ordered† orthogonal unit eigenvectors $\boldsymbol{\phi}_i$ $i = 1, \ldots, n$ of $\mathbf{M}$. Thus

$$\checkmark \quad \mathbf{U} = (\boldsymbol{\phi}_1, \ldots, \boldsymbol{\phi}_n). \tag{4.3–5}$$

Moreover, it can be shown that $\mathbf{U}^T\mathbf{U} = \mathbf{I}$ (and that $\mathbf{U}^T = \mathbf{U}^{-1}$) so that Equation 4.3–4 can be written as

$$\checkmark \quad \mathbf{U}^T\mathbf{M}\mathbf{U} = \mathbf{\Lambda}. \tag{4.3–6}$$

Discussion. Matrices $\mathbf{U}$, which satisfy $\mathbf{U}^T\mathbf{U} = \mathbf{I}$, are called *unitary.* They have the property of *distance preservation* in the following sense: Consider a vector $\mathbf{x} = (x_1, \ldots, x_n)^T$. The Euclidean distance of $\mathbf{x}$ from the origin is

$$\|\mathbf{x}\| \triangleq (\mathbf{x}^T\mathbf{x})^{1/2},$$

where $\|\mathbf{x}\|$ is called the *norm* of $\mathbf{x}$. Now consider the transformation $\mathbf{y} = \mathbf{U}\mathbf{x}$ where $\mathbf{U}$ is unitary. Then

$$\|\mathbf{y}\|^2 = \mathbf{y}^T\mathbf{y} = \mathbf{x}^T\mathbf{U}^T\mathbf{U}\mathbf{x} = \|\mathbf{x}\|^2.$$

Thus the new vector $\mathbf{y}$ has the same distance from the origin as the old vector $\mathbf{x}$ under the transformation $\mathbf{y} = \mathbf{U}\mathbf{x}$.

Since a covariance matrix $\mathbf{K}$ is real symmetric it can be easily diagonalized according to Equation 4.3–6 once $\mathbf{U}$ is known. The columns of $\mathbf{U}$ are just the eigenvectors of $\mathbf{K}$ and these can be obtained once the eigenvalues are known. The diagonalization of covariance matrices is a very important procedure in applied probability theory. It is used to transform correlated r.v.'s into uncorrelated r.v.'s; in statistical pattern recognition (for example, computing the Fisher discriminant—defined in Section 4.4); in solving problems involving the multidimensional normal pdf; and numerous other places.

If $\boldsymbol{\phi}_1, \ldots, \boldsymbol{\phi}_n$ are the orthogonal unit eigenvectors of a real symmetric matrix $\mathbf{M}$ then the system of equations

$$\begin{aligned} \mathbf{M}\boldsymbol{\phi}_1 &= \lambda_1\boldsymbol{\phi}_1 \\ &\vdots \\ \mathbf{M}\boldsymbol{\phi}_n &= \lambda_n\boldsymbol{\phi}_n \end{aligned}$$

can be compactly written as

$$\mathbf{M}\mathbf{U} = \mathbf{U}\mathbf{\Lambda}. \tag{4.3–7}$$

The next Theorem establishes a relation between the eigenvalues of a r.s. matrix and its positive definite character.

Theorem 4.3–4. A real symmetric matrix $\mathbf{M}$ is positive definite if and only if all its eigenvalues are positive.

† That is, $\boldsymbol{\phi}_i$ goes with λ_i $i = 1, \ldots, n$.

Proof. First let $\lambda_i > 0$, $i = 1, \ldots, n$. Then with the linear transformation $\mathbf{x} \triangleq \mathbf{U}\mathbf{y}$ we can write for any vector $\mathbf{x}$

$$\begin{aligned}\mathbf{x}^T\mathbf{M}\mathbf{x} &= (\mathbf{U}\mathbf{y})^T\mathbf{M}(\mathbf{U}\mathbf{y}) \\ &= \mathbf{y}^T\mathbf{U}^T\mathbf{M}\mathbf{U}\mathbf{y} \\ &= \mathbf{y}^T\mathbf{\Lambda}\mathbf{y} \\ &= \sum_{i=1}^{n} \lambda_i y_i^2 > 0 \end{aligned} \tag{4.3–8}$$

unless $\mathbf{y} = 0$. But if $\mathbf{y} = 0$ then from $\mathbf{x} = \mathbf{A}\mathbf{y}$, $\mathbf{x} = 0$ as well. Hence we have shown that $\mathbf{M}$ is p.d. if $\lambda_i > 0$ for all i. Conversely, we must show that if $\mathbf{M}$ is p.d. then all $\lambda_i > 0$. Thus for any $\mathbf{x} \neq 0$

$$0 < \mathbf{x}^T\mathbf{M}\mathbf{x}. \tag{4.3–9}$$

In particular, Equation 4.3–9 must hold for $\boldsymbol{\phi}_1, \ldots, \boldsymbol{\phi}_n$. But

$$0 < \boldsymbol{\phi}_i^T\mathbf{M}\boldsymbol{\phi}_i = \lambda_i \qquad i = 1, \ldots, n.$$

Hence $\lambda_i > 0$, $i = 1, \ldots, n$. Thus a p.d. covariance matrix will have all positive eigenvalues.

The next section deals with a problem of great importance in a branch of applied probability called *statistical pattern recognition.*

4.4 SIMULTANEOUS DIAGONALIZATION OF TWO COVARIANCE MATRICES

Theorem 4.4–1. Let $\mathbf{P}$ and $\mathbf{Q}$ be $n \times n$ real symmetric matrices. If $\mathbf{P}$ is positive definite, then there exists a $n \times n$ matrix $\mathbf{V}$ which achieves

$$\mathbf{V}^T\mathbf{P}\mathbf{V} = \mathbf{I} \tag{4.4–1}$$

and

$$\mathbf{V}^T\mathbf{Q}\mathbf{V} = \mathbf{\Lambda} = \text{diag}(\lambda_1, \ldots, \lambda_n).$$

The real numbers $\lambda_1, \ldots, \lambda_n$ satisfy the generalized eigenvalue equation

$$\mathbf{Q}\mathbf{v}_i = \lambda_i\mathbf{P}\mathbf{v}_i. \tag{4.4–2}$$

The numbers λ_i and vectors $\mathbf{v}_i$ $i = 1, \ldots, n$ are sometimes called generalized eigenvalues and eigenvectors.

Proof. Let $\mathbf{U} = (\boldsymbol{\phi}_1, \ldots, \boldsymbol{\phi}_n)$ where $\boldsymbol{\phi}_i$, γ_i $i = 1, \ldots, n$ are the unit eigenvectors and eigenvalues of $\mathbf{P}$. Then

$$\mathbf{U}^T\mathbf{P}\mathbf{U} = \text{diag}(\gamma_1, \ldots, \gamma_n) \triangleq \mathbf{M}. \tag{4.4–3}$$

Since $\mathbf{P}$ is positive definite, all the $\gamma_i > 0$ and the diagonal matrix $\mathbf{Z} \triangleq \text{diag}(\gamma_1^{-1/2}, \ldots, \gamma_n^{-1/2})$ exists and is real. Now

$$\mathbf{Z}^T\mathbf{U}^T\mathbf{P}\mathbf{U}\mathbf{Z} = \mathbf{Z}^T\mathbf{M}\mathbf{Z} = \mathbf{I}. \tag{4.4–4}$$

Thus by the similarity transformation $(\mathbf{U}\mathbf{Z})^T\mathbf{P}(\mathbf{U}\mathbf{Z})$ we have not only diagonalized $\mathbf{P}$ but also reduced it to an identity matrix. This process is called *whitening.* Next we inquire whether the similarity transformation

$(\mathbf{UZ})^T\mathbf{Q}(\mathbf{UZ})$ has produced a matrix that retains a real symmetric structure. But

$$[\mathbf{Z}^T\mathbf{U}^T\mathbf{QUZ}]^T = \mathbf{Z}^T\mathbf{U}^T\mathbf{QUZ} \tag{4.4-5}$$

so that $\mathbf{Z}^T\mathbf{U}^T\mathbf{QUZ} \triangleq \mathbf{A}$ is r.s. Since $\mathbf{A}$ is r.s. there exists a unitary similarity transformation $\mathbf{W}^T\mathbf{AW} = \text{diag}(\lambda_1, \ldots, \lambda_n) \triangleq \mathbf{\Lambda}$ where $\mathbf{W}^T\mathbf{W} = \mathbf{I}$. Now we ask does applying $\mathbf{W}$ to $\mathbf{Z}^T\mathbf{MZ} = \mathbf{I}$ (Equation 4.4–4), that is,

$$\mathbf{W}^T(\mathbf{Z}^T\mathbf{MZ})\mathbf{W} \tag{4.4-6}$$

affect the whitening transformation? The answer is no since substituting Equation 4.4–4 in 4.4–6 results in $\mathbf{W}^T\mathbf{IW} = \mathbf{I}$.

Comparing $\mathbf{W}^T\mathbf{AW} \triangleq (\mathbf{UZW})^T\mathbf{P}(\mathbf{UZW})$ with Equation 4.4–1 we see that the matrix $\mathbf{V}$, which achieves the simultaneous diagonalization of $\mathbf{P}$ and $\mathbf{Q}$, is given by

$$\mathbf{V} \triangleq \mathbf{UZW}. \tag{4.4-7}$$

Using Equation 4.4–1 we obtain

$$\det(\mathbf{V}^T\mathbf{PV}) = \det(\mathbf{I}) = 1.$$

From matrix theory we know that for any real $n \times n$ matrices $\mathbf{A}, \mathbf{B}, \mathbf{C}$ $\det(\mathbf{ABC}) = \det(\mathbf{A})\det(\mathbf{B})\det(\mathbf{C})$ and $\det(\mathbf{A}^T) = \det(\mathbf{A})$. Thus

$$\det(\mathbf{V}^T\mathbf{PV}) = \det(\mathbf{V})\det(\mathbf{P})\det(\mathbf{V}) = 1.$$

Hence $\det(\mathbf{V}) \neq 0$ and $\mathbf{V}$ is invertible, that is, $\mathbf{V}^{-1}$ exists. From

$$\mathbf{V}^T\mathbf{PV} = \mathbf{I}$$

we obtain

$$\mathbf{PV} = (\mathbf{V}^T)^{-1}. \tag{4.4-8}$$

Thus from $\mathbf{V}^T\mathbf{QV} = \mathbf{\Lambda}$ we obtain

$$\mathbf{QV} = [\mathbf{V}^T]^{-1}\mathbf{\Lambda}$$

and, using Equation 4.4–8,

$$\mathbf{QV} = \mathbf{PV\Lambda}, \tag{4.4-9}$$

which the reader will recognize as compact notation for the system of generalized eigenvalue equations

$$\mathbf{Qv}_i = \lambda_i\mathbf{Pv}_i \qquad i = 1, \ldots, n \tag{4.4-10}$$

given in Equation 4.4–2. This completes the proof of Theorem 4.4–1.

We observe that Equation 4.4–10 can also be written as (since $\mathbf{P}$ is p.d. it is invertible)

$$\mathbf{P}^{-1}\mathbf{Qv}_i = \lambda_i\mathbf{v}_i. \tag{4.4-11}$$

Thus the $\mathbf{v}_i$ can be viewed as the eigenvalues of the matrix $\mathbf{P}^{-1}\mathbf{Q}$. However, the normalization of $\mathbf{v}_i$ must be such as to satisfy $\mathbf{v}_i^T\mathbf{Pv}_i = 1$, which implies $\mathbf{V}^T\mathbf{PV} = \mathbf{I}$. The procedure for diagonalizing two matrices $\mathbf{P}$ and $\mathbf{Q}$ simultaneously is the following:

1. Calculate the eigenvalues of Equation 4.4–11; calculate unnor-

malized eigenvectors $\mathbf{v}'_i$ $i = 1, \ldots, n$ by solving

$$(\mathbf{P}^{-1}\mathbf{Q} - \lambda\mathbf{I})\mathbf{v}'_i = 0;$$

2. Find constants K_i, $i = 1, \ldots, n$ such that $\mathbf{v}_i \triangleq K_i\mathbf{v}'_i$ satisfies $\mathbf{v}_i^T\mathbf{P}\mathbf{v}_i = 1, i = 1, \ldots, n$.

As a final comment we point out that an important result can be obtained from Equation 4.4–4. Since $\mathbf{P}$ is positive definite both $\mathbf{U}$ and $\mathbf{Z}$ are invertible. Moreover, since $\mathbf{U}$ is unitary $\mathbf{U}^{-1} = \mathbf{U}^T$ and $(\mathbf{Z}^{-1})^T = \mathbf{Z}^{-1}$. Then if we premultiply

$$(\mathbf{UZ})^T\mathbf{P}(\mathbf{UZ}) = \mathbf{I}$$

by $[(\mathbf{UZ})^T]^{-1}$ and post-multiply by $(\mathbf{UZ})^{-1}$, we obtain

$$\mathbf{P} = \mathbf{CC}^T, \tag{4.4–12}$$

where $\mathbf{C} \triangleq \mathbf{UZ}^{-1}$. Thus we see that any real-symmetric, positive definite matrix $\mathbf{P}$ can be factored as in Equation 4.4–12. In terms of $\mathbf{C}$ we obtain

$$\mathbf{C}^{-1}\mathbf{P}[\mathbf{C}^{-1}]^T = \mathbf{I} \tag{4.4–13a}$$

and

$$\mathbf{C}^T\mathbf{P}^{-1}\mathbf{C} = \mathbf{I}. \tag{4.4–13b}$$

Example 4.4–1: (Simultaneous diagonalization of two covariance matrices $\mathbf{P}$ and $\mathbf{Q}$). We are given

$$\mathbf{P} = \begin{bmatrix} 2 & 1 \\ 1 & 2 \end{bmatrix} \qquad \mathbf{Q} = \begin{bmatrix} 3 & -1 \\ -1 & 3 \end{bmatrix}$$

We compute

$$\mathbf{P}^{-1} = \frac{1}{3}\begin{bmatrix} 2 & -1 \\ -1 & 2 \end{bmatrix} \qquad \mathbf{P}^{-1}\mathbf{Q} = \begin{bmatrix} \frac{7}{3} & \frac{-5}{3} \\ \frac{-5}{3} & \frac{7}{3} \end{bmatrix}.$$

The eigenvalues are computed from $\det(\mathbf{P}^{-1}\mathbf{Q} - \lambda\mathbf{I}) = 0$. They are $\lambda_1 = 4$, $\lambda_2 = 2/3$. Next we compute unnormalized eigenvectors from

$$(\mathbf{P}^{-1}\mathbf{Q} - \lambda\mathbf{I})\mathbf{v}'_i = 0 \qquad i = 1, 2.$$

This yields

$$\mathbf{v}'_1 = K_1(1, -1)^T \qquad \mathbf{v}'_2 = K_2(1, 1)^T.$$

The eigenvectors in this case are orthogonal. In general they will not be.† Finally, we must find the constants K_1 and K_2 such that $\mathbf{v}_i^T\mathbf{P}\mathbf{v}_i = 1$, $i = 1, 2$. Hence

$$[1 \quad -1]\begin{bmatrix} 2 & 1 \\ 1 & 2 \end{bmatrix}\begin{bmatrix} 1 \\ -1 \end{bmatrix}K_1^2 = 1$$

and

$$[1 \quad 1]\begin{bmatrix} 2 & 1 \\ 1 & 2 \end{bmatrix}\begin{bmatrix} 1 \\ 1 \end{bmatrix}K_2^2 = 1.$$

† The reader should not conclude from this example that the product of two real-symmetric matrices is always real symmetric. In general, the product of two r.s. matrices is not r.s. See Problem 4.10(b).

Such constants are found to be $K_1 = 1/\sqrt{2}\ K_2 = 1/\sqrt{6}$. Thus the matrix $\mathbf{V} = (\mathbf{v}_1, \mathbf{v}_2)$ is given by

$$\mathbf{V} = \begin{bmatrix} \frac{1}{\sqrt{2}} & \frac{1}{\sqrt{6}} \\ -\frac{1}{\sqrt{2}} & \frac{1}{\sqrt{6}} \end{bmatrix}.$$

It is easily verified that $\mathbf{V}^T\mathbf{PV} = \mathbf{I}$ and $\mathbf{V}^T\mathbf{QV} = \text{diag}(4, 2/3)$.

Projection. Given two real nonzero vectors $\mathbf{x} = (x_1, \ldots, x_n)^T$ and $\mathbf{a} = (a_1, \ldots, a_n)^T$ we define the projection of $\mathbf{x}$ onto $\mathbf{a}$ as the inner product $\mathbf{x}^T\mathbf{a} = \mathbf{a}^T\mathbf{x}$. The angle θ between $\mathbf{x}$ and $\mathbf{a}$ is implicitly defined by

$$\cos\theta = \frac{\mathbf{a}^T\mathbf{x}}{\|\mathbf{x}\|\,\|\mathbf{a}\|} = \frac{a_1x_1 + \ldots + a_nx_n}{\left[\sum_i x_i^2\right]^{1/2}\left[\sum_i a_i^2\right]^{1/2}}.$$

If $\|\mathbf{a}\| = 1$, we can think of $\mathbf{a}$ is a *unit direction vector* in which case $\mathbf{x}^T\mathbf{a}$ can be regarded as the projection of $\mathbf{x}$ *along the direction* $\mathbf{a}$ (Figure 4.4–1).

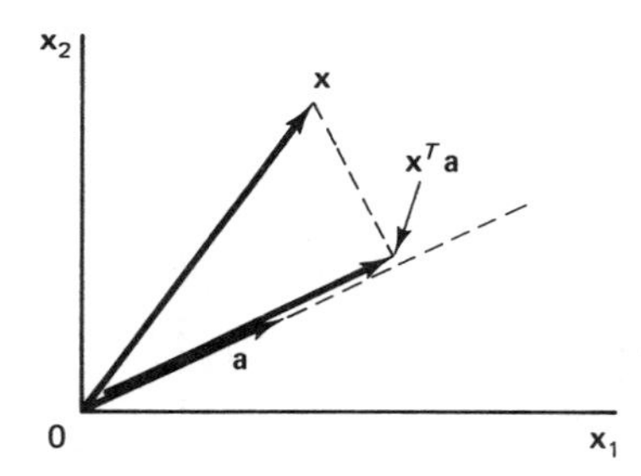

Figure 4.4–1 Projection of $\mathbf{x}$ along the direction $\mathbf{a}$ ($\|\mathbf{a}\| = 1$).

From the Schwarz inequality we have $|\mathbf{a}^T\mathbf{x}| \leq \|\mathbf{a}\|\,\|\mathbf{x}\|$ with equality if and only if $\mathbf{a}$ and $\mathbf{x}$ are colinear, that is, $\mathbf{a} = K\mathbf{x}$ (K a constant).

Example 4.4–2: Let $\boldsymbol{\mu}_i = E[\mathbf{X}_i]\ i = 1, 2$. Along what direction $\mathbf{a} = (a_1, a_2)^T$ is the projection of $\boldsymbol{\mu}_1 - \boldsymbol{\mu}_2$ a maximum?

Solution: To the astute reader, this problem may seem somewhat trivial but we shall proceed as if we didn't know the Schwarz inequality. We wish to maximize $|\mathbf{a}^T(\boldsymbol{\mu}_1 - \boldsymbol{\mu}_2)|$ subject to the constraint that $\|\mathbf{a}\|^2 = a_1^2 + a_2^2 = 1$. By the method of Lagrange multipliers we write

$$q(\mathbf{a}) \triangleq \mathbf{a}^T(\boldsymbol{\mu}_1 - \boldsymbol{\mu}_2) + \lambda\,\|\mathbf{a}\|^2$$

and set

$$\frac{\partial q(\mathbf{a})}{\partial a_i} = 0 \qquad i = 1, 2$$

subject to

$$a_1^2 + a_2^2 = 1.$$

The solution is

$$\mathbf{a} = \pm \frac{\boldsymbol{\mu}_1 - \boldsymbol{\mu}_2}{\|\boldsymbol{\mu}_1 - \boldsymbol{\mu}_2\|}. \tag{4.4–14}$$

Actually, the solution could have been obtained immediately from the Schwarz inequality: $|\mathbf{a}^T(\boldsymbol{\mu}_1 - \boldsymbol{\mu}_2)| \leq \|\mathbf{a}\| \, \|\boldsymbol{\mu}_1 - \boldsymbol{\mu}_2\|$ with equality if

$$\mathbf{a} = K(\boldsymbol{\mu}_1 - \boldsymbol{\mu}_2) \quad (K \text{ a constant}).$$

Since $\|\mathbf{a}\| = 1$, we obtain

$$|K| = \|\boldsymbol{\mu}_1 - \boldsymbol{\mu}_2\|^{-1},$$

whence follows Equation 4.4–14.

Maximization of Quadratic Forms. In solving problems involving random vectors we sometimes have to maximize quadratic forms. The following three theorems deal with this important problem.

Theorem 4.4–2. Let $\mathbf{M}$ be a r.s. matrix with largest eigenvalue λ_1. Suppose we want to maximize the quadratic form $\mathbf{u}^T\mathbf{M}\mathbf{u}$ on the unit sphere, that is, $\|\mathbf{u}\| = 1$. Then the maximum value of $\mathbf{u}^T\mathbf{M}\mathbf{u}$ subject to $\|\mathbf{u}\| = 1$ is λ_1 and the maximum occurs when $\mathbf{u} = \boldsymbol{\phi}_1$, that is, the unit eigenvector associated with λ_1.

Theorem 4.4–3. If $\mathbf{M}$ is a r.s. matrix with largest eigenvalue λ_1 then

$$\lambda_1 = \max \frac{\mathbf{x}^T\mathbf{M}\mathbf{x}}{\|\mathbf{x}\|^2} \quad (\mathbf{x} \neq 0) \tag{4.4–15}$$

and the maximum is achieved whenever $\mathbf{x} = K\boldsymbol{\phi}_1$ where $\boldsymbol{\phi}_1$ is the unit eigenvector associated with λ_1 and K is any real constant.

Theorem 4.4–4. Let $\mathbf{M}$ be a r.s. matrix with eigenvalues $\lambda_1 \geq \lambda_2 \geq \ldots \geq \lambda_n$. Let $\boldsymbol{\phi}_1, \boldsymbol{\phi}_2, \ldots, \boldsymbol{\phi}_n$ be the eigenvectors associated with $\lambda_1, \lambda_2, \ldots, \lambda_n$ respectively. Suppose we want to maximize the quadratic form $\mathbf{u}^T\mathbf{M}\mathbf{u}$ on the unit sphere, that is, $\|\mathbf{u}\| = 1$ subject to the constraints $\mathbf{u}^T\boldsymbol{\phi}_1 = \ldots = \mathbf{u}^T\boldsymbol{\phi}_{i-1} = 0$. Then the maximum value of $\mathbf{u}^T\mathbf{M}\mathbf{u}$ subject to all i constraints is λ_i and the maximum is achieved when $\mathbf{u} = \boldsymbol{\phi}_i$.

The proofs of these theorems can be found in books on the subject of linear algebra (for example, see pages 140–145 of the book by Franklin, Reference 4–2). For the reader's convenience and also because we shall use the result in the example to follow we furnish the proof of Theorem 4.4–2 below.

Proof of Theorem 4.4–2. The complete set of eigenvectors $\boldsymbol{\phi}_i$, $i = 1, \ldots, n$ form a *basis* for n-dimensional Euclidean space E^n. Therefore any vector $\mathbf{u}$ can be represented as

$$\mathbf{u} = c_1\boldsymbol{\phi}_1 + \ldots + c_n\boldsymbol{\phi}_n,$$

where the constants c_i are determined from $c_i = \boldsymbol{\phi}^T\mathbf{u}$. Now $\|\mathbf{u}\| = 1$ implies

$$\|\mathbf{u}\|^2 = 1 = \left(\sum_{i=1}^{n} c_i \boldsymbol{\phi}_i^T\right)\left(\sum_{j=1}^{n} c_j \boldsymbol{\phi}_j\right)$$

$$= \sum_{i=1}^{n} c_i^2 \quad \text{since } \boldsymbol{\phi}_i^T \boldsymbol{\phi}_j = \delta_{ij}.\dagger$$

Furthermore

$$\mathbf{Mu} = \mathbf{M}\left(\sum_{i=1}^{n} c_i \boldsymbol{\phi}_i\right)$$

$$= \sum_{i=1}^{n} c_i \mathbf{M} \boldsymbol{\phi}_i = \sum_{i=1}^{n} c_i \lambda_i \boldsymbol{\phi}_i.$$

Hence

$$\mathbf{u}^T\mathbf{Mu} = \left(\sum_{j=1}^{n} c_j \boldsymbol{\phi}_j^T\right)\left(\sum_{i=1}^{n} c_i \lambda_i \boldsymbol{\phi}_i\right)$$

$$= \sum_{i=1}^{n} \lambda_i c_i^2 \quad \text{(again since } \boldsymbol{\phi}_j^T \boldsymbol{\phi}_i = \delta_{ij})$$

$$\leq \lambda_1 \sum_{i=1}^{n} c_i^2 = \lambda_1$$

with equality if and only if $c_2 = \ldots = c_n = 0$. Thus

$$\mathbf{u}^T\mathbf{Mu} \leq \lambda_1$$

with equality when $\mathbf{u} = \boldsymbol{\phi}_1$ (in which case $c_1 = 1$ and all other c_i are zero).

■

The application of much of the material in this section will be illustrated in the following example from pattern recognition.

Example 4.4–3: (Automatic recognition of black-lung disease by computer [4–3].) We return to Example 4.1–3 that dealt with computer recognition of black-lung disease (BLD) from scanning X-ray films of the suspect population. We suppose that the X-rays fall into two categories: those that exhibit the BLD syndrome (say class ω_1) and those free of the BLD syndrome (say class ω_2). When an X-ray is presented to the computer, it extracts an n-component measurement vector $\mathbf{X}$ from which it must decide whether the X-ray should be assigned to class ω_1 or ω_2. We assume that the computer does not have available the conditional pdf's $f_{\mathbf{X}|\omega_i}(\mathbf{x} \mid \omega_i)$ $i = 1, 2$ but knows the conditional class means and covariances.‡ These are defined as

$$\boldsymbol{\mu}_i \triangleq E[\mathbf{X} \mid \text{given that } \mathbf{X} \text{ comes from class } \omega_i]$$

$$\triangleq E[\mathbf{X} \mid \omega_i] \qquad i = 1, 2 \tag{4.4–16a}$$

$$\mathbf{K}_i \triangleq E[(\mathbf{X} - \boldsymbol{\mu}_i)(\mathbf{X} - \boldsymbol{\mu}_i)^T \mid \text{given that } \mathbf{X} \text{ comes from class } \omega_i]$$

$$\triangleq E[(\mathbf{X} - \boldsymbol{\mu}_i)(\mathbf{X} - \boldsymbol{\mu}_i)^T \mid \omega_i] \qquad i = 1, 2. \tag{4.4–16b}$$

† The Kronecker δ_{ij} has value 1 when $i = j$ and value zero otherwise.

‡ If the conditional pdf's were known then Bayes' decision theory could be used to decide on class membership. Bayesian theory, which can be regarded as optimum in a rather broad sense for problems of this type, is discussed in Chapter 5.

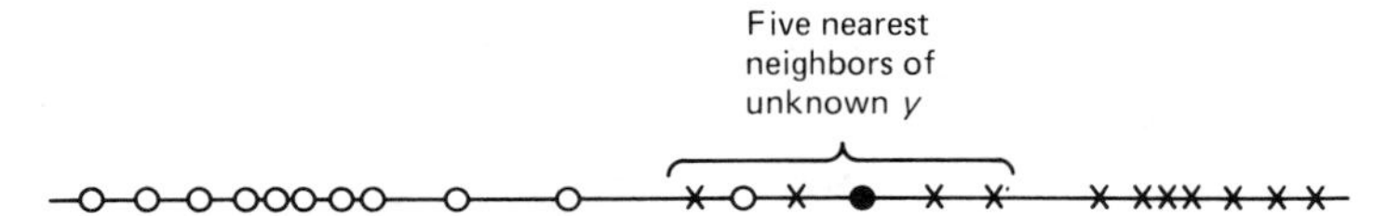

Figure 4.4–2 Classifying an unknown y by the N-nearest-neighbor rule. In the example shown, $N = 5$. Four out of five nearest neighbors belong to ω_1, hence y gets assigned to ω_1.

The decision whether any given value of **X** belongs to ω_1 or ω_2 will be done as follows: From **X**, a single scalar r.v. Y known as a feature is extracted by projecting **X** along a direction **a**, that is,

$$Y = \mathbf{a}^T\mathbf{X} \qquad \|\mathbf{a}\| = 1. \tag{4.4–17}$$

Next we assume that from *training data* we have available two sets $\mathscr{R}_1 \triangleq \{y_1^{(1)}, \ldots, y_n^{(1)}, \text{ all from } \omega_1\}$ and $\mathscr{R}_2 \triangleq \{y_1^{(2)}, \ldots, y_n^{(2)}, \text{ all from } \omega_2\}$.† To be specific, let the feature Y assume the value y. Our problem is then to assign to y a class membership, that is, should y be assigned to ω_1 or ω_2? We shall do this using the *N-nearest-neighbor rule* [4–4, pp. 103–105]. This is realized as follows: The N nearest neighbors of the unassigned y are found from the training sets. Being from the training sets, these nearest neighbors have an assigned class membership. If the majority of the N nearest neighbors belong to ω_1 then y gets assigned to ω_1; otherwise y gets assigned to ω_2. For this rule to work well, the data in $\mathscr{R}_1$ and $\mathscr{R}_2$ should cluster in different regions of the real line. To be specific, refer to Figure 4.4–2. There, a particular realization y of the r.v. Y is shown by the solid dot. We take $N = 5$. Since four of the nearest neighbors belong to ω_1 and only one belongs to ω_2, y gets assigned to ω_1. With high probability, the ω_1 classification for y is correct, since the *projected training data nicely separate.*

Assuming that there exists at least one direction **a** which will cluster $\mathscr{R}_1$ and $\mathscr{R}_2$ in different regions, how do we find **a**? To answer this question refer to Figure 4.4–3. There we see that $\mathbf{a}_1$ is a poor direction for projecting **X** because although the projected difference in means is relatively large, the projected variances are also large and there is a significant region of confusion (the crosshatched region). Thus $\mathbf{a}_1$ is not a good direction for projection of the data. Along $\mathbf{a}_2$, the projected variances are less but so is the difference in projected means; once again there is manifest a region of confusion. However, along $\mathbf{a}_3$ the projected difference in means is large in relation to the projected variances; the region of confusion is small or nonexistent. Thus $\mathbf{a}_3$ is a good projection direction. If we call ξ_i and σ_i^2 $i = 1, 2$ the *projected* means and variances respectively then an excellent indicator of how good is any

† The y's in $\mathscr{R}_1$ and $\mathscr{R}_2$ represent particular outcomes of the r.v. Y. In practice the raw training data $\{\mathbf{X}_i^{(j)}, i = 1, \ldots, n; j = 1, 2\}$ from which the $\{Y_i^{(j)}, i = 1, \ldots, n; j = 1, 2\}$ would be computed would be obtained by having expert radiologists carefully examine X-rays from *known* sufferers of BLD and *known* healthy humans. This training data would then be made available to the computer, in effect converting the computer to an "expert" system.

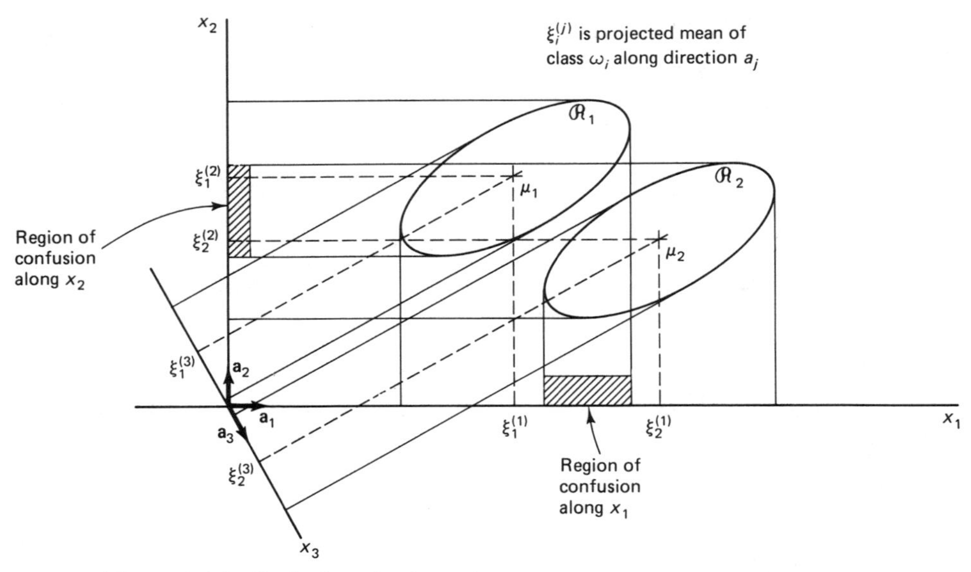

Figure 4.4–3 Projecting the data along $\mathbf{a}_1$ produces a region of confusion because the projected variances along $\mathbf{a}_1$ are too large. Projecting the data along $\mathbf{a}_2$ results in a large region of confusion because the projected difference in means $|\xi_1^{(2)} - \xi_1^{(2)}|$ is too small. Along $\mathbf{a}_3$, however, the projected difference in means is large in relation to the projected variances. Hence there is no region of confusion and $\mathbf{a}_3$ is the preferred direction.

particular projection direction $\mathbf{a}$ is the size of the criterion function $J(\mathbf{a})$ defined by

$$J(\mathbf{a}) \triangleq \frac{(\xi_1 - \xi_2)^2}{\sigma_1^2 + \sigma_2^2}. \tag{4.4–18}$$

If for a particular $\mathbf{a}$, $J(\mathbf{a})$ is large, the regions of confusion will be small; if $J(\mathbf{a})$ is small, the converse is true. Thus having identified a measure of cluster separation, how do we now proceed to find the value of $\mathbf{a}$ that maximizes $J(\mathbf{a})$?

Solution: We are looking to maximize $J(\mathbf{a})$ in Equation 4.4–18 by a suitable choice of $\mathbf{a}$. Note that under the transformation

$$Y = \mathbf{a}^T\mathbf{X} \tag{4.4–19}$$

we obtain

$$\begin{aligned}\xi_i &= E[Y \mid \omega_i] \\ &= \mathbf{a}^T\boldsymbol{\mu}_i \qquad i = 1, 2\end{aligned} \tag{4.4–20}$$

and

$$\begin{aligned}\sigma_i^2 &= E[(\mathbf{a}^T(\mathbf{X} - \boldsymbol{\mu}_i))^2 \mid \omega_i] \\ &= \mathbf{a}^T\mathbf{K}_i\mathbf{a} \qquad i = 1, 2.\end{aligned} \tag{4.4–21}$$

Thus

$$J(\mathbf{a}) = \frac{(\mathbf{a}^T(\boldsymbol{\mu}_1 - \boldsymbol{\mu}_2))^2}{\mathbf{a}^T(\mathbf{K}_1 + \mathbf{K}_2)\mathbf{a}} \tag{4.4–22}$$

$$= \frac{\mathbf{a}^T\mathbf{Q}\mathbf{a}}{\mathbf{a}^T\mathbf{P}\mathbf{a}}, \tag{4.4–23}$$

where

$$\mathbf{Q} \triangleq (\boldsymbol{\mu}_1 - \boldsymbol{\mu}_2)(\boldsymbol{\mu}_1 - \boldsymbol{\mu}_2)^T$$
$$\mathbf{P} \triangleq \mathbf{K}_1 + \mathbf{K}_2 \quad \text{(assumed full rank).} \tag{4.4–24}$$

Now we simultaneously diagonalize $\mathbf{Q}$ and $\mathbf{P}$ with the transformation

$$\mathbf{a} = \mathbf{V}\mathbf{b}, \tag{4.4–25}$$

where the columns $\mathbf{v}_i$ $i = 1, \ldots, n$ of $\mathbf{V}$ are the eigenvectors of the equations (the λ_i are the eigenvalues)

$$\mathbf{P}^{-1}\mathbf{Q}\mathbf{v}_i = \lambda_i \mathbf{v}_i$$

and satisfy the generalized orthogonality $\mathbf{v}_i^T\mathbf{P}\mathbf{v}_i = \mathbf{I}$. Use of Equation 4.4–25 in Equation 4.4–23 yields

$$J(\mathbf{a}) = \frac{\mathbf{b}^T\mathbf{V}^T\mathbf{Q}\mathbf{V}\mathbf{b}}{\mathbf{b}^T\mathbf{V}^T\mathbf{P}\mathbf{V}\mathbf{b}}$$
$$= \frac{\mathbf{b}^T\boldsymbol{\Lambda}\mathbf{b}}{\|\mathbf{b}\|^2} \tag{4.4–26}$$

where $\mathbf{V}^T\mathbf{P}\mathbf{V} = \mathbf{I}$ and $\mathbf{V}^T\mathbf{Q}\mathbf{V} = \boldsymbol{\Lambda} = \text{diag}(\lambda_1, \ldots, \lambda_n)$. Next we use Theorem 4.4–3 to find the maximum of Equation 4.4–26. The maximum is λ_1, and it occurs when $\mathbf{b} = \boldsymbol{\phi}_1$, that is, the eigenvector associated with λ_1. This eigenvector is $\boldsymbol{\phi}_1 = (1, 0, 0, \ldots, 0)^T$ since $\boldsymbol{\Lambda}$ is diagonal. Also from $\mathbf{a} = \mathbf{V}\mathbf{b}$ we find that

$$\mathbf{a} = \mathbf{V}\boldsymbol{\phi}_1 = \mathbf{v}_1.$$

Hence $\mathbf{a}$ satisfies

$$\mathbf{P}^{-1}\mathbf{Q}\mathbf{a} = \lambda_1\mathbf{a}$$

or

$$\mathbf{P}^{-1}(\boldsymbol{\mu}_1 - \boldsymbol{\mu}_2)(\boldsymbol{\mu}_1 - \boldsymbol{\mu}_2)^T\mathbf{a} = \lambda_1\mathbf{a}.$$

Since $(\boldsymbol{\mu}_1 - \boldsymbol{\mu}_2)^T\mathbf{a}$ is a scalar, say k, we obtain

$$\mathbf{a} = \frac{k}{\lambda_1}\mathbf{P}^{-1}(\boldsymbol{\mu}_1 - \boldsymbol{\mu}_2). \tag{4.4–27}$$

Since we are interested only in direction, we can ignore the generalized orthogonalization implied by Equation 4.4–1, that is, that $\mathbf{a}^T\mathbf{P}\mathbf{a} = 1$. Instead we normalize $\mathbf{a}$ to unity by choosing k so that $\|\mathbf{a}\| = 1$. It is easy to establish that this requires

$$k = \lambda_1\|\mathbf{P}^{-1}(\boldsymbol{\mu}_2 - \boldsymbol{\mu}_1)\|^{-1}.$$

Equation 4.4–27 is a result of principal importance in the field of applied probability called statistical pattern recognition. The vector $\mathbf{a}$ is called the Fisher† linear discriminant.

Having found the best direction for separating the two classes, we generate the scalar r.v. from

$$Y = \mathbf{a}^T\mathbf{X}.$$

When the dimension of $\mathbf{X}$ is large and the a_iX, $i = 1, \ldots, n$ satisfy the Lindenberg conditions, then from the *central limit theorem* (Section 3.6), Y can be taken to be a normal r.v. The assumption that Y is normal—often made even when the above conditions are not quite satisfied—enables an optimum partitioning of the real line

† Ronald Aylmer Fisher (1890–1962) British mathematician who along with Richard von Mises and Andrei Kolmogorov developed modern probability theory.

into regions Q_1^* and Q_2^* such that (1) the probability of a correct decision is maximized and/or (2) the cost of an incorrect decision is minimized. We only mention this approach here, deferring the actual computations until the next chapter.

You will have observed that Example 4.4–2 made use of many of the results of this chapter, for example, covariance matrices, their simultaneous diagonalization, maximization of quadratic forms, and others.

4.5 THE MULTIDIMENSIONAL GAUSSIAN LAW

The general n-dimensional Gaussian law has a rather forbidding mathematical appearance upon first acquaintance but is fortunately rather easily extended from the one-dimensional Gaussian pdf. We already know that if X is a (scalar) Gaussian r.v. with mean μ and variance σ^2, its pdf is

$$f_X(x) = \frac{1}{\sqrt{2\pi}\sigma}\exp\left(-\frac{1}{2}\left(\frac{x-\mu}{\sigma}\right)^2\right).$$

Now consider the random vector $\mathbf{X} = (X_1, \ldots, X_n)^T$ with *independent* components X_i, $i = 1, \ldots, n$. The pdf of $\mathbf{X}$ is the point pdf of the $X_1, \ldots, X_n$, that is,

$$\begin{aligned} f_{\mathbf{X}}(x_1, \ldots, x_n) &= \prod_{i=1}^{n} f_{X_i}(x_i) \\ &= \frac{1}{(2\pi)^{n/2}\sigma_1 \ldots \sigma_n}\exp\left[-\frac{1}{2}\sum_{i=1}^{n}\left(\frac{x_i-\mu_i}{\sigma_i}\right)^2\right], \end{aligned} \quad (4.5\text{–}1)$$

where μ_i, σ_i^2 are the mean and variance, respectively, of X_i, $i = 1, \ldots, n$. Equation 4.5–1 can be written compactly as

$$f_{\mathbf{X}}(\mathbf{x}) = \frac{1}{(2\pi)^{n/2}[\det(\mathbf{K})]^{1/2}}\exp[-\tfrac{1}{2}(\mathbf{x}-\boldsymbol{\mu})^T\mathbf{K}^{-1}(\mathbf{x}-\boldsymbol{\mu})], \quad (4.5\text{–}2)$$

where

$$\mathbf{K} \triangleq \begin{bmatrix} \sigma_1^2 & & 0 \\ & \ddots & \\ 0 & & \sigma_n^2 \end{bmatrix}, \quad (4.5\text{–}3)$$

$\boldsymbol{\mu} = (\mu_1, \ldots, \mu_n)^T$, and $\det(\mathbf{K}) = \prod_{i=1}^{n}\sigma_i^2$. Note that $\mathbf{K}^{-1}$ is merely

$$\mathbf{K}^{-1} = \begin{bmatrix} \sigma_1^{-2} & & 0 \\ & \ddots & \\ 0 & & \sigma_n^{-2} \end{bmatrix}.$$

The fact that Equations 4.5–1 and 4.5–2 are identical is left as an exercise to the reader. Note that because the X_i, $i = 1, \ldots, n$ are independent, the covariance matrix $\mathbf{K}$ is diagonal, since

$$E[(X_i - \mu_i)^2] \triangleq \sigma_i^2 \qquad i = 1, \ldots, n. \quad (4.5\text{–}4)$$

$$E[(X_i - \mu_i)(X_j - \mu_j)] = 0 \qquad i \neq j. \quad (4.5\text{–}5)$$

Next we ask, what happens if $\mathbf{K}$ is a positive definite covariance matrix that is not necessarily diagonal? Does Equation 4.5–2 with *arbitrary* p.d. covariance $\mathbf{K}$ still obey the requirements of a pdf? If it does, we shall call $\mathbf{X}$ a *normal* random vector and $f_{\mathbf{X}}(\mathbf{x})$ the *multidimensional normal* pdf. To show that $f_{\mathbf{X}}(\mathbf{x})$ is indeed a pdf we must show that

$$f_{\mathbf{X}}(\mathbf{x}) \geq 0 \tag{4.5–6a}$$

and

$$\int_{-\infty}^{\infty} f_{\mathbf{X}}(\mathbf{x})\, d\mathbf{x} = 1. \tag{4.5–6b}$$

(We use the vector notation $d\mathbf{x} \triangleq dx_1\, dx_2 \ldots dx_n$ for a volume element.) We assume as always that $\mathbf{X}$ is real, that is, $X_1, \ldots, X_n$ are real r.v.'s. To show that Equation 4.5–2 with arbitrary p.d. covariance matrix $\mathbf{K}$ satisfies Equation 4.5–6a is simple and left as an exercise; to prove Equation 4.5–6b is more difficult, and follows here.

Proof of Equation 4.5–6*b when* $f_{\mathbf{X}}(\mathbf{x})$ *is as in Equation* 4.5–2 *and* $\mathbf{K}$ *is an arbitrary p.d. covariance matrix.* We note that with $\mathbf{z} \triangleq \mathbf{x} - \boldsymbol{\mu}$, Equation 4.5–2 can be written as

$$f_{\mathbf{X}}(\mathbf{x}) \triangleq \frac{1}{(2\pi)^{n/2}(\det(\mathbf{K}))^{1/2}}\, \phi(\mathbf{z}),$$

where

$$\phi(\mathbf{z}) \triangleq \exp(-\tfrac{1}{2}\mathbf{z}^T\mathbf{K}^{-1}\mathbf{z}). \tag{4.5–7a}$$

With

$$\alpha \triangleq \int_{-\infty}^{\infty} \phi(\mathbf{z})\, d\mathbf{z} \tag{4.5–7b}$$

we see that

$$\int_{-\infty}^{\infty} f_{\mathbf{X}}(\mathbf{x})\, d\mathbf{x} = \frac{\alpha}{(2\pi)^{n/2}(\det(\mathbf{K}))^{1/2}}.$$

Hence we need only evaluate α to prove (or disprove) Equation 4.5–6b.

From Equations 4.4–12 and 4.4–13b we know that there exists an $n \times n$ matrix $\mathbf{C}$ such that $\mathbf{K} = \mathbf{C}\mathbf{C}^T$ and $\mathbf{C}^T\mathbf{K}^{-1}\mathbf{C} = \mathbf{I}$ (the identity matrix). Now consider the linear transformation

$$\mathbf{z} = \mathbf{C}\mathbf{y} \tag{4.5–8}$$

for use in Equations 4.5–7. To understand the effect of this transformation we note first that

$$\mathbf{z}^T\mathbf{K}^{-1}\mathbf{z} = \mathbf{y}^T\mathbf{C}^T\mathbf{K}^{-1}\mathbf{C}\mathbf{y} = \|\mathbf{y}\|^2 = \sum_{i=1}^{n} y_i^2$$

so that $\phi(\mathbf{z})$ is given by

$$\phi(\mathbf{z}) = \prod_{i=1}^{n} \exp[-\tfrac{1}{2}y_i^2].$$

Next we use a result from advanced calculus (see Kenneth Miller, Reference [4–5, p. 16]) that for a linear transformation such as in Equation 4.5–8 volume elements are related as

$$d\mathbf{z} = |\det(\mathbf{C})|\, d\mathbf{y},$$

where $d\mathbf{z} \triangleq dz_1 \ldots dz_n$ and $d\mathbf{y} = dy_1 \ldots dy_n$. Hence Equation 4.5–7b is transformed to

$$\begin{aligned}\alpha &= \int_{-\infty}^{\infty} \exp\left(-\frac{1}{2}\sum_{i=1}^{n} y_i^2\right) dy_1 \ldots dy_n\, |\det(\mathbf{C})| \\ &= \left[\int_{-\infty}^{\infty} e^{-y^2/2}\, dy\right]^n |\det(\mathbf{C})| \\ &= [2\pi]^{n/2}\, |\det(\mathbf{C})| .\end{aligned}$$

But since $\mathbf{K} = \mathbf{C}\mathbf{C}^T$, $\det(\mathbf{K}) = \det(\mathbf{C})\det(\mathbf{C}^T) = [\det(\mathbf{C})]^2$ or

$$|\det(\mathbf{C})| = |\det(\mathbf{K})|^{1/2} = (\det(\mathbf{K}))^{1/2}.$$

Hence

$$\alpha = (2\pi)^{n/2}(\det(\mathbf{K}))^{1/2}$$

and

$$\frac{\alpha}{[2\pi]^{n/2}[\det(\mathbf{K})]^{1/2}} = 1,$$

which proves Equation 4.5–6b. ■

Having established that

$$f_{\mathbf{X}}(\mathbf{x}) = \frac{1}{(2\pi)^{n/2}(\det(\mathbf{K}))^{1/2}} \exp(-\tfrac{1}{2}(\mathbf{x} - \boldsymbol{\mu})^T\mathbf{K}^{-1}(\mathbf{x} - \boldsymbol{\mu})) \tag{4.5–9}$$

indeed satisfies the requirements of a pdf and is a generalization of the univariate normal pdf we now ask what is the pdf of the random vector $\mathbf{Y}$ given by

$$\mathbf{Y} \triangleq \mathbf{A}\mathbf{X}, \tag{4.5–10}$$

where $\mathbf{A}$ is a nonsingular $n \times n$ transformation. The answer is furnished by the following theorem.

Theorem 4.5–1. Let $\mathbf{X}$ be an n-dimensional normal random vector with positive definite covariance matrix $\mathbf{K}$ and mean vector $\boldsymbol{\mu}$. Let $\mathbf{A}$ be a nonsingular linear transformation in n dimensions. Then $\mathbf{Y} \triangleq \mathbf{A}\mathbf{X}$ is an n-dimensional normal random vector with covariance matrix $\mathbf{Q} \triangleq \mathbf{A}\mathbf{K}\mathbf{A}^T$ and mean vector $\boldsymbol{\beta} \triangleq \mathbf{A}\boldsymbol{\mu}$.

Proof. We use the n-dimensional form of Equation 2.11–15, that is,

$$f_{\mathbf{Y}}(\mathbf{y}) = \sum_{i=1}^{l} \frac{f_{\mathbf{X}}(\mathbf{x}_i)}{|J_i|}, \tag{4.5–11}$$

where $\mathbf{Y}$ is some function of $\mathbf{X}$, that is, $\mathbf{Y} = \mathbf{g}(\mathbf{X}) \triangleq (g_1(\mathbf{X}), \ldots, g_n(\mathbf{X}))^T$, the $\mathbf{x}_i$, $i = 1, \ldots, l$ are the roots of the equation $\mathbf{g}(\mathbf{x}_i) - \mathbf{y} = \mathbf{0}$ and J_i is the

Jacobian evaluated at the ith root, that is,

$$J_i = \det\left(\frac{\partial \mathbf{g}}{\partial \mathbf{x}}\right)\Bigg|_{\mathbf{x}=\mathbf{x}_i} = \begin{vmatrix} \frac{\partial g_1}{\partial x_1} & \cdots & \frac{\partial g_1}{\partial x_n} \\ \vdots & & \\ \frac{\partial g_n}{\partial x_1} & \cdots & \frac{\partial g_n}{\partial x_n} \end{vmatrix}_{\mathbf{x}=\mathbf{x}_i}. \quad (4.5\text{–}12)$$

Since we are dealing with a nonsingular *linear* transformation, the only solution to

$$\mathbf{A}\mathbf{x} - \mathbf{y} = \mathbf{0} \quad \text{is} \quad \mathbf{x} = \mathbf{A}^{-1}\mathbf{y}. \quad (4.5\text{–}13)$$

Also

$$J_i = \det\left(\frac{\partial(\mathbf{A}\mathbf{x})}{\partial \mathbf{x}}\right) = \det(\mathbf{A}). \quad (4.5\text{–}14)$$

Hence

$$f_{\mathbf{Y}}(\mathbf{y}) = \frac{1}{(2\pi)^{n/2}(\det(\mathbf{K}))^{1/2}\,|\det(\mathbf{A})|} \exp(-\tfrac{1}{2}(\mathbf{A}^{-1}\mathbf{y} - \boldsymbol{\mu})^T\mathbf{K}^{-1}(\mathbf{A}^{-1}\mathbf{y} - \boldsymbol{\mu})). \quad (4.5\text{–}15)$$

Can this formidable expression be put in the form of Equation 4.5–9? First we note that

$$(\det(\mathbf{K}))^{1/2}\,|\det(\mathbf{A})| = (\det(\mathbf{A}\mathbf{K}\mathbf{A}^T))^{1/2} = (\det(\mathbf{Q}))^{1/2}. \quad (4.5\text{–}16)$$

Next we note that

$$(\mathbf{A}^{-1}\mathbf{y} - \boldsymbol{\mu})^T\mathbf{K}^{-1}(\mathbf{A}^{-1}\mathbf{y} - \boldsymbol{\mu}) = (\mathbf{y} - \mathbf{A}\boldsymbol{\mu})^T(\mathbf{A}\mathbf{K}\mathbf{A}^T)^{-1}(\mathbf{y} - \mathbf{A}\boldsymbol{\mu}). \quad (4.5\text{–}17)$$

(We leave the algebraic manipulations as an exercise.) But $\mathbf{A}\boldsymbol{\mu} \triangleq \boldsymbol{\beta} = E[\mathbf{Y}]$ and $\mathbf{A}\mathbf{K}\mathbf{A}^T = E[(\mathbf{Y} - \boldsymbol{\beta})(\mathbf{Y} - \boldsymbol{\beta})^T]$. Hence Equation 4.5–15 can be rewritten as

✓ $$f_{\mathbf{Y}}(\mathbf{y}) = \frac{1}{(2\pi)^{n/2}[\det(\mathbf{Q})]^{1/2}} \exp[-\tfrac{1}{2}(\mathbf{y} - \boldsymbol{\beta})^T\mathbf{Q}^{-1}(\mathbf{y} - \boldsymbol{\beta})]. \quad (4.5\text{–}18)$$

■

The next question that arises quite naturally as an extension of the previous result is: Does $\mathbf{Y}$ remain a normal random vector under more general (nontrivial) linear transformation? The answer is given by the following theorem, which is a generalization of Theorem 4.5–1.

Theorem 4.5–2. Let $\mathbf{X}$ be an n-dimensional normal random vector with positive definite covariance matrix $\mathbf{K}$ and mean vector $\boldsymbol{\mu}$. Let $\mathbf{A}_{mn}$ be an $m \times n$ matrix of rank m. Then the random vector generated by

$$\mathbf{Y} = \mathbf{A}_{mn}\mathbf{X}$$

has an m-dimensional normal pdf with p.d. covariance matrix $\mathbf{Q}$ and mean vector $\boldsymbol{\beta}$ given respectively by

$$\mathbf{Q} \triangleq \mathbf{A}_{mn}\mathbf{K}\mathbf{A}_{mn}^T \quad (4.5\text{–}19)$$

$\underline{\beta} = \underline{A}_{mn}\underline{\mu}$

and

$$\boldsymbol{\beta} = \mathbf{A}_{mn}\boldsymbol{\mu}. \tag{4.5–20}$$

The proof of this theorem is quite similar to the proof of Theorem 4.5–1; it is given by Miller in Reference [4–5, p. 22].

Some examples involving transformations of normal random variables are given below.

Example 4.5–1: A zero-mean normal random vector $\mathbf{X} = (X_1, X_2)^T$ has covariance matrix $\mathbf{K}$ given by

$$\mathbf{K} = \begin{bmatrix} 3 & -1 \\ -1 & 3 \end{bmatrix}.$$

Find a transformation $\mathbf{Y} = \mathbf{DX}$ such that $\mathbf{Y} = (Y_1, Y_2)^T$ is a normal random vector with uncorrelated (and therefore independent) components of unity variance.

Solution: Write

$$E[\mathbf{YY}^T] = E[\mathbf{DXX}^T\mathbf{D}^T] = \mathbf{DKD}^T = \mathbf{I}.$$

The last equality on the right follows from the requirement that the covariance of $\mathbf{Y}$, $\mathbf{K_Y}$, satisfies

$$\mathbf{K_Y} = \begin{bmatrix} 1 & 0 \\ 0 & 1 \end{bmatrix} \triangleq \mathbf{I}.$$

From Equation 4.4–13a, the matrix $\mathbf{D}$ must be $\mathbf{D} = \mathbf{C}^{-1} = \mathbf{ZU}^T$, where $\mathbf{Z}$ is the normalizing matrix

$$\mathbf{Z} = \begin{bmatrix} \lambda_1^{-1/2} & 0 \\ 0 & \lambda_2^{-1/2} \end{bmatrix} \quad (\lambda_i,\ i = 1, 2 \text{ are eigenvalues of } \mathbf{K})$$

and $\mathbf{U}$ is the matrix whose columns are the unit eigenvectors of $\mathbf{K}$ (recall $\mathbf{U}^{-1} = \mathbf{U}^T$). From $\det(\mathbf{K} - \lambda\mathbf{I}) = 0$ we find $\lambda_1 = 4$, $\lambda_2 = 2$. Hence

$$\boldsymbol{\Lambda} = \begin{bmatrix} 4 & 0 \\ 0 & 2 \end{bmatrix} \qquad \mathbf{Z} = \boldsymbol{\Lambda}^{-1/2} = \begin{bmatrix} \frac{1}{2} & 0 \\ 0 & \frac{1}{\sqrt{2}} \end{bmatrix}.$$

Next from

$$(\mathbf{K} - \lambda_1\mathbf{I})\boldsymbol{\phi}_1 = 0 \quad \text{with } \|\boldsymbol{\phi}_1\| = 1$$

and

$$(\mathbf{K} - \lambda_2\mathbf{I})\boldsymbol{\phi}_2 = 0 \quad \text{with } \|\boldsymbol{\phi}_2\| = 1$$

we find $\boldsymbol{\phi}_1 = (1/\sqrt{2}, -1/\sqrt{2})^T$ $\boldsymbol{\phi}_2 = (1/\sqrt{2}, 1/\sqrt{2})^T$. Thus

$$\mathbf{U} = (\boldsymbol{\phi}_1, \boldsymbol{\phi}_2) = \frac{1}{\sqrt{2}}\begin{bmatrix} 1 & 1 \\ -1 & 1 \end{bmatrix}$$

and

$$\mathbf{D} = \mathbf{ZU}^T = \frac{1}{\sqrt{2}}\begin{bmatrix} \frac{1}{2} & -\frac{1}{2} \\ \frac{1}{\sqrt{2}} & \frac{1}{\sqrt{2}} \end{bmatrix}.$$

As a check to see if $\mathbf{DKD}^T$ is indeed an identity covariance matrix we compute

$$\frac{1}{2}\begin{bmatrix} \frac{1}{2} & -\frac{1}{2} \\ \frac{1}{\sqrt{2}} & \frac{1}{\sqrt{2}} \end{bmatrix}\begin{bmatrix} 3 & -1 \\ -1 & 3 \end{bmatrix}\begin{bmatrix} \frac{1}{2} & \frac{1}{\sqrt{2}} \\ -\frac{1}{2} & \frac{1}{\sqrt{2}} \end{bmatrix} = \begin{bmatrix} 1 & 0 \\ 0 & 1 \end{bmatrix}.$$

In many situations we might want to generate correlated samples of a random vector $\mathbf{X}$ whose covariance matrix $\mathbf{K}$ is not diagonal. Unfortunately most Gaussian random number generators to be found in computer libraries generate independent normal samples whose covariance is the identity matrix. From Example 4.5–1 we see that the transformation

$$\mathbf{X} = \mathbf{D}^{-1}\mathbf{Y}, \tag{4.5–21}$$

where $\mathbf{D} = \mathbf{ZU}^T$ produces a normal random vector whose covariance is $\mathbf{K}$. Thus one way of obtaining correlated from uncorrelated samples is to use the transformation given in Equation 4.5–21 on the computer-generated samples. This procedure is the reverse of what we did in Example 4.5–1.

Example 4.5–2: Two jointly normal r.v.'s X_1 and X_2 have joint pdf given by

$$f_{X_1X_2}(x_1, x_2) = \frac{1}{2\pi\sigma^2\sqrt{1-\rho^2}}\exp\left(\frac{-1}{2\sigma^2(1-\rho^2)}(x_1^2 - 2\rho x_1x_2 + x_2^2)\right).$$

Let the correlation coefficient ρ be -0.5. From X_1, X_2 find two jointly normal r.v.'s Y_1 and Y_2 such that Y_1 and Y_2 are independent. Avoid the trivial case of $Y_1 = Y_2 = 0$.

Solution: Define $\mathbf{x} \triangleq (x_1, x_2)^T$ and $\mathbf{y} = (y_1, y_2)^T$. Then with $\rho = -0.5$, the quadratic in the exponent can be written as

$$x_1^2 + x_1x_2 + x_2^2 = \mathbf{x}^T\begin{bmatrix} a & b \\ c & d \end{bmatrix}\mathbf{x} = ax_1^2 + (b+c)x_1x_2 + dx_2^2,$$

where the a, b, c, d are to be determined. We immediately find that $a = d = 1$ and—because of the real symmetric requirement—we find $b = c = 0.5$. We can rewrite $f_{X_1,X_2}(x_1, x_2)$ in standard form as

$$f_{X_1X_2}(x_1, x_2) = \frac{1}{2\pi(\det(\mathbf{K}))^{1/2}}\exp(-\tfrac{1}{2}(\mathbf{x}^T\mathbf{K}^{-1}\mathbf{x}))$$

whence

$$\mathbf{K}^{-1} = \frac{1}{\sigma^2(1-\rho^2)}\begin{bmatrix} a & b \\ c & d \end{bmatrix} = \frac{4}{3\sigma^2}\begin{bmatrix} 1 & 0.5 \\ 0.5 & 1 \end{bmatrix}.$$

Our task is now to find a transformation that diagonalizes $\mathbf{K}^{-1}$. This will enable the joint pdf of Y_1 to Y_2 to be factored, thereby establishing that Y_1 and Y_2 are independent.

The factor $4/3\sigma^2$ affects the eigenvalues of $\mathbf{K}^{-1}$ but not the eigenvectors. To compute a set of orthonormal eigenvectors of $\mathbf{K}^{-1}$ we need only consider $\mathbf{K}^{-1}$ given by

$$\tilde{\mathbf{K}}^{-1} \triangleq \begin{bmatrix} 1 & 0.5 \\ 0.5 & 1 \end{bmatrix}$$

for which we obtain $\tilde{\lambda}_1 = 3/2$ $\tilde{\lambda}_2 = 1/2$. The corresponding unit eigenvectors are $\boldsymbol{\phi}_1 = 1/\sqrt{2}\,(1, 1)^T$ and $\boldsymbol{\phi}_2 = 1/\sqrt{2}\,(1, -1)^T$. Thus with

$$\tilde{\mathbf{U}} \triangleq \begin{bmatrix} 1 & 1 \\ 1 & -1 \end{bmatrix}$$

(normalization by $1/\sqrt{2}$ is not needed to obtain a diagonal covariance matrix so we dispense with these factors) we find that

$$\tilde{\mathbf{U}}^T \tilde{\mathbf{K}}^{-1} \tilde{\mathbf{U}} = \text{diag}(1, 3).$$

Hence a transformation that will work is†

$$\mathbf{Y} = \tilde{\mathbf{U}}^T \mathbf{X},$$

that is,

$$Y_1 = X_1 + X_2$$

$$Y_2 = X_1 - X_2.$$

To find $f_{Y_1Y_2}(y_1, y_2)$ we use Equation 2.11–15 of Chapter 2:

$$f_{Y_1Y_2}(y_1, y_2) = \sum_{i=1}^{l} f_{X_1X_2}(\mathbf{x}_i)/|J_i|$$

where the $\mathbf{x}_i \triangleq (x_1^{(i)}, x_2^{(i)})^T$ $i = 1, \ldots, l$ are the l solutions to $\mathbf{y} - \tilde{\mathbf{U}}^T\mathbf{x} = 0$ and J_i is the Jacobian. The only solution, that is $l = 1$, to $\mathbf{y} - \tilde{\mathbf{U}}^T\mathbf{x} = 0$ is

$$x_1 = \frac{y_1 + y_2}{2}$$

$$x_2 = \frac{y_1 - y_2}{2}$$

and, dispensing with subscripts there being only one root,

$$J = \det\left(\frac{\partial \mathbf{g}}{\partial \mathbf{x}}\right) = \det\begin{bmatrix} 1 & 1 \\ 1 & -1 \end{bmatrix} = -2.$$

Hence

$$f_{Y_1Y_2}(y_1, y_2) = \tfrac{1}{2} f_{X_1X_2}\left(\frac{y_1 + y_2}{2}, \frac{y_1 - y_2}{2}\right)$$

$$= \frac{1}{\sqrt{2\pi\sigma^2}} \exp\left[-\frac{y_1^2}{2\sigma^2}\right] \cdot \frac{1}{\sqrt{2\pi\sigma'^2}} \exp\left[-\frac{y_2^2}{2\sigma'^2}\right],$$

where $\sigma' \triangleq \sqrt{3}\,\sigma$.

Examples 4.5–1 and 4.5–2 are special cases of the following theorem:

Theorem 4.5–3. Let X be a normal zero-mean (for convenience) random vector with positive definite covariance matrix $\mathbf{K}$. Then there exists a nonsingular $n \times n$ matrix $\mathbf{C}$ such that under the transformation

$$\mathbf{Y} = \mathbf{C}^{-1}\mathbf{X},$$

the components $Y_1, \ldots, Y_n$ of $\mathbf{Y}$ are independent.

Proof. Use the $\mathbf{C}$ from Equation 4.4–12, that is,

$$\mathbf{K} = \mathbf{C}\mathbf{C}^T. \qquad \blacksquare$$

† There is no requirement to whiten the covariance matrix as in Example 4.5–1. Also, diagonalizing $\mathbf{K}^{-1}$ is equivalent to diagonalizing $\mathbf{K}$.

Example 4.5–3: (The generalized Rayleigh law.) Let $\mathbf{X} = (X_1, X_2, X_3)^T$ be a normal random vector with covariance matrix

$$\mathbf{K} = \sigma^2\mathbf{I}.$$

Compute the pdf of $R_3 \triangleq \|\mathbf{X}\| = \sqrt{X_1^2 + X_2^2 + X_3^2}$.

Solution: The probability of the event $\{R_3 \leq r\}$ is the probability distribution function $F_{R_3}(r)$ of R_3. Thus

$$F_{R_3}(r) = \frac{1}{(2\pi)^{3/2}[\sigma^2]^{3/2}} \iiint_{\mathcal{S}} \exp\left[-\frac{1}{2\sigma^2}(x_1^2 + x_2^2 + x_3^2)\right] dx_1\, dx_2\, dx_3,$$

where $\mathcal{S} = \{(x_1, x_2, x_3) : \sqrt{x_1^2 + x_2^2 + x_3^2} \leq r\}$. Now let

$$x_1 \triangleq \xi \cos\phi$$
$$x_2 \triangleq \xi \sin\phi \cos\theta$$
$$x_3 = \xi \sin\phi \sin\theta,$$

that is, a rectangular-to-spherical coordinate transformation. The Jacobian of this transformation is $\xi^2 \sin\phi$. Using this transformation in the expression for $F_{R_3}(r)$ we obtain for $r > 0$

$$\begin{aligned} F_{R_3}(r) &= \frac{1}{(2\pi)^{3/2}(\sigma^2)^{3/2}} \int_0^r \int_{\theta=0}^{2\pi} \int_{\phi=0}^{\pi} \exp\left[-\frac{\xi^2}{2\sigma^2}\right] \xi^2 \sin\phi\, d\xi\, d\theta\, d\phi \\ &= \frac{4\pi}{(2\pi)^{3/2}[\sigma^2]^{3/2}} \int_0^r \xi^2 \exp\left[-\frac{\xi^2}{2\sigma^2}\right] d\xi. \end{aligned}$$

To obtain $f_{R_3}(r)$, we differentiate $F_{R_3}(r)$ with respect to r. This yields

$$f_{R_3}(r) = \frac{2r^2}{\Gamma(\frac{3}{2})[2\sigma^2]^{3/2}} \exp\left[-\frac{r^2}{2\sigma^2}\right] \cdot u(r), \tag{4.5–22}$$

where $u(r)$ is the unit step and $\Gamma(3/2) = \sqrt{\pi}/2$. Equation 4.5–22 is an extension of the ordinary two-dimensional Rayleigh introduced in Chapter 2. The general n-dimensional Rayleigh is the pdf associated with $R_n \triangleq \|\mathbf{X}\| = \sqrt{X_1^2 + \ldots + X_n^2}$ and is given by

$$f_{R_n}(r) = \frac{2r^{n-1}}{\Gamma\left(\frac{n}{2}\right)[2\sigma^2]^{n/2}} \exp\left[-\frac{r^2}{2\sigma^2}\right] \cdot u(r). \tag{4.5–23}$$

The proof of Equation 4.5–23 requires the use of n-dimensional spherical coordinates. Such generalized spherical coordinates are well known in the mathematical literature [4–4, p. 9]. The demonstration of Equation 4.5–23 is left as a challenging problem.

4.6 CHARACTERISTIC FUNCTIONS OF RANDOM VECTORS

In Equation 3.6–1 we defined the characteristic function (c.f.) of a random variable as

$$\Phi_X(\omega) \triangleq E[e^{j\omega X}].$$

The extension to random vectors is completely straightforward. Let $\mathbf{X} = (X_1, \ldots, X_n)^T$ be a real n-component random vector. Let $\boldsymbol{\omega} = (\omega_1, \ldots, \omega_n)^T$ be a real n-component parameter vector. The c.f. of $\mathbf{X}$ is defined as

$$\Phi_{\mathbf{X}}(\boldsymbol{\omega}) \triangleq E[e^{j\boldsymbol{\omega}^T\mathbf{X}}]. \tag{4.6–1}$$

The similarity to the scalar case is obvious. The actual evaluation of Equation 4.6–1 is done through

$$\Phi_{\mathbf{X}}(\boldsymbol{\omega}) = \int_{-\infty}^{\infty} f_{\mathbf{X}}(\mathbf{x})e^{j\boldsymbol{\omega}^T\mathbf{x}}\, d\mathbf{x}. \tag{4.6–2}$$

In Equation 4.6–2 we use the usual compact notation that $d\mathbf{x} = dx_1 \ldots dx_n$ and the integral sign refers to an n-fold integration. If $\mathbf{X}$ is a discrete random vector, $\Phi_{\mathbf{X}}(\boldsymbol{\omega})$ can be computed from the joint probability mass function (PMF) as

$$\Phi_{\mathbf{X}}(\boldsymbol{\omega}) = \sum_{i_1} \ldots \sum_{i_n} \exp[j(\omega_1 x_{i_1} + \ldots + \omega_n x_{i_n})] \cdot P[X_1 = x_{i_1}, \ldots, X_n = x_{i_n}] \tag{4.6–3}$$

From Equation 4.6–2 we see that $\Phi_{\mathbf{X}}(\boldsymbol{\omega})$ is, except for a sign reversal in the exponent, the n-dimensional Fourier transform of $f_{\mathbf{X}}(\mathbf{x})$. This being the case, we can recover the pdf by an inverse n-dimensional Fourier transform (again with a sign reversal). Thus

$$f_{\mathbf{X}}(\mathbf{x}) = \frac{1}{(2\pi)^n}\int_{-\infty}^{\infty} \Phi_{\mathbf{X}}(\boldsymbol{\omega})e^{-j\boldsymbol{\omega}^T\mathbf{x}}\, d\boldsymbol{\omega}. \tag{4.6–4}$$

The c.f. is very useful for computing joint moments. We illustrate with an example.

Example 4.6–1: Let $\mathbf{X} = (X_1, X_2, X_3)^T$ and $\boldsymbol{\omega} = (\omega_1, \omega_2, \omega_3)$. Compute $E[X_1X_2X_3]$.

Solution: Since

$$\Phi_{\mathbf{X}}(\omega_1, \omega_2, \omega_3) = \int_{-\infty}^{\infty}\int_{-\infty}^{\infty}\int_{-\infty}^{\infty} f_{\mathbf{X}}(x_1, x_2, x_3)e^{j[\omega_1x_1+\omega_2x_2+\omega_3x_3]}\, dx_1\, dx_2\, dx_3$$

we obtain by partial differentiation

$$\frac{1}{j^3}\frac{\partial^3\Phi_{\mathbf{X}}(\omega_1, \omega_2, \omega_3)}{\partial\omega_1\,\partial\omega_2\,\partial\omega_3}\bigg|_{\omega_1=\omega_2=\omega_3=0} = \int_{-\infty}^{\infty}\int_{-\infty}^{\infty}\int_{-\infty}^{\infty} x_1x_2x_3 f_{\mathbf{X}}(x_1, x_2, x_3)\, dx_1\, dx_2\, dx_3$$
$$\triangleq E[X_1X_2X_3].$$

Any moment—provided that it exists—can be computed by the method used in Example 4.6–1, that is, by partial differentiation. Thus

$$E[X_1^{k_1} \ldots X_n^{k_n}] = j^{-(k_1+\ldots+k_n)}\frac{\partial^{k_1+\ldots+k_n}\Phi_{\mathbf{X}}(\omega_1, \ldots, \omega_n)}{\partial\omega_1^{k_1} \ldots \partial\omega_n^{k_n}}\bigg|_{\omega_1=\ldots=\omega_n=0} \tag{4.6–5}$$

By writing $E[\exp(j\boldsymbol{\omega}^T\mathbf{X})] = E\left[\exp\left(j\sum_{i=1}^{n}\omega_iX_i\right)\right] = E\left[\prod_{i=1}^{n}\exp(j\omega_iX_i)\right]$ and expanding each term in the product into a power series we readily obtain the rather cumbersome formula

$$\Phi_{\mathbf{X}}(\boldsymbol{\omega}) = \sum_{k_1=0}^{\infty} \cdots \sum_{k_n=0}^{\infty} E[X_1^{k_1} \ldots X_n^{k_n}] \frac{(j\omega_1)^{k_1}}{k_1!} \cdots \frac{(j\omega_n)^{k_n}}{k_n!}, \tag{4.6-6}$$

which has the advantage of explicitly revealing the relationship between the c.f. and the joint moments of the X_i, $i = 1, \ldots, n$. Of course Equation 4.6–6 has meaning only if

$$E[X_1^{k_1} \ldots X_n^{k_n}]$$

exists for all values of the nonnegative integers $k_1, \ldots, k_n$.

From Equation 4.6–2 we make the rather trivial observations that

1. $|\Phi_{\mathbf{X}}(\boldsymbol{\omega})| \leq \Phi_{\mathbf{X}}(\mathbf{0}) = 1$ and
2. $\Phi_{\mathbf{X}}^*(\boldsymbol{\omega}) = \Phi_{\mathbf{X}}(-\boldsymbol{\omega})$ (* indicates conjugation).
3. All c.f.'s of *subsets* of $\mathbf{X}$ can be obtained once $\Phi_{\mathbf{X}}(\boldsymbol{\omega})$ is known.

The last property is easily demonstrated with the following example. Suppose $\mathbf{X} = (X_1, X_2, X_3)^T$ has c.f.† $\Phi_{\mathbf{X}}(\omega_1, \omega_2, \omega_3) = E[\exp j(\omega_1 X_1 + \omega_2 X_2 + \omega_3 X_3)]$. Then

$$\Phi_{X_1X_2}(\omega_1, \omega_2) = \Phi_{X_1X_2X_3}(\omega_1, \omega_2, 0)$$
$$\Phi_{X_1X_3}(\omega_1, \omega_3) = \Phi_{X_1X_2X_3}(\omega_1, 0, \omega_3)$$
$$\Phi_{X_1}(\omega_1) = \Phi_{X_1X_2X_3}(\omega_1, 0, 0).$$

As already pointed out in Chapter 3, characteristic functions are also extremely useful in solving problems involving sums of independent random variables. Thus suppose $\mathbf{X} = (X_1, \ldots, X_n)^T$ where the X_i are *independent* r.v.'s with marginal pdf's $f_{X_i}(x_i)$, $i = 1, \ldots, n$. Then the pdf of the sum

$$Z = X_1 + \ldots + X_n$$

can be obtained from

$$f_Z(z) = f_{X_1}(z) * \ldots * f_{X_n}(z). \tag{4.6-7}$$

However, the actual carrying out of the n-fold convolution in Equation 4.6–7 can be quite tedious. The computation of $f_Z(z)$ is often done more advantageously from the c.f.'s as follows. In

$$\Phi_{\mathbf{X}}(\omega_1, \ldots, \omega_n) = E[e^{j\boldsymbol{\omega}^T\mathbf{X}}]$$

choose $\boldsymbol{\omega} = (\omega, \ldots, \omega)^T$. Then

$$\begin{aligned}\Phi_{\mathbf{X}}(\omega, \ldots, \omega) &= E[e^{j\omega(X_1+\ldots+X_n)}] \\ &= E\left[\prod_{i=1}^{n} e^{j\omega X_i}\right] \\ &= \prod_{i=1}^{n} E[e^{j\omega X_i}] = \prod_{i=1}^{n} \Phi_{X_i}(\omega) \\ &= \Phi_Z(\omega).\end{aligned} \tag{4.6-8}$$

† We use $\Phi_{\mathbf{X}}(\cdot)$ and $\Phi_{X_1X_2X_3}(\cdot)$ interchangeably if $\mathbf{X} = (X_1, X_2, X_3)^T$.

In this development, line three follows from the fact that if $X_1, \ldots, X_n$ are n independent r.v.'s then $Y_i = g_i(X_i)$, $i = 1, \ldots, n$ will also be n independent r.v.'s and $E[Y_1 \ldots Y_n] = E[Y_1] \ldots E[Y_n]$. We leave the demonstration of this result as an exercise. The inverse Fourier transform of Equation 4.6–8 yields the pdf $f_Z(z)$. This approach works equally well when the X_i are discrete. Then the PMF and the discrete Fourier transform can be used. We illustrate this approach to computing the pdf's of sums of random variables with an example.

Example 4.6–2: Let $\mathbf{X} = (X_1, \ldots, X_n)^T$ where the X_i, $i = 1, \ldots, n$ are i.i.d. Poisson r.v.'s with Poisson parameter λ. Let $Z = X_1 + \ldots + X_n$. Then

$$P_{X_i}(k) = \frac{\lambda^k e^{-\lambda}}{k!} \tag{4.6–9}$$

and

$$\begin{aligned} \Phi_{X_i}(\omega) &= \sum_{k=0}^{\infty} \frac{\lambda^k \exp(j\omega k)}{k!} e^{-\lambda} \\ &= e^{\lambda(\exp(j\omega)-1)}. \end{aligned} \tag{4.6–10}$$

Hence, from the third line leading up to Equation 4.6–8 we obtain

$$\begin{aligned} \Phi_Z(\omega) &= \prod_{i=1}^{n} e^{\lambda(\exp(j\omega)-1)} \\ &= e^{n\lambda(\exp(j\omega)-1)}. \end{aligned} \tag{4.6–11}$$

Comparing Equation 4.6–11 with Equation 4.6–10 we see by inspection that $\Phi_Z(z)$ is the c.f. of the PMF

$$P_Z(k) = \frac{\xi^k e^{-\xi}}{k!} \qquad k = 0, 1 \ldots \tag{4.6–12}$$

where $\xi \triangleq n\lambda$. Thus the PMF of the sum of n i.i.d. Poisson r.v.'s is Poisson with parameter $n\lambda$.

The Characteristic Function of the Normal Law. Let $\mathbf{X}$ be a real normal random vector with nonsingular covariance matrix $\mathbf{K}$. Then from Equation 4.4–12 both $\mathbf{K}$ and $\mathbf{K}^{-1}$ can be factored as

$$\mathbf{K} = \mathbf{C}\mathbf{C}^T \tag{4.6–13}$$

$$\mathbf{K}^{-1} = \mathbf{D}\mathbf{D}^T, \qquad \mathbf{D} = [\mathbf{C}^T]^{-1} \tag{4.6–14}$$

where $\mathbf{C}$ and $\mathbf{D}$ are nonsingular. This observation will be put to good use shortly. The c.f. of $\mathbf{X}$ is by definition

$$\Phi_{\mathbf{X}}(\boldsymbol{\omega}) = \frac{1}{(2\pi)^{n/2}(\det(\mathbf{K}))^{1/2}} \int_{-\infty}^{\infty} \exp(-\tfrac{1}{2}(\mathbf{x} - \boldsymbol{\mu})^T \mathbf{K}^{-1}(\mathbf{x} - \boldsymbol{\mu})) \cdot \exp(j\boldsymbol{\omega}^T\mathbf{x})\, d\mathbf{x}. \tag{4.6–15}$$

Now introduce the transformation

$$\mathbf{z} \triangleq \mathbf{D}^T(\mathbf{x} - \boldsymbol{\mu}) \tag{4.6–16}$$

so that

$$\begin{aligned} \mathbf{z}^T\mathbf{z} &= (\mathbf{x} - \boldsymbol{\mu})^T\mathbf{D}\mathbf{D}^T(\mathbf{x} - \boldsymbol{\mu}) \\ &= (\mathbf{x} - \boldsymbol{\mu})^T\mathbf{K}^{-1}(\mathbf{x} - \boldsymbol{\mu}). \end{aligned} \tag{4.6–17}$$

The Jacobian of this transformation is $\det(\mathbf{D}^T) = \det(\mathbf{D})$. Thus under the transformation in Equation 4.6–16, Equation 4.6–15 is transformed to

$$\Phi_{\mathbf{X}}(\boldsymbol{\omega}) = \frac{\exp(j\boldsymbol{\omega}^T\boldsymbol{\mu})}{(2\pi)^{n/2}(\det(\mathbf{K}))^{1/2}\,|\det(\mathbf{D})|}\int_{-\infty}^{\infty}\exp(-\tfrac{1}{2}\mathbf{z}^T\mathbf{z})\cdot\exp(j\boldsymbol{\omega}^T(\mathbf{D}^T)^{-1}\mathbf{z})\,d\mathbf{z}. \quad (4.6\text{–}18)$$

We can complete the squares in the integrand as follows:

$$\exp[-\{\tfrac{1}{2}[\mathbf{z}^T\mathbf{z} - 2j\boldsymbol{\omega}^T(\mathbf{D}^T)^{-1}\mathbf{z}]\}] = \exp(-\tfrac{1}{2}\boldsymbol{\omega}^T(\mathbf{D}^T)^{-1}(\mathbf{D})^{-1}\boldsymbol{\omega}) \cdot \exp(-\tfrac{1}{2}\|\mathbf{z} - j\mathbf{D}^{-1}\boldsymbol{\omega}\|^2). \quad (4.6\text{–}19)$$

Equations 4.6–18 and 4.6–19 will be greatly simplified if we use the following easily proven results: (A) If $\mathbf{K}^{-1} = \mathbf{D}\mathbf{D}^T$ then $\mathbf{K} = [\mathbf{D}^T]^{-1}\mathbf{D}^{-1}$; (B) $\det(K^{-1}) = \det(\mathbf{D})\det(\mathbf{D}^T) = [\det(\mathbf{D})]^2 = [\det(\mathbf{K})]^{-1}$. Hence $|\det(\mathbf{D})|^{-1} = [\det(\mathbf{K})]^{1/2}$. It then follows that

$$\Phi_{\mathbf{X}}(\boldsymbol{\omega}) = \exp[j\boldsymbol{\omega}^T\boldsymbol{\mu} - \tfrac{1}{2}\boldsymbol{\omega}^T\mathbf{K}\boldsymbol{\omega}]\cdot\frac{1}{(2\pi)^{n/2}}\int_{-\infty}^{\infty} e^{-\frac{1}{2}\|\mathbf{z}-j\mathbf{D}^{-1}\boldsymbol{\omega}\|^2}\,d\mathbf{z}.$$

Finally we recognize that the n-fold integral is the product of n identical integrals, each of unit variance. Hence the value of the integral is merely $(2\pi)^{n/2}$, which cancels the factor $(2\pi)^{-n/2}$ and yields as the c.f. for the normal random vector:

$$\Phi_{\mathbf{X}}(\boldsymbol{\omega}) = \exp[j\boldsymbol{\omega}^T\boldsymbol{\mu} - \tfrac{1}{2}\boldsymbol{\omega}^T\mathbf{K}\boldsymbol{\omega}], \quad (4.6\text{–}20)$$

where $\boldsymbol{\mu}$ is the mean vector, $\boldsymbol{\omega} = (\omega_1, \ldots, \omega_n)^T$, and $\mathbf{K}$ is the covariance. We observe in passing that $\Phi_{\mathbf{X}}(\boldsymbol{\omega})$ has a multidimensional complex Gaussian form as a function of $\boldsymbol{\omega}$. Thus the Gaussian pdf has mapped into a Gaussian c.f. a result that should not be too surprising since we already know that the one-dimensional Fourier transform maps a Gaussian function into a Gaussian function.

4.7 SUMMARY

In this chapter we studied the calculus of multiple random variables. We found it convenient to organize multiple random variables into random vectors and treat these as single entities. Because in practice it is often difficult to describe the joint probability law of n random variables we argued that in the case of random vectors we often settle for a less complete but more computable characterization than that furnished by the pdf. We focused on the characterizations furnished by the moments, especially the mean and covariance. In particular, because of the great importance of covariance matrices in signal processing, communication theory, pattern recognition, multiple regression analysis, and many other areas of engineering and science, we made use of numerous results from matrix theory and linear algebra to reveal the properties of these matrices. To illustrate the usefulness of the mathematical techniques introduced in this chapter (the techniques generally requiring somewhat greater mathematical exposure than was required for Chapters 1 to 3) we solved an important problem in the field of applied probability called statistical-pattern recognition (Example 4.4–3).

The last third of the chapter dealt with the multidimensional Gaussian (normal) law and characteristic functions of random vectors. We demonstrated that under linear transformations Gaussian random vectors map into Gaussian random vectors. We showed how to derive a transformation that can convert correlated random variables into uncorrelated ones. The characteristic function (c.f.) of random vectors in general was defined and shown to be useful in computing moments and solving problems involving the sums of independent random variables; these assertions were illustrated with examples. Finally, using vector and matrix techniques we derived the c.f. for the Gaussian random vector and showed that it too had a Gaussian shape.

PROBLEMS

4.1. Let $f_{\mathbf{X}}(\mathbf{x})$ be given as

$$f_{\mathbf{X}}(\mathbf{x}) = Ke^{-\mathbf{x}^T\mathbf{\Lambda}}u(\mathbf{x}),$$

where $\mathbf{\Lambda} = (\lambda_1, \ldots, \lambda_n)^T$ with $\lambda_i > 0$ all i, $\mathbf{x} = (x_1, \ldots, x_n)^T u(\mathbf{x}) = 1$ if $x_i \geq 0$ $i = 1, \ldots, n$ and zero otherwise, and K is a constant to be determined. What value of K will enable $f_{\mathbf{X}}(\mathbf{x})$ to be a pdf?

4.2. Let B_i, $i = 1, \ldots, n$ be n disjoint and exhaustive events. Show that the probability distribution function (PDF) of $\mathbf{X}$ can be written as

$$F_{\mathbf{X}}(\mathbf{x}) = \sum_{i=1}^{n} F_{\mathbf{X}|B_i}(\mathbf{x} \mid B_i)P[B_i].$$

4.3. Let

$$f_{\mathbf{X}}(\mathbf{x}) = \frac{1}{(2\pi)^{n/2}\sigma_1 \ldots \sigma_n} \exp\left\{-\frac{1}{2}\left(\sum_{i=1}^{n}\left(\frac{x_i}{\sigma_i}\right)^2\right)\right\}.$$

Show that all the marginals pdf's are Gaussian.

4.4. Show that any matrix $\mathbf{M}$ generated by an outer product of two vectors, that is, $\mathbf{M} = \mathbf{X}\mathbf{X}^T$ has rank at most unity. Explain why $\mathbf{R} \triangleq E[\mathbf{X}\mathbf{X}^T]$ can be of full rank.

4.5. Show that the two r.v.'s X_1 and X_2 with joint pdf

$$f_{X_1X_2}(x_1, x_2) = \begin{cases} \frac{1}{16}, & |x_1| < 4, \quad 2 < x_2 < 4 \\ 0, & \text{otherwise} \end{cases}$$

are independent and orthogonal.

4.6. Let $\mathbf{X}_i$, $i = 1, \ldots, n$ be n mutually orthogonal random vectors. Show that

$$E\left[\left\|\sum_{i=1}^{n} \mathbf{X}_i\right\|^2\right] = \sum_{i=1}^{n} E[\|\mathbf{X}_i\|^2].$$

Hint: Use the definition $\|\mathbf{X}\|^2 \triangleq \mathbf{X}^T\mathbf{X}$.

4.7. Let $\mathbf{X}_i$, $i = 1, \ldots, n$ be n mutually uncorrelated random vectors with means $\boldsymbol{\mu}_i \triangleq E[\mathbf{X}_i]$. Show that

$$E\left[\left\|\sum_{i=1}^{n} (\mathbf{X}_i - \boldsymbol{\mu}_i)\right\|^2\right] = \sum_{i=1}^{n} E[\|\mathbf{X}_i - \boldsymbol{\mu}_i\|^2].$$

4.8. Let $\mathbf{X}_i$, $i = 1, \ldots, n$ be n mutually uncorrelated random vectors with $E[\mathbf{X}_i] = \boldsymbol{\mu}_i$, $i = 1, \ldots, n$. Show that

$$E\left[\sum_{i=1}^{n} (\mathbf{X}_i - \boldsymbol{\mu}_i) \sum_{j=1}^{n} (\mathbf{X}_j - \boldsymbol{\mu}_j)^T\right] = \sum_{i=1}^{n} \mathbf{K}_i,$$

where $\mathbf{K}_i \triangleq E[(\mathbf{X}_i - \boldsymbol{\mu}_i)(\mathbf{X}_i - \boldsymbol{\mu}_i)^T]$.

4.9. Explain why none of the following matrices can be covariance matrices associated with real random vectors.

$$\begin{bmatrix} 2 & -4 & 0 \\ -4 & 3 & 1 \\ 0 & 1 & 2 \end{bmatrix} \begin{bmatrix} 4 & 0 & 0 \\ 0 & 6 & 0 \\ 0 & 0 & -2 \end{bmatrix} \begin{bmatrix} 6 & 1+j & 2 \\ 1-j & 5 & -1 \\ 2 & -1 & 6 \end{bmatrix} \begin{bmatrix} 4 & 6 & 2 \\ 6 & 9 & 3 \\ 8 & 12 & 16 \end{bmatrix}$$

(a) (b) (c) (d)

4.10. (a) Let a vector $\mathbf{X}$ have $E[\mathbf{X}] = \mathbf{0}$ with covariance $\mathbf{K_X}$ given by

$$\mathbf{K_X} = \begin{bmatrix} 3 & \sqrt{2} \\ \sqrt{2} & 4 \end{bmatrix}$$

Find a linear transformation $\mathbf{C}$ such that $\mathbf{Y} = \mathbf{CX}$ will have

$$\mathbf{K_Y} = \begin{bmatrix} 1 & 0 \\ 0 & 1 \end{bmatrix}$$

Is $\mathbf{C}$ a unitary transformation?

(b) Consider the two real symmetric matrices $\mathbf{A}$ and $\mathbf{A}'$ given by

$$\mathbf{A} \triangleq \begin{bmatrix} a & b \\ b & c \end{bmatrix} \qquad \mathbf{A}' \triangleq \begin{bmatrix} a' & b' \\ b' & c' \end{bmatrix}.$$

Show that when $a = c$ and $a' = c'$, the product $\mathbf{AA}'$ is real symmetric. More generally, show that if $\mathbf{A}$ and $\mathbf{A}'$ are any real symmetric matrices then $\mathbf{AA}'$ will be symmetric if $\mathbf{AA}' = \mathbf{A}'\mathbf{A}$.

4.11. (K. Fukunaga [4–6, p. 35].) Let $\mathbf{K}_1$ and $\mathbf{K}_2$ be positive definite covariance matrices and form

$$\mathbf{K} = a_1\mathbf{K}_1 + a_2\mathbf{K}_2 \qquad \text{where } a_1, a_2 > 0.$$

Let $\mathbf{A}$ be a transformation that achieves

$$\mathbf{A}^T\mathbf{K}\mathbf{A} = \mathbf{I} \qquad \mathbf{A}^T\mathbf{K}_1\mathbf{A} = \boldsymbol{\Lambda}^{(1)} = \text{diag}(\lambda_1^{(1)}, \ldots, \lambda_n^{(1)}).$$

(a) Show that $\mathbf{A}$ satisfies

$$\mathbf{K}^{-1}\mathbf{K}_1\mathbf{A} = \mathbf{A}\boldsymbol{\Lambda}^{(1)}.$$

(b) Show that $\mathbf{A}^T\mathbf{K}_2\mathbf{A} \triangleq \boldsymbol{\Lambda}^{(2)}$ is also diagonal, that is, $\boldsymbol{\Lambda}^{(2)} \triangleq \text{diag}(\lambda_1^{(2)}, \ldots, \lambda_n^{(2)})$.

(c) Show that $\mathbf{A}^T\mathbf{K}_1\mathbf{A}$ and $\mathbf{A}^T\mathbf{K}_2\mathbf{A}$ share the same eigenvectors.

(d) Show that the eigenvalues of $\mathbf{\Lambda}^{(2)}$ are related to the eigenvalues of $\mathbf{\Lambda}^{(1)}$ as

$$\lambda_i^{(2)} = \frac{1}{a_2}[1 - a_1\lambda_i^{(1)}]$$

and therefore are in inverse order from those of $\mathbf{\Lambda}^{(1)}$.

4.12. (J. A. McLaughlin [4–7].) Consider the m vectors $\mathbf{X}_i = (X_{i1}, \ldots, X_{in})^T$, $i = 1, \ldots, m$ where $n > m$. Consider the $n \times n$ matrix $\mathbf{S} = \frac{1}{m}\sum_{i=1}^{m} \mathbf{X}_i\mathbf{X}_i^T$.

(a) Show that with $\mathbf{W} \triangleq (\mathbf{X}_1 \ldots \mathbf{X}_m)$, $\mathbf{S}$ can be written as

$$\mathbf{S} = \frac{1}{m}\mathbf{W}\mathbf{W}^T.$$

(b) What is the maximum rank of $\mathbf{S}$?

(c) Let $\mathbf{S}' \triangleq \frac{1}{m}\mathbf{W}^T\mathbf{W}$. What is the size of $\mathbf{S}'$? Show that the first m nonzero eigenvalues of $\mathbf{S}$ can be computed from

$$\mathbf{S}'\mathbf{\Phi} = \mathbf{\Phi}\mathbf{\Lambda},$$

where $\mathbf{\Phi}$ is the eigenvector matrix of $\mathbf{S}'$ and $\mathbf{\Lambda}$ is the matrix of eigenvalues. What are the relations between the eigenvectors and eigenvalues of $\mathbf{S}$ and $\mathbf{S}'$?

(d) What is the advantage of computing the eigenvectors from $\mathbf{S}'$ rather than $\mathbf{S}$?

4.13. (K. Fukunaga [4–6], p. 45–46.)

(a) Let $\mathbf{K}$ be an $n \times n$ covariance matrix and let $\Delta\mathbf{K}$ be a real symmetric perturbation matrix. Let λ_i, $i = 1, \ldots, n$ be the eigenvalues of $\mathbf{K}$ and $\boldsymbol{\phi}_i$ the associated eigenvectors. Show that the first order approximation to the eigenvalues λ_i' of $\mathbf{K} + \Delta\mathbf{K}$ yield

$$\lambda_i' = \boldsymbol{\phi}_i^T(\mathbf{K} + \Delta\mathbf{K})\boldsymbol{\phi}_i, \qquad i = 1, \ldots, n.$$

(b) Show that the first-order approximation to the eigenvectors is given by

$$\Delta\boldsymbol{\phi}_i = \sum_{j=1}^{n} b_{ij}\boldsymbol{\phi}_j,$$

where $b_{ij} = \boldsymbol{\phi}_j^T\Delta\mathbf{K}\boldsymbol{\phi}_i/(\lambda_i - \lambda_j) i \neq j$ and $b_{ii} = 0$.

4.14. (a) Let $\lambda_1 \geq \lambda_2 \geq \ldots \geq \lambda_n$ be the eigenvalues of a real symmetric matrix $\mathbf{M}$. For $i \geq 2$, let $\boldsymbol{\phi}_1, \boldsymbol{\phi}_2, \ldots, \boldsymbol{\phi}_{i-1}$ be mutually orthogonal unit eigenvectors belonging to $\lambda_1, \ldots, \lambda_{i-1}$. Prove that the *maximum value* of $\mathbf{u}^T\mathbf{M}\mathbf{u}$ subject to $\|\mathbf{u}\| = 1$ and $\mathbf{u}^T\boldsymbol{\phi}_1 = \ldots = \mathbf{u}^T\boldsymbol{\phi}_{i-1} = 0$ is λ_i, that is, $\lambda_i = \max(\mathbf{u}^T\mathbf{M}\mathbf{u})$.

(b) Prove Theorem 4.4–3 using Theorem 4.4–2.

4.15. Let $\mathbf{X} = (X_1, X_2, X_3)^T$ be a random vector with $\boldsymbol{\mu} \triangleq E[\mathbf{X}]$ given by

$$\boldsymbol{\mu} = (5, -5, 6)^T$$

and covariance given by

$$\mathbf{K} = \begin{bmatrix} 5 & 2 & -1 \\ 2 & 5 & 0 \\ -1 & 0 & 4 \end{bmatrix}.$$

Calculate the mean and variance of

$$Y = \mathbf{A}^T\mathbf{X} + B,$$

where $\mathbf{A} = (2, -1, 2)^T$ and $B = 5$.

4.16. Two jointly normal r.v.'s X_1 and X_2 have joint pdf given by

$$f_{X_1X_2}(x_1, x_2) = \frac{2}{\pi\sqrt{7}} \exp[-\tfrac{8}{7}(x_1^2 + \tfrac{3}{2}x_1x_2 + x_2^2)].$$

ex. 4.5-2
Th. 4.5-3

Find a nontrivial transformation $\mathbf{A}$ in

$$\begin{pmatrix} Y_1 \\ Y_2 \end{pmatrix} = \mathbf{A}\begin{pmatrix} X_1 \\ X_2 \end{pmatrix}$$

such that Y_1 and Y_2 are independent. Compute the joint pdf of Y_1, Y_2.

4.17. Show that if $\mathbf{X} = (X_1, \ldots, X_n)^T$ has mean $\boldsymbol{\mu} = (\mu_1, \ldots, \mu_n)^T$ and covariance

$$\mathbf{K} = \{K_{ij}\}_{n\times n}$$

then the scalar r.v. Y given by

$$Y \triangleq p_1X_1 + \ldots + p_nX_n$$

has mean

$$E[Y] = \sum_{i=1}^{n} p_i\mu_i$$

and variance

$$\sigma_Y^2 = \sum_{i=1}^{n}\sum_{j=1}^{n} p_ip_jK_{ij}.$$

4.18. Compute the joint characteristic function of $\mathbf{X} = (X_1, \ldots, X_n)^T$, where the X_i, $i = 1, \ldots, n$ are mutually independent and identically distributed Cauchy r.v.'s, that is,

$$f_{X_i}(x) = \frac{\alpha}{\pi(x^2 + \alpha^2)}.$$

Use this result to compute the pdf of $Y = \sum_{i=1}^{n} X_i$.

4.19. Compute the joint characteristic function of $\mathbf{X} = (X_1, \ldots, X_n)^T$, where the X_i, $i = 1, \ldots, n$ are mutually independent and identically distributed binomial r.v.'s. Use this result to compute the PMF of $Y = \sum_{i=1}^{n} X_i$.

4.20. Let $\mathbf{X} = (X_1, \ldots, X_4)$ be a Gaussian random vector with $E[\mathbf{X}] = 0$. Show that

$$E[X_1X_2X_3X_4] = K_{12}K_{34} + K_{13}K_{24} + K_{14}K_{23},$$

where the K_{ij} are elements of the covariance matrix $\mathbf{K} = \{K_{ij}\}_{4\times 4}$ of $\mathbf{X}$.

REFERENCES

4-1. W. B. Davenport, Jr., *Probability and Random Processes.* New York; McGraw-Hill, 1970.

4-2. J. N. Franklin, *Matrix Theory.* Englewood-Cliffs, N.J.: Prentice-Hall, 1968.

4-3. H. Stark and R. O'Toole, "Statistical Pattern Recognition Using Optical

Fourier Transform Features," Chapter 11 in *Applications of Optical Fourier Transforms*, H. Stark, ed. New York: Academic Press, 1982.
4–4. R. O. Duda and P. E. Hart, *Pattern Classification and Scene Analysis*. New York: John Wiley, 1973.
4–5. K. S. Miller, *Multidimensional Gaussian Distributions*. New York: John Wiley, 1964.
4–6. K. Fukunaga, *Introduction to Statistical Pattern Recognition*. New York: Academic, 1972.
4–7. J. A. McLaughlin and J. Raviv, "*N*th Order Autocorrelations in Pattern Recognition," *Information and Control*, 12, pp. 121–142, Chapter 2, 1968.

5

ESTIMATION AND DECISION THEORY I

In Chapter 4 a large number of results involved manipulations of objects such as the vector of means $\boldsymbol{\mu}$ and the covariance matrix $\mathbf{K}$. Sometimes these quantities are known from theoretical considerations but very often they have to be estimated from measurement data. How should this data be organized to furnish a good estimator for $\boldsymbol{\mu}$?; for $\mathbf{K}$? More generally suppose we want to estimate a parameter vector $\boldsymbol{\theta} = (\theta_1, \ldots, \theta_n)^T$. Suppose further that our measurements are corrupted by random measurement errors that we call noise. What is a good estimator for $\boldsymbol{\theta}$ that is based only on the acquired data? Finally, suppose we want to estimate not a parameter $\boldsymbol{\theta}$ but rather a random vector $\mathbf{Y}$ by observing another random vector, say $\mathbf{X}$. How should this be done? This class of problems falls within the realm of estimation theory. A second class of problems closely related to estimation are those involved with making decisions in a random, that is, probabilistic environment. For example, in a digital communication system a *zero* and a *one* are sent with equal probability. In the absence of noise the *zero* and *one* signals are easily distinguishable. However, when channel noise is present (as it invariably is) the *zero* and *one* signals are distorted and it may not be so obvious which is which. We sample and observe the received voltage level at the receiver at some instant t_0. How should we decide on what was sent?

The material in this chapter is the first of two dealing with problems in estimation and decision. As such, the chapter covers essentially introductory material. Chapter 10 continues and extends the discussion to a more advanced level. The basic mathematical tools required to understand the material covered here are calculus (as always) and matrix algebra.

5.1 PARAMETER ESTIMATION

The problem of estimating the parameters μ and σ^2 or $\boldsymbol{\mu}$ and $\mathbf{K}$ is a problem of parameter estimation. We illustrate what is meant by parameter estimation with two examples.

Example 5.1–1: Suppose θ is an unknown scalar that we wish to estimate. We make n measurements called observations and collect n samples (sometimes called realizations) $x_1, \ldots, x_n$ where

$$x_i = \theta + \varepsilon_i \qquad i = 1, \ldots, n. \tag{5.1–1}$$

In Equation 5.1–1 x_i is the value of the measurement and ε_i is the value of the measurement noise on the ith observations. After completing the n measurements and collecting all the data, a reasonable estimate $\hat{\theta}$ of θ is furnished by

$$\hat{\theta} = \frac{1}{n}\sum_{i=1}^{n} x_i. \tag{5.1–2}$$

Example 5.1–2: We observe the values $x_1^{(1)}, \ldots, x_n^{(1)}$ that a normal r.v. takes in n trials. We wish to estimate the mean μ of the pdf. An estimate of μ is furnished by

$$\hat{\mu}^{(1)} = \frac{1}{n}\sum_{i=1}^{n} x_i^{(1)}.$$

Note that if we collected a second set of samples $x_1^{(2)}, \ldots, x_n^{(2)}$ then the estimate of μ based on this second set, that is,

$$\hat{\mu}^{(2)} = \frac{1}{n}\sum_{i=1}^{n} x_i^{(2)}$$

would probably be different from $\hat{\mu}_1$.

In Examples 5.1–1 and 5.1–2 we can expect different sets of n sample values to yield different estimates. This suggests that we should view the estimate as a particular value of a random variable (r.v.) called an *estimator* and the n sample values $x_1, \ldots, x_n$ for a particular measurement as the values assumed by a set of n random variables $X_1, \ldots, X_n$ where X_i is the outcome of the ith measurement. We can think of each measurement as an observation on a generic r.v. X with some underlying pdf $f_X(x)$. Then the X_i $i = 1, \ldots, n$ represent n independent, identically distributed (i.i.d.) observations on X, and each has pdf $f_{X_i}(x_i) = f_X(x_i)$, $i = 1, \ldots, n$. Thus the estimate $\hat{\theta}$ is a particular value of the *estimator*

$$\hat{\Theta} \triangleq \frac{1}{n}\sum_{i=1}^{n} X_i. \tag{5.1–3}$$

The estimator in Equation 5.1–3 is often used to estimate $E[X]$.† Other estimators are used to estimate $\text{Var}[X]$, $\mathbf{K}$, etc. Some estimators are better than others. To evaluate estimators we need some definitions.‡

† The validity of estimating parameters as well as other objects, for instance probabilities, from repeated observations is based, fundamentally, on the law of large numbers and the Chebyshev inequality.

‡ The definitions are given for scalar estimators. They are easily extended to vector estimators.

Definition 5.1–1. An estimator $\hat{\Theta}$ is a function of the observation vector $\mathbf{X} = (X_1, \ldots, X_n)^T$ that estimates θ but is not dependent on θ.

Definition 5.1–2. An estimator $\hat{\Theta}$ for θ is said to be unbiased if and only if $E[\hat{\Theta}] = \theta$. The bias in estimating θ with $\hat{\Theta}$ is†

$$|E[\hat{\Theta}] - \theta|.$$

Definition 5.1–3. An estimator $\hat{\Theta}$ is said to be a *linear estimator* of θ if it is a linear function of the observation vector $\mathbf{X} \triangleq (X_1, \ldots, X_n)^T$, that is,

$$\hat{\Theta} = \mathbf{b}^T\mathbf{X}. \tag{5.1–4}$$

The vector $\mathbf{b}$ is an nxl vector of coefficients that do not depend on $\mathbf{X}$.

Definition 5.1–4. Let $\hat{\Theta}_n$ be an estimator computed from n samples $X_1, \ldots, X_n$ for every $n \geq 1$. Then $\hat{\Theta}_n$ is said to be *consistent* if

$$\lim_{n\to\infty} P[|\hat{\Theta}_n - \theta| > \varepsilon] = 0 \quad \text{for every} \quad \varepsilon > 0. \tag{5.1–5}$$

The condition in Equation 5.1–5 is often referred to as *convergence in probability*.

Definition 5.1–5. An estimator $\hat{\Theta}$ is called *minimum-variance unbiased* if

$$E[(\hat{\Theta} - \theta)^2] \leq E[(\hat{\Theta}' - \theta)^2] \tag{5.1–6}$$

where $\hat{\Theta}'$ is any other estimator and $E[\hat{\Theta}'] = E[\hat{\Theta}] = \theta$.

Definition 5.1–6. An estimator $\hat{\Theta}$ is called a *minimum mean-square error* (MMSE) estimator if

$$E[(\hat{\Theta} - \theta)^2] \leq E[(\hat{\Theta}' - \theta)^2], \tag{5.1–7}$$

where $\hat{\Theta}'$ is any other estimator.

There are several other properties of estimators that are deemed desirable such as efficiency, completeness, and invariance. These properties are discussed in books on statistics‡ and will not be discussed further here.

Estimation of E[X]. Let X be a r.v. with pdf $f_X(x)$ and finite variance σ^2. We wish to estimate $\mu = E[X]$ from n i.i.d. observations on X, that is, we repeat the experiment n times with X_i denoting the ith outcome. Sometimes it is said that the X_i are *drawn independently* from $f_X(x)$. Then $f_{X_i}(x) = f_X(x)$, $i = 1, \ldots, n$.

The sample mean estimator is

$$\hat{\Theta} \triangleq \frac{1}{n}\sum_{i=1}^{n} X_i. \tag{5.1–8}$$

We now show that $\hat{\Theta}$ is an unbiased, consistent estimator of μ.

† The bias is often defined without the magnitude sign.

‡ See, for example, Reference [5–1].

Unbiasedness of $\hat{\Theta}$ of Equation 5.1–8. Since $E[X_i] = \mu, i = 1, \ldots, n$ we have

$$E[\hat{\Theta}] = E\left[\frac{1}{n}\sum_{i=1}^{n} X_i\right]$$
$$= \frac{1}{n}\sum_{i=1}^{n} E[X_i]$$
$$= \frac{1}{n}(n)\mu = \mu.$$

Consistency of $\hat{\Theta}$. By the Chebyshev inequality, Equation 3.4–1, we obtain

$$P[|\hat{\Theta} - \mu| > \varepsilon] \leq \frac{\text{Var}[\hat{\Theta}]}{\varepsilon^2}. \tag{5.1–9}$$

The variance of $\hat{\Theta}$ is obtained as

$$\text{Var}[\hat{\Theta}] = E\left[\left(\frac{1}{n}\sum_{i=1}^{n} X_i\right) - \mu\right]^2$$
$$= E\left[\frac{1}{n^2}\sum_{i=1}^{n} X_i^2 + \frac{1}{n^2}\sum_{i\neq j}^{n}\sum^{n} X_i X_j + \mu^2 - 2\mu\left(\frac{1}{n}\sum_{i=1}^{n} X_i\right)\right]$$
$$= \frac{1}{n^2}(n\sigma^2 + n\mu^2) + \frac{1}{n^2}n(n-1)\mu^2 + \mu^2 - 2\mu^2$$
$$= \frac{\sigma^2}{n}. \tag{5.1–10}$$

Now using Equation 5.1–10 in Equation 5.1–9 we obtain

$$P[|\hat{\Theta} - \mu| > \varepsilon] \leq \frac{\sigma^2}{n\varepsilon^2}, \tag{5.1–11}$$

which goes to zero for $n \to \infty$ and every $\varepsilon > 0$.†

Estimation of Var[X]. Let X be a random variable with pdf $f_X(x)$ with mean μ and variance σ^2. We wish to estimate σ^2 from n i.i.d. observations $X_1, \ldots, X_n$ on X. Consider the estimator

$$\hat{\Theta} \triangleq \frac{1}{n-1}\sum_{i=1}^{n} (X_i - \hat{\mu})^2, \tag{5.1–12}$$

where

$$\hat{\mu} \triangleq \frac{1}{n}\sum_{i=1}^{n} X_i. \tag{5.1–13}$$

The following shows that $\hat{\Theta}$ in Equation 5.1–12 is an unbiased, consistent estimator of Var(X).

† This result, or variations thereof, is essentially a statement of the weak *law of large numbers*. See Section 6.5 for additional discussion of the law of large numbers.

Unbiasedness of $\hat{\Theta}$ of Equation 5.1–12. Consider.

$$E\left[\sum_{i=1}^{n}\left(X_i - \frac{1}{n}\sum_{j=1}^{n} X_j\right)^2\right]$$
$$= E\left[\sum_{i=1}^{n}\left\{X_i^2 - \frac{2}{n}X_i^2 - \frac{2}{n}\sum_{\substack{j=1\\ j\neq i}}^{n} X_iX_j + \frac{1}{n^2}\sum_{j=1}^{n} X_j^2 + \frac{1}{n^2}\sum_{j\neq l}^{n}\sum^{n} X_jX_l\right\}\right]$$
$$= (n-1)\sigma^2. \tag{5.1–14}$$

In obtaining Equation 5.1–14, we used the fact that $E[X_i^2] = \sigma^2 + \mu^2$ $i = 1, \ldots, n$. Clearly if

$$E\left[\sum_{i=1}^{n}(X_i - \hat{\mu})^2\right] = (n-1)\sigma^2$$

then

$$E\left[\frac{1}{n-1}\sum_{i=1}^{n}(X_i - \hat{\mu})^2\right] = \sigma^2. \tag{5.1–15}$$

But the quantity inside the square brackets is $\hat{\Theta}$ of Equation 5.1–12. Hence $\hat{\Theta}$ is unbiased for σ^2.

Consistency of $\hat{\Theta}$ of Equation 5.1–12. To indicate the dependence of $\hat{\Theta}$ on the sample size n we add the subscript n. The variance of $\hat{\Theta}_n$ is given by

$$\mathrm{Var}[\hat{\Theta}_n] = E[\hat{\Theta}_n - \sigma^2]^2$$
$$= E\left[\frac{1}{(n-1)^2}\left\{\sum_{i=1}^{n}(X_i - \hat{\mu})^4 + \sum_{i\neq j}^{n}\sum^{n}(X_i - \hat{\mu})^2(X_j - \hat{\mu})^2\right\}\right.$$
$$\left. + \sigma^4 - \frac{2\sigma^2}{n-1}\sum_{i=1}^{n}(X_i - \hat{\mu})^2\right].$$

A straightforward calculation shows that for $n \gg 1$

$$\mathrm{Var}[\hat{\Theta}_n] \simeq \frac{1}{n} m_4, \tag{5.1–16}$$

where $m_4 \triangleq E[(X_i - \mu)^4]$ (see Equation 3.3–2). Assuming that m_4 (the fourth-order central moment) exists, we once again use the Chebyshev inequality to write that

$$P(|\hat{\Theta}_n - \sigma^2| > \varepsilon) \leq \frac{\mathrm{Var}[\hat{\Theta}_n]}{\varepsilon^2}$$
$$\simeq \frac{m_4}{n\varepsilon^2}, \tag{5.1–17}$$

which goes to zero as $n \to \infty$ for every $\varepsilon > 0$. Hence $\hat{\Theta}_n$ is consistent.

5.2 ESTIMATION OF VECTOR MEANS AND COVARIANCE MATRICES

Let $\mathbf{X} \triangleq (X_1, \ldots, X_p)^T$ be a p-component random vector with pdf $f_{\mathbf{X}}(\mathbf{x})$. Let $\mathbf{X}_1, \ldots, \mathbf{X}_n$ be n observations on $\mathbf{X}$, that is, the $\mathbf{X}_i$, $i = 1, \ldots, n$ are drawn from $f_{\mathbf{X}}(\mathbf{x})$. Then $\mathbf{X}_i$, $i = 1, \ldots, n$ are i.i.d. random vectors with pdf $f_{\mathbf{X}}(\mathbf{x}_i)$. We show below how to estimate

$$\text{(i)}\ \boldsymbol{\mu} \triangleq E[\mathbf{X}] = (\mu_1, \ldots, \mu_p)^T,$$

where

$$\mu_j \triangleq E[X_j] \qquad j = 1, \ldots, p$$

and

$$\text{(ii)}\ \mathbf{K} \triangleq E[(\mathbf{X} - \boldsymbol{\mu})(\mathbf{X} - \boldsymbol{\mu})^T].$$

Estimation of $\boldsymbol{\mu}$. Consider the p-vector estimator $\hat{\boldsymbol{\Theta}}$ given by

$$\hat{\boldsymbol{\Theta}} \triangleq \frac{1}{n} \sum_{i=1}^{n} \mathbf{X}_i. \tag{5.2–1}$$

We shall show that $\hat{\boldsymbol{\Theta}}$ is unbiased and consistent for $\boldsymbol{\mu}$. We arrange the observations as in Table 5.2–1.

Table 5.2–1 Observed Data

	$\mathbf{X}_1$	$\cdots$	$\mathbf{X}_i$	$\cdots$	$\mathbf{X}_n$	
$\mathbf{Y}_1$	X_{11}		X_{i1}		X_{n1}	
$\vdots$	$\vdots$		$\vdots$		$\vdots$	
$\mathbf{Y}_j$	$\cdot$		X_{ij}		$\cdot$	p rows
$\vdots$	$\vdots$		$\vdots$		$\vdots$	
$\mathbf{Y}_p$	X_{1p}				X_{np}	
	n columns					

In Table 5.2–1 X_{ij} is the jth component of the random vector $\mathbf{X}_i$. The vectors $\mathbf{Y}_j$, $j = 1, \ldots, p$ represent n i.i.d. observations on the jth component of the random vector $\mathbf{X}$. From the scalar case we already know that

$$\hat{\Theta}_j \triangleq \frac{1}{n} \sum_{i=1}^{n} X_{ij} \triangleq \hat{\mu}_j \qquad j = 1, \ldots, p \tag{5.2–2}$$

is unbiased and consistent for $\mu_j \triangleq E[X_{ij}]$ $i = 1, \ldots, n$. It follows therefore that the vector estimator $\hat{\boldsymbol{\Theta}} \triangleq (\hat{\Theta}_1, \ldots, \hat{\Theta}_p)^T$ is unbiased and consistent for $\boldsymbol{\mu}$.

When $\mathbf{X}$ is normal, $\hat{\boldsymbol{\Theta}}$ is normal. Even when $\mathbf{X}$ is not normal, $\hat{\boldsymbol{\Theta}}$ tends to the normal for large n by the central limit theorem (Theorem 3.6–1).

Estimation of the Covariance K. If the mean $\boldsymbol{\mu}$ is known, then the estimator

$$\hat{\boldsymbol{\Theta}} \triangleq \frac{1}{n} \sum_{i=1}^{n} (\mathbf{X}_i - \boldsymbol{\mu})(\mathbf{X}_i - \boldsymbol{\mu})^T \tag{5.2–3}$$

is unbiased for $\mathbf{K}$. However, since the mean is generally estimated from the *sample mean* $\hat{\boldsymbol{\mu}}$, the estimator

$$\hat{\boldsymbol{\Theta}} \triangleq \frac{1}{n-1}\sum_{i=1}^{n}(\mathbf{X}_i - \hat{\boldsymbol{\mu}})(\mathbf{X}_i - \hat{\boldsymbol{\mu}})^T \tag{5.2–4}$$

is unbiased for $\mathbf{K}$. To prove this result requires some effort. First observe that the diagonal elements of $\hat{\boldsymbol{\Theta}}$ are of the form

$$S_{jj} \triangleq \frac{1}{n-1}\sum_{i=1}^{n}(X_{ij} - \hat{\mu}_j)^2, \tag{5.2–5}$$

which we already know from the univariate case are unbiased for $\sigma_j^2 \triangleq E[(X_j - \mu_j)^2]$. Next consider the sequence $(l \neq m)$

$$X_{1l} + X_{1m}, X_{2l} + X_{2m}, \ldots, X_{nl} + X_{nm}, \tag{5.2–6}$$

which are n i.i.d. observations on a univariate r.v. $Z_{lm} \triangleq X_l + X_m$ with mean $\mu_l + \mu_m$ and variance

$$\begin{aligned}\text{Var}[Z_{lm}] &= E[(X_l - \mu_l) + (X_m - \mu_m)]^2 \\ &= \sigma_l^2 + \sigma_m^2 + 2K_{lm},\end{aligned} \tag{5.2–7}$$

where $K_{lm} \triangleq E[(X_l - \mu_l)(X_m - \mu_m)]$ is the lmth element of $\mathbf{K}$. Finally, consider

$$\hat{\Theta}_{lm} \triangleq \frac{1}{n-1}\sum_{i=1}^{n}[Z_{lm}^{(i)} - (\hat{\mu}_l + \hat{\mu}_m)]^2, \tag{5.2–8}$$

which, by Equation 5.1–15, is unbiased for $\sigma_l^2 + \sigma_m^2 + 2K_{lm}$. If we expand Equation 5.2–8 and use the fact that $Z_{lm} \triangleq X_l + X_m$, we obtain

$$\begin{aligned}\hat{\Theta}_{lm} &\triangleq \frac{1}{n-1}\sum_{i=1}^{n}((X_{il} - \hat{\mu}_l) + (X_{im} - \hat{\mu}_m))^2 \\ &= \frac{1}{n-1}\sum_{i=1}^{n}(X_{il} - \hat{\mu}_l)^2 + \frac{1}{n-1}\sum_{i=1}^{n}(X_{im} - \hat{\mu}_m)^2 \\ &\quad + \frac{2}{n-1}\sum_{i=1}^{n}(X_{il} - \hat{\mu}_l)(X_{im} - \hat{\mu}_m).\end{aligned} \tag{5.2–9}$$

In Equation 5.2–9, the first term is unbiased for σ_l^2, the second is unbiased for σ_m^2, and the sum of all three is unbiased by Equation 5.2–8 for $\sigma_l^2 + \sigma_m^2 + 2K_{lm}$. We therefore conclude that

$$S_{lm} \triangleq \frac{1}{n-1}\sum_{i=1}^{n}(X_{il} - \hat{\mu}_l)(X_{im} - \hat{\mu}_m) \tag{5.2–10}$$

is unbiased for $K_{lm}(= K_{ml})$. Hence every term of $\hat{\boldsymbol{\Theta}}$ in Equation 5.2–4 is unbiased for every corresponding term in $\mathbf{K}$. In this sense $\hat{\boldsymbol{\Theta}} \triangleq \hat{\mathbf{K}}$ is unbiased for $\mathbf{K}$.

By resorting again to the univariate case and assuming that all moments up to the fourth order exist, we can show consistency for every term in the estimator for $\mathbf{K}$, that is, equation 5.2–4. Hence without specifying the distribution, Equations 5.2–1 and 5.2–4 are unbiased and consistent estimators for $\boldsymbol{\mu}$ and $\mathbf{K}$ respectively.

When $\mathbf{X}$ is normal, $\hat{\mathbf{K}}$ obeys a complex probability law called the Wishart distribution (see Fukunaga, Reference [5–2, p. 126]). More generally, when the pdf of $\mathbf{X}$ is known, one can use another method of estimating such parameters as μ, σ^2, $\boldsymbol{\mu}$ and $\mathbf{K}$ called the method of *maximum likelihood.* Maximum likelihood estimators have several desirable properties as estimators and by some measures can be regarded as "best" estimators. However, the reader should be cautioned that *best* implies a criterion of performance and what may be best by one criterion may be far from best by another. For instance, the sample mean estimator $\hat{\Theta}$ given in Equation 5.1–8 may be best in the sense of being unbiased and consistent but is not best according to a criterion called minimum mean-square-error (MMSE) (see Definition 5.1–6). We illustrate with an Example.

Example 5.2–1: (Kendall and Stuart, [5–3], p. 21.) Consider the sample mean estimator from Equation 5.1–3, that is,

$$\hat{\mu} = \frac{1}{n}\sum_{i=1}^{n} X_i.$$

What constant a in $\hat{\Theta} \triangleq a\hat{\mu}$ will generate the MMSE estimator of μ? Recall the X_i $i = 1, \ldots, n$ are i.i.d. r.v.'s with $E[X_i] = \mu$ and $\text{Var}[X_i] = \sigma^2$.

Solution: We are seeking the value of a such that

$$E[a\hat{\mu} - \mu]^2 \tag{5.2–11}$$

is a minimum. Clearly $\hat{\mu}$ is unbiased for μ, and it seems hard to believe that there may exist an $\hat{\Theta}$ with $a \neq 1$ that—though yielding a biased estimator—gives a lower MSE than $\hat{\Theta} = \mu$.

For *any* estimator $\hat{\Theta}$, the mean square error in estimating μ is

$$\begin{aligned} E[(\hat{\Theta} - \mu)^2] &= E[\{(\hat{\Theta} - E[\hat{\Theta}]) + (E[\hat{\Theta}] - \mu)\}^2] \\ &= \text{Var}[\hat{\Theta}] + (E[\hat{\Theta}] - \mu)^2. \end{aligned} \tag{5.2–12}$$

If $\hat{\Theta}$ is unbiased then the last term, which is the square of the bias (Definition 5.1–2), is zero. For the case at hand, $\hat{\Theta} = a\hat{\mu}$; thus

$$\begin{aligned} E[(\hat{\Theta} - \mu)^2] &= a^2\,\text{Var}[\hat{\mu}] + (a\mu - \mu)^2 \\ &= \frac{a^2\sigma^2}{n} + (a - 1)^2\mu^2. \end{aligned} \tag{5.2–13}$$

To find the MMSE estimator, we differentiate Equation 5.2–13 with respect to a and set to zero. This yields the optimum value of $a = a_0$, that is,

$$a_0 = \frac{\mu^2}{(\sigma^2/n) + \mu^2} = \frac{n}{(\sigma^2/\mu^2) + n}, \tag{5.2–14}$$

which is not unity. As $n \to \infty$, $a_0 \to 1$ and the estimator becomes unbiased but for any finite n, $a_0 \neq 1$. Thus the MMSE estimator is biased.

The question as to why not use the MMSE approach all the time is a valid one. An examination of Equation 5.2–14 shows that we need to know σ^2 and μ^2 or at least their ratio. But these quantities are not generally known.

5.3 MAXIMUM LIKELIHOOD PARAMETER ESTIMATION

For the sake of convenience we deal with a univariate r.v. X whose pdf $f_X(x)$ involves a parameter θ that we wish to estimate. We assume that θ is constant but unknown.† To make the point more explicitly we write for the pdf of X: $f_X(x; \theta)$. Let X_i, $i = 1, \ldots, n$ be n i.i.d. observations on X drawn from $f_X(x; \theta)$. The joint pdf of $X_1, X_2, \ldots, X_n$ is

$$\begin{aligned} f_{X_1 \ldots X_n}(x_1, \ldots, x_n; \theta) &= f_X(x_1; \theta) f_X(x_2; \theta) \ldots f_X(x_n; \theta) \\ &\triangleq L(x_1, \ldots, x_n; \theta), \end{aligned} \tag{5.3–1}$$

where $L(x_1, \ldots, x_n; \theta)$ is called the *likelihood function* (LF). The *maximum likelihood estimator* (MLE) $\hat{\Theta}_0$ of θ, is that value of $\hat{\Theta}_0$ that maximizes the LF, that is,

$$L(x_1, \ldots, x_n; \hat{\Theta}_0) \geq L(x_1, \ldots, x_n; \hat{\Theta}), \tag{5.3–2}$$

where $\hat{\Theta}$ is any other estimator. Any particular value of the maximum likelihood estimator $\hat{\Theta}_0$, say $\hat{\theta}_0$, is interpreted as the parameter value most likely to be responsible for the observed data values $x_1, \ldots, x_n$. In the literature on statistics (for example, see Kendall and Stuart [5–3]) it is shown that MLE's have numerous desirable qualities, not the least of which is consistency. In the normal case they are also minimum variance unbiased, which allows ML estimators some claim to general optimality. However, in order to use the ML method *we need to know the form of the pdf* (*or* PMF) of X. Needless to say, this is not always known.

Often we can compute $\hat{\Theta}_0$ by differentiating the LF and setting the derivative equal to zero; other times we must use a different approach. We illustrate with two examples.

Example 5.3–1: Assume X: $N(\mu, \sigma^2)$ where σ is known. Compute the MLE of the mean μ.

Solution: The LF for n realizations of X is

$$L(\mathbf{x}; \mu) = \left(\frac{1}{\sqrt{2\pi\sigma^2}}\right)^n \exp\left(-\frac{1}{2\sigma^2} \sum_{i=1}^{n} (x_i - \mu)^2\right). \tag{5.3–3}$$

Since the log function is monotonic, the maximum of $L(\mathbf{x}; \mu)$ is also that of $\log L(\mathbf{x}; \mu)$. Hence

$$\log L(\mathbf{x}; \mu) = -\frac{n}{2} \log(2\pi\sigma^2) - \frac{1}{2\sigma^2} \sum_{i=1}^{n} (x_i - \mu)^2$$

and set

$$\frac{\partial \log L(\mathbf{x}; \mu)}{\partial \mu} = 0.$$

This yields

$$\sum_{i=1}^{n} (x_i - \mu) = 0.$$

† Another approach called Bayesian estimation assumes that θ is a random variable. (See Fukunaga [5–2], p. 132.)

Thus the value of μ, say μ_0 that maximizes $L(\mathbf{x}; \mu)$ is

$$\mu_0 = \frac{1}{n} \sum_{i=1}^{n} x_i$$

This implies that our MLE of μ should be

$$\hat{\Theta}_0 = \frac{1}{n} \sum_{i=1}^{n} X_i \triangleq \hat{\mu} \qquad \text{(see Equation 5.1–8).} \tag{5.3–4}$$

Thus we see that in the normal case, the MLE for μ can be computed by differentiation and that it turns out to be the sample mean.

Example 5.3–2: Assume X is uniform in $(0, \theta)$, that is,

$$f_X(x) = \begin{cases} \dfrac{1}{\theta}, & 0 < x \leq \theta \\ 0, & x > \theta, \end{cases}$$

and we wish to compute the MLE for θ. Let a particular realization of the n observations $X_1, \ldots, X_n$ be $\mathbf{x} = (x_1, \ldots, x_n)^T$ and let $x_m \triangleq \max(x_1, \ldots, x_n)$. The likelihood function is

$$L(\mathbf{x}; \theta) = \begin{cases} \dfrac{1}{\theta^n}, & x_m \leq \theta \\ 0, & \text{otherwise.} \end{cases}$$

Clearly to maximize L we must make the estimate $\hat{\theta}$ as small as possible. But $\hat{\theta}$ cannot be smaller than x_m. Hence $\hat{\theta}_0$ is x_m and the MLE is

$$\hat{\Theta}_0 = \max(X_1, \ldots, X_n). \tag{5.3–5}$$

The probability distribution function (PDF) of $\hat{\Theta}_0$ for $n = 2$ is

$$F_{\hat{\Theta}_0}(\alpha) = F_{X_1}(\alpha) F_{X_2}(\alpha) = F_X^2(\alpha). \tag{5.3–6}$$

We leave the computation of the PDF and pdf of $\hat{\Theta}_0$ for arbitrary n as an exercise for the reader.

Maximum Likelihood Estimation in the Multidimensional Normal pdf. The most important multidimensional pdf is the normal, and the ML estimation of its parameters is given below. Let $\mathbf{X} \triangleq (X_1, \ldots, X_p)^T$ be a p-component Gaussian random vector with covariance

$$\mathbf{K} \triangleq \begin{bmatrix} K_{11} & \ldots & K_{1p} \\ \vdots & & \vdots \\ K_{p1} & & K_{pp} \end{bmatrix}, \tag{5.3–7}$$

where $K_{ij} = K_{ji} = E[(X_i - \mu_i)(X_j - \mu_j)]$ and $K_{ii} \triangleq \sigma_i^2 = \text{Var}(X_i)$ $i, j = 1, \ldots, p$. Let the mean vector $\boldsymbol{\mu}$ be

$$\boldsymbol{\mu} = (\mu_1, \ldots, \mu_p)^T. \tag{5.3–8}$$

Let $\mathbf{X}_1, \ldots, \mathbf{X}_n$ be n i.i.d. observation on $\mathbf{X}$. Then, using the notation of Section 5.2 we obtain

$$\hat{\Theta}_j^{(1)} = \frac{1}{n} \sum_{i=1}^{n} X_{ij} \qquad j = 1, \ldots, p$$

$$\triangleq \hat{\mu}_j \text{ (the sample mean),} \tag{5.3–9}$$

which is the MLE of μ_j;

$$\hat{\Theta}_j^{(2)} = \frac{1}{n} \sum_{i=1}^{n} (X_{ij} - \hat{\mu}_j)^2 \qquad j = 1, \ldots, p$$
$$\triangleq S'_{jj}, \tag{5.3–10}$$

which is the MLE of σ_j^2; and

$$\hat{\Theta}_{lm} = \frac{1}{n} \sum_{i=1}^{n} (X_{il} - \hat{\mu}_l)(X_{im} - \hat{\mu}_m) \qquad l = 1, \ldots, p \qquad m = 1, \ldots, p$$
$$\triangleq S'_{lm} \tag{5.3–11}$$

is the MLE of K_{lm} in $\mathbf{K}$. Note that Equation 5.3–9 is identical with Equation 5.2–2, which is the unbiased, consistent estimator of $\boldsymbol{\mu}$ based on non-parametric methods.† Also observe how similar Equations 5.3–10 and 5.3–11 are to the nonparametric, unbiased, consistent estimators of σ_j^2 and K_{lm} given in Equations 5.2–5 and 5.2–10, respectively. The MLE estimators for σ_j^2 and K_{lm} are slightly biased since

$$E[S'_{jj}] = \frac{n-1}{n} \sigma_j^2$$

and

$$E[S'_{lm}] = \frac{n-1}{n} K_{lm}.$$

To obtain the results given in Equations 5.3–9 to 5.3–11 we write the likelihood function for $\mathbf{X}_1, \ldots, \mathbf{X}_n$ and differentiate the log-likelihood with respect to $\boldsymbol{\mu}$ and $\mathbf{K} = \{K_{ij}\}$. The procedure requires some nontrivial matrix calculus but is quite straightforward, see for example, Reference 5–4.

Aside from the matrix calculus, the principal idea is the *simultaneous* ML *estimation* of several parameters. We illustrate how this is done with an example.

Example 5.3–3: Consider the normal pdf

$$f_X(x; \mu, \sigma^2) = \frac{1}{\sqrt{2\pi}\sigma} \exp\left(-\frac{1}{2\sigma^2}(x - \mu)^2\right) \qquad -\infty < x < \infty.$$

The log likelihood function is

$$\log L(x_1, \ldots, x_n; \mu, \sigma) \triangleq \log L = -\frac{n}{2} \log 2\pi - n \log \sigma - \frac{1}{2\sigma^2} \sum_{i=1}^{n} (x_i - \mu)^2.$$

Now set

$$\frac{\partial L}{\partial \mu} = 0 \qquad \frac{\partial L}{\partial \sigma} = 0$$

and obtain the simultaneous equations

$$\sum_{i=1}^{n} (x_i - \mu) = 0 \tag{5.3–12}$$

$$-\frac{n}{\sigma} + \frac{1}{\sigma^3} \sum_{i=1}^{n} (x_i - \mu)^2 = 0. \tag{5.3–13}$$

† That is, no *a priori* knowledge of the pdf was required.

From Equation 5.3–12 we infer that

$$\hat{\mu} = \frac{1}{n} \sum_{i=1}^{n} X_i. \tag{5.3–14}$$

From Equation 5.3–13 we infer that, using the result from Equation 5.3–12,

$$\hat{\sigma} = \frac{1}{n} \sum_{i=1}^{n} (X_i - \hat{\mu})^2. \tag{5.3–15}$$

5.4 LINEAR ESTIMATION OF VECTOR PARAMETERS

A great many measurement problems in the real world are described by the following model

$$y(t) = \int_T h(t, \tau)\theta(\tau)\, d\tau + n(t), \tag{5.4–1}$$

where $y(t)$ is the *observation* or *measurement*, T is the integration set, $\theta(\tau)$ is the unknown *parameter* function, and $h(t, \tau)$ is a function that is characteristic of the system and links the parameter function to the measurement but is itself independent of $\theta(\tau)$, and $n(t)$ is the inevitable error in the measurement due to noise. For computational purposes Equation 5.4–1 must be reduced to its discrete form

$$\mathbf{Y} = \mathbf{H}\boldsymbol{\theta} + \mathbf{N}, \tag{5.4–2}$$

where $\mathbf{Y}$ is an $n \times 1$ vector of observations, $\mathbf{H}$ is an $n \times k$ matrix $(n > k)$, $\boldsymbol{\theta}$ is a $k \times 1$ parameter vector, and $\mathbf{N}$ is an $n \times 1$ random vector whose components N_i, $i = 1, \ldots, n$ are the errors or noise associated with the ith observation Y_i. We shall assume without loss of generality that $E[\mathbf{N}] = \mathbf{0}$.†

Equation 5.4–2 is known as the *linear model.* We now ask the following question: How do we extract a "good" estimate $\hat{\boldsymbol{\Theta}}$ of $\boldsymbol{\theta}$ from the observed values of $\mathbf{Y}$ if we restrict our estimator $\hat{\boldsymbol{\Theta}}$ be a linear function of $\mathbf{Y}$? By a linear function we mean

$$\hat{\boldsymbol{\Theta}} = \mathbf{B}\mathbf{Y}, \tag{5.4–3}$$

where $\mathbf{B}$, which *does not* depend on $\mathbf{Y}$, is to be determined. The problem posed here is of great practical significance. It is one of the most fundamental problems in parameter estimation theory and covered in great detail in numerous books, for example, Kendall and Stuart [5–3] and Lewis and Odell [5–5]. It also is an immediate application of the probability theory of random vectors and is fundamental for understanding numerous topics in the second half of this book, especially Chapter 10.

Before computing the matrix $\mathbf{B}$ in Equation 5.4–3 for various cases, we must first develop some results from matrix calculus.

† The symbol $\mathbf{0}$ here stands for the zero vector, that is, the vector whose components are all zero.

Derivative of a scalar with respect to a vector. Let $q(\mathbf{x})$ be a scalar function of the vector $\mathbf{x} = (x_1, \ldots, x_n)^T$. Then

$$\frac{dq(\mathbf{x})}{d\mathbf{x}} \triangleq \left(\frac{\partial q}{\partial x_1}, \ldots, \frac{\partial q}{\partial x_n}\right)^T. \qquad (5.4\text{–}4)$$

Thus the derivate of $q(\mathbf{x})$ with respect to $\mathbf{x}$ is a *column vector* whose ith component is the partial derivative of $q(\mathbf{x})$ with respect to x_i.

Derivative of quadratic forms. Let $\mathbf{A}$ be a real-symmetric $n \times n$ matrix and let $\mathbf{x}$ be an arbitrary n-vector. Then the derivative of the quadratic form

$$q(\mathbf{x}) \triangleq \mathbf{x}^T\mathbf{A}\mathbf{x}$$

with respect to $\mathbf{x}$ is

$$\frac{dq(\mathbf{x})}{d\mathbf{x}} = 2\mathbf{A}\mathbf{x}. \qquad (5.4\text{–}5)$$

The proof of Equation 5.4–5 is obtained by writing

$$\begin{aligned} q(\mathbf{x}) &= \sum_{i=1}^{n} \sum_{j=1}^{n} x_i a_{ij} x_j \\ &= \sum_{i=1}^{n} x_i^2 a_{ii} + \sum_{i \neq j}\sum a_{ij} x_i x_j. \end{aligned}$$

Hence

$$\begin{aligned} \frac{\partial q(\mathbf{x})}{\partial x_k} &= 2x_k a_{kk} + 2\sum_{i \neq k} a_{ki} x_i \\ &= 2\sum_{i=1}^{n} a_{ki} x_i \end{aligned}$$

or

$$\frac{dq(\mathbf{x})}{d\mathbf{x}} = \mathbf{A}\mathbf{x}. \qquad (5.4\text{–}6)$$

Derivative of scalar products.† Let $\mathbf{a}$ and $\mathbf{x}$ be two n-vectors. Then with $y = \mathbf{a}^T\mathbf{x}$, we obtain

$$\frac{dy}{d\mathbf{x}} = \mathbf{a}. \qquad (5.4\text{–}7)$$

Let $\mathbf{x}$, $\mathbf{y}$, and $\mathbf{A}$ be two n-vectors and an $n \times n$ matrix respectively. Then with $q \triangleq \mathbf{y}^T\mathbf{A}\mathbf{x}$,

$$\frac{\partial q}{\partial \mathbf{x}} = \mathbf{A}^T\mathbf{y}. \qquad (5.4\text{–}8)$$

We return now to Equation 5.4–2:

$$\mathbf{Y} = \mathbf{H}\boldsymbol{\theta} + \mathbf{N}$$

† The *scalar or inner product* of two n-vectors, say $\mathbf{a}$ and $\mathbf{b}$, is $\mathbf{a}^T\mathbf{b}(=\mathbf{b}^T\mathbf{a})$. For the relation between scalar product and norm see Section 4.3. For the definitions of scalar product and norm of square-integrable functions see Section 3.4.

and assume that (recall $E[\mathbf{N}] = \mathbf{0}$)

$$\mathbf{K} \triangleq E[\mathbf{N}\mathbf{N}^T] = \sigma^2\mathbf{I} \tag{5.4–9}$$

where $\mathbf{I}$ is the identity matrix. Equation 5.4–9 is equivalent to stating that the measurement errors N_i, that is, $i = 1, \ldots, n$ are uncorrelated, and their variances are the same and equal to σ^2. This situation is sometimes called *white* noise.

We have not yet defined what we mean by a *good* or *best* estimator of θ. A reasonable choice is to find a $\hat{\mathbf{\Theta}}$ that *minimizes* the sum squares S defined by

$$S \triangleq (\mathbf{Y} - \mathbf{H}\hat{\mathbf{\Theta}})^T(\mathbf{Y} - \mathbf{H}\hat{\mathbf{\Theta}}) \triangleq \|\mathbf{Y} - \mathbf{H}\hat{\mathbf{\Theta}}\|^2. \tag{5.4–10}$$

Note that by finding $\hat{\mathbf{\Theta}}$ that best fits the measurement $\mathbf{Y}$ in the sense of minimizing $\|\mathbf{Y} - \mathbf{H}\hat{\mathbf{\Theta}}\|^2$, we are realizing what is commonly called a *least-squares* fit to the data. For this reason finding $\hat{\mathbf{\Theta}}$ that minimizes S in Equation 5.4–10 is called the least-squares (LS) method. To find the minimum of S with respect to $\hat{\mathbf{\Theta}}$, write

$$S = \mathbf{Y}^T\mathbf{Y} + \hat{\mathbf{\Theta}}^T\mathbf{H}^T\mathbf{H}\hat{\mathbf{\Theta}} - \hat{\mathbf{\Theta}}^T\mathbf{H}^T\mathbf{Y} - \mathbf{Y}^T\mathbf{H}\hat{\mathbf{\Theta}}$$

and compute

$$\frac{\partial S}{\partial \hat{\mathbf{\Theta}}} = 0 = 2[\mathbf{H}^T\mathbf{H}]\hat{\mathbf{\Theta}} - 2\mathbf{H}^T\mathbf{Y},$$

whence (assuming $\mathbf{H}^T\mathbf{H}$ has an inverse)

$$\hat{\mathbf{\Theta}}_{LS} = (\mathbf{H}^T\mathbf{H})^{-1}\mathbf{H}^T\mathbf{Y}. \tag{5.4–11}$$

Comparing our result with Equation 5.4–3 we see that $\mathbf{B}$ in Equation 5.4–3 is given by $\mathbf{B}_0 \triangleq (\mathbf{H}^T\mathbf{H})^{-1}\mathbf{H}^T$ in the LS method. Equation 5.4–11 is the LS estimator of $\boldsymbol{\theta}$ based on the measurement $\mathbf{Y}$.

The astute reader will have noticed that we never involved the fact that $\mathbf{K} = \sigma^2\mathbf{I}$. Indeed, in arriving at Equation 5.4–11 we essentially treated $\mathbf{Y}$ as deterministic and merely obtained $\hat{\mathbf{\Theta}}_{LS}$ as the *generalized inverse* (see Lewis and Odell [5–5, p. 6]) of the system of equations $\mathbf{Y} = \mathbf{H}\boldsymbol{\theta}$. As it stands, the estimator $\hat{\mathbf{\Theta}}_{LS}$ given in Equation 5.4–11 has no claim to being optimum. However, when the covariance of the noise $\mathbf{N}$ is as in Equation 5.4–9 then $\hat{\mathbf{\Theta}}_{LS}$ does indeed have optimal properties in an important sense. However, before discussing this point further in Section 5.5, we give some examples.

Example 5.4–1: We are given the following data

$$6.2 = 3\theta + n_1$$
$$7.8 = 4\theta + n_2$$
$$2.2 = \theta + n_3.$$

Find the LS estimate of θ.

Solution: The data can be put in the form

$$\mathbf{y} = \mathbf{H}\theta + \mathbf{n},$$

where $\mathbf{y} = (6.2, 7.8, 2.2)^T$ is a realization of $\mathbf{Y}$, $\mathbf{H} = (3, 4, 1)^T$ and $\mathbf{n} = (n_1, n_2, n_3)^T$ is a realization of $\mathbf{N}$. Hence $\mathbf{H}^T\mathbf{H} = \sum H_i^2 = 26$ and $\mathbf{H}^T\mathbf{y} = \sum_{i=1}^{3} H_i y_i = 52$. Thus

$$\hat{\theta}_{LS} = (\mathbf{H}^T\mathbf{H})^{-1}\mathbf{H}^T\mathbf{y} = \frac{\sum_{i=1}^{3} H_i y_i}{\sum_{i=1}^{3} H_i^2} = \frac{52}{26} = 2.$$

Example 5.4–2: (Reference 5–3, p. 77.) Let $\boldsymbol{\theta} = (\theta_1, \theta_2)^T$ be a two-component parameter vector to be estimated, and let $\mathbf{H}$ be a $n \times 2$ matrix of coefficients partitioned into column vectors as $\mathbf{H} = (\mathbf{H}_1\mathbf{H}_2)$ where $\mathbf{H}_i$, $i = 1, 2$ is an n-vector. Then with the n-vector $\mathbf{Y}$ representing the observation data, the linear model assumes the form

$$\mathbf{Y} = (\mathbf{H}_1\mathbf{H}_2)\boldsymbol{\theta} + \mathbf{N}$$

and the LS estimator of $\boldsymbol{\theta}$ is

$$\hat{\boldsymbol{\Theta}}_{LS} = \begin{bmatrix} \mathbf{H}_1^T\mathbf{H}_1 & \mathbf{H}_1^T\mathbf{H}_2 \\ \mathbf{H}_2^T\mathbf{H}_1 & \mathbf{H}_2^T\mathbf{H}_2 \end{bmatrix}^{-1} \begin{bmatrix} \mathbf{H}_1^T\mathbf{Y} \\ \mathbf{H}_2^T\mathbf{Y} \end{bmatrix}.$$

5.5 OPTIMAL PROPERTIES OF LEAST-SQUARES ESTIMATORS; THE GAUSS-MARKOV THEOREM

The LS estimator has a number of properties that account for its widespread use in estimation problems. It is simple to construct, does not require knowledge of the pdf of $\mathbf{N}$, is unbiased, and has a minimum variance property, which we discuss below.

Unbiasedness of the LS Estimator. To show that $\hat{\boldsymbol{\Theta}}_{LS}$ in Equation 5.4–11 is unbiased we must show that $E[\hat{\boldsymbol{\Theta}}_{LS}] = \boldsymbol{\theta}$. We write

$$\begin{aligned} \hat{\boldsymbol{\Theta}}_{LS} &= (\mathbf{H}^T\mathbf{H})^{-1}\mathbf{H}^T\mathbf{Y} \\ &= (\mathbf{H}^T\mathbf{H})^{-1}\mathbf{H}^T(\mathbf{H}\boldsymbol{\theta} + \mathbf{N}) \\ &= (\mathbf{H}^T\mathbf{H})^{-1}\mathbf{H}^T\mathbf{H}\boldsymbol{\theta} + (\mathbf{H}^T\mathbf{H})^{-1}\mathbf{H}^T\mathbf{N} \\ &= \boldsymbol{\theta} + \mathbf{B}_0\mathbf{N} \quad (\mathbf{B}_0 \triangleq (\mathbf{H}^T\mathbf{H})^{-1}\mathbf{H}^T). \end{aligned} \tag{5.5–1}$$

Hence

$$E[\hat{\boldsymbol{\Theta}}_{LS}] = \boldsymbol{\theta} + \mathbf{B}_0E[\mathbf{N}] = \boldsymbol{\theta}$$

since, by assumption, $E[\mathbf{N}] = \mathbf{0}$.

The covariance of $\hat{\boldsymbol{\Theta}}_{LS}$ is easily computed from Equation 5.5–1. Thus

$$\begin{aligned} E[(\hat{\boldsymbol{\Theta}}_{LS} - \boldsymbol{\theta})(\boldsymbol{\Theta}_{LS} - \boldsymbol{\theta})^T] &= E[\mathbf{B}_0\mathbf{N}\mathbf{N}^T\mathbf{B}_0^T] \\ &= \mathbf{B}_0E[\mathbf{N}\mathbf{N}^T]\mathbf{B}_0^T \\ &= \sigma^2\mathbf{B}_0\mathbf{B}_0^T \quad (\text{since } E[\mathbf{N}\mathbf{N}^T] = \sigma^2\mathbf{I}) \\ &= \sigma^2(\mathbf{H}^T\mathbf{H})^{-1}. \end{aligned} \tag{5.5–2}$$

Minimum Variance Property of $\hat{\boldsymbol{\Theta}}_{LS}$. We now demonstrate one of the most important properties of $\hat{\boldsymbol{\Theta}}_{LS}$, namely, its minimum variance property.

Because of this property, the LS estimator is sometimes considered a best estimator and given the acronym BLUE (Best Linear Unbiased Estimator).

To begin with, consider any linear unbiased estimator $\hat{\boldsymbol{\Theta}} = (\hat{\Theta}_1, \ldots, \hat{\Theta}_n)^T$ of $\boldsymbol{\theta}$ of the form $\hat{\boldsymbol{\Theta}} = \mathbf{BY}$. If $\hat{\boldsymbol{\Theta}}$ is to be unbiased then

$$\begin{aligned} E[\hat{\boldsymbol{\Theta}}] &= E[\mathbf{BY}] = E[\mathbf{B}(\mathbf{H}\boldsymbol{\theta} + \mathbf{N})] \\ &= \mathbf{BH}\boldsymbol{\theta} \\ &= \boldsymbol{\theta} \quad \text{(by unbiasedness of } \hat{\boldsymbol{\Theta}}\text{).} \end{aligned}$$

The result $\mathbf{BH}\boldsymbol{\theta} = \boldsymbol{\theta}$ implies that

$$\mathbf{BH} = \mathbf{I}. \tag{5.5-3}$$

Equation 5.5–3 is a necessary and sufficient condition for $\mathbf{BY}$ to be an unbiased estimator of $\boldsymbol{\theta}$. Now consider the covariance matrix $\mathbf{K}_{\hat{\boldsymbol{\Theta}}}$ of the error $\hat{\boldsymbol{\Theta}} - \boldsymbol{\theta}$:

$$\begin{aligned} \mathbf{K}_{\hat{\boldsymbol{\Theta}}} &= E[(\hat{\boldsymbol{\Theta}} - \boldsymbol{\theta})(\hat{\boldsymbol{\Theta}} - \boldsymbol{\theta})^T] \\ &= \sigma^2 \mathbf{BB}^T. \end{aligned} \tag{5.5-4}$$

To arrive at Equation 5.5–4 we used the facts that (1) $\mathbf{BH} = \mathbf{I}$, (2) that $\hat{\boldsymbol{\Theta}} = \boldsymbol{\theta} + \mathbf{BN}$ and (3) that $\mathbf{K} \triangleq E[\mathbf{NN}^T] = \sigma^2 \mathbf{I}$. We now ask what matrix $\mathbf{B}$ will minimize the diagonal elements of Equation 5.5–4, which are the variance $\sigma^2_{\hat{\Theta}_i}$ of the estimators $\hat{\Theta}_i$, $i = 1, \ldots, n$. We can solve this problem by using the following identity:

$$\mathbf{BB}^T = \mathbf{B}_0\mathbf{B}_0^T + (\mathbf{B} - \mathbf{B}_0)(\mathbf{B} - \mathbf{B}_0)^T. \tag{5.5-5}$$

To prove Equation 5.5–5 we need only substitute $\mathbf{B}_0 = (\mathbf{H}^T\mathbf{H})^{-1}\mathbf{H}^T$ and use the fact that $\mathbf{BH} = \mathbf{I}$. Indeed, observing that

$$\mathbf{B}_0\mathbf{B}_0^T = (\mathbf{H}^T\mathbf{H})^{-1}$$

and

$$\mathbf{BB}_0^T = (\mathbf{H}^T\mathbf{H})^{-1} = \mathbf{B}_0\mathbf{B}^T$$

and expanding Equation 5.5–5 to

$$\mathbf{BB}^T = 2\mathbf{B}_0\mathbf{B}_0^T + \mathbf{BB}^T - \mathbf{B}_0\mathbf{B}^T - \mathbf{BB}_0^T \tag{5.5-6}$$

enables us to demonstrate the identity.

Finding the minimum variance estimator is readily accomplished once we realize that for any real matrix $\mathbf{A}$ the diagonal terms of $\mathbf{AA}^T$ are always sums of squares and hence nonnegative. In Equation 5.5–5 the diagonal terms of $\mathbf{BB}^T$ are minimized when the diagonal terms of $(\mathbf{B} - \mathbf{B}_0)(\mathbf{B} - \mathbf{B}_0)^T$ are minimized. But the smallest that the latter can get is zero since they are sums of squares. Thus $\mathbf{BB}^T$ has strictly minimum diagonal elements when $\operatorname{tr}[(\mathbf{B} - \mathbf{B}_0)(\mathbf{B} - \mathbf{B}_0)^T] = 0$, which is achieved only when $\mathbf{B} = \mathbf{B}_0$. Hence we have shown that the LS estimator

$$\hat{\boldsymbol{\Theta}}_{\text{LS}} = \mathbf{B}_0\mathbf{Y} \quad (\mathbf{B}_0 \triangleq (\mathbf{H}^T\mathbf{H})^{-1}\mathbf{H}^T) \tag{5.5-7}$$

is also the *minimum variance*, unbiased linear estimator of θ. Henceforth such estimators will have the subscript 0 rather than LS to emphasize their optimum, that is, minimum-variance property. The fact that the LS estimator is minimum variance when $\mathbf{K} = \sigma^2\mathbf{I}$ is a special case of the Gauss-Markov theorem. A slightly more general form of this special case is given below.

Theorem 5.5–1. Consider the model

$$\mathbf{Y} = \mathbf{H}\boldsymbol{\theta} + \mathbf{N},$$

where $E[\mathbf{N}] = \mathbf{0}$ $E[\mathbf{N}\mathbf{N}^T] \triangleq \mathbf{K} = \sigma^2\mathbf{I}$. Suppose we want to estimate a linear function of $\boldsymbol{\theta}$, say $\boldsymbol{\phi} = \mathbf{D}\boldsymbol{\theta}$ where $\mathbf{D}$ is a matrix of known coefficients. Then the minimum-variance, unbiased, linear estimator $\hat{\boldsymbol{\Phi}}_0$ of $\boldsymbol{\phi}$ is given by

$$\hat{\boldsymbol{\Phi}}_0 = \mathbf{D}[\mathbf{H}^T\mathbf{H}]^{-1}\mathbf{H}^T\mathbf{Y}. \tag{5.5–8}$$

Comment. We have already proven the validity of Equation 5.5–8 for $\mathbf{D} = \mathbf{I}$ in the discussion leading up to Equation 5.5–7. The demonstration of Equation 5.5–8 for $\mathbf{D}$ arbitrary is a straightforward extension of the case when $\mathbf{D} = \mathbf{I}$. We leave this as well as the demonstration that any linear unbiased estimator of $\boldsymbol{\phi}$ of the form $\boldsymbol{\Phi} = \mathbf{L}\mathbf{Y}$ where $\mathbf{L}$ is to be determined must satisfy

$$\mathbf{L}\mathbf{H} = \mathbf{D} \tag{5.5–9}$$

as exercises to the reader.

So far we have assumed that the noise $\mathbf{N}$ has covariance matrix $\mathbf{K} = \sigma^2\mathbf{I}$. What is the minimum variance, unbiased, linear estimator in the general case, that is, when $\mathbf{K}$ is an arbitrary covariance matrix? The answer follows.

Consider again the model

$$\mathbf{Y} = \mathbf{H}\boldsymbol{\theta} + \mathbf{N}, \tag{5.5–10}$$

where all quantities are defined as before including $E[\mathbf{N}] = \mathbf{0}$, except the covariance matrix $\mathbf{K}$ of $\mathbf{N}$ is arbitrary positive definite. Then from Equation 4.4–12 we know that there exists a factorization

$$\mathbf{K}^{-1} = \mathbf{C}\mathbf{C}^T \tag{5.5–11}$$

such that $\mathbf{C}^T\mathbf{K}\mathbf{C} = \mathbf{I}$ and we know how to compute $\mathbf{C}$. Now consider the transformation

$$\mathbf{W} = \mathbf{C}^T\mathbf{Y}. \tag{5.5–12}$$

Then premultiplying Equation 5.5–10 by $\mathbf{C}^T$ yields the matrix equation

$$\begin{aligned} \mathbf{W} &= \mathbf{C}^T\mathbf{H}\boldsymbol{\theta} + \mathbf{C}^T\mathbf{N} \\ &= \mathbf{H}'\boldsymbol{\theta} + \mathbf{N}', \end{aligned} \tag{5.5–13}$$

where $\mathbf{H}' \triangleq \mathbf{C}^T\mathbf{H}$ and $\mathbf{N}' = \mathbf{C}^T\mathbf{N}$. Now observe that

$$\begin{aligned} \mathbf{K}_{\mathbf{N}'} &= E[\mathbf{N}'\mathbf{N}'^T] \\ &= E[\mathbf{C}^T\mathbf{N}\mathbf{N}^T\mathbf{C}] \\ &= \mathbf{C}^T E[\mathbf{N}\mathbf{N}^T]\mathbf{C} \\ &= \mathbf{C}^T\mathbf{K}\mathbf{C} \\ &= \mathbf{I}, \text{ the identity matrix.} \end{aligned} \tag{5.5–14}$$

Hence from Equation 5.5–7, the LS, minimum-variance, unbiased, linear estimator of θ is given by

$$\begin{aligned}\hat{\mathbf{\Theta}}_0 &= (\mathbf{H}'^T\mathbf{H}')^{-1}\mathbf{H}'^T\mathbf{W} \\ &= (\mathbf{H}^T\mathbf{C}\mathbf{C}^T\mathbf{H})^{-1}\mathbf{H}^T\mathbf{C}\mathbf{C}^T\mathbf{Y} \\ &= (\mathbf{H}^T\mathbf{K}^{-1}\mathbf{H})^{-1}\mathbf{H}^T\mathbf{K}^{-1}\mathbf{Y}. \end{aligned} \tag{5.5–15}$$

Equation 5.5–15 is what is generally taken to be as the statement of the celebrated Gauss-Markov theorem. To show that $\hat{\mathbf{\Theta}}_0$ is unbiased we write

$$\begin{aligned}E[\hat{\mathbf{\Theta}}_0] &= E[(\mathbf{H}^T\mathbf{K}^{-1}\mathbf{H})^{-1}\mathbf{H}^T\mathbf{K}^{-1}(\mathbf{H}\boldsymbol{\theta} + \mathbf{N})] \\ &= \boldsymbol{\theta}.\end{aligned}$$

The covariance matrix of $\hat{\mathbf{\Theta}}_0$ is

$$\begin{aligned}E[(\hat{\mathbf{\Theta}}_0 - \boldsymbol{\theta})(\hat{\mathbf{\Theta}}_0 - \boldsymbol{\theta})^T] &= (\mathbf{H}^T\mathbf{K}^{-1}\mathbf{H})^{-1}\mathbf{H}^T\mathbf{K}^{-1}\mathbf{K} \\ &\quad \times \mathbf{K}^{-1}\mathbf{H}(\mathbf{H}^T\mathbf{K}^{-1}\mathbf{H})^{-1} \\ &= (\mathbf{H}^T\mathbf{K}^{-1}\mathbf{H})^{-1} \end{aligned} \tag{5.5–16}$$

and from the theory advanced earlier, its diagonal elements, that is, the variances $\sigma^2_{\hat{\Theta}_i}$ of the individual estimators $\hat{\Theta}_i$, $i = 1, \ldots, n$ in $\hat{\mathbf{\Theta}} \triangleq (\hat{\Theta}_1, \ldots, \hat{\Theta}_n)$ are minimal. Let us summarize this result in the following theorem.

Theorem 5.5–2. (The Gauss–Markov theorem.) Consider the model

$$\mathbf{Y} = \mathbf{H}\boldsymbol{\theta} + \mathbf{N},$$

where $\mathbf{Y}$ is an $n \times 1$ vector of observations, $\mathbf{H}$ is an $n \times k$ $(n > k)$ matrix of known coefficients, $\boldsymbol{\theta}$ is a $k \times 1$ vector of parameters, and $\mathbf{N}$ is an $n \times 1$ random vector consisting of "measurement" noise with

$$E[\mathbf{N}] = \mathbf{0}$$

and

$$E[\mathbf{N}\mathbf{N}^T] \triangleq \mathbf{K}.$$

Then the minimum-variance, unbiased, linear estimate of $\boldsymbol{\theta}$ is

$$\hat{\mathbf{\Theta}}_0 = (\mathbf{H}^T\mathbf{K}^{-1}\mathbf{H})^{-1}\mathbf{H}^T\mathbf{K}^{-1}\mathbf{Y}.$$

Extension. The minimum-variance, unbiased, linear estimate of a linear function of $\boldsymbol{\theta}$, that is, $\boldsymbol{\phi} = \mathbf{D}\boldsymbol{\theta}$ is given by

$$\hat{\mathbf{\Phi}}_0 = \mathbf{D}(\mathbf{H}^T\mathbf{K}^{-1}\mathbf{H})^{-1}\mathbf{H}^T\mathbf{K}^{-1}\mathbf{Y}. \tag{5.5–17}$$

The proof of Equation 5.5–17 follows directly from the facts that (1) any unbiased estimator $\hat{\mathbf{\Phi}}$ of $\boldsymbol{\phi}$ that is a linear function of $\mathbf{Y}$, that is, $\hat{\mathbf{\Phi}} = \mathbf{L}\mathbf{Y}$ must satisfy

$$\mathbf{L}\mathbf{H} = \mathbf{D}$$

for unbiasedness and (2) the following identify with $\mathbf{L}_0 \triangleq \mathbf{D}(\mathbf{H}^T\mathbf{K}_N^{-1}\mathbf{H})^{-1}\mathbf{H}^T\mathbf{K}_N^{-1}$ holds:

$$\mathbf{L}\mathbf{L}^T = \mathbf{L}_0\mathbf{L}_0^T + (\mathbf{L} - \mathbf{L}_0)(\mathbf{L} - \mathbf{L}_0)^T. \tag{5.5–18}$$

We leave the details as an exercise to the reader.

5.6 ESTIMATION OF RANDOM VARIABLES

In the previous sections we considered the estimation of a parameter θ (or $\boldsymbol{\theta}$) by observing random variables and forming a function of these random variables called an estimator. This estimator was then used to estimate the unknown parameter. We now consider a different problem, namely that of estimating one random variable with another or one random vector with another. We introduce the basic ideas in this section; subsequent development and application of these ideas will be taken up in Chapter 10.

To make clear what we mean by estimating one random variable with another, consider the following example: Let X_1 denote the barometric pressure (BP) and X_2 denote the rate of change of the BP at $t = 0$. Let Y denote the relative humidity one hour after measuring $\mathbf{X} = (X_1, X_2)^T$. Clearly, in this case, $\mathbf{X}$ and Y are dependent r.v.'s; then using $\mathbf{X}$ to estimate Y is a case of estimating one r.v. with another (actually $\mathbf{X}$ here is a random vector).†

In terms of the axiomatic theory we can describe the problem of estimating one r.v. with another in the following terms: Consider an underlying experiment $\mathscr{H}$ with probability space $\mathscr{P} \triangleq (\Omega, \mathscr{F}, P)$. Let X and Y be two r.v.'s defined on $\mathscr{P}$. For every $\zeta \in \Omega$, we generate the numbers $X(\zeta)$, $Y(\zeta)$. Suppose we can observe only $X(\zeta)$; how do we proceed to estimate $Y(\zeta)$ in some optimum fashion?

At this point the reader may wonder why observing $X(\zeta)$ doesn't uniquely specify $Y(\zeta)$. After all, since X and Y are deterministic functions, why can't we reason that $X(\zeta)$ specifies ζ specifies $Y(\zeta)$? The answer is that observing $X(\zeta)$ does not, in general, uniquely specify the outcome ζ and therefore does not uniquely specify $Y(\zeta)$. For example, let $\Omega = \{-2, -1, 0, 1, 2\}$, $X(\zeta) \triangleq \zeta^2$ and $Y(\zeta) \triangleq \zeta$. Then the observation $X(\zeta) = 4$ is associated with the outcomes $\zeta = 2$ *or* $\zeta = -2$ (of course, these may not be equally probable) and $Y(2) = 2$ while $Y(-2) = -2$. Hence all we can say about $Y(\zeta)$ after observing $X(\zeta)$ is that $Y(\zeta)$ has value 2 or -2. If all outcomes $\zeta \in \Omega$ are equally likely then the *a priori* probability $P[Y = 2]$ is $\frac{1}{5}$ and $P[Y = 2 \mid X = 4] = \frac{1}{2}$.

Assume, at first, for simplicity that we are constrained to estimate Y‡ by the linear function aX. Assume $E[X] = E[Y] = 0$. Note that a generalization of this problem has already been treated in Example 3.3–4. The mean square error (MSE) in estimating Y by aX is given by

$$\begin{aligned} \varepsilon &\triangleq E[(Y - aX)^2] \\ &= \sigma_Y^2 - 2a\,\mathrm{Cov}(X, Y) + a^2\sigma_X^2 \qquad (5.6\text{–}1) \end{aligned}$$

Setting the first derivative with respect to a equal to zero to find the

† The abbreviation r.v. can mean random variable or random vector without ambiguity.

‡ All random variables in this section are initially assumed to be real. However, later in the discussion we shall generalize to include complex r.v.'s.

minimum, we obtain

$$a_o = \frac{\text{Cov}(X, Y)}{\sigma_X^2}. \tag{5.6–2}$$

Equation 5.6–2 furnishes the value of a, which yields a minimum mean square error (MMSE) if we restrict ourselves to linear estimates of the form $Y = aX$. We note in passing that the inner product†

$$\begin{aligned}(Y - a_o X, X) &\triangleq E[(Y - a_o X)X] \\ &= 0,\end{aligned} \tag{5.6–3}$$

which suggests that the random error $\mathscr{E} \triangleq Y - a_o X$ is orthogonal to the datum. This interesting result will shortly be generalized.

Let us now remove the constraints of linear estimation and consider the more general problem of estimating Y with a (possibly nonlinear) function of X, that is, $g(X)$ so as to *minimize the* MSE. Thus we seek the function $g_o(X)$, which minimizes

$$\varepsilon = E[(Y - g(X))^2].$$

The answer to this problem is surprisingly easy, although its implementation is, except for the Gaussian case, generally very difficult. The result is given in Theorem 5.6–1.

Theorem 5.6–1. The MMSE estimator of Y based on observing the r.v. X is the conditional mean, that is, $g_o(X) \triangleq E[Y \mid X]$.‡

Proof. We write $g(X)$ as $g(X) = g_o(X) + \delta g$, that is, as a variation about the assumed optimal value. Then

$$\begin{aligned}\varepsilon &= E[(Y - E[Y \mid X] - \delta g)^2] \\ &= E[(Y - E[Y \mid X])^2] - 2E[(Y - E[Y \mid X])\,\delta g] \\ &\quad + E[(\delta g)^2].\end{aligned}$$

Now regarding the cross-term observe that

$$\begin{aligned}&E[(Y - E[Y \mid X])\delta g] \\ &\quad = E[Y\delta g] - E[E[Y \mid X]\delta g] \\ &\quad = E[Y\delta g] - E[Y\delta g] \\ &\quad = 0.\end{aligned}$$

■

We leave it as an exercise to the reader to show how line 3 was obtained from line 2. With the result just obtained we write

$$\begin{aligned}\varepsilon &= E[(Y - E[Y \mid X])^2] + E[(\delta g)^2] \\ &\geq E[(Y - E[Y \mid X])^2] \\ &= \varepsilon_{\min}.\end{aligned} \tag{5.6–4}$$

† For a discussion of inner products involving random variables see the last portion of Section 3.4.

‡ The conditional mean estimator is a function of X and hence is itself a random variable. See the discussion on the conditional mean as a r.v. in Section 3.2.

So to make ε a minimum, we set $\delta g = 0$, which implies that the MMSE estimator is $g_o(X) = E[Y \mid X]$.

The preceding theorem readily generalizes to random vectors† $\mathbf{X}$ and $\mathbf{Y}$, where we wish to minimize the individual component MSE. It follows readily from Theorem 5.6–1 (the actual demonstration is left as an exercise) that

$$\varepsilon_{\min}^{(i)} = \sum_{i=1}^{N} E[|Y_i - g_o^{(i)}(\mathbf{X})|^2], \tag{5.6–5}$$

where

$$g_o^{(i)}(\mathbf{X}) \triangleq E[Y_i \mid \mathbf{X}].$$

Thus the MMSE estimate of a random vector $\mathbf{Y}$ after observing a random vector $\mathbf{X}$ is also the conditional mean that, in vector notation, becomes

$$\mathbf{g}_o(\mathbf{X}) \triangleq E[\mathbf{Y} \mid \mathbf{X}]. \tag{5.6–6}$$

Before leaving this section we generalize the orthogonality property observed in Equation 5.6–3. We show that this is a property of the conditional mean.

Property 5.6–1. The MMSE *error vector* $\mathscr{E}$ of the vector $\mathbf{Y}$ given the random vector $\mathbf{X}$, that is,

$$\mathscr{E} \triangleq \mathbf{Y} - E[\mathbf{Y} \mid \mathbf{X}]$$

is orthogonal to any measurable function $h(\mathbf{X})$ of the data, that is,

$$E[(\mathbf{Y} - E[\mathbf{Y} \mid \mathbf{X}]h^*(\mathbf{X})] = \mathbf{0} \quad \text{(the asterisk denotes conjugation).} \tag{5.6–7}$$

Proof. Use the same method as in Theorem 5.6–1, which showed that the error was orthogonal to δg. Then generalize the result to the vector case.

Theorem 5.6–2. Let $\mathbf{X} = (X_1, \ldots, X_N)^T$ and Y be jointly Gaussian distributed with *zero means*. The MMSE estimate is the conditional mean given as

$$E[Y \mid \mathbf{X}] = \sum_{i=1}^{N} a_i X_i,$$

where the a_i's are chosen such that

$$E\left[\left(Y - \sum_{i=1}^{N} a_i X_i\right) X_k^*\right] = 0 \quad \text{for } k = 1, \ldots, N, \tag{5.6–8}$$

which is called the *orthogonality condition* and is written

$$\left(Y - \sum_{i=1}^{N} a_i X_i\right) \perp X_k \qquad k = 1, \ldots, N.$$

We see that this condition is a special case of Property 5.6–1.

† In anticipation of the material in Chapter 10, we let $\mathbf{X}$, $\mathbf{Y}$ be complex random vectors. A complex r.v. X is written $X = X_r + jX_i$; where X_r and X_i are real r.v.'s and represent the real and imaginary components of X, respectively, and $j = \sqrt{-1}$. The PDF of X is the joint PDF of X_r and X_i, that is, $F_{X_r X_i}(x_r, x_i)$.

Proof. The random variables

$$\left(Y - \sum_{i=1}^{N} a_i X_i\right), X_1, X_2, \ldots, X_N$$

are jointly Gaussian. Hence, since the first one is uncorrelated with all the rest, it is independent of them. Thus the error

$$Y - \sum_{i=1}^{N} a_i X_i$$

is independent of the random vector $\mathbf{X} = (X_1, \ldots, X_N)^T$, so

$$\begin{aligned} E\left[\left(Y - \sum_{i=1}^{N} a_i X_i\right)\middle| \mathbf{X}\right] &= E\left[Y - \sum_{i=1}^{N} a_i X_i\right] \\ &= E[Y] - \sum_{i=1}^{N} a_i E[X_i] \\ &= 0 \quad \text{since } E[Y] = E[X_i] = 0 \end{aligned}$$

But

$$\begin{aligned} 0 &= E\left[\left(Y - \sum_{i=1}^{N} a_i X_i\right)\middle| \mathbf{X}\right] \\ &= E[Y \mid \mathbf{X}] - \sum_{i=1}^{N} a_i E[X_i \mid \mathbf{X}] \\ &= E[Y \mid \mathbf{X}] - \sum_{i=1}^{N} a_i X_i. \end{aligned}$$

Hence

$$E[Y \mid \mathbf{X}] = \sum_{i=1}^{N} a_i X_i. \qquad (5.6\text{–}9)$$

■

Theorem 5.6–2 points out a great simplification of the jointly Gaussian case, that is, that the conditional mean is linear and easily determined with linear algebraic methods as the solution to the *orthogonality equations*, which can be put into matrix form as

$$(Y - \mathbf{a}^T \mathbf{X}) \perp \mathbf{X}$$

or

$$E[(Y - \mathbf{a}^T\mathbf{X})\mathbf{X}^\dagger] = \mathbf{0}^T \quad (\dagger \text{ denotes transpose conjugate}).$$

Hence the optimum value of $\mathbf{a}$, denoted by $\mathbf{a}_o$, is given by

$$\mathbf{a}_o^T = \mathbf{k}_{Y\mathbf{X}} \mathbf{K}_{\mathbf{XX}}^{-1}, \qquad (5.6\text{–}10)$$

where $\mathbf{k}_{Y\mathbf{X}} \triangleq E[Y\mathbf{X}^\dagger]$ and $\mathbf{K}_{\mathbf{XX}} \triangleq E[\mathbf{X}\mathbf{X}^\dagger]$.

If the means are not zero, the answer is slightly more complicated. If $\mathbf{X}$ and Y are jointly Gaussian with means $\boldsymbol{\mu}_\mathbf{X}$ and μ_Y respectively, we can define the zero-mean random variables

$$\begin{aligned} \mathbf{X}_c &\triangleq \mathbf{X} - \boldsymbol{\mu}_\mathbf{X} \\ Y_c &\triangleq Y - \mu_Y. \end{aligned}$$

Then Theorem 5.6–2 applies to them directly, and we can write

$$E[Y_c \mid \mathbf{X}_c] = \sum_{i=1}^{N} a_i X_{ci} = E[(Y - \mu_Y) \mid \mathbf{X}_c]. \tag{5.6–11}$$

But the conditional expectation is a linear operation so that

$$E[(Y - \mu_Y) \mid \mathbf{X}_c] = E[Y \mid \mathbf{X}_c] - \mu_Y. \tag{5.6–12}$$

Let us observe next that

$$E[Y \mid \mathbf{X}_c] = E[Y \mid \mathbf{X}], \tag{5.6–13}$$

a result that is intuitively agreeable since we do not expect the conditional expectation of Y to depend on whether the average value of $\mathbf{X}$ is included or not. A formal demonstration of Equation 5.6–13 is readily obtained by writing out the definition of conditional expectation using pdf's and considering the transformation $\mathbf{X}_c = \mathbf{X} - \boldsymbol{\mu}_\mathbf{X}$; this is left as an exercise. Now using Equation 5.6–13 in Equation 5.6–12 and using the latter in Equation 5.6–11 we obtain our final result as

$$\begin{aligned} E[Y \mid \mathbf{X}] &= \sum_{i=1}^{N} a_i X_{ci} + \mu_Y \\ &= \sum_{i=1}^{N} a_i (X_i - \mu_{X_i}) + \mu_Y, \end{aligned} \tag{5.6–14}$$

which is the general expression for the jointly Gaussian case when the means are nonzero. We see that the estimate is in the form of a linear transformation plus a bias. In passing, we note that the a_i's would be determined from Equation 5.6–10 using the correlation matrix and cross-correlation vector of the zero-mean random variables, which is the same as the covariance matrices of the original random variables $\mathbf{X}$ and Y.

5.7 DECISION THEORY: THE BAYESIAN APPROACH

Consider the state of mind of a physician examining a chest X-ray of a patient and finding a vague shadow that might be, on the one hand, harmless scar tissue or, on the other hand, a cancer. Assume for simplicity that the only logical courses of action are to operate or not.

To operate carries its own risks including a long period of debilitation, the risks of incurring infections, and a reduced life span if a lung is removed. Not to operate, however, will mean almost certain death *if a cancer is present.* Thus either course of action carries with it a penalty or *cost* that could be measured in terms of years of reduced life span from the norm. If the physician knew for sure what the state of nature was, for example, that the shadow was a cancer, he could make a rational decision based on minimizing the cost. Thus if the cost of not operating is 30 years of reduced life span versus 5 years for removing the lung, the choice is clearly to operate. However, the physician does not know categorically the state of nature. He may know, based on the data $\mathbf{X}$ extracted from the X-ray the probability that the radiopacity is scar tissue or cancer, but he doesn't know for sure. What should he do? According to Bayesian decision theory he should make that decision that minimizes the *average cost* or *risk*.† In this example there are four costs to consider: the cost of removing a lung with no cancer present; the cost of removing a lung and a cancer being present; the cost of not operating and no cancer being present; and the cost of not operating and a cancer being present.

The foregoing example contains all of the elements of Bayesian decision theory. We now discuss these elements in a more general setting. First, we assume that there exists a set Ω partitioned into M states of nature ω_i $i = 1, \ldots, M$ and that one of these states prevails when the observations are made. The prevailing state of nature is not known to the observer. Second, we assume that the observer obtains real vector data $\mathbf{X} = (X_1, \ldots, X_n)^T$, which is probabilistic in nature and represents a manifestation of the state of nature.‡ Third, we postulate the existence of a set of costs that are known to the observer. Fourth, we assume that the *a priori* probabilities of ω_i, $i = 1, \ldots, M$ as well as the conditional probability functions of observing $\mathbf{X}$ given ω_i, $i = 1, \ldots, M$ are known to the observer. And fifth, we assume a decision function $d(\mathbf{X})$ whose domain is the set of all real valued n-tuples and range is Ω. We define

$$P[\hat{\omega}_i, \omega_j] \triangleq P[d(\mathbf{X}) = \omega_i, \omega_j]; \tag{5.7-1}$$

that is, $P[\hat{\omega}_i, \omega_j]$ is the joint probability that the state of nature is ω_j and that—based on $\mathbf{X}$ and the function $d(\cdot)$—our estimated or *perceived* state of nature is ω_i. If $d(\mathbf{X}) = \omega_j$ our decision is a correct one; that is, our perception agrees with the true state of nature. If $d(\mathbf{X}) = \omega_i$, $i \neq j$ then an error has been made and a penalty (cost) is incurred.

As stated earlier, a rational basis for making a decision is to minimize the average cost. We define a cost matrix

$$\mathbf{C} = \begin{bmatrix} C_{11}, & C_{12}, & \ldots, & C_{1M} \\ \vdots & \vdots & & \vdots \\ C_{M1}, & C_{M2}, & \ldots, & C_{MM} \end{bmatrix}, \tag{5.7-2}$$

where C_{ij} is the cost, that is, penalty of deciding that the state of nature is ω_i when in fact the true state of nature is ω_j, and where elements C_{ii},

† Thus *risk* in Bayesian theory is a technical term, and that is how it shall be used in this section.

‡ The elements of ω_i consist of the totality of outcomes that we associate with the state-of-nature ω_i. Thus in the above example, the subset ω_i might represent the state of malignancy and $\zeta_i \in \omega_i$ would be a particular type of malignancy. The observation vector $\mathbf{X}$ is an n-dimensional r.v. whose domain is Ω and range is R_n. As $\mathbf{X}$ roams over ω_i, it generates numerical data according to the probability law $f(\mathbf{x} \mid \omega_i)$. For a particular value $\zeta_i \in \omega_i$, $\mathbf{X}$ assumes the value $\mathbf{X}(\zeta_i) = (X_1(\zeta_i), \ldots, X_n(\zeta_i))^T$.

$i = 1, \ldots, M$, represent the costs of correct decisions and are often made zero. However, since some correct decisions may be better than others, one sometimes assigns nonzero values to some of the C_{ii} as well. Nevertheless, the cost of making an error will always be assumed to be larger than the cost of a correct decision if C_{ij} is to reflect the fact that a bad decision carries with it a higher penalty than a correct one.

In the model considered here it is assumed that the actual state of nature is a random outcome, that is, that we can think of an urn holding all the states of nature and ω_j being picked from the urn with probability P_j. P_j is called the *a priori* probability that the state of nature is ω_j. The observation $\mathbf{X}$ is associated with an outcome in one of the ω_j's $\subset \Omega$; which one we don't know. But we know the conditional pdf's (or PMF's)

$$f(\mathbf{x} \mid \omega_j) \quad j = 1, \ldots, M, \tag{5.7–3}$$

which lend themselves to the interpretation

$$f(\mathbf{x} \mid \omega_j)\, d\mathbf{x} \simeq P[\mathbf{x} < \mathbf{X} < \mathbf{x} + d\mathbf{x} \mid \omega_j]. \tag{5.7–4}$$

Decisions are made in accordance with a decision rule. Such a rule is essentially a partition of all possible observations into disjoint decision regions $\mathscr{X}_1, \mathscr{X}_2, \ldots, \mathscr{X}_M$ such that if the realization $\mathbf{x}$ falls into $\mathscr{X}_i$, the decision is that ω_i prevails (Figure 5.7–1). We illustrate with an example.

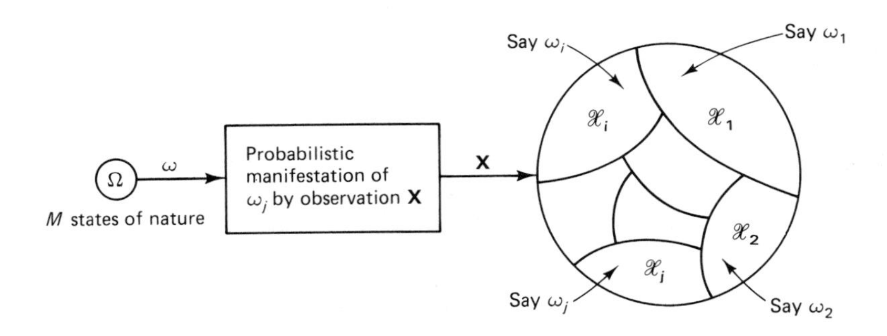

Figure 5.7–1 The optimum decision process viewed as an optimum partitioning of the space of observations $\mathscr{X}$.

Example 5.7–1: (Digital communication system.) Assume that a digital source transmits a *zero* or a *one*. A *zero* is transmitted by sending a negative dc voltage $-A$, for a time interval τ while a *one* is transmitted by sending a positive dc voltage $+A$, for a similar interval. Let ω_1 represent the sending of a *zero*; ω_2 the sending of a *one*. Because of random channel noise a negative voltage sometimes goes positive and a positive voltage sometimes goes negative (Figure 5.7–2).

Let $x(t)$, $0 < t < \tau$, represent the received signal and let $x_i \triangleq x(t_i)$, $i = 1, \ldots, n$ represent the sampled received signal at the instants t_i where $0 \le t_1 < \ldots < t_n \le \tau$. The received sample vector is $\mathbf{x} \triangleq (x_1, \ldots, x_n)^T$. Suppose we decide on

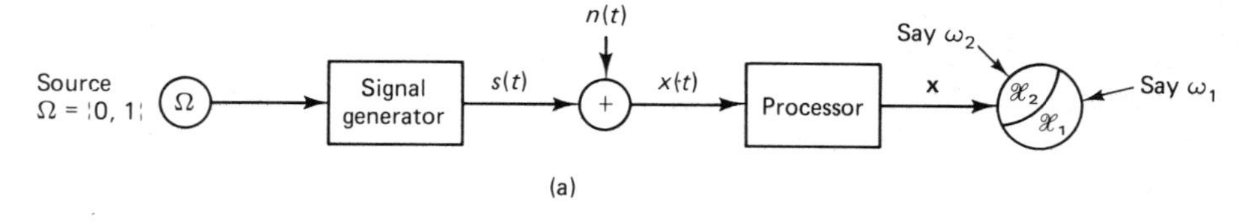

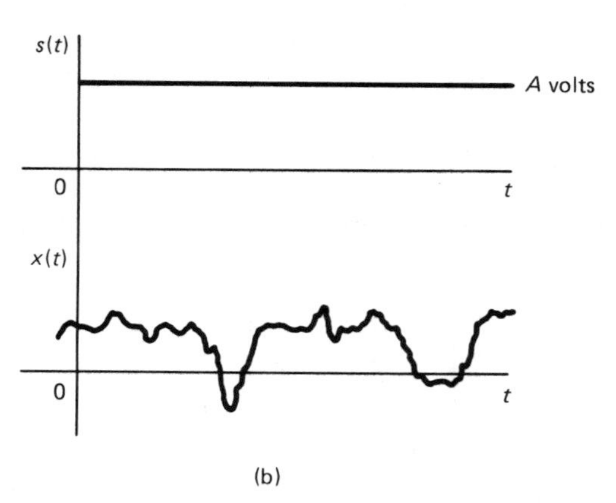

5.7–2 (a) Block diagram of digital communication link showing decision regions for ω_1, ω_2; (b) the distortion due to noise of a *one* can lead to an error in decoding.

ω_1 when

$$\frac{1}{n}\sum_{i=1}^{n} x_i < 0 \quad \text{(negative sample mean)}$$

and decide on ω_2 when

$$\frac{1}{n}\sum_{i=1}^{n} x_i \geq 0. \quad \text{(positive sample mean)}$$

Although this particular strategy seems reasonable, it is not necessarily optimal. An optimal strategy would depend on the statistics of the random channel noise.

In this example the regions $\mathscr{X}_1$ and $\mathscr{X}_2$ would contain all vectors x having negative and positive sample mean values, respectively. Note that to find out in which region x belongs involves some processing, that is, sampling and taking the average.

There are generally very many decision rules, some reasonable and others not. For instance, we could decide always to choose ω_2, independent of what is observed. Or we might decide to pick ω_2 whenever any one element of the vector $\mathbf{x}$ exceeds some positive number.

We should mention that the decision process is sometimes generalized to include random decision rules. Such rules are defined in terms of the probability of deciding in favor of ω_i given that $\mathbf{x}$ is observed. This generalization has some mathematical advantages, but we need not consider it here.

Optimum Partitioning of Observation Space $\mathscr{X}$. In terms of Equation 5.7–1 and the elements of the cost matrix $\mathbf{C}$, the risk $\mathscr{R}$ is defined by

$$\mathscr{R} \triangleq \sum_{i=1}^{M} \sum_{j=1}^{M} P[\hat{\omega}_i, \omega_j]C_{ij} \tag{5.7–5}$$

$$= \sum_{i=1}^{M} \sum_{j=1}^{M} P[\hat{\omega}_i \mid \omega_j]P_jC_{ij}$$

$$= \sum_{i=1}^{M} \sum_{j=1}^{M} P_jC_{ij} \int_{\mathscr{X}_i} f(\mathbf{x} \mid \omega_j)\, d\mathbf{x}, \tag{5.7–6}$$

where $P[\hat{\omega}_i \mid \omega_j]$ is defined as the conditional probability that we choose ω_i to be the state of nature given that ω_j is the (true) state of nature. The integral is n-fold, and the fact that it is over the region $\mathscr{X}_i$ implies that the decision is in favor of ω_i. The probability density $f(\mathbf{x} \mid \omega_j)$ indicates that the observations are actually due to ω_j.

We now seek to adjust the partitioning $\{\mathscr{X}_i\}$ so as to minimize $\mathscr{R}$. The optimum partitioning can be found with the help of Bayes' rule:

$$f(\mathbf{x} \mid \omega_j) = \frac{P[\omega_j \mid \mathbf{x}]f(\mathbf{x})}{P_j}$$

or

$$P_jf(\mathbf{x} \mid \omega_j) = f(\mathbf{x})P[\omega_j \mid \mathbf{x}], \tag{5.7–7}$$

where $f(\mathbf{x})$ is the unconditional pdf of $\mathbf{x}$. Use of Equation 5.7–7 in Equation 5.7–6 furnishes

$$\mathscr{R} = \sum_{i=1}^{M} \sum_{j=1}^{M} C_{ij} \int_{\mathscr{X}_i} P[\omega_j \mid \mathbf{x}]f(\mathbf{x})\, d\mathbf{x} \tag{5.7–8}$$

$$= \sum_{i=1}^{M} \int_{\mathscr{X}_i} \beta_i(\mathbf{x})f(\mathbf{x})\, d\mathbf{x}, \tag{5.7–9}$$

where

$$\beta_i(\mathbf{x}) \triangleq \sum_{j=1}^{M} C_{ij}P[\omega_j \mid \mathbf{x}], \qquad i = 1, \ldots, M. \tag{5.7–10}$$

We are now in a position to find the optimum regions $\{\mathscr{X}^*\}$ which partition $\mathscr{X}$ so as to minimize the risk. For ease of illustration we consider a univariate, that is, scalar r.v. X whose range is the real line. Let $M = 2$ and refer to Figure 5.7–3. In this case $\mathscr{R}$ is given by

$$\mathscr{R} = \int_{\mathscr{X}_1} \beta_1(x)f(x)\, dx + \int_{\mathscr{X}_2} \beta_2(x)f(x)\, dx. \tag{5.7–11}$$

Since $f(x)$ is everywhere nonnegative, $\mathscr{R}$ is minimized by choosing the partitions $\mathscr{X}_1^*$ and $\mathscr{X}_2^*$ such that x is in $\mathscr{X}_1^*$ whenever $\beta_1(x) < \beta_2(x)$ and in $\mathscr{X}_2^*$ whenever $\beta_2(x) < \beta_1(x)$. Any other choice of $\mathscr{X}_1$ and $\mathscr{X}_2$ would increase $\mathscr{R}$ (for example, try the obvious $\mathscr{X}_1 = \mathscr{X}_2^*$ and $\mathscr{X}_2 = \mathscr{X}_1^*$).

The result is easily generalized to M decision regions (Figure 5.7–4). We find that the optimum partitioning is equivalent to the following (Bayes)

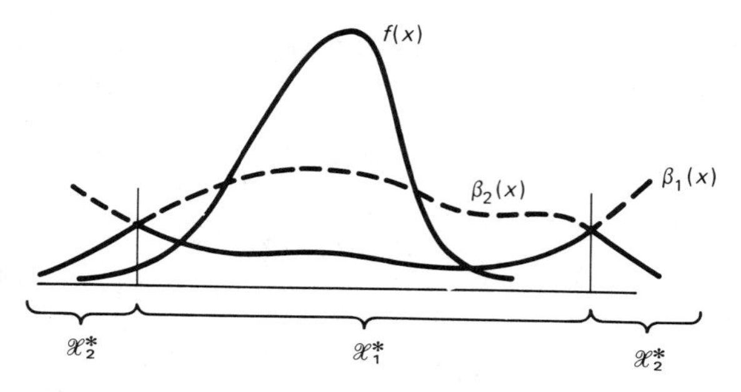

Figure 5.7–3 Optimum partitioning of $\mathscr{X}$ into two decision regions.

decision rule: For every observed $\mathbf{x}$, compute

$$\beta_i(\mathbf{x}) = \sum_{j=1}^{M} C_{ij} P[\omega_j \mid \mathbf{x}] \qquad i = 1, \ldots, M, \tag{5.7–12}$$

and choose that ω_i for which β_i is smallest.

The function $P[\omega_j \mid \mathbf{x}]$ is the probability that ω_j is the state of nature if $\mathbf{x}$ was observed. It is referred to as the *a posteriori* probability of the state-of-nature ω_j given $\mathbf{x}$, and it contains all the information available to the observer after reception of $\mathbf{x}$. Therefore, it forms a natural basis for making the decision. The function β_i is the *a posteriori* average cost, or *a posteriori risk*, associated with the state-of-nature ω_i and the particular received signal $\mathbf{x}$. Hence the Bayes decision rule is to decide in such a way as to always minimize the *a posteriori* risk.

Equation 5.7–7 can be written in the equivalent form

$$\begin{aligned} P[\omega_j \mid \mathbf{x}] &= \frac{P_j f(\mathbf{x} \mid \omega_j)}{f(\mathbf{x})} \\ &= \frac{P_j f(\mathbf{x} \mid \omega_j)}{\sum_{j=1}^{M} P_j f(\mathbf{x} \mid \omega_j)}. \end{aligned} \tag{5.7–13}$$

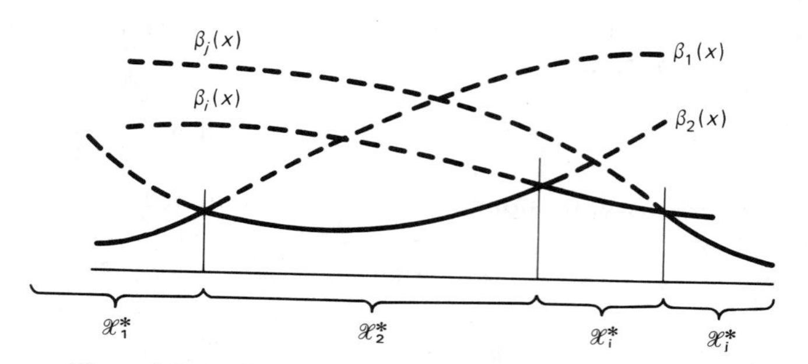

Figure 5.7–4 Optimum partitioning in the general case.

The second line is the form in which Bayes' theorem is usually given. We see that the computation of the *a posteriori* risk involves the *a priori* probabilities P_j, the probability densities $f(\mathbf{x} \mid \omega_j)$, and the cost matrix $\mathbf{C}$. Not all of these are always known. In these cases the Bayes strategy cannot be implemented, and the decision rule must be based on other criteria ([5–6], Chap. 2).

The Zero-One Cost Function. An important special case occurs when the cost function is

$$C_{ij} = 1 - \delta_{ij}, \tag{5.7–14}$$

where δ_{ij} is the Kronecker delta. This gives zero cost for correct decisions and unity cost for any incorrect decision; that is, any error is as bad as any other. Then Equation 5.7–10 becomes

$$\beta_i(\mathbf{x}) = \sum_{j \neq i}^{M} P[\omega_j \mid \mathbf{x}] = 1 - P[\omega_i \mid \mathbf{x}]. \tag{5.7–15}$$

Hence the optimum decision rule, that is,

$$\min_i \beta_i(\mathbf{x}) \quad \text{for each } \mathbf{x},$$

in this case is equivalent to

$$\max_i P[\omega_i \mid \mathbf{x}] \quad \text{for each } \mathbf{x}.$$

Since the optimum decision rule maximizes the *a posteriori* probability, it is sometimes called a *m*aximum *a* posteriori *p*robability strategy or MAP strategy for short. The risk function for the zero-one loss is given by

$$\begin{aligned}
\mathcal{R} &= \sum_{i=1}^{M} \sum_{j \neq i}^{M} P_j \int_{\mathcal{X}_i} f(\mathbf{x} \mid \omega_j)\, d\mathbf{x} \\
&= \sum_{i=1}^{M} \sum_{j \neq i}^{M} \int_{\mathcal{X}_i} P[\omega_j \mid \mathbf{x}] f(\mathbf{x})\, d\mathbf{x} \\
&= 1 - \sum_{i=1}^{M} \int_{\mathcal{X}_i} P[\omega_i \mid \mathbf{x}] f(\mathbf{x})\, d\mathbf{x} \\
&= 1 - \text{probability of a correct decision.}
\end{aligned} \tag{5.7–16}$$

Hence minimizing the risk in this case is equivalent to maximizing the probability of making a correct decision. Since $f(\mathbf{x}) \geq 0$, the probability of making a correct decision is maximized by using the MAP decision rule.

Another way of describing the MAP decision rule is as follows: $d(\mathbf{x}) = \omega_i$ if

$$P[\omega_i \mid \mathbf{x}] > P[\omega_k \mid \mathbf{x}] \quad \text{for } k = 1, \ldots, M, \quad k \neq i. \tag{5.7–17}$$

Equivalently, the Bayes rule enables us to write the following: $d(\mathbf{x}) = \omega_i$ if

$$\frac{f(\mathbf{x} \mid \omega_i) P_i}{f(\mathbf{x})} > \frac{f(\mathbf{x} \mid \omega_k) P_k}{f(\mathbf{x})} \quad \text{for every } k \neq i \tag{5.7–18}$$

or, since $f(\mathbf{x})$ is irrelevant, $d(\mathbf{x}) = \omega_i$ if

$$f(\mathbf{x} \mid \omega_i)P_i > f(\mathbf{x} \mid \omega_k)P_k \quad \text{for every } k \neq i. \tag{5.7–19}$$

If the $\{P_i\}_{i=1}^M$ are not known or are assumed equal, that is, $P_1 = P_2 = \ldots = P_M = 1/M$, the optimum rule chooses $d(\mathbf{x}) = \omega_i$ if $f(\mathbf{x} \mid \omega_i)$ is the maximum conditional pdf from the set $\{f(\mathbf{x} \mid \omega_j)\}_{j=1}^M$. This kind of rule is known as a maximum likelihood (ML) strategy. It minimizes the probability of error if all *a priori* probabilities $\{P_j\}$ are equal.

Note that if the *a priori* probabilities are not equal, there generally exists a MAP rule having a smaller error probability than the ML rule. However, if this same MAP rule is faced with a different set of *a priori* probabilities, its error probability could be larger than that of the ML receiver; in fact for certain unfavorable *a priori* distributions its performance might be much worse. Since *a priori* probabilities are frequently unavailable, one would want a strategy that guards against the possibility of a particularly unfavorable *a priori* distribution resulting in very large errors. The strategy whose worst-case performance is better than that of any other rule is called a *minimax strategy*.† For a discussion of such strategies, see Reference 5–7, p. 264. It turns out that under conditions frequently encountered in practice the ML rule is, in fact, minimax.

The Likelihood Ratio. The computation that implements the optimum decision rule can be done serially. For example, Equation 5.7–19 enables us to write the following: If

$$\frac{f(\mathbf{x} \mid \omega_i)}{f(\mathbf{x} \mid \omega_k)} > \frac{P_k}{P_i}, \tag{5.7–20}$$

reject ω_k for consideration and test $f(\mathbf{x} \mid \omega_{k+1})$. Otherwise, that is, if Equation 5.7–20 is not true, reject ω_i, replace $f(\mathbf{x} \mid \omega_i)$ with $f(\mathbf{x} \mid \omega_k)$, and test the next pdf.‡ The ratio

$$\text{LR}_{ik}(\mathbf{x}) \equiv \frac{f(\mathbf{x} \mid \omega_i)}{f(\mathbf{x} \mid \omega_k)}. \tag{5.7–21}$$

is known as the likelihood ratio, and a test of the form of Equation 5.7–20 is known as a likelihood-ratio test.

Example 5.7–2: Let ω_i, $i = 1, 2, 3$, represent three source messages in a communication system, and let C_{ij} be the zero-one cost function; that is,

$$C_{ij} = \begin{cases} 1, & i \neq j \\ 0, & i = j. \end{cases}$$

† Minimax is a contraction of minimizing the maximum risk or probability of error, whichever the case may be.

‡ In other words, if Equation 5.7–20 doesn't hold the new test will be: If $f(\mathbf{x} \mid \omega_k)/f(\mathbf{x} \mid \omega_{k+1}) > P_{k+1}/P_k$ reject ω_{k+1}.

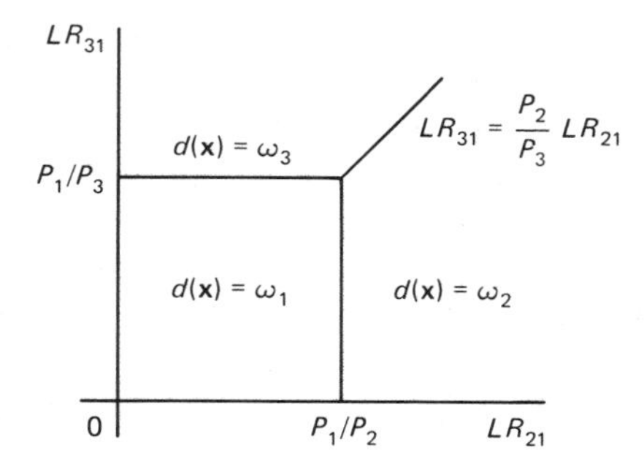

Figure 5.7–5 Optimum decision regions in likelihood-ratio space for the case of three messages.

The *a priori* probability of ω_i is P_i, $i = 1, 2, 3$.

$$\text{LR}_{21}(x) = \frac{f(x \mid \omega_2)}{f(x \mid \omega_1)}$$

$$\text{LR}_{31}(x) = \frac{f(x \mid \omega_3)}{f(x \mid \omega_1)}$$

$$\text{LR}_{23}(x) = \frac{\text{LR}_{21}(x)}{\text{LR}_{31}(x)} = \frac{f(x \mid \omega_2)}{f(x \mid \omega_3)}.$$

The decision boundaries are established when the right and left side of Equation 5.7–20 are set equal; that is,

$$\text{LR}_{21}(x) = \frac{P_1}{P_2}; \quad \text{i.e., if } \text{LR}_{21}(x)\begin{cases} > P_1/P_2, \text{ don't choose } \omega_1 \\ < P_1/P_2, \text{ don't choose } \omega_2 \end{cases}$$

$$\text{LR}_{31}(x) = \frac{P_1}{P_3}; \quad \text{i.e., if } \text{LR}_{31}(x)\begin{cases} > P_1/P_3, \text{ don't choose } \omega_1 \\ < P_1/P_3, \text{ don't choose } \omega_3 \end{cases}$$

$$\text{LR}_{23}(x) = \frac{P_3}{P_2}; \quad \text{i.e., if } \text{LR}_{23}(x)\begin{cases} > P_3/P_2, \text{ don't choose } \omega_3 \\ < P_3/P_2, \text{ don't choose } \omega_2. \end{cases}$$

The decision space is shown in Figure 5.7–5. Note that the optimum partitioning is unambiguous and can be represented in a two-dimensional coordinate system even though three messages are involved. This is true more generally; in the case of M decision possibilities, the minimum dimension of the decision space is $M - 1$.

5.8 SUMMARY

Decision and estimation theory (DET) are closely related topics in applied probability that have great practical significance. One can think of estimation as the process of making decisions in the presence of an infinite number of choices. In this chapter we discussed the basic principles of DET; more advanced topics are discussed in Chapter 10. We began the chapter by defining properties of estimators and furnishing estimators for the mean and covariance of random vectors. We discussed such properties as *unbiasedness* and *consistency* and showed why it was desirable for an estimator to have them. We then focused on a particular class of estimators called maximum likelihood estimators (MLE) that have many desirable properties and computed the MLE's of $\boldsymbol{\mu}$ and $\mathbf{K}$ (the mean and covariance, respectively) for the multidimensional normal probability law.

In Section 5.4 we turned our attention to a very general problem, that is, the estimation of a parameter vector $\boldsymbol{\theta}$ from a set of linear observations corrupted by additive noise. We restricted ourselves to linear estimators because of their simplicity. We derived the least-squares and minimum variance estimators and derived the famous Gauss–Markov theorem that furnishes the minimum variance estimator for arbitrary observation noise.

In Section 5.6 we considered the problem of estimating one random variable by observing another. We extended our results to the case of estimating one random vector with another. The main result, subsequently extended to random vectors, was the demonstration that the minimum mean square estimate of Y, based on observing X, is the condition mean $E[Y \mid X]$.

We then undertook a discussion of the fundamentals of Bayesian decision theory. We introduced the notions of *cost* and *risk* and showed that Bayesian decision theory has as its aim the *minimization of risk*. We then derived the Bayesian decision function implicitly by partitioning the observation space in a manner that minimized the risk. We considered special cases of the Bayesian model such as the *zero-one* cost function and showed that the Bayesian strategy in that case led to the celebrated maximum *a posteriori* (MAP) rule. We showed how the MAP rule can be easily implemented using likelihood ratios. We closed the chapter by finding the optimum decision regions in an example involving deciding among three alternatives.

PROBLEMS

5.1. Explain the difference between an estimate and an estimator. Explain the unbiasedness, consistency, and minimum mean-square error properties of estimators.

5.2. Show that if $\mu \triangleq E[X_i]$, $i = 1, \ldots, n$ is known then

$$\hat{\Theta} = \frac{1}{n} \sum_{i=1}^{n} (X_i - \mu)^2$$

is unbiased and consistent for estimating σ^2. The X_i, $i = 1, \ldots, n$ are i.i.d.† r.v.'s with $\text{Var}(X_i) = \sigma^2$.

5.3. Let X_i, $i = 1, \ldots, n$ be i.i.d. r.v.'s. Let $\hat{\mu}$ be the sample mean estimator given by

$$\hat{\mu} \triangleq \frac{1}{r} \sum_{i=1}^{n} X_i.$$

† As in the text, i.i.d. r.v.'s stands for independent, identically distributed random variables.

Show that $\hat{\Theta}_n$, given by

$$\hat{\Theta}_n \triangleq \frac{1}{n-1}\sum_{i=1}^{n}(X_i - \hat{\mu})^2$$

is unbiased and consistent for $\text{Var}(X_i) \triangleq \sigma^2$.

5.4. Consider the r.v. X that satisfies the binomial law, that is,

$$P_X(k) = \binom{n}{k}p^n(1-p)^{n-k}.$$

Show that

$$\hat{\Theta} \triangleq \frac{X}{n}$$

is unbiased and consistent for p.

5.5. In Problem 5.4 show that

$$\hat{\Theta} \triangleq \frac{X(X-1)}{n(n-1)}$$

is unbiased for p^2.

5.6. Let X_i, $i = 1, \ldots, n$ be i.i.d. r.v.'s with

$$f_{X_i}(x) = \frac{1}{\pi}\left(\frac{1}{1+(x-\mu)^2}\right) \qquad -\infty < x < \infty.$$

Consider

$$\hat{\Theta} = \frac{1}{n}\sum_{i=1}^{n} X_i \triangleq \hat{\mu}$$

as an estimator for μ. Is $\hat{\mu}$ consistent for μ? Hint: Show with the help of characteristic functions that the pdf of $\hat{\mu}$ is

$$f_{\hat{\mu}}(\alpha) = \frac{1}{\pi}\left(\frac{1}{1+(\alpha-\mu)^2}\right)$$

and does not depend on n.

5.7. Let X_i, $i = 1, \ldots, n$ be n i.i.d. normal r.v.'s with

$$f_{X_i}(x) = \frac{1}{\sqrt{2\pi}}e^{-\frac{1}{2}(x-\mu)^2}.$$

Consider the estimator for μ:

$$\hat{\mu} \triangleq \frac{1}{n}\sum_{i=1}^{n} X_i.$$

How large should n be so that

$$P[|\hat{\mu} - \mu| < 0.1] \geq 0.95?$$

5.8. Compute the maximum likelihood estimates (MLE) of μ and σ^2 in

$$f_X(x) = \frac{1}{\sqrt{2\pi}\sigma}\exp\left(-\frac{1}{2\sigma^2}(x-\mu)^2\right)$$

based on n i.i.d. observations. Prove Equations 5.3–14 and 5.3–15 by working out the details.

5.9. In Problem 5.8 what are the MLE's of μ and σ^2 when $n = 1$? What conclusions do you draw from this result?

5.10. Let X_i, $i = 1, \ldots, n$ be n i.i.d. observations with

$$f_{X_i}(x) = \begin{cases} \dfrac{1}{b - a}, & a \leq x \leq b \qquad b > a \\ 0, & \text{otherwise.} \end{cases}$$

Compute the MLE's of a and b. Compute the probability distribution functions of the estimators.

5.11. Let $\mathbf{Z}_i = (X_i, Y_i)^T$, $i = 1, \ldots, n$ be n i.i.d. observations with

$$f_{\mathbf{Z}_i}(\mathbf{z}) = \frac{1}{2\pi(1 - \rho^2)^{1/2}\sigma_1\sigma_2} e^{-Q(\mathbf{z})},$$

where $\mathbf{z} = (x, y)^T$ and

$$Q(\mathbf{z}) \triangleq \frac{1}{2(1 - \rho^2)}\left[\left(\frac{x}{\sigma_1}\right)^2 - 2\rho\frac{xy}{\sigma_1\sigma_2} + \left(\frac{y}{\sigma_2}\right)^2\right]$$

and $-\infty < x_1, x_2 < \infty$, $\sigma_1, \sigma_2 > 0$, $|\rho| < 1$. Compute the MLE of ρ assuming σ_1 and σ_2 are known.

5.12. Let $\mathbf{X}$ and $\mathbf{Y}$ be real random n-vectors with $E[\mathbf{X}] = E[\mathbf{Y}] = \mathbf{0}$ and $\mathbf{K}_1 \triangleq E[\mathbf{X}\mathbf{X}^T]$, $\mathbf{K}_2 \triangleq E[\mathbf{Y}\mathbf{Y}^T]$ $\mathbf{K}_{12} \triangleq E[\mathbf{X}\mathbf{Y}^T] = \mathbf{K}_{21}^T \triangleq (E[\mathbf{Y}\mathbf{X}^T])^T$. It is desired to estimate the value of $\mathbf{Y}$ from observing the value of $\mathbf{X}$ according to the rule

$$\hat{\mathbf{Y}} = \mathbf{A}\mathbf{X}.$$

Show that with

$$\mathbf{A} = \mathbf{A}_0 \triangleq \mathbf{K}_{21}\mathbf{K}_1^{-1}$$

the estimator $\hat{\mathbf{Y}}_0$ is a minimum-variance estimator of $\mathbf{Y}$. By minimum variance is meant that the diagonal terms of $E[(\hat{\mathbf{Y}} - \mathbf{Y})(\hat{\mathbf{Y}} - \mathbf{Y})^T]$ are at a minimum.

5.13. Let $\mathbf{X}$ and $\mathbf{Y}$ be random vectors with $E[\mathbf{X}] = \boldsymbol{\mu}_1$, $E[\mathbf{Y}] = \boldsymbol{\mu}_2$, $E[(\mathbf{X} - \boldsymbol{\mu}_1)(\mathbf{X} - \boldsymbol{\mu}_1)^T] \triangleq \mathbf{K}_1$ $E[(\mathbf{Y} - \boldsymbol{\mu}_2)(\mathbf{Y} - \boldsymbol{\mu}_2)^T] \triangleq \mathbf{K}_2$ and $E[(\mathbf{X} - \boldsymbol{\mu}_1)(\mathbf{Y} - \boldsymbol{\mu}_2)^T] \triangleq \mathbf{K}_{12}$. Show that

$$\hat{\mathbf{Y}} = \boldsymbol{\mu}_2 + \mathbf{K}_{21}\mathbf{K}_1^{-1}(\mathbf{X} - \boldsymbol{\mu}_1)$$

is a minimum variance estimator for $\mathbf{Y}$.

5.14. Consider the model $\mathbf{Y} = \mathbf{H}\boldsymbol{\theta} + \mathbf{N}$ where $E[\mathbf{N}] = \mathbf{0}$ and $E[\mathbf{N}\mathbf{N}^T] \triangleq \mathbf{K} = \sigma^2\mathbf{I}$. Let a linear function of $\boldsymbol{\theta}$ be given by $\boldsymbol{\phi} = \mathbf{D}\boldsymbol{\theta}$ where $\mathbf{D}$ is a matrix of known coefficients. Show that if $\hat{\boldsymbol{\Phi}} = \mathbf{L}\mathbf{Y}$ is to be an unbiased estimator of $\boldsymbol{\phi}$, it must satisfy

$$\mathbf{L}\mathbf{H} = \mathbf{D}.$$

Show that the minimum variance unbiased estimator of $\boldsymbol{\phi}$ is

$$\hat{\boldsymbol{\Phi}}_0 = \mathbf{D}(\mathbf{H}^T\mathbf{H})^{-1}\mathbf{H}^T\mathbf{Y}.$$

5.15. Let the set of nature-states be $\Omega = \{\omega_1, \omega_2, \omega_3\}$ with $P[\omega_1] = \frac{1}{6}$, $P[\omega_2] = \frac{2}{6}$ and $P[\omega_3] = \frac{3}{6}$. The conditional pdf's of the observations are $f(x \mid \omega_1) = 2\exp(-2x)u(x)$, $f(x \mid \omega_2) = 3\exp(-3x)u(x)$ and $f(x \mid \omega_3) = 4\exp(-4x) \cdot u(x)$ where $u(x)$ is the unit step function.

(a) Find the optimum decision regions in likelihood ratio space when the zero-one cost function is used.

(b) Find the optimum partitioning of the observation space $\mathscr{X}$.

5.16. Let $\Omega = \{\omega_1, \ldots, \omega_M\}$ be M distinct states. Associated with each state is a unique vector signal (that is, the manifestation of that state) $\mathbf{s}_i = (s_{i1}, \ldots, s_{iK})^T$ $i = 1, \ldots, M$. Let $P_i \triangleq P[\omega_i]$, $i = 1, \ldots, M$ be the *a priori* probabilities of ω_i. Because of additive Gaussian measurement noise $\mathbf{N} = (N_1, \ldots, N_K)^T$ the signal perceived by the observer is not one of the $\{\mathbf{s}_i\}$ but rather $\mathbf{s}_i + \mathbf{N}$. Let $E[\mathbf{N}] = \mathbf{0}$, $E[\mathbf{N}\mathbf{N}^T] = \sigma^2\mathbf{I}$. The conditional pdf's of the observations are

$$f_{\mathbf{X}}(\mathbf{x} \mid \mathbf{s}_i) = \frac{1}{\sqrt{2\pi\sigma^2}} \exp\left(\frac{-1}{2\sigma^2} \sum_{k=1}^{K} (x_k - s_{ik})^2\right).$$

Show that the Bayes strategy for the zero-one loss function is to choose that value of ω, say ω_j, if for any realization $\mathbf{x}$

$$\tilde{d}(\omega_j \mid \mathbf{x}) = \min_i \tilde{d}(\omega_i \mid \mathbf{x})$$

where

$$\tilde{d}(\omega_i \mid \mathbf{x}) \triangleq \|\mathbf{x} - \mathbf{s}_i\|^2 - 2\sigma^2 \ln P_i.$$

Note $\tilde{d}(\omega_i \mid \mathbf{x})$ is known as a *discriminant* function.

5.17. In Problem 5.16, let $P_i = \dfrac{1}{M}$ and $\|\mathbf{s}_i\|^2 = E$, $i = 1, \ldots, M$. Show that the optimum strategy for the Bayes zero-one loss is to choose ω_j if

$$\mathbf{x}^T\mathbf{s}_j \geq \mathbf{x}^T\mathbf{s}_i \qquad i = 1, \ldots, M.$$

5.18. In a hospital ward for lung diseases, a surgeon has discovered a suspicious shadow $\mathbf{x}$ in the lung X-ray of his patient. Two possibilities exist: The shadow represents a cancer or it represents harmless scar tissue. The surgeon has two choices: to operate (O) or not (N). Let M denote the state of cancer (malignancy) and B denote the state of *benign* scar tissue. The following costs are measured in number of years subtracted from a normal life span: $C_{OM} = 5$, $C_{NM} = 30$, $C_{NB} = 0$, $C_{OB} = 5$. Assume that the contents of an X-ray can, upon analysis, be quantified into a vector $\mathbf{X}$. For this X-ray assume that $\mathbf{X} = \mathbf{x}$ for which it is known that $P[M \mid \mathbf{X} = \mathbf{x}] = 0.1$ and $P[B \mid \mathbf{X} = \mathbf{x}] = 0.9$. Based on all this data and his desire to minimize the risk, should the surgeon operate?

REFERENCES

5–1. F. A. Graybill, *An Introduction to Linear Statistical Models, Vol.* 1. New York: McGraw-Hill, 1961, pp. 34–197.

5–2. K. Fukunaga, *Introduction to Statistical Pattern Recognition.* New York: Academic Press, 1972.

5–3. M. G. Kendall and A. Stuart, *The Advanced Theory of Statistics*, Vol. 2, 3rd Ed. London, England: Charles Griffin and Co., 1951.

5–4. C. R. Rao, *Linear Statistical Inference and Its Application.* New York: John Wiley, 1965, pp. 447–449.

5–5. T. O. Lewis and P. L. Odell, *Estimation in Linear Models.* Englewood Cliffs, N.J.: Prentice Hall, 1971.

5–6. H. L. van Trees, *Detection, Estimation and Modulation Theory.* New York: John Wiley, 1968.

5–7. J. M. Wozencraft and I. W. Jacobs, *Principles of Communication Engineering.* New York: John Wiley, 1965.

6

RANDOM SEQUENCES

Random sequences are used as models of sampled data arising in signal and image processing, digital control, and communications. They also arise as nonsampled data such as economic variables, the content of a register in a digital computer, something as simple as coin flipping, or stochastic languages in syntactic pattern recognition. In this chapter we will introduce the concept of a random sequence and study some of its important properties. As we will see, a random sequence or stochastic sequence can be thought of as an infinite-dimensional vector of random variables. As such it stands between the finite dimensional vectors studied in Chapter 4 and the continuous-time random functions to be studied in the next chapter on stochastic processes.

6.1 BASIC CONCEPTS

In the course of developing this material we will have need to review and extend some of the basic material presented in Chapter 1 on the axioms of probability. We start out by defining a random sequence and then give some examples.

Definition 6.1–1. Let $(\Omega, \mathscr{F}, P)$ be a probability space. Let $\zeta \in \Omega$. If $X[n, \zeta]$, is a random variable for each fixed integer n in an index set I, then $X[n, \zeta]$ is a random or stochastic sequence. The index set I may be all the integers or a subset of the integers.

We see that $X[n, \zeta]$ for a *fixed* ζ is an ordinary sequence of numbers that is, a deterministic function of the discrete parameter n. For n *fixed* and ζ variable, $X[n, \zeta]$ is a random variable. Thus the collection of all these r.v.'s, $-\infty < n < +\infty$, *is* the random sequence. We often denote the random sequence as just $X[n]$. We can give the following specific examples:

Example 6.1–1a: $X[n, \zeta] = X(\zeta)f[n]$ where X is a random variable and $f[n]$ is a deterministic sequence. We also write $X[n] = Xf[n]$.

Example 6.1–1b: $X[n, \zeta] = A(\zeta)\sin(\pi n/10 + \Theta(\zeta))$ where A and Θ are random variables, alternately written $X[n] = A\sin(\pi n/10 + \Theta)$.

The above random sequences are "deterministic" in that their future values are exactly predictable from their present and past values. In Example, 6.1–1a, once we observe $X[0]$ then, since the sequence $f[n]$ is assumed to be known and nonrandom, all of $X[n]$ becomes known. We say that this X is *conditionally known* given its value at $n = 0$. The situation in Example 6.1–1b is a little more complicated but the same approach suffices to show that given two observations, say at $n = 0$ and $n = 5$, one can determine the values taken on by the random variables A and Θ; then the sequence X becomes conditionally known given these observations at $n = 0$ and $n = 5$.

In the next example we see how a more general but still "deterministic" random sequence can be made out of a random vector.

Example 6.1–2: (Random sequence with finite support.) Let $X[n, \zeta]$ be given by:

$$X[n, \zeta] \triangleq \begin{cases} X_1(\zeta) & n = 1 \\ X_2(\zeta) & n = 2 \\ \vdots & \\ X_N(\zeta) & n = N \\ 0 & \text{elsewhere,} \end{cases}$$

which may alternately be written:

$$X[n] \triangleq \begin{cases} X_n, & 1 \le n \le N \\ 0, & \text{else,} \end{cases}$$

where $(X_1, \ldots, X_N)^T$ is a random vector.

In order to deal with general random sequences, we will have to be able to compute the probabilities of infinite intersections of events, for example, the event $\{X < 5 \text{ for all positive } n\}$, which can be written as either $\bigcap_{n=1}^{+\infty} \{X[n] < 5\}$ or in terms of the infinite union $\left(\bigcup_{n=1}^{\infty} \{X[n] \ge 5\}\right)^c$. (See Section 1.3 for the definition of infinite unions and intersections.) This requires that we can define probabilities of infinite collections of events, which presents a problem with Axiom 3 of the probability measure: that is, for $AB = \phi$,

$$P[A \cup B] = P[A] + P[B] \quad \text{(Axiom 3).} \tag{6.1–1}$$

By iteration we could only build this result up to the result

$$P\left[\bigcup_{n=1}^{N} A_n\right] = \sum_{n=1}^{N} P[A_n]$$

for any finite positive N, assuming $A_iA_j = \phi$ for all $i \neq j$. This is called *finite additivity*. It will permit us to evaluate $\lim_{N\to\infty} P\left[\bigcup_{n=1}^{N} A_n\right]$ but what we need above is $P\left[\bigcup_{n=1}^{\infty} A_n\right]$ where $A_n \triangleq \{X[n] \geq 5\}$. In general they might not be the same. Hence we need some kind of continuity built into P. This can be achieved by augmenting or replacing Axiom 3 by the stronger Axiom 4 given as

Axiom 4 (Countable Additivity)

$$P\left[\bigcup_{n=1}^{\infty} A_n\right] = \sum_{n=1}^{\infty} P[A_n], \tag{6.1-2}$$

for an infinite collection of events satisfying $A_iA_j = \phi$ for $i \neq j$.

Fortunately, in measure theory [6–1] it is shown that it is always possible to construct probability measures satisfying such an axiom. Moreover if one has defined a probability measure P satisfying Axiom 3, that is, it is finitely additive, then Kolmogorov [6–2] has shown that it is always possible to extend P to satisfy Axiom 4, the property of being countably additive. We pause now for an example, after which we will show that Axiom 4 is equivalent to the desired continuity of the probability measure.

Example 6.1–3: (Infinite-length Bernoulli trials.) Let $\Omega = \{H, T\}$ with $P[H] = p$ and $P[T] = 1 - p$. Define the random variable W by $W(H) \triangleq 1$ and $W(T) \triangleq 0$, indicative of successes and failures in coin flipping. Let Ω_n be the sample space on the nth flip ($=\Omega$). Then define a new event space as $\Omega_\infty \triangleq \bigtimes_{n=1}^{\infty} \Omega_n$. This would be the sample space associated with an infinite sequence of flips, each with sample space Ω_n. Here the infinite cross product $\bigtimes_{n=1}^{\infty} \Omega_n$ simply means that the points in Ω_∞ consist of all the infinite-length sequences of events, each one in Ω_n for some n. Thus if $\boldsymbol{\zeta} \in \Omega_\infty$ then $\boldsymbol{\zeta} = (\zeta_1, \zeta_2, \zeta_3, \ldots)$, where ζ_n is in Ω_n for each $n \geq 1$. (The finite-length case was treated in Section 1.8.) We also define the random sequence $W[n, \zeta] \triangleq W(\zeta_n)$, thus generating a Bernoulli random sequence $W[n]$, $n \geq 1$.

We next consider the probability measure for Ω_∞. Letting A_n denote an event at trial n, that is, $A_n \in \mathscr{F}_n$ where $\mathscr{F}_n$ is the field of events in the probability space $(\Omega_n, \mathscr{F}_n, P)$ of trial n, then it is reasonable to consider $\bigcap_{n=1}^{\infty} A_n$ as an event in $\mathscr{F}_\infty$, the σ-field of events in Ω_∞. To complete this field of events we will have to augment it with all the countable intersections and unions of such events. For example, we may want to calculate the probability of the event

$$\{W[1] = 1, W[2] = 0\} \cup \{W[1] = 0, W[2] = 1\},$$

which can be interpreted as the union of two events of the form $\bigcap_{n=1}^{\infty} A_n$; for example $\{W[1] = 1,\ W[2] = 0\} = \bigcap_{n=1}^{\infty} A_n$ with $A_1 = \{W[1] = 1\}$, $A_2 = \{W[2] = 0\}$, and $A_n = \Omega_n$ for $n \geq 3$. Hence $\mathscr{F}_\infty$ must include all such events.

To construct a probability measure on Ω_∞, we start with sets of the form $A_\infty = \bigcap_{n=1}^{\infty} A_n$ and define in the case of independent trials,

$$P_\infty[A_\infty] \triangleq \prod_{n=1}^{\infty} P[A_n] \quad \text{(where we have prematurely used the continuity of } P\text{)}.$$

We then extend this probability measure to all of $\mathscr{F}_\infty$ by using Axiom 4 and the fact that every member of $\mathscr{F}_\infty$ is expressible as the countable union and intersection of events of the form $\bigcap_{n=1}^{\infty} A_n$.

We have thus constructed the probability space $(\Omega_\infty, \mathscr{F}_\infty, P_\infty)$ corresponding to the infinite-length Bernoulli trials, with associated Bernoulli random sequence

$$W[n, \zeta] = W(\zeta_n), \qquad n \geq 1.$$

Continuity of Probability Measure. When dealing with an infinite number of events, we have seen that continuity of the probability measure can be quite useful. Fortunately, the desired continuity can be shown to be a direct consequence of the four axioms.

Theorem 6.1–1. Consider an *increasing sequence* of events B_n, that is, $B_n \subset B_{n+1}$ for all $n \geq 1$ as shown in Fig. 6.1–1. Define $B_\infty \triangleq \bigcup_{n=1}^{\infty} B_n$, then $\lim_{n\to\infty} P[B_n] = P[B_\infty]$.

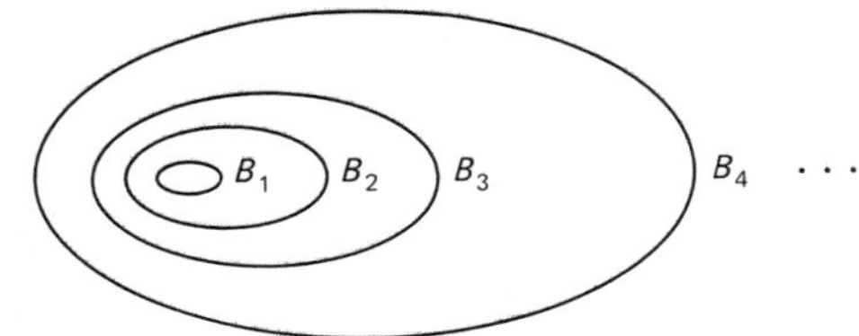

Figure 6.1–1 Increasing sequence of events.

Proof. Define the sequence of events A_n as follows:

$$A_1 \triangleq B_1$$

$$A_n \triangleq B_n B_{n-1}^c, \qquad n > 1.$$

The A_n are disjoint and $\bigcup_{n=1}^{N} A_n = \bigcup_{n=1}^{N} B_n$ for all N. Also $B_N = \bigcup_{n=1}^{N} B_n$ because the B_n are increasing. So

$$P[B_N] = P\left[\bigcup_{n=1}^{N} B_n\right] = P\left[\bigcup_{n=1}^{N} A_n\right] = \sum_{n=1}^{N} P[A_n],$$

and

$$\lim_{N\to\infty} P[B_N] = \lim_{N\to\infty} \sum_{n=1}^{N} P[A_n]$$

$$= \sum_{n=1}^{\infty} P[A_n] \quad \text{by definition of the limit of a sum,}$$

$$= P\left[\bigcup_{n=1}^{\infty} A_n\right] \quad \text{by Axiom 4,}$$

$$= P[B_\infty] \quad \text{by definition of the } A_n.$$

This last step is so because $\bigcup_{n=1}^{\infty} A_n = \bigcup_{n=1}^{\infty} B_n \triangleq B_\infty$. ■

Corollary. Let B_n be a *decreasing* sequence of events, that is, $B_n \supset B_{n+1}$ for all $n \geq 1$. Then

$$\lim_{n\to\infty} P[B_n] = P[B_\infty],$$

where

$$B_\infty \triangleq \bigcap_{n=1}^{\infty} B_n.$$

Proof. Similar to proof of Theorem 6.1–1 (omitted). ■

We next use the continuity of P to prove an elementary fact about probability distribution functions (PDF's).

Example 6.1–4: The PDF is continuous from the *right*; that is, for $F_X(x) = P[X(\zeta) \leq x]$ (cf. Property (iii) of F_X in Section 2.3) we have

$$\lim_{n\to\infty} F_X\left(x + \frac{1}{n}\right) = F_X(x).$$

To show this, we define:

$$B_n \triangleq \left\{\zeta : X(\zeta) \leq x + \frac{1}{n}\right\}$$

and note that B_n is a decreasing sequence of events, where $B_\infty \triangleq \bigcap_{n=1}^{\infty} B_n = \{\zeta : X(\zeta) \leq x\}$ and

$$F_X\left(x + \frac{1}{n}\right) = P[B_n].$$

By application of the corollary, we then have

$$\lim_{n\to\infty} F_X\left(x + \frac{1}{n}\right) = \lim_{n\to\infty} P[B_n] = P[B_\infty]$$

$$= F_X(x).$$

A random sequence is said to be *statistically specified* by its nth-order distribution functions for all $n \geq 1$, and all times $k, k+1, \ldots, k+n-1$, that is, if we are given

$$F_X(x_k, x_{k+1}, x_{k+2}, \ldots, x_{k+n-1}) \\ = P[X[k] \leq x_k, X[k+1] \leq x_{k+1}, \ldots, X[k+n-1] \leq x_{k+n-1}]. \tag{6.1–3}$$

We note that this specification requires an infinite set of PDF's of all orders. To make the observation times explicit when F_X is evaluated at a particular sequence $x[n] = a_n$, we may also write the PDF as

$$F_X(a_k, a_{k+1}, \ldots, a_{k+n-1}; k, k+1, \ldots, k+n-1). \tag{6.1–4}$$

It may seem that this statistical specification is some distance from a complete description of the entire random sequence since no one distribution function in this infinite set of distribution functions describes the entire random sequence. Nevertheless, if we specify all the joint distributions at all finite times and use the continuity of the probability measure that we have just shown, we can calculate the probabilities of events involving infinite numbers of random variables.

Thus, in summary, we have seen two ways to specify the random sequence: the statistical approach (Equation 6.1–4) and the direct specification in terms of the random functions $X[n, \zeta]$. We use the word *statistical* to indicate that the former information can be obtained, at least conceptually, by estimating the nth order PDF's for $n = 1, 2, 3, \ldots$ and so forth.

The nth-order pdf's, if they exist, are given as

$$f_X(x_k, x_{k+1}, \ldots, x_{k+n-1}) \\ = \partial^n F_X(x_k, x_{k+1}, \ldots, x_{k+n-1}) / \partial x_k \, \partial x_{k+1} \ldots \partial x_{k+n-1}, \tag{6.1–5}$$

for every integer k and positive integer n. Sometimes we will omit the subscript X when no confusion can arise. As in Equation 6.1–4, we will sometimes find it convenient to make the times more explicit by also writing the pdf as

$$f_X(a_k, a_{k+1}, \ldots, a_{k+n-1}; k, k+1, \ldots, k+n-1).$$

We will sometimes want to deal with complex random variables and sequences. By this we mean an ordered pair of real random variables, that is, $X = X_R + jX_I$ with PDF

$$F_X(x_R, x_I) \triangleq P[X_R \leq x_R, X_I \leq x_I].$$

The corresponding pdf is then

$$f_X(x_R, x_I) = \frac{\partial^2 F_X(x_R, x_I)}{\partial x_R \, \partial x_I}.$$

To simplify notation we will write $f_X(x)$ for $f_X(x_R, x_I)$ in what follows, with the understanding that the respective integrals (or sums in the case of discrete complex random variables) are really double integrals on the (x_R, x_I) plane if the random variable is complex.

The moments of a random sequence play an important role in most applications. The first moment or *mean function* of a random sequence is defined as

$$\mu_X[n] \triangleq E[X[n]] = \int_{-\infty}^{+\infty} x_n f(x_n)\, dx_n$$

$$= \int_{-\infty}^{+\infty} x f_X(x; n)\, dx \tag{6.1–6}$$

for a continuous-valued random sequence X. The mean function for a discrete-valued random sequence is evaluated as

$$\mu_X[n] = E[X[n]] = \sum_{i=-\infty}^{+\infty} iP[X[n] = i]. \tag{6.1–7}$$

In the case of a mixed random sequence, as in the case of mixed random variables, it is convenient to write

$$\mu_X[n] = \int_{-\infty}^{+\infty} x_n f(x_n)\, dx_n + \sum_{i=-\infty}^{+\infty} iP[X[n] = i]. \tag{6.1–8}$$

Actually using the concept of the Stieltjes integral both can be written in the one form,

$$\mu_X[n] = \int_{-\infty}^{+\infty} x_n\, dF(x_n),$$

in terms of the PDF F. In fact, this form also covers the case where the discrete-valued part of the random variable is not integer valued, all with a relatively simple notation. For more on Stieltjes integrals see Reference 6–3.

The expected value of the product $X[k]X^*[l]$ is called the *autocorrelation function*, a function of k and l,

$$R_X[k, l] \triangleq E[X[k]X^*[l]]$$

$$= \iint_{-\infty}^{+\infty} x_k x_l^* f(x_k, x_l)\, dx_k\, dx_l. \tag{6.1–9}$$

Later we shall see that the conjugate on the second factor results in some notational simplicities in the case of complex valued random sequences, which find application in the study of certain bandpass signals and noises. We can also define the *centered random sequence* $X_c[n] \triangleq X[n] - \mu_X[n]$, which is zero-mean, and consider its autocorrelation function, which is called the *autocovariance function* of the original sequence X. It is defined as

$$K_X[k, l] \triangleq E[(X[k] - \mu_X[k])(X[l] - \mu_X[l])^*]. \tag{6.1–10}$$

We note that, dropping the subscript X,

$$R[k, l] = R^*[l, k] \tag{6.1–11a}$$

$$K[k, l] = K^*[l, k], \tag{6.1–11b}$$

called *hermitian symmetry*, and also note

$$K[k, l] = R[k, l] - \mu[k]\mu^*[l]. \tag{6.1–11c}$$

The variance function is denoted $\sigma_X^2[n] \triangleq K[n, n]$.

Example 6.1–1a (*cont'd.*): The mean function of $X[n]$ as given in Example 6.1–1a is

$$\mu_X[n] = E[X[n]] = E[Xf[n]] = \mu_X f[n],$$

where μ_X is the mean of the random variable X. The autocorrelation function is

$$\begin{aligned} R[k, l] &= E[X[k]X^*[l]] = E[Xf[k]X^*f^*[l]] \\ &= E[|X|^2]f[k]f^*[l], \end{aligned}$$

and so the autocovariance function is given as,

$$\begin{aligned} K[k, l] &= E[|X|^2]f[k]f^*[l] - |\mu_X|^2 f[k]f^*[l] \\ &= (E[|X|^2 - |\mu_X|^2])f[k]f^*[l], \\ &= E[|X - \mu_X|^2]f[k]f^*[l] \\ &= \sigma_X^2 f[k]f^*[l]. \end{aligned}$$

We thus see that the variance $\sigma_X^2[n]$ is just $\sigma_X^2 |f[n]|^2$.

We look at a more random sequence in the next example.

Example 6.1–5: Consider the random sequence consisting of independent random variables $\Upsilon[n]$ for $n \geq 1$, each identically distributed with the exponential pdf of Equation 2.4–12, that is,

$$f_\Upsilon(t; n) = \lambda \exp(-\lambda t)u(t), \qquad n = 1, 2, \ldots.$$

We also need the running sum of the $\Upsilon[k]$ up to time n, defined as

$$T[n] \triangleq \sum_{k=1}^{n} \Upsilon[k] \tag{6.1–12}$$

and consider $T[n]$ as a new random sequence for $n = 1, 2, \ldots$.

The arrival of random events in time is often modeled in this way. We say that $T[n]$ is the *time to the nth arrival* and we call the $\Upsilon[n]$ the *interarrival times*. Later, in Chapter 7, we shall see that the important Poisson random process can be described in this way. Here we want to determine the density of $T[n]$ at each n based on the definition of Equation 6.1–12. Using the fact that the $\Upsilon[k]$ are independent, we can use Equation 3.6–3 and conclude that the probability density function of $T[n]$ will be the $(n - 1)$-fold convolution product of exponential pdf's. Using convolution to determine the pdf of $T[2]$ we get

$$f_T(t; 2) = \lambda^2 t \exp(-\lambda t)u(t).$$

Convolving this result with the exponential pdf a second time we get

$$f_T(t; 3) = \tfrac{1}{2}\lambda^3 t^2 \exp(-\lambda t)u(t).$$

If we now assume the general result

$$f_T(t; n) = \frac{(\lambda t)^{n-1}}{(n - 1)!}\lambda \exp(-\lambda t)u(t), \tag{6.1–13}$$

we can establish this result by the *principle of mathematical induction*: That is, first show the formula is correct at $n = 1$, then show that if the formula is true at $n - 1$, it must also be true at n.

We see that $f_T(t; 1)$ in Equation 6.1–13 is correct, so we proceed by assuming Equation 6.1–13 is true at $n - 1$. By convolving with the exponential, we can show that it is true at n as follows:

$$
\begin{aligned}
f_T(t; n) &= f_T(t; n-1) * \lambda \exp(-\lambda t) u(t) \\
&= \int_0^t \lambda \exp(-\lambda t) \frac{(\lambda \tau)^{n-2}}{(n-2)!} \lambda \exp(-\lambda(t - \tau))\, d\tau\, u(t) \\
&= \lambda^n \exp(-\lambda t) \int_0^t \frac{\tau^{n-2}}{(n-2)!}\, d\tau\, u(t) \\
&= \lambda^n \exp(-\lambda t) \frac{t^{n-1}}{(n-1)!} u(t).
\end{aligned}
$$

Using the independence of the $\Upsilon[n]$, we can also compute the mean as

$$\mu_T[n] = n\mu_\Upsilon = n(1/\lambda) = n/\lambda$$

and variance of the sum $T[n]$ by repeated use of property (A) on p. 87

$$\text{Var}[T[n]] = n\,\text{Var}[\Upsilon] = n/\lambda.$$

We next introduce the most widely used random model in electrical engineering, communications, and control: the Gaussian random sequence. Its wide popularity stems from two important facts: (1) the Central Limit Theorem 3.6–2 assures that many processes occurring in practice are approximately Gaussian; and (2) the mathematics is especially tractable in problems involving estimation, detection, and control.

Definition 6.1–2. A random sequence $X[n]$ is called a *Gaussian random sequence* if all its nth-order PDF's are jointly Gaussian.

We note that the mean and covariance function will specify a Gaussian random sequence in the same way that the mean vector and covariance matrix determine a Gaussian random vector (see Section 4.5). This is because each nth-order distribution function is just the PDF of a Gaussian random vector whose mean vector and covariance matrix are in turn specified by the mean and covariance functions of the Gaussian random sequence.

Example 6.1–6: Let $W[n]$ be a Gaussian random sequence with mean $\mu_W[n] = 0$ for all n and autocorrelation function $R[k, l] = \sigma^2 \delta[k - l]$, where δ is the *discrete-time impulse*

$$\delta[n] \triangleq \begin{cases} 1 & n = 0 \\ 0 & n \neq 0. \end{cases}$$

Then each nth-order distribution function factors into a product of n first-order distribution functions. Hence the elements of the random sequence are jointly independent. Such a sequence is said to be an *independent random sequence*. Next define

$$X[n] = W[n] + W[n-1] \quad \text{for } -\infty < n < +\infty.$$

Then $X[n]$ is also Gaussian in all its nth-order distributions (since a linear transformation of a Gaussian random vector produces a Gaussian vector); hence $X[n]$ is also a Gaussian random sequence. We can easily evaluate the mean of $X[n]$ as

$$\begin{aligned}\mu_X[n] = E[X[n]] &= E[W[n]] + E[W[n-1]] \\ &= 0,\end{aligned}$$

and its correlation function as

$$\begin{aligned}R_X[k,l] &= E[X[k]X[l]^*] \\ &= E[(W[k] + W[k-1])(W[l] + W[l-1])^*] \\ &= E[W[k]W[l]^*] + E[W[k]W[l-1]^*] \\ &\quad + E[W[k-1]W[l]^*] + E[W[k-1]W[l-1]^*] \\ &= R_W[k,l] + R_W[k,l-1] + R_W[k-1,l] + R_W[k-1,l-1] \\ &= \sigma^2(\delta[k-l] + \delta[k-l+1] + \delta[k-l-1] + \delta[k-l]).\end{aligned}$$

We can plot this autocorrelation in the (k, l) plane as shown in Figure 6.1–2 to see the extent of the dependence of the random sequence X.

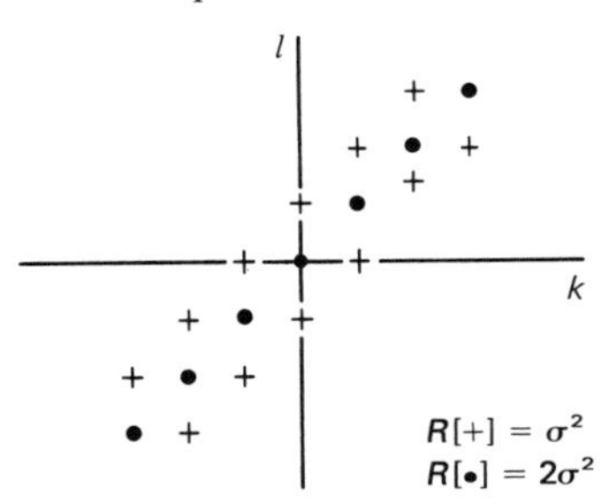

Figure 6.1–2 Plot of correlation function.

From Figure 6.1–2 we see that the autocorrelation has value $2\sigma^2$ on the diagonal line $l = k$ and has value σ^2 on the diagonal lines $l = k \pm 1$. It should be clear from Figure 6.1–2 that $X[n]$ is not an independent random sequence. However, the banded support of this covariance function signifies that dependence is limited to shift $(k - l) = \pm 1$.

Example 6.1–3 (cont'd): (The random walk.) Continuing the example of Bernoulli trials, we define a random sequence $X[n]$ as the number of successes minus the number of failures in n trials multiplied times a step factor s. The resulting sequence then models a random walk on the integers starting at 0 at $n = 0$. At each succeeding time unit a step of size s is taken either to the right or to the left. After n steps we will be at a position rs for some integer r. This is illustrated in Figure 6.1–3.

If there are k successes and necessarily $(n - k)$ failures, then we have the following relation:

$$\begin{aligned}rs &= ks - (n-k)s \\ &= (2k - n)s,\end{aligned}$$

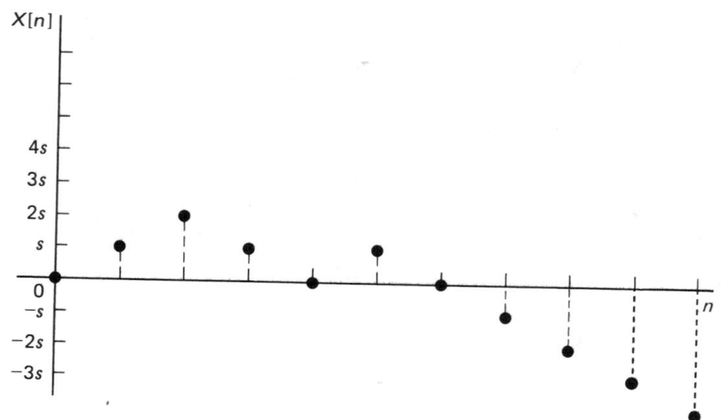

Figure 6.1–3 Sample sequence of random walk.

which implies that $k = (r + n)/2$, for those values of r that make this value an integer. Then with $P[s] = P[f] = \frac{1}{2}$, we have

$$P[X[n] = rs] = P[(r + n)/2 \text{ successes}]$$

$$= \begin{cases} \binom{n}{\frac{n+r}{2}} 2^{-n}, & \frac{n+r}{2} \text{ an integer} \\ 0, & \text{else.} \end{cases}$$

Using the fact that $X[n] = W[1] + W[2] + \ldots + W[n]$ and that the W's are jointly independent, we can compute the mean and variance of the random walk as follows:

$$E[X[n]] = \sum_{k=1}^{n} E[W[k]] = \sum_{k=1}^{n} 0 = 0,$$

and

$$E[X^2[n]] = \sum_{k=1}^{n} E[W^2[k]]$$

$$= \sum_{k=1}^{n} 0.5[(+s)^2 + (-s)^2]$$

$$= ns^2.$$

If we normalize $X[n]$ by dividing $\sqrt{n}$ and define

$$\tilde{X}[n] \triangleq \frac{1}{\sqrt{n}} X[n],$$

then by the Central Limit Theorem 3.6–2 we have that the PDF of $\tilde{X}[n]$ converges to $N(0, s^2)$. Thus for n large enough, we can approximate the probabilities

$$P[a < \tilde{X}[n] \leq b] = P\left[a\sqrt{n} < X[n] \leq b\sqrt{n}\right] \simeq \text{erf}(b/s) - \text{erf}(a/s).$$

Note, however, that when this probability is small, very large values of n might be required to keep the percentage error small because small errors in the distribution function may be comparable to the required probability value. In practice this means that the normal approximation will not be dependable on the tails of the distribution but only in the central part, hence the name Central Limit Theorem.

Note also that while $X[n]$ can never be considered approximately Gaussian for any n, (for example, if n is even $X[n]$ can only be an even multiple of s) still we can approximately calculate the probability

$$P[(r-2)s < X[n] < rs] = P\left[\frac{(r-2)s}{\sqrt{n}} < \tilde{X}[n] \leq \frac{rs}{\sqrt{n}}\right]$$

$$= \frac{1}{\sqrt{2\pi}} \int_{(r-2)/\sqrt{n}}^{r/\sqrt{n}} \exp(-0.5v^2)\, dv$$

$$\approx 1/\sqrt{\pi(n/2)} \exp(-r^2/2n),$$

when r is not large with respect to $\sqrt{n}$. See Section 1.9 for a similar result.

The waiting-time sequence in Example 6.1–5 and the random walk in Example 6.1–3(*cont'd*) both have the property that they build up over time from independent increments. More generally we can define an independent-increments property.

10-22
10-24

Definition 6.1–3. A random sequence is said to have *independent-increments* if for all $N > 1$ and all $n_1 < n_2 < \ldots < n_N$, the increments $X[n_1]$, $X[n_2] - X[n_1], \ldots, X[n_N] - X[n_{N-1}]$ are jointly independent.

If a random sequence has independent increments one can build up its nth-order probabilities (PMF's and pdf's) as products of the probabilities of its increments. (For an application see Problem 6.7.)

Many practical applications of random sequences involve the important case where the underlying statistical properties are constant or invariant with respect to the parameter n, which is normally time or space. This simplifies the random model in two ways: First, the model can be specified with many fewer parameters than would be the case if the statistics were varying with time, and second, these few parameters can then be more reliably estimated from the data.

Definition 6.1–4. If for all positive n the nth-order PDF's of a random sequence do not depend on the shift parameter k, then the random sequence is said to be *stationary*. In symbols, for all $n \geq 1$

$$f(a_0, a_1, \ldots, a_{n-1}; k, k+1, \ldots, k+n-1)$$
$$= f(a_0, a_1, \ldots, a_{n-1}; 0, 1, \ldots, n-1) \qquad (6.1\text{–}14)$$

for all $-\infty < k < +\infty$ and for all a_0 thru a_{n-1}.

If we look back at Example 6.1–6 we see that $X[n]$ and $W[n]$ are both stationary random sequences. The same was true of T$[n]$ in Example 6.1–5, but the random arrival time sequence $T[n]$ was clearly nonstationary.

It is often desirable to partially characterize a random sequence based on knowledge of only its first two moments, that is, its mean function and its covariance function. This has already been mentioned in Chapter 4 in connection with linear estimation of random vectors. We will encounter this for random sequences when we continue our discussion of linear estimation in Chapter 10. Thus we define a weakened kind of stationarity that involves only the mean and covariance (or correlation) functions. Specifically, if these functions are consistent with stationarity, then we say that the random sequence is *wide-sense stationary*, or just WSS.

Definition 6.1–5. If the mean function is constant for all n and the covariance function is shift-invariant for all k, l, and shift n

$$\mu[n] = \mu[0] \quad \text{and} \quad K[k, l] = K[k + n, l + n], \tag{6.1–15}$$

then the random sequence is called *wide-sense stationary*.

It should be noticed that all stationary sequences are wide-sense stationary but not the reverse. For example, the third moment could be shift-variant in a manner not consistent with stationarity even though the first moment is constant and the second moment is shift-invariant. Then the random sequence would be wide-sense stationary but not stationary. To distinguish these cases, sometimes the stationary case is called *strict-sense stationary* to avoid confusion with the weaker concept of wide-sense stationarity. Going the other way one can establish that all stationary random sequences are in fact also wide-sense stationary. This will then establish the claim made earlier that wide-sense stationary is a weaker form of stationarity.

Theorem 6.1–2. Wide-sense stationarity is implied by stationarity.

Proof. We first show that the mean is constant for a stationarity random sequence. Let n be arbitrary

$$\mu_X[n] = E[X[n]] = \int_{-\infty}^{+\infty} x f_X(x; n)\, dx = \int_{-\infty}^{+\infty} x f_X(x; 0)\, dx = \mu_X[0],$$

since $f_X(x; n)$ does not depend on n. We show that the covariance function is shift-invariant by first showing that the correlation function is shift-invariant:

$$\begin{aligned} R[k, l] &= E[X[k]X^*[l]] \\ &= \iint_{-\infty}^{+\infty} x_k x_l^* f(x_k, x_l)\, dx_k\, dx_l \\ &= \iint_{-\infty}^{+\infty} x_{n+k} x_{n+l}^* f(x_{n+k}, x_{n+l})\, dx_{n+k}\, dx_{n+l}, \\ &= R_X[n + k, n + l], \end{aligned}$$

since $f(x_k, x_l)$ doesn't depend on the shift n, and the x's are dummy variables. Finally, we use Equation 6.1–11c and the result on the means to conclude that the covariance function is also shift-invariant. ■

Since the covariance function is shift-invariant for any wide-sense stationary random sequence, we can define a one-parameter covariance function to simplify the notation for WSS sequences

$$K[n] \triangleq E[X_c[k+n]X_c^*[k]]. \tag{6.1–16}$$

We also do the same for correlation functions.

Example 6.1–6 (*cont'd.*): The covariance function of Example 6.1–6 is shift-invariant and so can take advantage of the simplified notation. We write

$$K_X[n] = \sigma^2(2\delta[n] + \delta[n-1] + \delta[n+1]).$$

In Section 6.2 we provide a review or summary of the theory of linear systems for sequences, that is, discrete-time linear system theory. Readers with adequate background may skip this section. In Section 6.3 we will apply this theory to study the effect of linear systems on random sequences, an area rich in applications in communications, signal processing, and control.

6.2 BASIC PRINCIPLES OF DISCRETE-TIME LINEAR SYSTEMS

In this section we present some fundamental material on discrete-time linear system theory. This will then be extended in the next section to the case of random sequence inputs and outputs. The material is very similar to the continuous-time linear system theory including the topics of differential equations, Fourier transforms, and Laplace transforms. The corresponding quantities in the discrete-time theory are difference equations, Fourier transforms (for discrete-time signals), and Z-transforms.

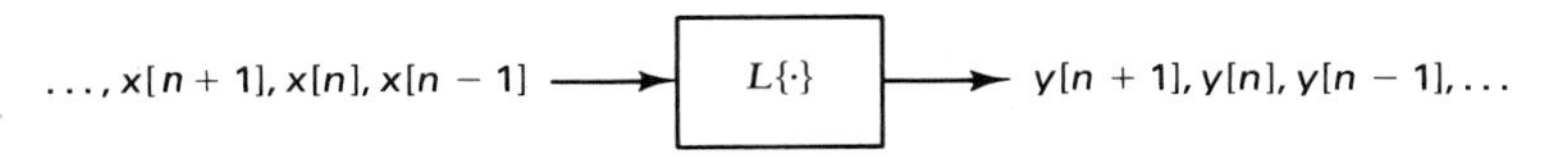

Figure 6.2–1 Discrete–time linear system.

With reference to Figure 6.2–1 we see that a linear system can be thought of as having an infinite length sequence $x[n]$ as input with a corresponding infinite-length sequence $y[n]$ as output. Representing this linear operation in equation form we have

$$y[n] = L\{x[n]\}, \tag{6.2–1}$$

where the linear operator L is defined to satisfy the following familiar definition adapted to the case of discrete-time signals.

Definition 6.2–1. We say a system with operator L is *linear* if for all permissible input sequences $x_1[n]$ and $x_2[n]$, and for all permissible scalar gains a_1 and a_2 we have

$$L\{a_1x_1[n] + a_2x_2[n]\} = a_1L\{x_1[n]\} + a_2L\{x_2[n]\}.$$

Examples of linear systems would include *moving averages* such as

$$y[n] = 0.33(x[n+1] + x[n] + x[n-1]), \qquad -\infty < n < +\infty,$$

and *autoregressions* such as,

$$y[n] = ay[n-1] + by[n-2] + cx[n], \qquad 0 \le n < +\infty,$$

when the initial conditions are zero. Both these equations are special cases of the more general *linear constant-coefficient difference equation* (LCCDE),

$$y[n] = \sum_{k=1}^{M} a_k y[n-k] + \sum_{k=0}^{N} b_k x[n-k]. \tag{6.2–2}$$

Example 6.2–1: (Solution of a difference equation.) Consider the following second-order LCCDE,

$$y[n] = 1.7y[n-1] - 0.72y[n-2] + u[n], \quad \text{with} \quad y[-1] = y[-2] = 0. \tag{6.2–3}$$

We first find the general solution to the *homogeneous equation*

$$y_h[n] = 1.7y_h[n-1] - 0.72y_h[n-2].$$

We try $y_h[n] = Ar^n$, where A and r are to be determined and obtain

$$A(r^n - 1.7r^{n-1} + 0.72r^{n-2}) = 0$$

or

$$Ar^{n-2}(r^2 - 1.7r + 0.72) = 0.$$

We thus see that any value of r satisfying the *characteristic equation*

$$r^2 - 1.7r + 0.72 = 0$$

will give a general solution to the homogeneous equation. In this case there are two roots at $r_1 = 0.8$ and $r_2 = 0.9$. By linear superposition the general *homogeneous solution* must be of the form

$$y_h[n] = A_1 r_1^n + A_2 r_2^n,$$

where the constants A_1 and A_2 will be determined using the initial conditions.

To obtain the *particular solution*, we first observe that the input equals 1 for $n \ge 0$. Thus we try as a particular solution a constant, that is,

$$y_p[n] = B \quad \text{for } n \ge 0$$

and obtain

$$B - 1.7B + 0.72B = 1$$

or

$$B = 1/(1 - 1.7 + 0.72) = 1/0.02 = 50.$$

More generally this method can be modified for any input of the form $C\rho^n$ over a time interval n_1 to n_2. One just assumes the corresponding form for the solution and determines the constant B as shown. In this approach, we solve the difference equation for each time interval separately, piecing the solution together at the boundaries by carrying across the initial conditions. We illustrate our approach here for the time interval starting at $n = 0$. The total solution is

$$\begin{aligned} y[n] &= y_h[n] + y_p[n] \\ &= A_1(0.8)^n + A_2(0.9)^n + 50 \quad \text{for } n \ge 0. \end{aligned}$$

To determine A_1 and A_2 we first evaluate Equation 6.2–3 at $n = 0$ and $n = 1$ using $y[-1] = y[-2] = 0$ to carry across the initial conditions to obtain $y[0] = 1$ and $y[1] = 2.7$, from which we obtain the linear equations

$$A_1 + A_2 + 50 = 1 \quad (\text{at } n = 0)$$

and

$$A_1(0.8) + A_2(0.9) + 50 = 2.7 \quad (\text{at } n = 1).$$

This can be put in matrix form

$$\begin{bmatrix} 1.0 & 1.0 \\ 0.8 & 0.9 \end{bmatrix} \begin{bmatrix} A_1 \\ A_2 \end{bmatrix} = \begin{bmatrix} -49.0 \\ -47.3 \end{bmatrix}$$

and solved to yield

$$\begin{bmatrix} A_1 \\ A_2 \end{bmatrix} = \begin{bmatrix} 32 \\ -81 \end{bmatrix}.$$

Thus the complete solution is

$$y[n] = 32(0.8)^n - 81(0.9)^n + 50, \qquad n \geq 0.$$

A system is called *linear time-invariant* (LTI) or *linear shift invariant* (LSI) if the response to a delayed (shifted) input is just the delayed (shifted) response. More precisely we have the following.

Definition 6.2–2. A linear system L is called *shift-invariant* if for all integers k with $-\infty < k < +\infty$ we have

$$y[n + k] = L\{x[n + k]\} \quad \text{for all } n. \tag{6.2–4}$$

An important property of LSI systems is that they are described by *convolution*, that is, L is a convolution operator,

$$y[n] = h[n] * x[n] = x[n] * h[n],$$

where

$$h[n] * x[n] = \sum_{k=-\infty}^{+\infty} h[k]x[n - k], \tag{6.2–5}$$

and

$$h[n] \triangleq L\{\delta[n]\},$$

which is called the *impulse response*.

In words we can say that—just as for continuous-time systems—if we know the impulse response of an LSI system, then we can compute the response to any other input by carrying out the convolution operation. In the discrete-time case this convolution operation is a summation rather than an integration, but the operation is otherwise the same.†

While in principle we could determine the output to any input, given knowledge of the impulse response, in practice the calculation of the convolution operation may be tedious and time consuming. To facilitate such calculations and also to gain added insight, we turn to a frequency-

† This is seen to be analogous to the case of discrete-valued and continuous-valued random variables as regards probability mass functions and density functions respectively, for the sum of two independent random variables.

domain characterization of LSI systems. We begin by defining the *Fourier Transform* (FT) for sequences as follows.

Definition 6.2–3. The *Fourier transform* for a discrete-time signal or sequence is defined by the infinite sum

$$X[\omega] = FT\{x[n]\} = \sum_{n=-\infty}^{+\infty} x[n]e^{-j\omega n}. \tag{6.2–6a}$$

The *inverse Fourier transform* is given as

$$x[n] = IFT\{X[\omega]\} = \frac{1}{2\pi}\int_{-\pi}^{+\pi} X[\omega]e^{j\omega n}\,d\omega. \tag{6.2–6b}$$

One can see that the Fourier transform and its inverse for sequences are really just the familiar Fourier series with the sequence x playing the role of the Fourier coefficients and the Fourier transform X playing the role of the periodic function. Thus the existence and uniqueness theorems of Fourier series are immediately applicable here to the Fourier transform for discrete-time signals.

For an LSI system the Fourier transform is particularly significant owing to the fact that the complex exponentials are the *eigenfunctions* of a discrete-time linear system, that is,

$$L\{e^{j\omega n}\} = H[\omega]e^{j\omega n}, \tag{6.2–7}$$

so long as the impulse response h is absolutely summable. For LSI systems this absolute summability can easily be seen to be equivalent to bounded-input bounded-output (BIBO) stability [6–4].

Just as in continuous-time system theory, multiplication of Fourier transforms corresponds to convolution in the time (or space) domain. The following theorem establishes this fact.

Theorem 6.2–1. The convolution,

$$y[n] = x[n] * h[n], \qquad -\infty < n < +\infty, \tag{6.2–8a}$$

is equivalent in the transform domain to

$$Y[\omega] = X[\omega]H[\omega], \qquad -\pi \le \omega \le +\pi. \tag{6.2–8b}$$

Proof.

$$\begin{aligned}
Y[\omega] &= \sum_{n=-\infty}^{+\infty} y[n]e^{-j\omega n} = \sum_{n=-\infty}^{+\infty} [x[n] * h[n]]e^{-j\omega n}, \\
&= \sum_n \sum_k x[k]h[n-k]e^{-j\omega n} = \sum_n \sum_k x[k]h[n-k]e^{-j\omega(n-k+k)}, \\
&= \sum_n \sum_k [x[k]e^{-j\omega k}h[n-k]e^{-j\omega(n-k)}] \\
&= \sum_k x[k]e^{-j\omega k}\left[\sum_n h[n-k]e^{-j\omega(n-k)}\right], \\
&= \sum_k x[k]e^{-j\omega k}H[\omega] \\
&= X[\omega]H[\omega]. \quad \blacksquare
\end{aligned}$$

Thus discrete-time linear shift invariant systems are easily understood in the frequency domain in analogy with the case of the continuous-time LSI system. Analogous to the Laplace transform for continuous-time signals, there is the Z-transform for discrete-time signals. It is defined as follows.

Definition 6.2–4. The *Z-transform* of a discrete-time signal or sequence is defined as the infinite summation

$$X[z] \triangleq \sum_{n=-\infty}^{+\infty} x[n]z^{-n}, \tag{6.2–9}$$

where z is a complex variable in the *region of absolute convergence* of the infinite sum.

Note that $X[z]$ is a function of a *complex* variable, while $X[\omega]$ is a function of a *real* variable. The two are related by $X[e^{j\omega}] = X[\omega]$. We thus see that the Fourier transform is just the restriction of the Z-transform to the unit circle in the z-plane. Similarly to the proof of Theorem 6.2–1, it is easy to show that Equation 6.2–8b is also true for Z-transforms. Analogous to continuous-time theory, the Z-transform of the impulse response of an LSI system is called the *system function.* For more information on discrete-time signals and systems, the reader is referred to Reference 6–4.

6.3 RANDOM SEQUENCES AND LINEAR SYSTEMS

In this section we look at the topic of linear systems with random sequence inputs. In particular we will look at how the mean and covariance functions are transformed by both linear and LSI systems. We will do this first for the general case of a nonstationary random sequence and then specialize to the more common case of a stationary sequence. The topics of this section are perhaps the most widely used concepts from the theory of random sequences. Applications arise in communications when analyzing signals and noise in linear filters, in digital signal processing for the analysis of quantization noise in digital filters, and in control theory to find the effect of disturbance inputs on an otherwise deterministic control system.

The first issue is the meaning of inputing a random sequence to a linear system. The problem is that a random sequence is not just one sequence but a whole family of sequences indexed by the parameter ζ, a point in the sample space. As such for each fixed ζ, the random sequence is just an ordinary sequence that may be a permissible input for the linear system. Thus when we talk about a linear system with a random sequence input, it is natural to say that for each point in the sample space Ω, we input the corresponding realization, that is, the sample sequence $x[n]$. We would therefore regard the corresponding output $y[n]$ as a sample sequence corresponding to the same point ζ in the sample space, thus defining an output random sequence $Y[n]$.

Definition 6.3–1. When we write $Y[n] = L\{X[n]\}$ for a random sequence, $X[n]$ and a linear system L mean that for each $\zeta \in \Omega$ we have

$$Y[n, \zeta] = L\{X[n, \zeta]\}. \tag{6.3–1}$$

This is the simplest way to treat systems with random inputs. A difficulty arises when the sample sequences do not "behave well," in which case it may not be possible to define the system operation for every one of them. In Chapter 8 we will generalize this definition and discuss a so-called mean-square description of the system operation, which avoids such problems, although of necessity it will be more abstract.

In most cases it is very hard to find the probability distribution of the output from the probabilistic description of the input to a linear system. The reason is that since the impulse response is often very long (or infinitely long), high-order distributions of the input sequence would be required to determine the output PDF. In other words, if $Y[n]$ depends on $X[n], \ldots, X[n-k+1]$, the kth-order pdf of X is required to compute the first-order pdf of Y. Moreover, the kth-order pdf of X may not even be known. The situation with moment functions is different. The moments of the output random sequence can be calculated from equal- or lower-order moments of the input random sequence. Partly for this reason, it is of considerable interest to determine the output moment functions in terms of the input moment functions. In the practical and important case of the Gaussian random sequence, we have seen that the entire probabilistic description depends only on the mean and covariance functions. In fact because the linear system is in effect performing a linear transformation on the infinite-dimensional vector that constitutes the input sequence, we can see that the output sequence will also obey the Gaussian law in its nth-order distributions if the input sequence is Gaussian. Thus the determination of the first- and second-order moment functions of the output is particularly important when the input sequence is Gaussian.

Theorem 6.3–1. For a linear system L and a random sequence $X[n]$, the mean of the output random sequence $Y[n]$ is

$$E[Y[n]] = L\{E[X[n]]\} \tag{6.3–2}$$

so long as both sides are well defined.

Proof (formal). Since L is a linear operator, we can write

$$y[n] = \sum_{k=-\infty}^{+\infty} h[n, k]x[k]$$

for each sample sequence input–output pair, or

$$Y[n, \zeta] = \sum_{k=-\infty}^{+\infty} h[n, k]X[k, \zeta].$$

If we operate on both sides with the expectation operator E we get

$$E[Y[n]] = E\left[\sum_{k=-\infty}^{+\infty} h[n,k]X[k]\right].$$

Now if we bring the operator E inside the right-hand sum we get

$$E[Y[n]] = \sum_{k=-\infty}^{+\infty} h[n,k]E[X[k]]$$

$$= L\{E[X[n]]\}. \quad \blacksquare$$

Some comments are necessary with regard to this interchange of the expectation and linear operators. This cannot always be done, even when the expectation of the input exists. For example, if the input is WSS with nonzero mean and the linear system is a summation into the infinite past, that is,

$$y[n] = \sum_{k=0}^{\infty} x[n-k].$$

In this case the mean of Y would be infinite. We will come back to this point when we study stochastic convergence in Section 6.4.

There are special cases of Equation 6.3–2 depending on whether the input sequence is WSS and whether the system is LSI. If the system is LSI and the input is at least WSS then the mean of the output is given as

$$E[Y[n]] = \sum_{k=-\infty}^{+\infty} h[n-k]\mu_X.$$

wide-sense stationary

Now because μ_X is a constant, we can take it out of the sum and obtain

$$E[Y[n]] = \left[\sum_{k=-\infty}^{+\infty} h[k]\right]\mu_X \qquad (6.3\text{–}3a)$$

$$= H[1+j0]\,\mu_X, \qquad (6.3\text{–}3b)$$

at least whenever $\sum_{k=-\infty}^{+\infty} |h[k]|$ exists, that is, for any BIBO stable system (see Section 6.2).

Thus we observe that in this case the mean of the output random sequence is a constant equal to the *dc gain* or *constant gain* of the LSI system times the mean of the input sequence.

Example 6.3–1: Let the system be a lowpass filter with system function

$$H[z] = 1/(1 + az^{-1}), \quad |z|>|a|$$

where $|a| < 1$ for stability of this assumed causal filter (that is, the region of convergence is $|z| > |a|$, which includes the unit circle). Then if a WSS sequence is the input to this filter, the mean of the output will be

$$E[Y[n]] = H[1+j0]E[X[n]]$$

$$= (1+a)^{-1}\mu_X.$$

We now turn to the problem of calculating the output covariance and correlation of the linear system L. We will find it convenient to introduce a cross-correlation function between the input and output,

$$R_{XY}[m, n] \triangleq E[X[m]Y[n]^*]. \tag{6.3–4}$$

Since there will be two times under consideration, that is, m and n, and since we will be using operator notation, that is, L, it will be necessary to distinguish which time index is being operated on. We will refer to the operator L as L_k meaning that the index k is considered the time index of the input and output of the operator (filter). This will become clear in what follows. Also we need the concept of *adjoint operator* L^*. This is the operator with impulse response $h^*(n, k)$, that is, the conjugate operator.

Theorem 6.3–2. Let $X[n]$ and $Y[n]$ be two random sequences that are the input and output respectively of the linear operator L_n. Let the input correlation function be $R_{XX}[m, n]$. Then the cross- and output correlation functions are given by

$$R_{XY}[m, n] = L_n^*\{R_{XX}[m, n]\} \tag{6.3–5a}$$

and

$$R_{YY}[m, n] = L_m\{R_{XY}[m, n]\}. \tag{6.3–5b}$$

Proof. Write $X[m]Y^*[n] = X[m]L_n^*\{X^*[n]\}$

$$= L_n^*\{X[m]X^*[n]\}.$$

Then
$$\begin{aligned} E[X[m]Y^*[n]] &= E\left[L_n^*\{X[m]X^*[n]\}\right], \\ &= L_n^*\{E[X[m]X^*[n]]\}, \\ &= L_n^*\{R_{XX}[m, n]\}, \end{aligned}$$

thus establishing the first part of the theorem. To show the second part, we proceed analogously by multiplying $Y[m]$ by $Y^*[n]$ to get

$$\begin{aligned} E[Y[m]Y^*[n]] &= E\left[L_m\{X[m]Y^*[n]\}\right], \\ &= L_m\{E[X[m]Y^*[n]]\}, \\ &= L_m\{R_{XY}[m, n]\}. \quad \blacksquare \end{aligned}$$

If we combine both parts of Theorem 6.3–2 we get an operator expression for the output correlation in terms of the input correlation function:

$$R_{YY}[m, n] = L_m\{L_n^*\{R_{XX}[m, n]\}\},$$

To find the corresponding results for covariance functions, we note that the centered output sequence is the output due to the centered input sequence, due to the linearity of the system and Equation 6.3–2. Then applying

Theorem 6.3–2 to these zero-mean sequences, we have immediately that

$$K_{XY}[m, n] = L_n^*\{K_{XX}[m, n]\} \tag{6.3–6a}$$

$$K_{YY}[m, n] = L_m\{K_{XY}[m, n]\} \tag{6.3–6b}$$

and

$$K_{YY}[m, n] = L_m\{L_n^*\{K_{XX}[m, n]\}\}.$$

Example 6.3–2: Let $Y[n] \triangleq X[n] - X[n-1] = L\{X[n]\}$, a real operator, which represents a first-order (backward) difference. It could be used to locate an impulse noise spike in some random data. The output mean is $E[Y[n]] = L\{E[X[n]]\} = \mu_X[n] - \mu_X[n-1]$. The cross-correlation function function is

$$\begin{aligned} R_{XY}[m, n] &= L_n\{R_{XX}[m, n]\} \\ &= R_{XX}[m, n] - R_{XX}[m, n-1]. \end{aligned}$$

The output autocorrelation function is

$$\begin{aligned} R_{YY}[m, n] &= L_m\{R_{XY}[m, n]\} \\ &= R_{XY}[m, n] - R_{XY}[m-1, n] \end{aligned}$$

$$- R_{XX}[m, n] - R_{XX}[m-1, n] - R_{XX}[m, n-1] + R_{XX}[m-1, n-1].$$

ıput random sequence were WSS with autocorrelation function,

$$R_{XX}[m, n] = a^{|m-n|}, \qquad 1 > a > 0,$$

above example would specialize to

$$\mu_Y[n] = 0,$$

$$R_{XY}[m, n] = a^{|m-n|} - a^{|m-n+1|}$$

$$R_{YY}[m, n] = 2a^{|m-n|} - a^{|m-1-n|} - a^{|m-n+1|},$$

involves only $m - n$. Hence the output sequence is WSS and we can with $k = m - n$)

$$R_{YY}[k] = 2a^{|k|} - a^{|k-1|} - a^{|k+1|},$$

s plotted in Figure 6.3–1.

e 6.3–3: (Solving for covariance functions of recursive systems.) With $|\alpha| < 1$

$$Y[n] = \alpha Y[n-1] + (1-\alpha) W[n] \tag{6.3–7}$$

0 subject to $Y[-1] = 0$. Since the i.c. is zero, the system is equivalently LSI 0, so we can represent L by convolution, where

$$h[n] = (1-\alpha)\, \alpha^n\, u[n].$$

ecializing Equation 6.3–2 we obtain

$$\mu_Y[n] = \sum_{k=0}^{\infty} (1-\alpha)\, \alpha^k\, \mu_W[n-k], \text{ where } \mu_W[n] = 0 \text{ for } n < 0.$$

ıg Equations 6.3–6 to this case enables us to write, for α real,

$$K_{WY}[m, n] = \sum_{k=0}^{\infty} (1-\alpha)\, \alpha^k\, K_{WW}[m, n-k]$$

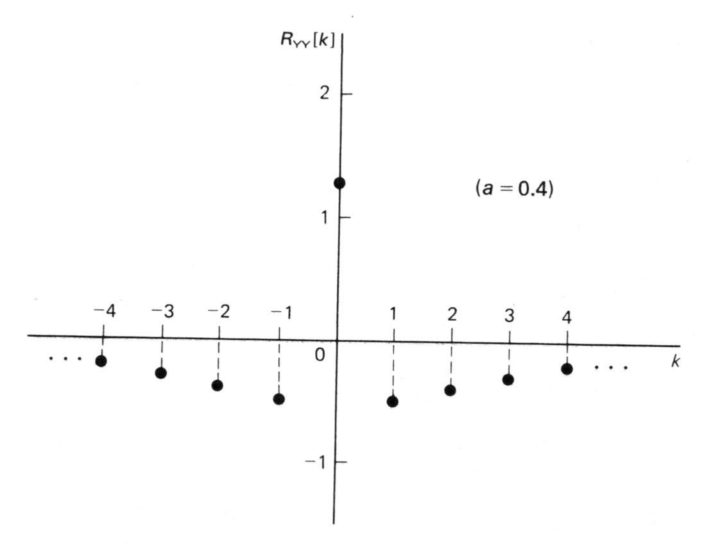

Figure 6.3–1 Plot of correlation function of Example 6.3–2.

and

$$K_{YY}[m, n] = \sum_{l=0}^{\infty} (1 - \alpha)\, \alpha^{l}\, K_{WY}[m - l, n],$$

which can be combined to yield

$$K_{YY}[m, n] = \sum_{k=0}^{\infty} \sum_{l=0}^{\infty} (1 - \alpha)^2\, \alpha^{k}\, \alpha^{l}\, K_{WW}[m - l, n - k].$$

Now if the input sequence $W[n]$ has covariance function

$$K_{WW}[m, n] = \sigma_W^2\, \delta[m - n]\, u(m)\, u(n),$$

then the output covariance is calculated as

$$\begin{aligned} K_Y[m, n] &= \sum_{k=0}^{n} (1 - \alpha)^2\, \alpha^k\, \alpha^{(m-n)+k} \sigma_W^2 \quad \text{for} \quad m \geq n, \\ &= \alpha^{(m-n)}(1 - \alpha)^2 \sum_{k=0}^{n} \alpha^{2k} \sigma_W^2 \\ &= \alpha^{(m-n)}(1 - \alpha)^2/(1 - \alpha^2)\sigma_W^2\, (1 - \alpha^{2n+2}) \\ &= [(1 - \alpha)/(1 + \alpha)]\alpha^{|m-n|}\sigma_W^2\, (1 - \alpha^{2\min(m,n)+2}) \quad \text{for all } m, n, \end{aligned}$$

where the last step follows from the required symmetry in (m, n).

As an alternative to this method of solution, one can take the expectation of Equation 6.3–7 to directly obtain a recursive equation for the mean which can be solved by the methods of Section 6.2:

$$\mu_Y[n] = \alpha \mu_Y[n - 1] + (1 - \alpha)\mu_W[n], \qquad n \geq 1,$$

with an initial condition specified at, say, $n = 0$. For example, if $\mu_Y[0] = 0$ and $\mu_W[n] = \mu_W$, a given constant, then the solution is

$$\mu_Y[n] = (1 - \alpha^n)\mu_W u[n].$$

We can also use this method to calculate the cross-correlation between input and output. First we conjugate Equation 6.3–7, then multiply by $W[m]$ and finally take the expectation to yield, for α real,

$$R_{WY}[m, n] = \alpha R_{WY}[m, n - 1] + (1 - \alpha)R_{WW}[m, n], \qquad (6.3\text{–}8)$$

which can be solved directly for R_{WY} in terms of R_{WW}. The partial difference equation for the output correlation R_{YY} is obtained by re-expressing Equation 6.3–7 as a function of m, multiplying by $Y^*[n]$ and then taking the expectation to yield

$$R_{YY}[m, n] = \alpha R_{YY}[m - 1, n] + (1 - \alpha)R_{WY}[m, n]. \qquad (6.3\text{–}9)$$

These two equations can be solved by the methods of Section 6.2 since they can each be seen to be one-dimensional difference equations with constant coefficients in one index, with the other index simply playing the role of a parameter. Thus, for example, one must solve Equation 6.3–8 as a function of n for each value of m in succession.

In the next section we shall continue our analysis of random sequences by investigating limits and convergence. This will serve as a good background for Chapter 7 where we investigate random functions with a continuous-time parameter, that is, *random processes.*

6.4 CONVERGENCE OF RANDOM SEQUENCES

Some nonstationary random sequences may converge to a limit as the sequence index goes to infinity. This asymptotic behavior is evidenced in probability theory by convergence of the fraction of successes in an infinite Bernoulli sequence, where the relevant theorems are called the laws of large numbers. Also, when we study random processes in Chapter 7 we will sometimes make a sequence of finer and finer approximations to the output of a random system at a given time, say, t_0, that is, $Y_n(t_0)$. The index n then defines a random sequence, which should converge in some sense to the true output. In this section we will look at several types of convergence for random sequences, that is, sequences of random variables.

We start by reviewing the concept of convergence for deterministic sequences. Let x_n be a sequence of complex (or real) numbers; then convergence is defined as follows.

Definition 6.4–1. A sequence of complex (or real) numbers x_n *converges* to the complex (or real) number x if given any $\varepsilon > 0$, there exists an integer n_0 such that whenever $n > n_0$, we have

$$|x_n - x| < \varepsilon.$$

Note that in this definition the value n_0 may depend on the value ε; that is, when ε is made smaller then most likely n_0 will need to be made larger. Sometimes this dependence is formalized by writing $n_0(\varepsilon)$ in place of n_0 in

this definition. A practical problem with this definition is that one
the limit x to test for convergence. For simple cases one can c
what the limit is and then use the definition to verify that this li
exists. Fortunately, for more complex situations there is an altern
Cauchy criterion for convergence, which we state as a theor
proof.

Theorem 6.4–1. (Cauchy criterion [6–5].) A sequence
(or real) numbers x_n converges to a limit if and only if (iff)

$$|x_n - x_m| \to 0 \text{ as both } n \text{ and } m \to \infty.$$

The reason that this works for complex (or real) numbers is th
all complex (or real) numbers is *complete*, meaning that it co
limit points. For example, the set $\{0 < x < 1\} = (0, 1)$ is not
the set $\{0 \leq x \leq 1\} = [0, 1]$ is complete because sequences
and tending to 0 or 1 have a *limit point* in the set $[0, 1]$ but
point in the set $(0, 1)$. In fact, the set of all complex (or re: is
complete as well as n-dimensional linear vector spaces over al
and complex number fields. Thus the Cauchy criterion fo e
applies in these cases also. For more on numerical converge].

Convergence for functions is defined using the concept ce
of sequences of numbers. We say the sequence of functions $f_n(x)$ converges
to the function $f(x)$ if the corresponding sequence of numbers converges for
each x. It is stated more formally in the following definition.

Definition 6.4–2. The sequence of functions $f_n(x)$ *converges* (point-wise) to the function $f(x)$ if for each x_0 the sequence of complex numbers $f_n(x_0)$ converges to $f(x_0)$.

The Cauchy criterion for convergence applies to pointwise convergence of functions if the set of functions under consideration is complete. The set of continuous functions is not complete because a sequence of continuous functions may converge to a discontinuous function. However, the set of bounded functions is complete [6–5].

The following are some examples of convergent sequences of numbers and functions. We leave the demonstration of these results as exercises for the reader.

Example 6.4–1: (Some convergent sequences.)

a. $x_n = (1 - 1/n)a + (1/n)b \longrightarrow a$ as $n \to \infty$.

b. $x_n = \sin(\omega + e^{-n}) \longrightarrow \sin \omega$

c. $f_n(x) = \sin[(\omega + 1/n)x] \longrightarrow \sin(\omega x)$, for all x.

d. $f_n(x) = \begin{Bmatrix} e^{-n^2 x} & \text{for } x > 0 \\ 1 & \text{for } x < 0 \end{Bmatrix} \to u(-x)$

Since a random variable is a function, a sequence of random variables (also called a random sequence) is a sequence of functions. Thus we can define the first and strongest type of convergence for random variables.

Definition 6.4–3. (Sure convergence.) The random sequence $X[n]$ *converges surely* to the random variable X if the sequence of functions $X[n, \zeta]$ converges to the function $X(\zeta)$ as $n \to \infty$ for all $\zeta \in \Omega$.

As a reminder, the functions $X(\zeta)$ are not arbitrary. They are random variables and thus satisfy the condition that the set $\{\zeta : X(\zeta) \leq x\} \subset \mathcal{F}$ for all x, that is, that this set be an event for all values of x. This is in fact necessary for the calculation of probability since the probability measure P is only defined for events. Such functions X are more generally called *measurable functions* and in a course on real analysis it is shown that the space of measurable functions is complete [6–1]. If we have a Cauchy sequence of measurable functions (random variables), then one can show that the limit function exists and is also measurable (a random variable). Thus the Cauchy convergence criterion also applies for random variables.

Most of the time we are not interested in precisely defining a random variable for sets in Ω of probability zero because it is thought that these events can never occur. In this case, we can weaken the concept of sure convergence to the still very strong concept of almost-sure convergence.

Definition 6.4–4. (Almost-sure convergence.) The random sequence $X[n]$ converges *almost surely* to the random variable X if the sequence of functions $X[n, \zeta]$ converges for all $\zeta \in \Omega$ except possibly on a set of probability zero.

This is the strongest type of convergence normally used in probability theory. It is also called *probability-1* convergence. It is sometimes written

$$P\left[\lim_{n\to\infty} X[n, \zeta] = X(\zeta)\right] = 1,$$

meaning simply that there is a set A such that $P[A] = 1$ and $X[n]$ converges to X for all $\zeta \in A$. In particular $A \triangleq \{\zeta : \lim_{n\to\infty} X[n, \zeta] = X(\zeta)\}$. Here the set A^c is the probability-zero set mentioned in Definition 6.4–4. As shorthand notation we also use

$$X[n] \to X \text{ a.s.} \quad \text{and} \quad X[n] \to X \text{ pr. 1},$$

where the abbreviation a.s. stands for *almost surely*.

An example of probability-1 convergence is the Strong Law of Large Numbers to be proved in Section 6.5. Three examples of random sequences are next evaluated for possible convergence.

Example 6.4–2: (Convergence of random sequences.) For each of the following three random sequences, we assume that the probability space $(\Omega, \mathcal{F}, P)$ is given as: $\Omega = [0, 1]$. $\mathcal{F}$ is the family of Borel subsets of Ω and the probability measure P is

Lebesgue measure, which on a real interval $(a, b]$, is just its length l, that is,

$$l(a, b] \triangleq b - a \quad \text{for } b \geq a.$$

(a) $X[n, \zeta] = n\zeta$

(b) $X[n, \zeta] = \sin(n\zeta)$

(c) $X[n, \zeta] = \exp[-n^2(\zeta - 1/n)]$

The sequence in (a) clearly diverges to $+\infty$ for any $\zeta \neq 0$. Thus this random sequence does not converge. The sequence in (b) does not diverge but it oscillates between -1 and $+1$ except for the one point $\zeta = 0$. Thus this random sequence does not converge either. Considering the random sequence in (c), the graph in Figure 6.4–1 shows that this sequence converges as follows:

$$\lim_{n\to\infty} X[n, \zeta] = \begin{cases} \infty & \text{for } \zeta = 0 \\ 0 & \text{for } \zeta > 0. \end{cases}$$

Thus we can say that the random sequence converges to the (degenerate) random variable $X = 0$ with probability-1. We simply take $A = (0, 1]$ and note that $P[A] = 1$ and that $X[n, \zeta] \simeq 0$ for every ζ in A. We write $X[n] \to 0$ a.s. However $X[n]$ clearly does not converge surely to zero.

Thus far we have been discussing pointwise convergence of sequences of functions and random sequences. This is the same as considering a space of *bounded* functions $\mathcal{F}$ with the norm

$$\|f\|_\infty \triangleq \sup_x |f(x)|.\dagger$$

When we write $f_n \to f$ in the function space $\mathcal{F}$, we mean that $\|f_n - f\|_\infty = \sup_x |f_n(x) - f(x)| \to 0$, giving us pointwise convergence. This space of bounded functions is denoted L^∞ and is complete [6–1].

Another type of function space of great practical interest uses the

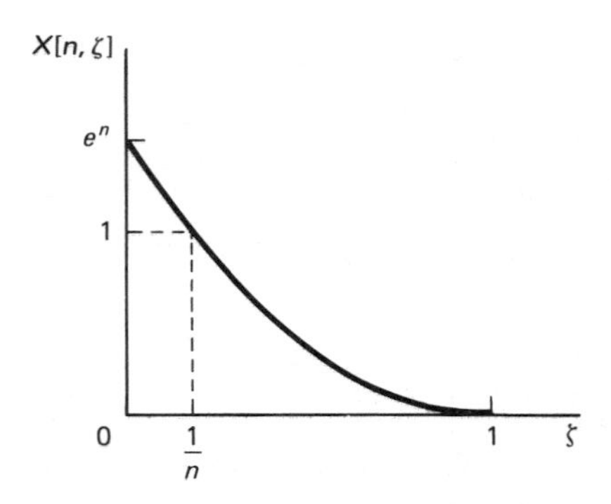

Figure 6.4–1 Random sequence plotted versus ζ.

† The supremum or sup operator is almost the same as the max operator. The supremum of a set of numbers is the smallest number greater than or equal to each number in the set, for example, $\sup\{0 < x < 1\} = 1$. Note the difficulty with max in this example since 1 is not included in the open interval $(0, 1)$.

energy norm (cf. Equation 3.4–6).

$$\|f\|_2 \triangleq \left(\int_{-\infty}^{+\infty} |f(x)|^2 \, dx\right)^{1/2}.$$

The space of integrable (measurable) functions with *finite energy norm* is denoted L^2. When we say a sequence of functions converges in L^2, that is, $\|f_n - f\|_2 \to 0$, we mean that

$$\left(\int_{-\infty}^{+\infty} |f_n(x) - f(x)|^2 \, dx\right)^{1/2} \to 0.$$

This space of integrable functions is also complete [6–1]. A corresponding concept for random sequences is given by mean-square convergence (cf. Equation 3.4–11).

Definition 6.4–5. (Mean-square convergence.) A random sequence $X[n]$ converges in the *mean-square sense* to the random variable X if $E[|X[n] - X|^2] \to 0$ as $n \to \infty$.

This type of convergence depends only on the second-order properties of the random variables and is thus often easier to calculate. A second benefit of the mean-square type of convergence is that it is closely related to the physical concept of power. If $X[n]$ converges to X in the mean-square sense, then we can expect that the variance of the error $\varepsilon[n] \triangleq X[n] - X$ will be small for large n. If we look back at Example 6.4–2(c), we can see that this random sequence does not converge in the mean-square sense, so that the error variance or power as defined here would not be expected ever to be small. To see this, consider possible mean-square convergence to zero (since $X[n] \to 0$ a.s.),

$$\begin{aligned}
E[|X[n] - 0|^2] &= E[X[n]^2] \\
&= \int_0^1 \exp(-2n^2\zeta)\exp 2n \, d\zeta \\
&= \exp(2n)\int_0^1 \exp(-2n^2\zeta)\, d\zeta \\
&= \exp(2n)\left[\frac{1 - \exp(-2n^2)}{2n^2}\right] \to \infty \quad \text{as } n \to \infty,
\end{aligned}$$

$e^{-n^2(\varphi - 1/n)} = X[n]$

hence $X[n]$ does not converge in the mean-square sense to 0.

Still another type of convergence that we will consider is called *convergence in probability*. It is weaker than probability-1 convergence and also weaker than mean-square convergence. This is the type of convergence displayed in the Weak Laws of Large Numbers to be discussed in Section 6.5. It is defined as follows:

Definition 6.4–6. (Convergence in probability.) Given the random sequence $X[n]$ and the limiting random variable X, we say that $X[n]$

narrower Gaussian still has outliers!

converges in probability to X if for every $\varepsilon > 0$,

$$\lim_{n\to\infty} P[|X[n] - X| > \varepsilon] = 0.$$

We sometimes write $X[n] \to X\ (p)$. Also convergence in probability is sometimes called p-convergence.

One can use Chebyshev's inequality (Theorem 3.4–1), $P[|Y| > \varepsilon] \le E[|Y|^2]/\varepsilon^2$ for $\varepsilon > 0$, to show that mean-square convergence implies convergence in probability. For example, let $Y \triangleq X[n] - X$; then the preceding inequality becomes

$$P[|X[n] - X| > \varepsilon] \le E[|X[n] - X|^2]/\varepsilon^2.$$

Now the right-hand side goes to zero as $n \to \infty$ given mean-square convergence, which implies that the left-hand side must also go to zero for any fixed $\varepsilon > 0$, which is the definition of convergence in probability. Thus we have proved the following result.

Theorem 6.4–2. Convergence of a random sequence in the mean-square sense implies convergence in probability.

The relation between convergence with probability-1 and convergence in probability is more subtle. The main difference between them can be seen by noting that the former talks about the probability of the limit while the latter talks about the limit of the probability. Further insight can be gained by noting that a.s. convergence is concerned with convergence of the entire sample sequences while p-convergence is concerned only with the convergence of the random variable at an individual n. That is to say, a.s. convergence is concerned with the joint events at an infinite number of times, while p-convergence is concerned with the simple event at time n. One can prove the following theorem.

Theorem 6.4–3. Convergence with probability-1 implies convergence in probability.

Proof. (adapted from Gnedenko [6–6].) Let $X[n] \to X$ a.s. and define the set A,

$$A \triangleq \bigcap_{k=1}^{\infty} \bigcup_{n=1}^{\infty} \bigcap_{m=1}^{\infty} \{\zeta : |X[n+m, \zeta] - X(\zeta)| < 1/k\}.$$

Then it must be that $P[A] = 1$. To see this we note that A is the set of ζ such that starting at some n and for all later n we have $|X[n, \zeta] - X(\zeta)| < 1/k$ and furthermore this must hold for all $k > 0$. Thus A is precisely the set of ζ on which $X[n, \zeta]$ is convergent. So $P[A]$ must be one. Eventually for n large enough and $1/k$ small enough we get $|X[n, \zeta] - X(\zeta)| < \varepsilon$, and the errors stays this small for all larger n. Thus

$$P\left[\bigcup_{n=1}^{\infty} \bigcap_{m=1}^{\infty} \{|X[n+m] - X| < \varepsilon\right] = 1 \qquad \text{for all } \varepsilon > 0,$$

which implies by the continuity of probability,

$$\lim_{n\to\infty} P\left[\bigcap_{m=1}^{\infty} \{|X[n+m]-X|<\varepsilon\}\right]=1 \qquad \text{for all } \varepsilon>0,$$

which in turn implies the greatly weakened result

$$\lim_{n\to\infty} P[|X[n]-X|<\varepsilon]=1 \qquad \text{for all } \varepsilon>0, \tag{6.4–1}$$

which is equivalent to the definition of p-convergence. ■

Because of the gross weakening of the a.s. condition, that is the enlargment of the set A in the foregoing proof, it can be seen that p-convergence does not imply a.s. convergence. We note in particular that Equation 6.4–1 may well be true even though no single sample sequence stays close to X for all $n+m>n$. This is in fact the key difference between these two types of convergence.

Example 6.4–3: Define a random pulse sequence $X[n]$ on $n \geq 0$ as follows: Set $X[0]=1$. Then for the *next two points* set exactly one of the $X[n]$'s to 1, the other to zero, equally likely. For the *next three points* set exactly one of the $X[n]$'s to 1 equally likely and set the others to zero. Continue this procedure for the *next four points*, setting exactly one of the $X[n]$'s to 1 equally likely, and so forth. A sample function would look like Figure 6.4–2.

Obviously this random sequence is slowly converging to zero in some sense as $n\to\infty$. In fact a simple calculation would show p-convergence and also m.s. convergence due to the growing distance between pulses as $n\to\infty$. In fact at $n \simeq \frac{1}{2}l^2$, the probability of a one (pulse) is only $1/l$. However, we do not have a.s. convergence, since *every* sample sequence has ones appearing arbitrarily far out on the n axis. Thus no sample sequences converge to zero.

One final type of convergence that we consider is not a convergence for random variables at all. Rather it is a type of convergence for distribution functions.

Definition 6.4–7. A random sequence $X[n]$ with distribution function $F_n(x)$ *converges in distribution* to the random variable X with distribution

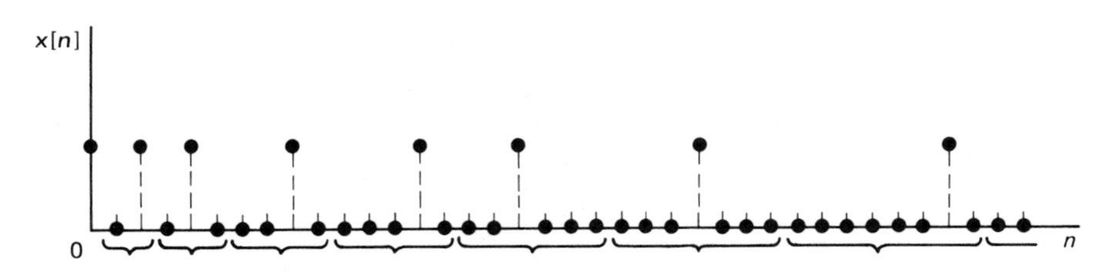

Figure 6.4–2 A sample sequence.

function $F(x)$ if

$$\lim_{n\to\infty} F_n(x) = F(x)$$

at all x for which F is continuous.

Note that in this definition we are not really saying anything about the random variables themselves, just their PDF's. Convergence in distribution just means that as n gets large the PDF's are converging or becoming alike. For example, the sequence $X[n]$ and the variable X can be jointly independent even though $X[n]$ converges to X in distribution. This is radically different from the four earlier types of convergence, where as n gets large the random variables $X[n]$ and X are becoming very dependent because some type of "error" between them is going to zero. Convergence in distribution is the type of convergence that occurs in the Central Limit Theorem (Section 3.7). The relationships between these five types of convergence are shown diagramatically in Figure 6.4–3, where we have used the fact that p-convergence implies convergence in distribution, which can be shown as follows.

Assume that the limiting random variable X is continuous so that it has a pdf. First we consider the conditional distribution function

$$F_{X[n]|X}(y \mid x) = P[X[n] \leq y \mid X = x].$$

From the definition of p-convergence, it should be clear that

$$F_{X[n]|X}(y \mid x) \to \begin{cases} 1 & \text{for } y > x \\ 0 & \text{for } y < x \end{cases} \quad \text{as } n \to \infty,$$

so that

$$F_{X[n]|X}(y \mid x) \to u(y - x)$$

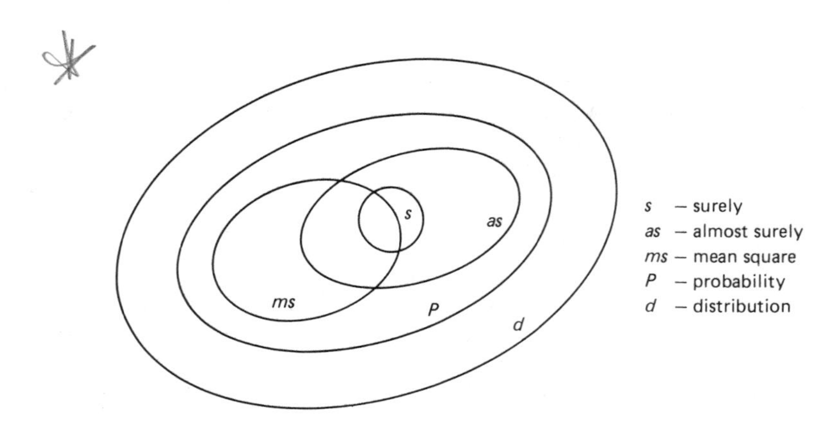

Figure 6.4–3 Venn diagram of relation of types of convergence.

and hence

$$
\begin{aligned}
F_{X[n]}(y) = P[X[n] \le y] &= \int_{-\infty}^{+\infty} F_{X[n]|X}(y \mid x) f_X(x)\, dx \\
&\to \int_{-\infty}^{+\infty} u(y - x) f_X(x)\, dx \\
&= \int_{-\infty}^{y} f_X(x)\, dx \\
&= F_X(y),
\end{aligned}
$$

as was to be shown. In the case where the limiting random variable X is not continuous, we must exercise more care but the result is still true at all points x for which $F_X(x)$ is continuous. (See Problem 6.18.)

6.5 LAWS OF LARGE NUMBERS

The Laws of Large Numbers have to do with the convergence of a sequence of estimates of the mean of a random variable. As such they concern the convergence of a random sequence to a constant. The Weak Laws obtain convergence in probability, while the Strong Laws yield convergence with probability-1. A version of the Weak Law has already been demonstrated in Example 3.4–3. We restate it here for convenience.

Theorem 6.5–1. (Weak Law of Large Numbers.) Let $X[n]$ be an independent random sequence with mean μ_X and variance σ_X^2 defined for $n \ge 1$. Define another random sequence as

$$\hat{\mu}_X[n] \triangleq (1/n) \sum_{k=1}^{n} X[k] \qquad \text{for } n \ge 1.$$

Then $\hat{\mu}_X[n] \to \mu_x \quad (p) \quad \text{as } n \to \infty$.

Another version of the Weak Law allows the random sequence to be of nonuniform variance.

Theorem 6.5–2. (Weak Law—Nonuniform Variance.) Let $X[n]$ be an independent random sequence with constant mean μ_X and variance $\sigma_X^2[n]$ defined for $n \ge 1$. Then if

$$\sum_{n=1}^{\infty} \sigma_X^2[n]/n^2 < \infty,$$

$$\hat{\mu}_X[n] \to \mu_X \quad (p) \quad \text{as } n \to \infty.$$

Both of these theorems are also true for convergence with probability-1, in which case they become Strong Laws. The theorems concerning convergence with probability-1 are best derived using the concept of Martingale sequences. By introducing this concept we can also get another useful result called the Martingale convergence theorem, which is helpful in estimation and decision/detection problems.

Definition 6.5–1.† A random sequence $X[n]$ defined for $n \geq 0$ is called a *Martingale* if the conditional expectation

$$E[X[n] \mid X[n-1], X[n-2], \ldots, X[0]] = X[n-1] \qquad \text{for all } n \geq 1.$$

Viewing the conditional expectation as an estimate of the future value of the sequence based on the past, then for a Martingale this estimate is just the most recent value. If we interpret $X[n]$ as an amount of capital in a betting game, then the Martingale condition can be regarded as necessary for fairness of the game, which in fact is how it was first introduced [6–1].

Example 6.5–1: (Binomial counting sequence.) Let $W[n]$ be a Bernoulli random sequence taking values ± 1 with equal probability and defined for $n \geq 0$. Let $X[n]$ be the corresponding Binomial counting sequence

$$X[n] \triangleq \sum_{k=0}^{n} W[k], \qquad n \geq 0.$$

Then $X[n]$ is a Martingale, which can be shown as follows:

$$\begin{aligned}
E[X[n] \mid X[n-1], \ldots, X[0]] &= E\left[\sum_{k=0}^{n} W[k] \mid X[n-1], \ldots, X[0]\right] \\
&= \sum_{k=0}^{n} E[W[k] \mid X[n-1], \ldots, X[0]] \\
&= \sum_{k=0}^{n} E[W[k] \mid W[n-1], \ldots, W[0]] \\
&= \sum_{k=0}^{n-1} W[k] + E[W[n]] \\
&= X[n-1].
\end{aligned}$$

The first equality follows from the definition of $X[n]$. The third equality follows the fact that knowledge of the first $n-1$ X's is equivalent to knowledge of the first $n-1$ W's. The next-to-last equality follows from $E[W \mid W] = W$.

Example 6.5–2: (Independent-increments sequences.) Let $X[n]$ be an independent-increments random sequence (see Definition 6.1–3) defined for $n \geq 0$. Then $X_c[n] = X[n] - \mu_X[n]$ is a Martingale. To show this we write $X_c[n] = [X_c[n] - X_c[n-1]] + X_c[n-1]$ and note that by independent increments and the fact that the mean of X_c is zero, we have

$$\begin{aligned}
E[X_c[n] \mid X_c[n-1], \ldots, X_c[0]] &= E[X_c[n] - X_c[n-1] \mid X_c[n-1], \ldots, X_c[0]] \\
&\quad + E[X_c[n-1] \mid X_c[n-1], \ldots, X_c[0]] \\
&= E[X_c[n] - X_c[n-1]] + X_c[n-1] \\
&= X_c[n-1].
\end{aligned}$$

The next theorem shows the connection between the strong laws, which have to do with the convergence of sample sequences, and Martingales. It provides a kind of Chebyshev inequality for the maximum term in an n-point Martingale sequence.

† The material dealing with Martingale sequences can be omitted on a first reading.

Theorem 6.5–3. Let $X[n]$ be a Martingale sequence defined on $n \geq 0$. Then for every $\varepsilon > 0$ and for any positive n,

$$P\Big[\max_{0 \leq k \leq n} |X[k]| \geq \varepsilon\Big] \leq E[X^2[n]]/\varepsilon^2.$$

Proof. For $0 \leq j \leq n$, define the mutually exclusive events,

$$A_j \triangleq \{|X[k] \geq \varepsilon \text{ for the } \textit{first time} \text{ at } j\}.$$

Then the event $\{\max_{0 \leq k \leq n} |X[k]| \geq \varepsilon\}$ is just a union of these events. Also define the random variables,

$$I_j \triangleq \begin{cases} 1 & \text{if } A_j \text{ occurs} \\ 0 & \text{otherwise,} \end{cases}$$

called the indicators of the events A_j. Then

$$E[X^2[n]] \geq \sum_{j=0}^{n} E[X^2[n]I_j] \tag{6.5–1}$$

since $\sum_{j=0}^{n} I_j \leq 1$. Also $X^2[n] = [X[j] + (X[n] - X[j])]^2$, so expanding and inserting into Equation 6.5–1 we get

$$\begin{aligned} E[X^2[n]] &\geq \sum_{j=0}^{n} E[X^2[j]I_j] + 2\sum_{j=0}^{n} E[X[j](X[n] - X[j])I_j] \\ &\quad + \sum_{j=0}^{n} E[(X[n] - X[j])^2 I_j] \\ &\geq \sum_{j=0}^{n} E[X^2[j]I_j] + 2\sum_{j=0}^{n} E[X[j](X[n] - X[j])I_j]. \end{aligned} \tag{6.5–2}$$

Letting $Z_j \triangleq X[j]I_j$, we can write the second term in Equation 6.5–2 as $E[Z_j(X[n] - X[j])]$ and noting that Z_j depends only on $X[0], \ldots, X[j]$, we then have

$$\begin{aligned} E[Z_j(X[n] - X[j])] &= E[E[Z_j(X[n] - X[j]) \mid X[0], \ldots, X[j]]] \\ &= E[Z_j E[X[n] - X[j] \mid X[0], \ldots, X[j]]] \\ &= E[Z_j(X[j] - X[j])] \\ &= 0. \end{aligned}$$

Thus Equation 6.5–2 becomes

$$\begin{aligned} E[X^2[n]] &\geq \sum_{j=0}^{n} E[X^2[j]I_j] \\ &\geq \varepsilon^2 E\Big[\sum_{j=0}^{n} I_j\Big] \\ &= \varepsilon^2 P\Big[\bigcup_{j=0}^{n} A_j\Big] \\ &= \varepsilon^2 P[\max_{0 \leq k \leq n} |X[k]| \geq \varepsilon]. \quad \blacksquare \end{aligned}$$

Theorem 6.5–4. (Martingale Convergence Theorem.) Let $X[n]$ be a Martingale sequence on $n \geq 0$, satisfying

$$E[X^2[n]] \leq C < \infty \quad \text{for all } n \text{ for some } C.$$

Then

$$X[n] \to X \quad \text{(a.s.)} \quad \text{as } n \to \infty.$$

Proof. Let $m \geq 0$ and define $Y[n] \triangleq X[n+m] - X[m]$ for $n \geq 0$. Then $Y[n]$ is a Martingale, so by Theorem 6.5–3

$$P\Big[\max_{0 \leq K \leq n} |X[m+k] - X[m]| \geq \varepsilon\Big] \leq \frac{1}{\varepsilon^2} E[Y^2[n]],$$

where

$$\begin{aligned} E[Y^2[n]] &= E[(X[n+m] - X[m])^2] \\ &= E[X^2[n+m]] - 2E[X[n+m]X[m]] + E[X^2[m]]. \end{aligned}$$

Rewriting the middle term, we have

$$\begin{aligned} E[X[m]X[n+m]] &= E\big[X[m]E[X[n+m] \,|\, X[m], \ldots, X[0]]\big] \\ &= E[X[m]X[m]] \\ &= E[X^2[m]] \quad \text{since } X \text{ is a Martingale.} \end{aligned}$$

So

$$E[Y^2[n]] = E\big[X^2[n+m]\big] - E\big[X^2[m]\big] \geq 0 \qquad \text{for all } m, n \geq 0. \tag{6.5–3}$$

Therefore $E[X^2[n]]$ must be monotonic nondecreasing. Since it is bounded from above by $C < \infty$, it must converge to a limit. Since it has a limit, then by Equation 6.5–3 the $E[Y^2[n]] \to 0$ as m and $n \to \infty$. Thus

$$\lim_{m \to \infty} P\Big[\max_{k \geq 0} |X[m+k] - X[m]| > \varepsilon\Big] = 0,$$

which implies $P\Big[\lim_{m \to \infty} \max_{k \geq 0} |X[m+k] - X[m]| > \varepsilon\Big] = 0$ by the continuity of the probability measure P (cf. Corollary to Theorem 6.1–1). Finally by the Cauchy convergence criteria, there exists a r.v. X such that

$$X[n] \to X \quad \text{(a.s.)}. \quad \blacksquare$$

Theorem 6.5–5. (Strong Law of Large Numbers.) Let $X[n]$ be a WSS independent random sequence with mean μ_X and variance σ_X^2 defined for $n \geq 1$. Then as $n \to \infty$

$$\hat{\mu}_X[n] = \frac{1}{n} \sum_{k=1}^{n} X[k] \to \mu_X \quad \text{(a.s.)}.$$

Proof. Let $Y[n] \triangleq \sum_{k=1}^{n} \frac{1}{k} X_c[n]$; then $Y[n]$ is a Martingale on $n \geq 1$.

Since

$$E[Y^2[n]] = \sum_{k=1}^{n} \frac{1}{k^2} \sigma_X^2 \leq \sigma_X^2 \sum_{k=1}^{\infty} \frac{1}{k^2} = C < \infty,$$

we can apply Theorem 6.5–4 to show that $Y[n] \to Y$ (a.s.) for some r.v. Y. Next noting that $X_c[k] = k(Y[k] - Y[k-1])$, we can write.

$$\frac{1}{n}\sum_{k=1}^{n} X_c[k] = \frac{1}{n}\left[\sum_{k=1}^{n} kY[k] - \sum_{k=1}^{n} kY[k-1]\right]$$

$$= -\frac{1}{n}\sum_{k=1}^{n} Y[k] + \frac{n+1}{n} Y[n]$$

$$\to -Y + Y = 0 \quad \text{(a.s.)}$$

so that

$$\hat{\mu}_X[n] \to \mu_X \quad \text{(a.s.).} \quad \blacksquare$$

6.6 SUMMARY

In this chapter we introduced the concept of random sequence and studied some of its properties. We defined the random sequence as an infinite-length sequence of random variables. We looked at a few important random sequences. Then we reviewed linear discrete-time theory and considered the practical problem of finding out how the mean and covariance function are transformed by a linear system. We looked at convergence of random sequences and learned to appreciate the variety of modes of convergence that are possible. We then applied some of these results to the laws of large numbers and used Martingale properties to prove the important strong law of large numbers.

In the next chapter we will discover that many of these results extend to the case of continuous time as we study random processes.

PROBLEMS

6.1. Prove the chain rule for the probability of the intersection of N events, $\{A_n\}_{n=1}^{N}$. For example, for $N = 3$ we have,

$$P[A_1A_2A_3] = P[A_1]P[A_2 \mid A_1]P[A_3 \mid A_1A_2].$$

Interpret this result for joint distribution functions and joint pdf's.

6.2. Often one is given a problem statement starting as follows: "Let X be a real valued random variable (r.v.) with pdf $f_X(x)$ . . .". Since an r.v. is a mapping from a sample space with a field of events and a probability measure, evidently the existence of an underlying probability space is assumed by such a problem statement. Show that a suitable underlying probability space can always be created, thus legitimizing problem statements such as the one above.

√ **6.3.** Let T be a continuous random variable denoting the time at which the first photon is emitted from a light source; T is measured from the instant the source is energized. Assume that the probability density function for T is $f_T(t) = \lambda e^{-\lambda t}u(t)$.

(a) What is the probability that at least one photon is emitted prior to time t_2 if it is known that none was emitted prior to time t_1, where $t_1 < t_2$?

(b) What is the probability that at least one photon is emitted prior to time t_2 if three independent sources of this type are energized simultaneously?

6.4. Let X be a conditionally normal random variable, with density function $N(\mu, \sigma^2)$, given the values of $M = \mu$ and $\Sigma^2 = \sigma^2$.

(a) Assume σ^2 is a known constant but that M is a random variable having the distribution function

$$F_M(m) = [l - e^{-\lambda m}]u(m)$$

where λ is a known positive value. Determine the characteristic function for X.

(b) Now assume both Σ^2 and M are independent random variables. Let their distributions be arbitrary, but assume both have a finite mean and variance. Determine the mean and variance for X in terms of those for Σ^2 and M.

6.5. Let X and Y be independent, identically distributed (i.i.d.) random variables with the exponential probability density function

$$f_X(w) = f_Y(w) = \lambda e^{-\lambda w}u(w).$$

(a) Determine the probability density function for the ratio

$$0 \le R \triangleq \frac{X}{X+Y} \le 1, \quad \text{i.e., } f_R(r), \qquad 0 < r \le 1.$$

(b) Let A be the event $X < \dfrac{1}{y}$. Determine the conditional probability density function of X given that A occurs and that $Y = y$; that is, determine

$$f_X(x \mid A, Y = y).$$

(c) Using the definitions of (b), what is the minimum mean-square error estimate of X when it is known that the event A occurs and that $Y = y$?

6.6. The random variables X_1, X_2, X_3, and X_4 are jointly Gaussian distributed with zero mean and covariances $K_{ij} = E[X_iX_j]$. Use the moment theorem of characteristic functions to conclude the useful formula

$$E[X_1X_2X_3X_4] = K_{12}K_{34} + K_{13}K_{24} + K_{14}K_{23}.$$

6.7. Let $\{X_i\}$ be a sequence of independent, identically distributed normal random variables with zero mean and unit variance. Let

$$S_k \triangleq X_1 + X_2 + \ldots + X_k \quad \text{for } k \ge 1.$$

Determine the joint probability density function for S_n and S_m where $1 \le m < n$.

6.8. In Example 6.1–4 we saw that PDF's are continuous from the right. Are they continuous from the left also? Either prove or give a counterexample.

6.9. Let $x[n]$ be a *deterministic* input to the linear, shift-invariant (LSI) discrete-time system H shown in Figure P-6.9.

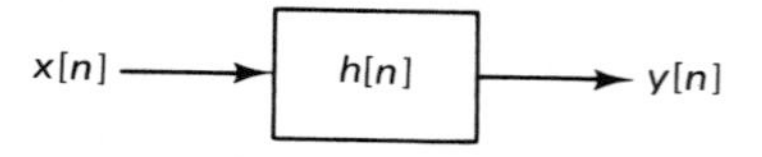

Figure P-6.9 LSI system H.

(a) Use linearity and shift-invariance properties to show that

$$y[n] = x[n] * h[n] \triangleq \sum_{k=-\infty}^{+\infty} x[k]h[n-k] = h[n] * x[n].$$

(b) Define the Fourier transform of a sequence $a(n)$ as

$$A[\omega] \triangleq \sum_{n=-\infty}^{\infty} a[n]e^{-j\omega n}, \qquad -\pi \leq \omega \leq +\pi$$

and show that the inverse Fourier transform is

$$a[n] = \frac{1}{2\pi}\int_{-\pi}^{+\pi} A[\omega]e^{+j\omega n}\, d\omega, \qquad -\infty < n < +\infty.$$

(c) Using the results in (a) and (b), show that

$$Y[\omega] = H[\omega]X[\omega], \qquad -\pi \leq \omega \leq +\pi$$

for an LSI discrete-time system.

6.10. Consider the difference equation:

$$y[n] + \alpha y[n-1] = x[n], \qquad -\infty < n < +\infty$$

where $-1 < \alpha < +1$.

(a) Let the input be $x[n] = \beta^n u[n]$ for $-1 < \beta < +1$. Find the solution for $y[n]$ assuming causality applies, i.e., $y(n) = 0$ for $n < 0$.

†(b) Let the input be $x[n] = \beta^{-n}u[-n]$ for $-1 < \beta < +1$. Find the solution for $y[n]$ assuming anticausality applies, i.e., $y(n) = 0$ for $n > 0$.

6.11. Let $W[n]$ be an independent random sequence with mean 0 and variance σ_W^2 defined for $-\infty < n < +\infty$. For appropriately chosen ρ, let a stationary random sequence $X[n]$ satisfy

$$X[n] = \rho X[n-1] + W[n], \qquad -\infty < n < +\infty$$

(a) Show that $X[n-1]$ and $W[n]$ are independent at time n.

(b) Derive the characteristic function equation

$$\Phi_X(\omega) = \Phi_X(\rho\omega)\Phi_W(\omega).$$

(c) Find the continuous solution to this functional equation for the unknown function Φ_X when $W[n]$ is assumed to be Gaussian. Note: $\Phi_X(0) = 1$.

(d) What is σ_X^2?

6.12. Consider the linear shift-invariant (LSI) system shown in Figure P-6.12, whose *deterministic input* $x[n]$ is contaminated by noise (a random sequence) $W[n]$. We wish to determine the properties of the output random sequence $Y[n]$. The noise $W[n]$ has mean $\mu_W[n] = 2$ and autocorrelation $E[W[m]W[n]] = \sigma_W^2\delta[m-n] + 4$. The impulse response is $h[n] = \rho^n u[n]$ with $|\rho| < 1$. The deterministic input $x[n]$ is given as $x[n] = 3$ for all n.

(a) Find the output mean $\mu_Y[n]$.

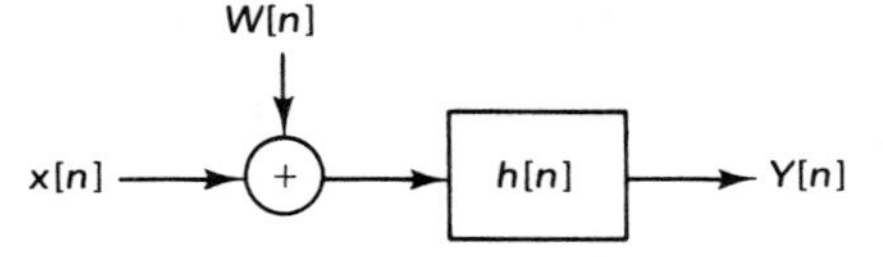

Figure P-6.12

† This part requires more detailed knowledge of the z-transform. See Reference 6–4.

(b) Find the output power $E[Y^2[n]]$.
(c) Find the output covariance $K_{YY}[m, n]$.

6.13. Let $W[n]$ be an independent random sequence with mean 0 and variance σ_W^2. Define a new random sequence $X[n]$ as follows:

$$X[0] \triangleq 0$$

$$X[n] = \rho X[n-1] + W[n] \quad \text{for} \quad n \geq 1.$$

(a) Find the mean value of $X[n]$ for $n \geq 0$.
(b) Find the covariance of $X[n]$, denoted $K_X[m, n]$.
(c) For what values of ρ does $K_X[m, n]$ tend to $g[m - n]$ (for some finite-valued function g) as m and n become large? This situation is called *asymptotic stationarity*.

6.14. Consider the probability space $(\Omega, \mathscr{F}, P)$ with $\Omega = [0, 1]$, $\mathscr{F}$ defined to be the Borel sets of Ω and $P[(0, \zeta]] = \zeta$ for $0 < \zeta \leq 1$.
(a) Show that $P[\{0\}] = 0$ by using the axioms of probability.
(b) Determine in what *senses* the following random sequences converge:
(i) $X[n, \zeta] = e^{-n\zeta}, n \geq 0$
(ii) $X[n, \zeta] = \sin\left(\zeta + \frac{1}{n}\right), n \geq 1$
(iii) $X[n, \zeta] = \cos^n(\zeta), n \geq 0$.
(c) If the preceding sequences converge, what are the limits?

6.15. In this chapter we used the Chebyshev inequality to prove that convergence in the mean-square sense implied convergence in probability. However, for many other applications we need much tighter bounds on probabilities, for example the probability of error in a digital communications system. In this problem we develop a popular bound called the *Chernoff bound*.

We are motivated by the need to calculate an upper bound to the probability:

$$P[X - \mu \geq k\sigma],$$

where X is a random variable with mean μ and variance σ^2. Assume for simplicity that X has a density function $f_X(x)$, then

$$P[X - \mu \geq k\sigma] = \int_{\mu+k\sigma}^{\infty} f_X(x)\,dx = \int_{k\sigma}^{\infty} f_X(x + \mu)\,dx.$$

(a) Now note that for any $\lambda \geq 0$ we have:

$$e^{\lambda(x-k\sigma)} \geq \begin{cases} 1 & \text{for } x \geq k\sigma \\ 0 & \text{for all } x. \end{cases}$$

Insert this function under the integral sign to conclude

$$P[X - \mu \geq k\sigma] \leq e^{-\lambda k\sigma} E[e^{\lambda(X-\mu)}] \quad \text{for } \lambda \geq 0.$$

(b) Derive the equation for the optimum value(s) of λ, say λ_0, by setting the derivative to zero. Verify that this is a true minimum.
(c) For the case where X is $N(\mu, \sigma^2)$, show that $\lambda_0 = k/\sigma$, and obtain the corresponding Chernoff bound. Compare to the Chebyshev bound.

6.16. The members of the sequence of jointly independent random variables $X[n]$

have pdf's of the form:

$$f_X(x; n) = \left(1 - \frac{1}{n}\right)\frac{1}{\sqrt{2\pi}\sigma}\exp\left[-\frac{1}{2\sigma^2}\left(x - \frac{n-1}{n}\sigma\right)^2\right] + \frac{1}{n}\sigma\exp(-\sigma x)u(x).$$

Determine whether or not the sequence $X[n]$ converges in:

(i) the mean-square sense,

(ii) probability,

(iii) distribution. Yes as $n \to \infty$ f_X's all $\approx \frac{1}{\sqrt{2\pi}\sigma}\exp\left(\frac{-1}{2\sigma^2}(x-\sigma)^2\right)$

6.17. The members of the random sequence $X[n]$ have joint pdf's of the form

$$f_X(\alpha, \beta; m, n) = \frac{mn}{2\pi\sqrt{1-\rho^2}}\exp\left(-\frac{1}{2(1-\rho^2)}[m^2\alpha^2 - 2\rho mn\alpha\beta + n^2\beta^2]\right)$$

for $m \geq 1$ and $n \geq 1$ where $-1 < \rho < +1$.

(a) Show that $X[n]$ converges in the mean-square sense as $n \to \infty$ for all $-1 < \rho < +1$.

(b) Specify the distribution function of the mean-square limit $X = \lim_{n\to\infty} X[n]$.

(c) State conditions under which the mean-square limit of a sequence of Gaussian random variables is also Gaussian.

6.18. This problem demonstrates that p-convergence implies convergence in distribution even when the limiting pdf does not exist. Assume the random sequence $X[n]$ converges to the random variable X in probability.

(a) For any real number x and any positive ε, show that

$$P[X \leq x - \varepsilon] \leq P[X[n] \leq x] + P[|X[n] - X| \geq \varepsilon].$$

(b) Similarly show that

$$P[X > x + \varepsilon] \leq P[X[n] > x] + P[|X[n] - X| \geq \varepsilon].$$

(c) Let $n \to \infty$ and conclude that

$$\lim_{n\to\infty} F_X(x; n) = F_X(x)$$

at points of continuity of F_X.

6.19. Let $X[n]$ be a second-order random sequence. Let $h[n]$ be the impulse response of an LSI system. We wish to define the output of the system $Y[n]$ as a mean-square limit.

(a) Show that we can define

$$Y[n] \triangleq \sum_{k=-\infty}^{+\infty} h[k]X[n-k], \qquad -\infty < n < +\infty, \quad \text{(m.s.)}$$

if

$$\sum_k \sum_l h[k]h^*[l]R_{XX}[n-k, n-l] < \infty \quad \text{for all } n.$$

Hint: Set $Y_N[n] \triangleq \sum_{k=-N}^{+N} h[k]X[n-k]$ and show the m.-s. limit of $Y_N[n]$ exists by using the Cauchy convergence criteria.

(b) Find a simpler condition for the case when $X[n]$ is wide-sense stationary.

(c) Find the condition on $h[n]$ when $X[n]$ is (stationary) white noise.

6.20. If $X[n]$ is a Martingale sequence on $n \geq 0$, show that

$$E[X[n+m] \mid X[m], \ldots, X[0]] = X[m] \qquad \text{for all } n \geq 0.$$

6.21. Let $Y[n]$ be a random sequence and X a random variable and consider the conditional expectation

$$E[X \mid Y[0], \ldots, Y[n]] \triangleq G[n].$$

Show that $G[n]$ is a Martingale.

6.22. We can enlarge the concept of Martingale sequence somewhat as follows. Let

$$G[n] = g(X[0], \ldots, X[n]) \quad \text{for each } n \geq 0 \text{ for measurable functions g.}$$

We say G is a *Martingale with respect to* (*wrt*) X if

$$E[G[n] \mid X[0], \ldots, X[n-1]] = G[n-1].$$

(a) Show that Theorem 6.5–3 holds for G a Martingale wrt X. Specifically, substitute G for X in the statement of the theorem. Then make necessary changes to the proof.

(b) Show that the Martingale convergence Theorem 6.5–4 holds for G a Martingale wrt X.

6.23. Consider the hypothesis-testing problem involving the $(n + 1)$ observations $X[0], \ldots, X[n]$ of the random sequence X. Define the likelihood ratio

$$L_X[n] \triangleq \frac{f_X(X[0], \ldots, X[n] \mid H_1)}{f_X(X[0], \ldots, X[n] \mid H_0)}, \qquad n \geq 0,$$

corresponding to two hypotheses, H_1 and H_0. Show that $L_X[n]$ is a Martingale wrt X under hypothesis H_0.

REFERENCES

6–1. R. B. Ash, *Real Analysis and Probability*. New York: Academic Press, 1972, pp. 1–53.

6–2. E. Wong and B. Hajek, *Stochastic Processes in Engineering Systems*, New York: Springer-Verlag, 1985, p. 3.

6–3. T. M. Apostol, *Mathematical Analysis*. Reading, Mass. Addison-Wesley, 1957, pp. 192–202.

6–4. A. V. Oppenheim and R. W. Schafer, *Digital Signal Processing*. Englewood Cliffs, N.J.: Prentice-Hall, 1975, Chapters 1 and 2.

6–5. W. Rudin, *Principles of Mathematical Analysis*. New York: McGraw-Hill, 1964, pp. 45–48.

6–6. B. V. Gnedenko, *The Theory of Probability* (translated by B. D. Seckler). New York: Chelsea, 1963, p. 237.

7

RANDOM PROCESSES

In the last chapter, we saw that it is possible to generalize the concept of random variable to the concept of random sequence. We did this by associating a sequence with each point $\zeta \in \Omega$, thereby generating a family of sequences collectively called a random sequence. In this chapter we generalize further by considering random functions of a *continuous* parameter. This continuous parameter is most often time, but could equally well be distance or some other continuous parameter. The collection of all these functions is called a random process. Random processes will be perhaps the most useful objects we study because they can be used to model physical processes directly without any intervening need to sample the data. Even when of necessity one is dealing with sampled data, the concept of random process will give us the ability to "know what we're missing" and thus be able to judge the adequacy of the sampling rate.

Random processes find a wide variety of applications. Perhaps the most common use is as a model for noise in physical systems, this modeling of the noise being the necessary first step in deciding on the best way to lessen its effects. A second class of applications concerns the modeling of random phenomena that are not noise but are nevertheless unknown to the system designer. An example would be an image or audio signal to be communicated to a distant point. The signal is not noise, but it can be considered as random from the viewpoint of the distant receiver. Thus it is reasonable to model such signals as random processes. Situations such as this arise in other contexts also, such as control systems, pattern recognition, and so forth. Indeed from information theory, it can be seen that any waveform that communicates information to a human must have at least some degree of randomness in it!

We start with a definition of random process and study some of the new difficulties to be encountered with continuous time. Then we look at the moment functions and define correlation and covariance functions. We next look at some basic random processes of practical importance. We then begin to investigate the effects of linear systems on random processes. Finally we present some classifications of random processes.

7.1 DEFINITIONS

It is most important to fully understand the basic concept of the random process and its associated moment functions. The situation is analogous to the discrete-time case treated in Chapter 6. The main new difficulty is that the time axis has now become uncountable. We start with the basic definition.

Definition 7.1–1. Let $(\Omega, \mathscr{F}, P)$ be a probability space. Then define a mapping X from the sample space Ω to a space of functions called *sample functions*. This mapping is called a *random process* if at each fixed time the mapping is a random variable, that is, $X(t, \zeta) \in \mathscr{F}$† for each fixed t on the real line $-\infty < t < +\infty$.

Thus we have a two-dimensional function $X(t, \zeta)$, which for each fixed ζ is an ordinary time function and for each fixed t is a random variable. This is shown diagrammatically in Figure 7.1–1 for the special case $\Omega = [0, \infty)$. We see a family of random variables indexed by t when we look along the time axis, and we see a family of time functions indexed by ζ when we look along the ζ-axis.

We have the following examples of random processes.

Example 7.1–1: $X(t, \zeta) = X(\zeta)f(t)$ where X is a random variable and f is a deterministic function of the parameter t. We also write $X(t) = Xf(t)$.

Example 7.1–2: $X(t, \zeta) = A(\zeta)\sin(\omega_0 t + \Theta(\zeta))$ where A and Θ are random variables. We also write $X(t) = A \sin(\omega_0 t + \Theta)$.

Example 7.1–3: $X(t) = \sum_n X[n]p_n(t - T[n])$ where $X[n]$ and $T[n]$ are random sequences and the p_n are deterministic waveforms that can take on various shapes. For example the p_n might be ideal unit step functions that could provide a model for a so-called *jump process.* In this interpretation the T would be the times of the *arrivals* and the $X[n]$ would be the *amplitudes* of the jumps. Then $X(t)$ would indicate the total amplitude up to time t. If all the $X[n]$'s were 1, we would have a *counting process* in that $X(t)$ would be a count of all the arrivals prior to time t.

If we sample the random process at n times t_1 through t_n, we get an n-dimensional random vector. If we know the probability distribution of this vector for any times t_1 through t_n and for all positive n, then clearly we

† $X \in \mathscr{F}$ is shorthand for $\{\zeta : X(\zeta) \leq x\} \subset \mathscr{F}$ for all x.

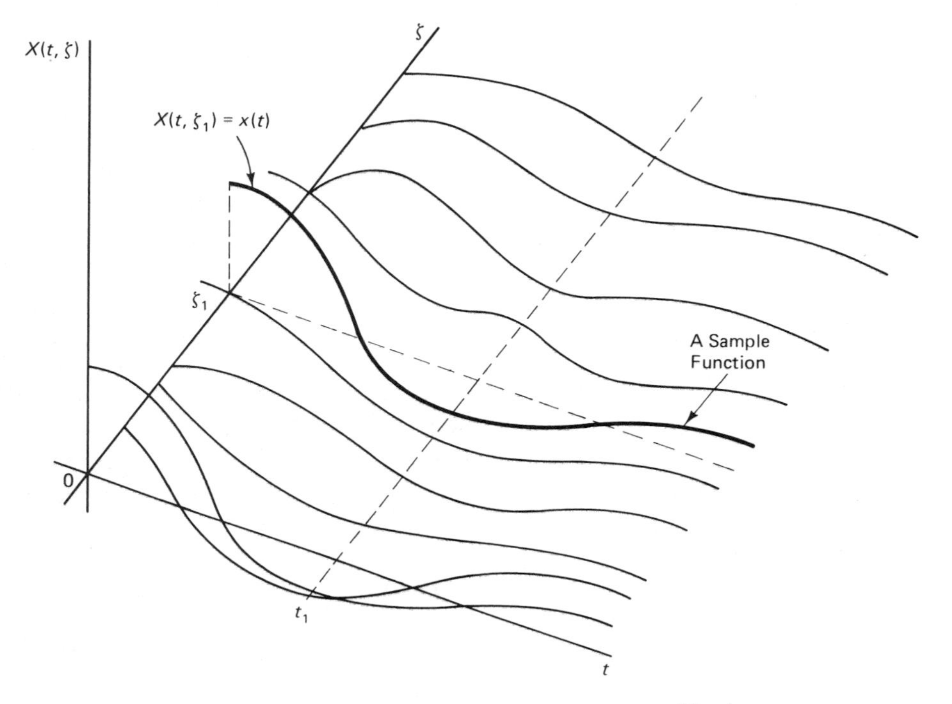

Figure 7.1–1 Random process depiction for $\Omega = [0, \infty)$.

know a lot about the random process. If we know all this information, we say that we have *statistically specified* (determined) the random process in a fashion that is analogous to the corresponding case for random sequences. The term statistical comes from the fact that this is the limit of the information that could be obtained from accumulating relative frequencies of events determined by the random process $X(t)$ at all finite collections of time instants. Clearly, this is all we could hope to determine by measurements on an observed process that we wish to model. However, the question arises: Is this enough information to completely determine the random process? Unfortunately the answer is no. We need to impose a continuity requirement on the sample functions $x(t)$. To see this the following simple example suffices.

Example 7.1–4: (From Karlin [7–1], p. 32.) Let U be a uniform random variable on $[0, 1]$ and define the random processes $X(t)$ and $Y(t)$ as follows:

$$X(t) \triangleq \begin{cases} 1 & \text{for } U = t \\ 0, & \text{otherwise,} \end{cases}$$

and

$$Y(t) \triangleq 0 \quad \text{for all } t.$$

Then $Y(t)$ and $X(t)$ have the same finite-order distributions, yet obviously the probability of the following two events is not the same:

$$\{X(t) \leq 0.5 \text{ for all } t\}$$

and

$$\{Y(t) \leq 0.5 \text{ for all } t\}.$$

To show that $Y(t)$ and $X(t)$ have the same nth-order pdf's, find the conditional nth-order pdf of X given $U = u$, then integrate out the conditioning on U. We leave this as an exercise to the reader.

The problem here is really that the complementary event $\{X(t) > 0.5$ for some $t \in [0, 1]\}$ involves an *uncountable* number of random variables. Yet the statistical determination and the extended additivity Axiom 4 only (see Section 6.1) allow us to evaluate probabilities corresponding to countable numbers of random variables. In what follows we will assume that we always have a process "continuous enough" that the family of finite-order distribution functions suffices to determine the process for all time. Such processes are called *separable*. The random process $X(t)$ of the above example is obviously not separable.

As in the case of random sequences, the moment functions play an important role in practical applications. The *mean function*, denoted $\mu_X(t)$ is given as

$$\mu_X(t) \triangleq E[X(t)], \qquad -\infty < t < +\infty. \tag{7.1-1}$$

Similarly the *correlation function* is defined as the expected value of the product,

$$R_X(t_1, t_2) \triangleq E[X(t_1)X^*(t_2)], \qquad -\infty < t_1, t_2 < +\infty. \tag{7.1-2}$$

The *covariance function* is defined as the expected value of the product of the *centered process* $X_c(t) \triangleq X(t) - \mu_X(t)$,

$$\begin{aligned} K_X(t_1, t_2) &\triangleq E[X_c(t_1)X_c^*(t_2)] \\ &= E[(X(t_1) - \mu_X(t_1))(X(t_2) - \mu_X(t_2))^*]. \end{aligned} \tag{7.1-3}$$

Clearly these three functions are not unrelated and in fact we have,

$$K_X(t_1, t_2) = R_X(t_1, t_2) - \mu_X(t_1)\mu_X^*(t_2). \tag{7.1-4}$$

We also define the *variance function* as $\sigma_X^2(t) \triangleq K_X(t, t) = E[|X_c(t)|^2]$. As in the discrete-time case, the correlation and covariance functions are Hermitian symmetric, that is,

$$R_X(t_1, t_2) = R_X^*(t_2, t_1),$$
$$K_X(t_1, t_2) = K_X^*(t_2, t_1),$$

which directly follow from the linearity of the expectation operator E.

If we sample the random process at N times t_1 through t_N, we form a random vector. We have already seen that the correlation or covariance matrix of this random vector must be positive semidefinite (Chapter 4). This, then, imposes certain requirements on the respective correlation and covariance function of the random process. Specifically, every correlation (covariance) matrix that can be formed from a correlation (covariance) function must be positive semidefinite.

Definition 7.1–2. The two-dimensional function $g(t, s)$ is *positive semidefinite* if for all $N > 0$ and all t_1 through t_N and for all complex constants a_1 through a_N we have

$$\sum_{i=1}^{N} \sum_{j=1}^{N} a_i a_j^* g(t_i, t_j) \geq 0.$$

Using this definition, we can thus say that all correlation and covariance functions must be positive semidefinite. Later we will see that this necessary condition is also sufficient. Although positive semidefiniteness is an important constraint, it is difficult to apply this condition in a test of the legitimacy of a proposed correlation function. Another fundamental property of correlation and covariance functions is *diagonal dominance*,

$$|R_{XX}(t, s)| \leq \sqrt{R_{XX}(t, t) R_{XX}(s, s)} \quad \text{for all } t, s,$$

which follows from the Cauchy-Schwarz inequality (Section 3.3, see Equation 3.3–17). Diagonal dominance is implied by positive semidefiniteness.

7.2 SOME IMPORTANT RANDOM PROCESSES

In this section we present examples of four important types of random processes. These are the Poisson-counting process; the phase-shift keying (PSK) random process, an example of digital modulation; the Wiener process, which is obtained as a continuous limit of a random walk sequence; and lastly the class of Markov random processes.

Poisson Counting Process. Let the process $N(t)$ represent the total number of counts (arrivals) up to time t. Then we can write

$$N(t) \triangleq \sum_{n=0}^{\infty} u[t - T[n]],$$

where $u(t)$ is the unit-step function and $T[n]$, *the time to the nth arrival,* is the random sequence of times considered in Example 6.1–5. There we showed that the $T[n]$ obeyed the nonstationary first-order density,

$$f_T(t; n) = \frac{(\lambda t)^{n-1}}{(n-1)!} \lambda e^{-\lambda t} u(t), \tag{7.2–1}$$

which was obtained as an n-fold convolution of exponential pdf's. A typical sample function is shown in Figure 7.2–1. Note that the time between the arrivals,

$$\Upsilon[n] \triangleq T[n] - T[n-1],$$

the *interarrival times,* are jointly independent and identically distributed,

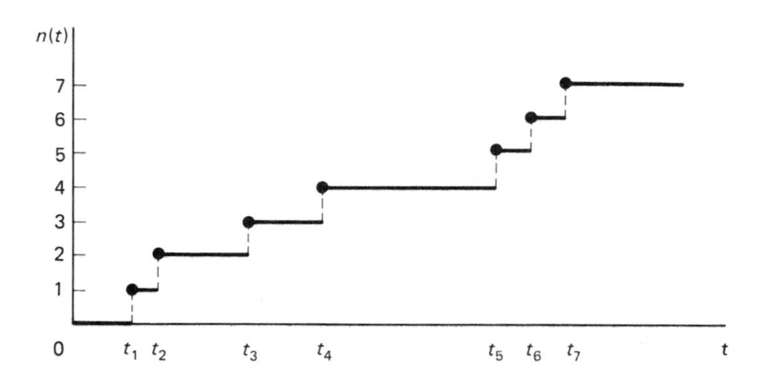

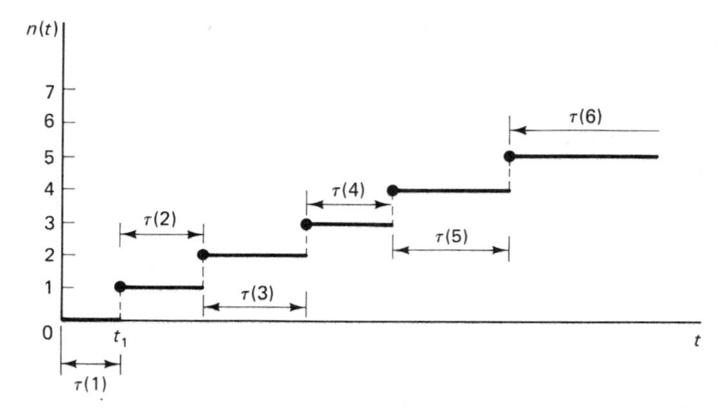

Figure 7.2–1 Two Poisson sample functions.

having the exponential density,

$$f_T(t) = \lambda e^{-\lambda t}u(t),$$

as in Example 6.1–5. Thus $T[n]$ denotes the total time until the nth arrival if we begin counting at the reference time $t = 0$.

Now by the construction involving the unit step function, the value $N(t)$ is the number of arrivals up to and including time t, so

$$P[N(t) = n] = P[T[n] \le t, T[n+1] > t],$$

because the only way that $N(t)$ can equal n is if the random variable $T[n]$ is less than or equal to t and the random variable $T[n + 1]$ is greater than t. If we bring in the independent interarrival times, we can re-express this probability as

$$P[T[n] \le t, \tau[n + 1] > t - T[n]],$$

which can be easily calculated on account of the statistical independence of the arrival time $T[n]$ and the interarrival time $\Upsilon[n+1]$ as follows,

$$\int_0^t f_T(\alpha; n)\left[\int_{t-\alpha}^{\infty} f_\Upsilon(\beta)\, d\beta\right] d\alpha = \int_0^t \frac{\lambda^n \alpha^{n-1} e^{-\lambda\alpha}}{(n-1)!}\left(\int_{t-\alpha}^{\infty} \lambda e^{-\lambda\beta}\, d\beta\right) d\alpha \cdot u(t)$$

$$= \left(\int_0^t \alpha^{n-1}\, d\alpha\right)\lambda^n e^{-\lambda t}/(n-1)! \quad u(t),$$

or

$$\checkmark \; P_N(n; t) = \frac{(\lambda t)^n}{n!} e^{-\lambda t} u(t) \quad \text{for } t \geq 0, \tag{7.2-2}$$

We have thus arrived at the PMF of the *Poisson counting process* and we note that it is the PMF of a Poisson random variable with mean λt, that is,

$$\checkmark \; E[N(t)] = \lambda t. \tag{7.2-3}$$

We call λ the *mean-arrival rate* (sometimes called *intensity*). It is intuitively satisfying that the average value of the process at time t is the mean arrival rate λ multiplied by the length of the time interval $(0, t]$. We leave it as an exercise for the reader to consider why this is so. It may be interesting to compare the constructive definition of the Poisson process in this section with the necessarily more elementary discussion in Section 1.10.

Since the random sequence $T[n]$ has independent increments (cf. Definition 6.1–3) and the unit step function used in the definition of the Poisson process is causal, it seems reasonable that the Poisson process $N(t)$ would also have independent increments. However, this result is not clear because one of the jointly independent interarrival times $\Upsilon[n]$ may be partially in two disjoint intervals, hence causing a dependency in neighboring increments. Nevertheless, using the *memoryless property* of the exponential density (See Problem 7.5), one can show that the independent increments property does hold for the Poisson process.

Using independent increments we can evaluate the PMF of the increment in the Poisson-counting process over an interval $(t_a, t_b]$ as

$$\checkmark \; P[N(t_b) - N(t_a) = n] = \frac{[\lambda(t_b - t_a)]^n}{n!} e^{-\lambda(t_b - t_a)} u(n), \tag{7.2-4}$$

where we have used the fact that the interarrival sequence is stationary, that is, that λ is a constant. We formalize this somewhat in the following definition.

Definition 7.2–1. A random process has *independent increments* when the set of n random variables,

$$X(t_1), X(t_2) - X(t_1), \ldots, X(t_n) - X(t_{n-1}),$$

are jointly independent for all $t_1 < t_2 < \ldots < t_n$ and for all $n \geq 1$.

This just says that the increments are statistically independent when the intervals do not overlap. Just as in the random sequence case, the independent-increment property makes it easy to get the higher-order distributions. For example, in the case at hand, the Poisson counting process, we can write for $t_2 > t_1$,

$$P_N(n_1, n_2; t_1, t_2) = P[N(t_1) = n_1]P[N(t_2) - N(t_1) = n_2 - n_1]$$

$$= \frac{(\lambda t_1)^{n_1}}{n_1!} e^{-\lambda t_1} \frac{[\lambda(t_2 - t_1)]^{n_2-n_1}}{(n_2 - n_1)!} e^{-\lambda(t_2-t_1)} u(n_1)u(n_2 - n_1),$$

which simplifies to

$$\frac{\lambda^{n_2} t_1^{n_1}(t_2 - t_1)^{n_2-n_1}}{n_1!(n_2 - n_1)!} e^{-\lambda t_2} u(n_1)u(n_2 - n_1), \qquad 0 \leq t_1 < t_2.$$

See also Problem 1.28. Using the independent-increments property we can formulate the following alternative definition of a Poisson counting process.

Definition 7.2–2. A Poisson counting process is the independent-increments process whose increments are Poisson distributed as in Equation 7.2–4.

We now proceed to calculate the moments for the Poisson process. The first-order moment has been shown to be λt. This is the mean function of the process. Let $t_2 \geq t_1$ and calculate the correlation function using the independent increments property as,

$$\begin{aligned} E[N(t_2)N(t_1)] &= E[(N(t_1) + [N(t_2) - N(t_1)])N(t_1)] \\ &= E[N^2(t_1)] + E[N(t_2) - N(t_1)]E[N(t_1)] \\ &= \lambda t_1 + \lambda^2 t_1^2 + \lambda(t_2 - t_1)\lambda t_1 \\ &= \lambda t_1 + \lambda^2 t_1 t_2. \end{aligned}$$

If $t_2 < t_1$ we merely interchange t_1 and t_2 in the preceding formula. Thus a general result for all t_1 and t_2 is

$$\begin{aligned} \checkmark\ R_N(t_1, t_2) &= E[N(t_1)N(t_2)] \\ &= \lambda \min(t_1, t_2) + \lambda^2 t_1 t_2. \end{aligned} \tag{7.2–5}$$

If we evaluate the covariance using Equations 7.2–3 and 7.2–5 we obtain

$$\checkmark\ K_N(t_1, t_2) = \lambda \min(t_1, t_2). \tag{7.2–6}$$

We thus see that the variance of the process is equal to λt and is the same as its mean, a property inherited from the Poisson random variable. Also we see that the covariance depends only on the earlier of the two times involved. The reason for this is seen by writing $N(t)$ as the value at an earlier time plus an increment, and then noting that the independence of this increment and N at the earlier time implies that the covariance between them must be zero. Thus the covariance of this independent-increments process is just the variance of the process at the earlier of the two times.

Example 7.2–1: (Radioactivity monitor.) In radioactivity monitoring, the particle-counting process can often be adequately modeled as Poisson. Let the counter start to monitor at some arbitrary time t and then count for T seconds. If the count is above a threshold, say N_0, an alarm will be sounded. Assuming the arrival rate to be λ, we want to know the probability that the alarm will not sound when radioactive material is present.

Since the process is Poisson, we know it has independent increments that satisfy the Poisson distribution. Thus the count ΔN in the interval $(t, t + T]$, that is, $\Delta N \triangleq N(t + T) - N(t)$, is Poisson distributed with mean λT independent of t. The probability of N_0 or fewer counts is thus

$$P[\Delta N \leq N_0] = \sum_{k=0}^{N_0} \frac{(\lambda T)^k}{k!} e^{-\lambda T}.$$

If N_0 is small we can calculate the sum directly. If $\lambda T \gg 1$, we can use the Gaussian approximation (Equation 1.9–10) to the Poisson distribution.

Example 7.2–2. (Sum of two independent Poisson processes.) Let $N_1(t)$ be a Poisson counting process with rate λ_1. Let $N_2(t)$ be a second Poisson counting process with rate λ_2, where N_2 is independent of N_1. The sum of the two processes, $N(t) \triangleq N_1(t) + N_2(t)$, could model the total number of failures of two separate machines. It is a remarkable fact that $N(t)$ is also a Poisson counting process with rate $\lambda \triangleq \lambda_1 + \lambda_2$.

To see this we use Definition 7.2–2 of the Poisson counting process and verify these conditions for $N(t)$. First, it is clear with a little reflection that the sum of two independent-increments processes will also be an independent-increments process *if* the processes are jointly independent. Second, for any increment $N(t_b) - N(t_a)$ with $t_b > t_a$, we can write

$$N(t_b) - N(t_a) = N_1(t_b) - N_1(t_a) + N_2(t_b) - N_2(t_a).$$

Thus the increment in N is the sum of two corresponding increments in N_1 and N_2. The result then follows from the fact that the sum of two independent Poisson random variables is also Poisson distributed with the sum of the two parameters (see Example 2.10–5). Thus the parameter of the increment in $N(t)$ is

$$\lambda_1(t_b - t_a) + \lambda_2(t_b - t_a) = (\lambda_1 + \lambda_2)(t_b - t_a)$$

as desired.

The Poisson counting process N can be generalized in several ways. We can let the arrival rate, sometimes called *intensity*, be a function of time. The arrival rate $\lambda(t)$ must satisfy $\lambda(t) \geq 0$. The average value of the resulting *nonuniform Poisson counting process* then becomes

$$\mu_X(t) = \int_0^t \lambda(\tau)\, d\tau, \qquad t \geq 0. \tag{7.2–7}$$

Another way to generalize the Poisson process is to use a different density for the independent interarrival times. With a nonexponential density the more general counting process is called a *renewal counting process* [7–3]. The word "renewal" comes from an interpretation of the arrival times as the failure times of certain equipment, thus the value of the counting process $N(t)$ models the number of renewals that have had to be made up to the present time.

Digital Modulation Using Phase-Shift Keying. Digital computers generate many binary sequences (data) to be communicated to other distant computers. Binary modulation methods frequency-shift this data to a region of the electromagnetic spectrum which is well-suited to the transmission media, for example, a telephone line. A basic method for modulating binary data is phase-shift keying (PSK). In this method binary data, modeled by the random sequence $B[n]$, are mapped bit-by-bit into a phase-angle sequence $\Theta[n]$, which is used to modulate the *carrier signal* $\cos(2\pi f_c t)$.

Specifically let $B[n]$ be a Bernoulli random sequence taking on the values 0 and 1 with equal probability. Then define the random phase sequence $\Theta[n]$ as follows:

$$\Theta[n] \triangleq \begin{cases} +\pi/2 & \text{if } B[n] = 1, \\ -\pi/2 & \text{if } B[n] = 0. \end{cases}$$

Using $\Theta_a(t)$ to denote the analog angle process,

$$\Theta_a(t) \triangleq \Theta[k] \quad \text{for } kT \leq t < (k+1)T,$$

we construct the modulated signal as

$$X(t) = \cos(2\pi f_c t + \Theta_a(t)). \tag{7.2–8}$$

Here T is a constant time for the transmission of one bit. Normally, T is chosen to be a multiple of $1/f_c$ so that there are an integral number of carrier cycles per bit time T. The reciprocal of T is called the *bit-rate*. The overall modulator is shown in Figure 7.2–2. The process $X(t)$ is the PSK process.

Our goal here is to evaluate the mean function and correlation function of the random PSK process. To help in the calculation we define two basis functions,

$$s_I(t) \triangleq \begin{cases} \cos(2\pi f_c t), & 0 \leq t < T \\ 0, & \text{else} \end{cases}$$

and

$$s_Q(t) \triangleq \begin{cases} \sin(2\pi f_c t), & 0 \leq t < T \\ 0, & \text{else} \end{cases}$$

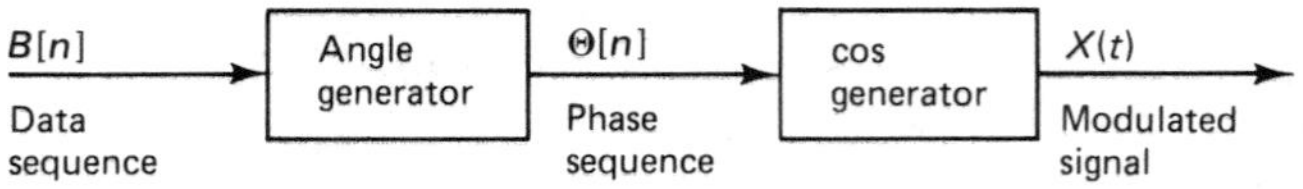

Figure 7.2–2 Random PSK modulator.

cos(a+b) = cos a cos b - sin a sin b

which together with Equation 7.2–8 imply

$$\begin{aligned}\cos[2\pi f_c t + \Theta(t)] &= \cos(\Theta_a(t))\cos 2\pi f_c t - \sin(\Theta_a(t))\sin 2\pi f_c t \\ &= \sum_{k=-\infty}^{+\infty} \cos(\Theta[k])s_I(t - kT) - \sum_{k=-\infty}^{+\infty} \sin(\Theta[k])s_Q(t - kT),\end{aligned} \tag{7.2–9}$$

by use of the sum of angles formula for cosines.

The mean of $X(t)$ can then be obtained in terms of the means of the random sequences $\cos\Theta[n]$ and $\sin\Theta[n]$. Because of the definition of $\Theta[n]$, in this particular case $\cos\Theta[n] = 0$ and $\sin\Theta[n] = \pm 1$ with equal probability so that mean of $X(t)$ is zero, that is, $\mu_X(t) = 0$.

Using Equation 7.2–9 we can calculate the correlation function

$$R_X(t_1, t_2) = \sum_{k,l} E[\sin\Theta[k]\sin\Theta[l]]s_Q(t_1 - kT)s_Q(t_2 - lT),$$

which involves the correlation function of the random sequence $\sin\Theta[n]$,

$$R_\Theta[k, l] = \delta[k - l].$$

Thus the overall correlation function then becomes

$$R_X(t_1, t_2) = \sum_{k=-\infty}^{+\infty} s_Q(t_1 - kT)s_Q(t_2 - kT). \tag{7.2–10}$$

Since the support of s_Q is only of width T, there is no overlap in (t_1, t_2) between product terms in Equation 7.2–10. So for any fixed (t_1, t_2), only one of the product terms in the sum can be nonzero. Also if t_1 and t_2 are not in the same period, then this term is zero also. More elegantly, we can write using the notation

$$((t)) \triangleq t \bmod T \quad \text{and} \quad \lfloor t/T \rfloor \triangleq \text{integer part } (t/T)$$

that

$$R_X(t_1, t_2) = \begin{cases} s_Q(((t_1)))s_Q(((t_2))) & \text{for } \left\lfloor \frac{t_1}{T} \right\rfloor = \left\lfloor \frac{t_2}{T} \right\rfloor \\ 0 & \text{otherwise.} \end{cases}$$

In particular for $0 \le t_1 \le T$ and $0 \le t_2 \le T$, we have

$$R_X(t_1, t_2) = s_Q(t_1)s_Q(t_2).$$

Stopped 11-7

Wiener Process or Brownian Motion. In Chapter 6 we considered a random sequence $X[n]$ called the random walk. See Example 6.1–3 (cont'd.). Here we construct a similar random process that is piecewise constant for intervals of length T as follows:

$$X_T(t) \triangleq \sum_{k=0}^{\infty} W[k]u(t - kT)$$

where

$$W[k] \triangleq \begin{cases} +s & \text{with } p = 0.5 \\ -s & \text{with } p = 0.5. \end{cases}$$

Then $X_T(nT) = X[n]$ the random walk sequence since

$$X_T(nT) = \sum_{k=0}^{n} W[k] = X[n].$$

Hence we can evaluate the PMF's and moments of this random process by employing the known results for the corresponding random walk sequence. Now the Wiener† process, sometimes also called *Wiener-Levy* or Brownian motion, is the process whose distribution is obtained as a limiting form of the distribution of the above piecewise constant process as the interval T shrinks to zero. We let s, the jump size, along with the interval T shrink to zero to obtain a *continuous random process* in the limit, that is, a process whose sample functions are continuous functions of time. In letting s and T tend to zero we must be careful to make sure that the limit of the variance stays finite and nonzero.

The main motivation for the Wiener process was to develop a model for the chaotic random motion of gas molecules. Modeling the basic discrete collisions with a random walk, one then finds the asymptotic process when an infinite (very large) number of molecules interact on an infinitesimal (very small) time scale.

Now the probability that $X_T(nT) = rs$ is the probability that there are $0.5(n + r)$ successes $(+s)$ and $0.5(n - r)$ failures $(-s)$ out of a total of n trials. Thus by the binomial PMF,

$$P[X_T(nT) = rs] = \binom{n}{\frac{n+r}{2}} 2^{-n} \quad \text{for } n + r \text{ even.}$$

The mean and variance can be most easily calculated by noting that the random variable $X[n]$ is the sum of n independent Bernoulli random variables as in Example 6.1–3(cont'd.). Thus

$$E[X_T(nT)] = 0$$

and

$$E[X_T^2(nT)] = ns^2.$$

On expressing the variance in terms of $t = nT$, we have

$$\text{Var}[X_T(t)] = E[X_T^2(t)] = t\frac{s^2}{T}.$$

Thus we need s^2 proportional to T, to get an interesting limiting distribution. We set $s^2 = \alpha T$, where α is a positive number. Now as T goes to zero we get the variance αt. Also, by an elementary application of the Central Limit Theorem (Section 3.6) we get a limiting Gaussian distribution. We take the limiting random process (convergence in the distribution sense) to be an independent-increments process since all the above random-walk

† After Norbert Wiener, American mathematician (1894–1964), a pioneer in the field of cybernetics.

processes had independent increments for all T, no matter how small. Hence we arrive at the following specifications for the limiting process, which is termed the *Wiener process*:

$$\mu_X(t) = 0, \qquad \mathrm{Var}[X(t)] = \alpha t \tag{7.2–11}$$

and

$$f_X(x; t) = \frac{1}{\sqrt{2\pi\alpha t}} \exp\left(-\frac{x^2}{2\alpha t}\right), \qquad t > 0. \tag{7.2–12}$$

The density of the increment $\Delta \triangleq X(t) - X(s)$ for all $t > s$ is given as

$$f_\Delta(\delta; t - s) = \frac{1}{\sqrt{2\pi\alpha(t - s)}} \exp\left(-\frac{\delta^2}{2\alpha(t - s)}\right), \tag{7.2–13}$$

since

$$E[X(t) - X(s)] = E[\Delta] = 0, \tag{7.2–14}$$

and

$$E[(X(t) - X(s))^2] = \alpha(t - s) \quad \text{for } t > s. \tag{7.2–15}$$

The sample function shown in Figure 7.2–3 illustrates the effect of the variance increasing with time.

From the first-order pdf of X and the density of the increment Δ, it is possible to calculate a complete set of consistent nth-order pdf's as we have seen before. It thus follows that all nth-order pdf's of a Wiener process are Gaussian.

Definition 7.2–3. If all the nth-order pdf's or more generally PDF's of a random process are Gaussian then the process is called a *Gaussian random process.*

The Wiener process is thus an example of a Gaussian random process. The covariance function of the Wiener process (which is also its correlation function because $\mu_X(t) = 0$) is given as

$$K_X(t, s) = \alpha \min(t, s). \tag{7.2–16}$$

To show this we take $t \geq s$, and noting that the increment $X(t) - X(s)$ is

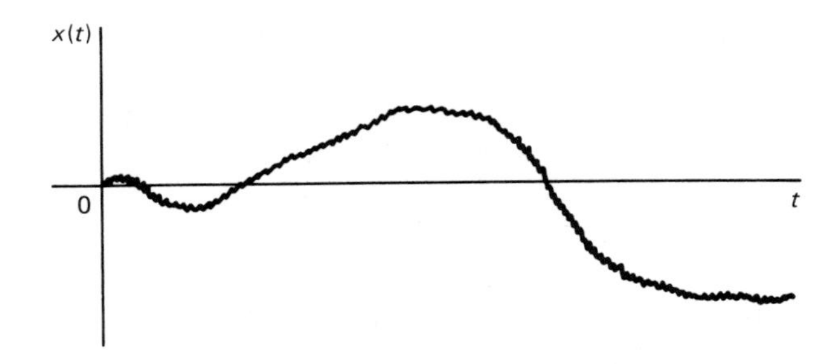

Figure 7.2–3 A sample function of the Wiener process.

independent of $X(s)$ and that they both have zero mean,

$$E[(X(t) - X(s))X(s)] = E[X(t) - X(s)]E[X(s)]$$
$$= 0$$

or

$$E[X(t)X(s)] = E[X^2(s)]$$
$$= \alpha s.$$

If $s > t$, we get $E[X(s)X(t)] = \alpha t$, thus establishing Equation 7.2–16.

Note that the Wiener process has the same variance function as the Poisson process, even though the two processes are dramatically different. While the Poisson process consists solely of jumps separated by constant values, the Wiener process has no jumps and can in fact be proven to be a.s. continuous, that is, the sample functions are continuous with probability-1. Later, we will show that the Wiener process is continuous in a weaker mean-square sense (to be specified precisely in Section 8.1).

Markov Random Processes. We have seen three random processes thus far. Of these, the Wiener and Poisson are fundamental in that many other rather general random processes can be obtained by transformations on these two basic processes. In both cases, the difficulty of specifying a consistent set of nth-order distributions for processes with dependence was overcome by use of the independent-increments property. In fact, this is quite a general approach in that we can start out with some arbitrary first-order distribution and then specify a distribution for the increment, thereby obtaining a consistent set of nth-order distributions that exhibit dependence.

Another way of going from the first-order probability to a consistent set of nth-order probabilities, which has proved quite useful, is the Markov process approach. Here we start with a first-order density (or PMF) and a conditional density (or PMF)

$$f_X(x; t) \quad \text{and} \quad f_X(x_2 \mid x_1; t_2, t), \qquad t_2 > t_1,$$

and then build up the nth-order density (or PMF) as the product,

$$f(x_1; t)f(x_2 \mid x_1; t_2, t_1) \dots f(x_n \mid x_{n-1}; t_n, t_{n-1}). \tag{7.2–17}$$

We ask the reader to show that this is a valid nth order pdf (that is, that this function is nonnegative and integrates to one) whenever the conditional and first-order pdf's are well-defined.

If we start with an arbitrary nth-order pdf and repeatedly use the definition of conditional probability we obtain,

$$f(x_1; t_1)f(x_2 \mid x_1; t_2, t_1)f(x_3 \mid x_2, x_1; t_3, t_2, t_1)$$
$$\dots f(x_n \mid x_{n-1}, \dots, x_1; t_n, \dots, t_1), \tag{7.2–18}$$

which can be made equivalent to Equation 7.2–17 if we constrain the conditional densities to depend only on the most recent conditioning value. This motivates the following definition of a Markov random process.

Definition 7.2–4. (Markov random process.)

(a) A *continuous-valued Markov process* $X(t)$ satisfies the conditional pdf expression

$$f_X(x_n \mid x_{n-1}, x_{n-2}, \ldots, x_1; t_n, \ldots, t_1) = f_X(x_n \mid x_{n-1}; t_n, t_{n-1}),$$

for all $x_1, x_2, \ldots, x_n$, for all $t_1 < t_2 < \ldots < t_n$ and for all $n > 0$.

(b) A *discrete-valued Markov random process* satisfies the conditional PMF expression

$$P_X(x_n \mid x_{n-1}, \ldots, x_1; t_n, \ldots, t_1) = P_X(x_n \mid x_{n-1}; t_n, t_{n-1})$$

for all $x_1, \ldots, x_n$ and for all $t_1 < \ldots < t_n$ and for all $n > 0$.

The value of the process X at a given time t thus determines the conditional probabilities for future values of the process. The values of the process are thus called the *state of the process*, and the conditional probabilities are thought of as *transition probabilities* between the states. If only a finite or countable set of values x_i is allowed, the discrete-valued Markov process is called a *Markov chain*. An example of a Markov chain would be the Poisson counting process studied earlier. The Wiener process is an example of a continuous-valued Markov process. Both these processes are Markov because of their independent increments property. In fact, any independent increment process is also Markov. To see this note that, for the discrete-valued case, for example,

$$\begin{aligned}
&P_X(x_n \mid x_{n-1}, \ldots, x_1; t_n, \ldots, t_1)\\
&= P[X(t_n) - X(t_{n-1}) = x_n - x_{n-1} \mid X(t_{n-1}) = x_{n-1}, \ldots, X(t_1) = x_1]\\
&= P[X(t_n) - X(t_{n-1}) = x_n - x_{n-1}] \quad \text{by independent increments}\\
&= P[X(t_n) - X(t_{n-1}) = x_n - x_{n-1} \mid X(t_{n-1}) = x_{n-1}]\\
&\qquad \text{by independent increments}\\
&= P_X(x_n \mid x_{n-1}; t_n, t_{n-1}).
\end{aligned}$$

Note, however, that the inverse argument is not true. A Markov random process does not necessarily have independent increments. (See Problem 7.9.)

Markov random processes find application in many areas including signal-processing and control systems. Markov chains are used in communications, computer networks, and reliability theory.

Example 7.2–3: ((Multiprocessor reliability.) Given a computer with two independent processors, we can model it as a 3-state system: 0—both processors down, 1—exactly one processor up, and 2—both processors up. We would like to know the probabilities of these three states. A reasonable probabilistic model is that the processors will fail randomly with time-to-failure, the *failure time*, exponentially distributed with some parameter $\lambda > 0$. Once a processor fails, the time to service it, the *service time*, will be assumed to be independently exponentially distributed with parameter $\mu > 0$. Furthermore, we assume that the processor's failures and servicing are independent; thus we make the failure and service times in our probabilistic model jointly independent.

If we define $X(t)$ as the state of the system at time t, then X is a continuous-time Markov chain. We can show this by first showing that the times between state transitions of X are exponentially distributed and then invoking the memoryless property of the exponential distribution (see Problem 7.5). Analyzing the *transition times* (either failure times or service times), we proceed as follows. The transition time for going from state $X = 0$ to $X = 1$ is the minimum of two exponentially distributed service times, which are assumed to be independent. By Problem 2.22, this time will be also exponentially distributed with parameter 2μ. The expected time for this transition will thus be $1/(2\mu) = 1/2\,(1/\mu)$ that is, one half the average time to service a single processor. This is quite reasonable since both processors are down in state $X = 0$ and hence both are being serviced independently and simultaneously. The rate parameter for the transition: 0 to 1 is thus 2μ. The transition: 1 to 2 awaits one exponential service time at rate μ. Thus its rate is also μ. Similarly, the state transition: 1 to 0 awaits only one failure at rate λ, while the transition: 2 to 1 awaits the minimum of two exponentially distributed failure times. Thus its rate is 2λ. Simultaneous transitions from 0 to 2 or 2 to 0 are of probability 0 and hence are ignored.

This Markov chain model is summarized in the *state-transition diagram* of Figure 7.2–4. In this diagram the directed branches represent transitions between the states. The transition times are assumed to be exponentially distributed with parameter given by the branch label. These transition times might be more properly called inter-transition times and are analogous to the interarrival times of the Poisson counting process, which are also exponentially distributed.

We can write the state probability at time $t + \Delta t$ in terms of the state probability at t in vector matrix form:

$$\begin{bmatrix} P_0(t+\Delta t) \\ P_1(t+\Delta t) \\ P_2(t+\Delta t) \end{bmatrix} = \begin{bmatrix} 1 - 2\mu\Delta t & \lambda\Delta t & 0 \\ +\,2\mu\Delta t & 1-(\lambda+\mu)\Delta t & 2\lambda\Delta t \\ 0 & \mu\Delta t & 1 - 2\lambda\Delta t \end{bmatrix} \begin{bmatrix} P_0(t) \\ P_1(t) \\ P_2(t) \end{bmatrix} + \mathbf{o}(\Delta t),$$

where $\mathbf{o}(\Delta t)$ denotes a quantity of order less than Δt.
Here $P_i(t) \triangleq P[X(t) = i]$ for $0 \le i \le 2$, and we have used the fact that for the exponential density with parameter a

$$P[t < \mathrm{T} \le t + \Delta t] = a\Delta t + \mathbf{o}(\Delta t).$$

Rearranging, we have

$$\begin{bmatrix} P_0(t+\Delta t) - P_0(t) \\ P_1(t+\Delta t) - P_1(t) \\ P_2(t+\Delta t) - P_2(t) \end{bmatrix} = \begin{bmatrix} -2\mu & +\lambda & 0 \\ +2\mu & -(\lambda+\mu) & +2\lambda \\ 0 & +\mu & -2\lambda \end{bmatrix} \begin{bmatrix} P_0(t) \\ P_1(t) \\ P_2(t) \end{bmatrix} \Delta t + \mathbf{o}(\Delta t).$$

Dividing both sides by Δt and using an obvious matrix notation we obtain

$$\frac{d\mathbf{P}(t)}{dt} = \mathbf{AP}(t). \tag{7.2–19}$$

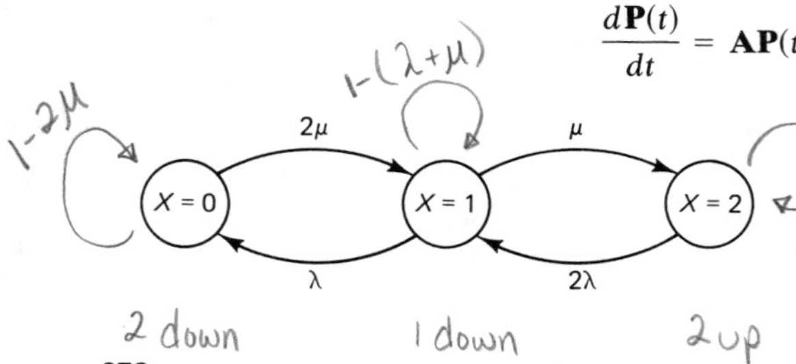

Figure 7.2–4 State-transition diagram for a continuous-time Markov chain.

The matrix $\mathbf{A}$ is called the *generator* of the Markov chain X. This first-order vector differential equation can be solved for an initial probability vector, $\mathbf{P}(0) \triangleq \mathbf{P}_0$, using methods of linear-system theory [7–2]. The solution is expressed in terms of the matrix exponential

$$e^{\mathbf{A}t} \triangleq I + \mathbf{A}t + \frac{1}{2!}(\mathbf{A}t)^2 + \frac{1}{3!}(\mathbf{A}t)^3 + \ldots,$$

which converges for all finite t. The solution $\mathbf{P}(t)$ is then given as

$$\mathbf{P}(t) = e^{\mathbf{A}t}\mathbf{P}_0, \qquad t \geq 0.$$

For details on this method as well as how to obtain an explicit solution see Reference 7–5.

Presently we content ourselves with the steady-state solution obtained by setting the time-derivative in Equation 7.2–19 to zero, thus yielding $\mathbf{AP} = \mathbf{0}$. From the first and last rows we get

$$-2\mu P_0 + \lambda P_1 = 0$$

and

$$+\mu P_1 - 2\lambda P_2 = 0.$$

From this we obtain $P_1 = (2\mu/\lambda)P_0$ and $P_2 = (\mu/2\lambda)P_1 = (\mu/\lambda)^2 P_0$. Then invoking $P_0 + P_1 + P_2 = 1$ we obtain $P_0 = \lambda^2/(\lambda^2 + 2\mu\lambda + \mu^2)$ and finally

$$\mathbf{P} = \frac{1}{\lambda^2 + 2\mu\lambda + \mu^2}[\lambda^2, 2\mu\lambda, \mu^2]^T.$$

Thus the steady-state probability of both processors being down is $P_0 = [\lambda/(\lambda + \mu)]^2$. Incidentally, if we had used only one processor modeled by a two-state Markov chain we would have obtained $P_0 = \lambda/(\lambda + \mu)$.

This is an example of a *queueing process*. Other applications are the number of toll booths busy on a highway, congestion and delay in a computer network or telephone network, and so forth. For more on queueing systems see Reference 7–3. An important point to notice in the last example is that the exponential transition times were crucial in showing the Markov property. In fact, any other distribution but exponential would not be memoryless, and the resulting state-transition process would not be a Markov chain.

We also define the Markov property, somewhat easier in fact, for random sequences.

Definition 7.2–5. (Markov random sequence.)

(a) A *contintinuous-valued Markov random sequence* $X[n]$, defined for $n \geq 0$, satisfies the conditional pdf expression

$$f_X(x_{n+k} \mid x_{n-1}, x_{n-2}, \ldots, x_0) = f_X(x_{n+k} \mid x_{n-1})$$

for all $x_0, \ldots, x_{n-1}, x_{n+k}$, for all $n > 0$ and for all $k \geq 0$.

(b) A *discrete-valued Markov random sequence* $X[n]$ defined for $n \geq 0$ satisfies the conditional PMF expression

$$P_X(x_{n+k} \mid x_{n-1}, \ldots, x_0) = P_X(x_{n+k} \mid x_{n-1})$$

for all $x_0, \ldots, x_{n-1}, x_{n+k}$, for all $n > 0$, and for all $k \geq 0$.

In the discrete-time case, it is sufficient for the above properties to hold for just $k = 0$. The discrete-*valued* Markov random sequence is also called a Markov chain. An example is provided in Problem 7.10. Next we present an example of a continuous-valued Markov random sequence which is also a Gaussian random sequence.

Example 7.2–4: (A Gaussian Markov random sequence.) Let $X[n]$ be a random sequence defined for $n \geq 0$, with initial pdf

$$f_X(x; 0) = N(0, \sigma_0^2)$$

for a given $\sigma_0 > 0$ and transition pdf

$$f_X(x_n \mid x_{n-1}; n, n-1) \sim N(\rho x_{n-1}, \sigma_W^2)$$

with $|\rho| < 1$ and $\sigma_W > 0$. We want to determine the unconditional density of $X[n]$ at an arbitrary time $n > 0$ and proceed as follows.

In general, one would have to advance recursively from the initial density by performing the integrals (c.f. Equation 2.7–45)

$$f_X(x; n) = \int_{-\infty}^{+\infty} f_X(x \mid \xi; n, n-1) f_X(\xi; n-1)\, d\xi \tag{7.2–20}$$

for $n = 1, 2, 3$, and so forth. However, in this example we know that the unconditional first-order density will be Gaussian because each of the pdf's in Equation 7.2–20 is Gaussian, and the Gaussian density "reproduces itself" in this context.† Hence the pdf $f_X(x; n)$ is determined by its first two moments. We first calculate the mean function

$$\begin{aligned} \mu_X[n] &= E[X[n]] \\ &= E\big[E[X[n] \mid X[n-1]]\big] \\ &= E[\rho X[n-1]] \\ &= \rho \mu_X[n-1], \end{aligned}$$

where the outer expectation in line 2 is over $X[n-1]$. We thus obtain the recursive equation,

$$\mu_X[n] = \rho \mu_X[n-1], \qquad n \geq 1,$$

with initial condition $\mu_X[0] = 0$. Hence $\mu_X[n] = 0$ for all n.

We also need the variance function $\sigma_X^2[n]$, which in this case is just $E[X^2[n]]$ since the mean is zero. Calculating, we obtain

$$\begin{aligned} E[X^2[n]] &= E\big[E[X^2[n] \mid X[n-1]]\big] \\ &= E[\sigma_w^2 + \rho^2 X^2[n-1]] \\ &= \sigma_w^2 + \rho^2 E[X^2[n-1]] \end{aligned}$$

or

$$\sigma_X^2[n] = \rho^2 \sigma_X^2[n-1] + \sigma_W^2, \qquad n \geq 1.$$

This is a first-order difference equation, which can be solved for $\sigma_X^2[n]$ given the

† For more information on reproducing densities, see Reference 7–4.

condition $\sigma_X^2[0] = \sigma_0^2$ supplied by the initial pdf. The solution then is

$$\sigma_X^2[n] = [1 + \rho^2 + \rho^4 + \ldots + \rho^{2(n-1)}]\sigma_W^2 + \rho^{2(n-1)}\sigma_0^2$$

$$\rightarrow \frac{1}{1 - \rho^2}\sigma_W^2 \quad \text{as } n \rightarrow \infty.$$

Chapman-Kolmogorov Equations. In Example 7.2–4 of a Markov random sequence, we specified the transition density as a one-step transition, that is, from $n - 1$ to n. More generally, we can specify the transition density from time n to time $n + k$ where $k \geq 0$, as in the general definition of a Markov random sequence. However, in this more general case we must make sure that this multistep transition density is consistent, that is, that there exists a one-step density that would sequentially yield the same results. This problem is even more important in the random process case, where due to continuous time one is always effectively considering multistep transition densities.

The Chapman-Kolmogorov equations supply both necessary and sufficient conditions for these more general transition densities. There is also another version of the Chapman-Kolmogorov equations for the discrete-valued case involving transition PMF's.

Consider three times $t_3 > t_2 > t_1$ and the Markov process r.v.'s at these three times $X(t_3)$, $X(t_2)$ and $X(t_1)$. We wish to compute the conditional density of $X(t_3)$ given $X(t_1)$. First, we write the joint pdf

$$f_X(x_3, x_1; t_3, t_1) = \int_{-\infty}^{+\infty} f_X(x_3 \mid x_2, x_1; t_3, t_2, t_1) f_X(x_2, x_1; t_2, t_1)\, dx_2.$$

If we now divide both sides of this equation by $f(x_1; t_1)$, we obtain

$$f_X(x_3 \mid x_1) = \int_{-\infty}^{-\infty} f_X(x_3 \mid x_2, x_1) f_X(x_2 \mid x_1)\, dx_2,$$

where we have suppressed the times t_i for notational simplicity. Then using the Markov property the above becomes

$$f_X(x_3 \mid x_1) = \int_{-\infty}^{+\infty} f_X(x_3 \mid x_2) f_X(x_2 \mid x_1)\, dx_2, \qquad (7.2\text{–}21)$$

which is known as the *Chapman-Kolmogorov equation* for the transition density $f_X(x_3 \mid x_1)$ of a Markov process. This equation must hold for all $t_3 > t_2 > t_1$ and for all values of x_3 and x_1. It can be proven that the Chapman-Kolmogorov condition expressed in Equation 7.2–21 is also sufficient for the existence of the transition density in question [7–5].

7.3 LINEAR SYSTEMS WITH RANDOM INPUTS

In this section we look at transformations of stochastic processes. We concentrate on the case of linear transformation with memory, since the memoryless case can be handled by the transformation of random variables method of Chapter 2. The definition of a linear continuous-time system is recalled first.

Definition 7.3–1. Let $x_1(t)$ and $x_2(t)$ be two deterministic time functions and let a_1 and a_2 be two scalar constants. Let the linear system be described by the operator equation $y = L\{x\}$. Then the system is linear if

$$L\{a_1x_1(t) + a_2x_2(t)\} = a_1L\{x_1(t)\} + a_2L\{x_2(t)\} \tag{7.3–1}$$

for all admissible functions x_1 and x_2 and all scalars a_1 and a_2.

In this definition we note that the inputs must be in the allowable input space for the system (operator) L. When we think of generalizing L to allow a random process input, the most natural choice is to input the sample functions of X and find the corresponding sample functions of the output, which thereby define a new random process Y. Just as the original random process X is a mapping from the sample space to a function space, the linear system in turn maps this function space to a new function space. The cascade or composition of the the two maps thus defines an output random process.

This is depicted graphically in Figure 7.3–1. Our goal in this section will be to find out how the first- and second-order moments, that is, the mean and correlation (and covariance) are transformed by a linear system.

Theorem 7.3.1. Let the random process $X(t)$ be the input to a linear system L with output process $Y(t)$. Then the mean function of the output is given as

$$\begin{aligned} E[Y(t)] &= L\{E[X(t)]\} \\ &= L\{\mu_X(t)\}. \end{aligned} \tag{7.3–2}$$

Proof (formal). By definition we have for each sample function

$$Y(t, \zeta) = L\{X(t, \zeta)\}$$

so

$$E[Y(t)] = E[L\{X(t)\}].$$

If we can interchange the two operators we get the result that the mean function of the output is just the result of L operating on the mean function of the input. This can be heuristically justified as follows, if we assume the operator L can be represented by the superposition integral

$$Y(t) = \int_{-\infty}^{+\infty} h(t, \tau)X(\tau)\, d\tau.$$

Then, taking the expectation, we obtain

$$\begin{aligned} E[Y(t)] &= E\left[\int_{-\infty}^{+\infty} h(t, \tau)X(\tau)\, d\tau\right] \\ &= \int_{-\infty}^{+\infty} h(t, \tau)E[X(\tau)]\, d\tau \\ &= L\{\mu_X(t)\}. \quad \blacksquare \end{aligned}$$

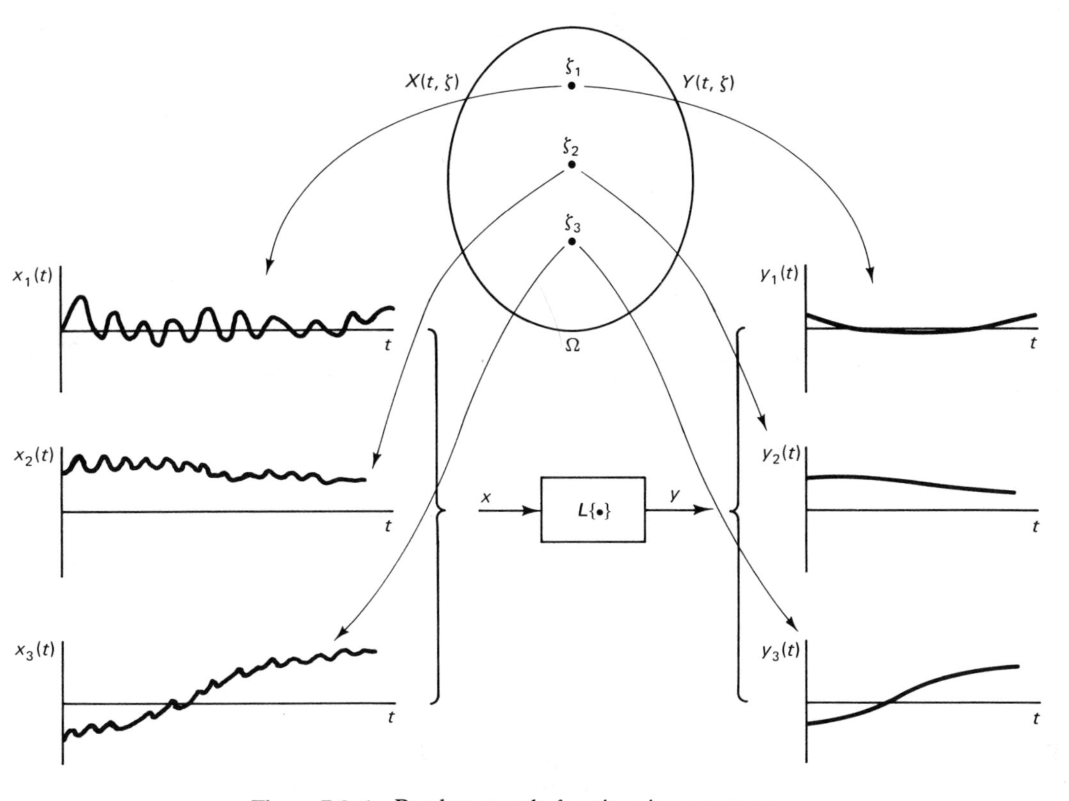

Figure 7.3–1 Random sample functions input to system.

We will present a rigorous proof of this theorem after we study the stochastic integral in Chapter 8. We now look at how the correlation function is transformed by a linear system. From the autocorrelation function of the input $R_{XX}(t_1, t_2)$ we first calculate the cross-correlation function $R_{XY}(t_1, t_2)$ and then the autocorrelation function of the output $R_{YY}(t_1, t_2)$. If the mean is zero for the input process, then by Theorem 7.3–1 the mean of the output process is also zero. Thus the following results can be seen also to hold for covariance functions by changing the input to $X_c(t) = X(t) - \mu_X(t)$, which produces the centered output $Y_c(t) = Y(t) - \mu_Y(t)$.

Theorem 7.3–2. Let $X(t)$ and $Y(t)$ be the input and output random processes of the linear operator L, then the following hold

$$R_{XY}(t_1, t_2) = L_2^*\{R_{XX}(t_1, t_2)\} \tag{7.3–3a}$$

$$R_{YY}(t_1, t_2) = L_1\{R_{XY}(t_1, t_2)\}, \tag{7.3–3b}$$

where L_i means the time variable of the operator L is t_i.

Proof. (formal). Write

$$\begin{aligned} X(t_1)Y^*(t_2) &= X(t_1)L_2^*\{X^*(t_2)\} \\ &= L_2^*\{X(t_1)X^*(t_2)\}, \end{aligned}$$

where we have used the *adjoint operator* L^* whose impulse response is $h^*(t, \tau)$. Then

$$\begin{aligned} E[X(t_1)Y^*(t_2)] &= E[L_2^*\{X(t_1)X^*(t_2)\}] \\ &= L_2^*\{E[X(t_1)X^*(t_2)]\} \text{ by interchanging } L_2^* \text{ and } E, \\ &= L_2^*\{R_{XX}(t_1, t_2)\}. \end{aligned}$$

which is Equation 7.3–3a.

Similarly, to prove Equation 7.3–3b, we multiply by $Y^*(t_2)$ and get

$$Y(t_1)Y^*(t_2) = L_1\{X(t_1)Y^*(t_2)\}$$

so that

$$\begin{aligned} E[Y(t_1)Y^*(t_2)] &= E[L_1\{X(t_1)Y^*(t_2)\}] \\ &= L_1\{E[X(t_1)Y^*(t_2)]\} \text{ by interchanging } L_1 \text{ and } E, \\ &= L_1\{R_{XY}(t_1, t_2)\}, \end{aligned}$$

which is Equation 7.3–3b. ■

If we combine Equation 7.3–3a and Equation 7.3–3b we get:

$$R_{YY}(t_1, t_2) = L_1L_2^*\{R_{XX}(t_1, t_2)\}. \tag{7.3–4}$$

Example 7.3–1: (Edge or "change" detector.) Let $Y(t) = L\{X(t)\} \triangleq X(t) - X(t-1)$ so

$$E[Y(t)] = L\{\mu_X(t)\} = \mu_X(t) - \mu_X(t-1).$$

Also

$$
\begin{aligned}
R_{XY}(t_1, t_2) &= L_2\{R_{XX}(t_1, t_2)\} \\
&= R_{XX}(t_1, t_2) - R_{XX}(t_1, t_2 - 1)
\end{aligned}
$$

and

$$
\begin{aligned}
R_{YY}(t_1, t_2) &= L_1\{R_{XY}(t_1, t_2)\} \\
&= R_{XY}(t_1, t_2) - R_{XY}(t_1 - 1, t_2) \\
&= R_{XX}(t_1, t_2) - R_{XX}(t_1 - 1, t_2) \\
&\quad - R_{XX}(t_1, t_2 - 1) + R_{XX}(t_1 - 1, t_2 - 1).
\end{aligned}
$$

To be specific, if we take

$$R_{XX}(t_1, t_2) \triangleq \sigma_X^2 \exp(-\alpha\,|t_1 - t_2|),$$

then

$$E[Y(t)] = 0 \qquad \text{since } \mu_X = 0,$$

and

$$
\begin{aligned}
R_{YY}(t_1, t_2) &= \sigma_X^2[2 \exp(-\alpha\,|t_1 - t_2|) - \exp(-\alpha\,|t_1 - t_2 - 1|) \\
&\quad - \exp(-\alpha\,|t_1 - t_2 + 1|)].
\end{aligned}
$$

The variance of Y is constant and is given as

$$\sigma_Y^2(t) = \sigma_Y^2 = 2\sigma_X^2[1 - \exp(-\alpha)].$$

We see that as α tends to zero, the variance of Y goes to zero. This is because as α tends to zero, $X(t)$ and $X(t - 1)$ become very positively correlated, and hence there is very little power in their difference.

7.4 CLASSIFICATION OF RANDOM PROCESSES

Here we look at several classes of random processes and pairs of processes. These classifications also apply to random sequences.

Definitions 7.4–1. Let X and Y be random processes. They are:

(a) *Uncorrelated* if $R_{XY}(t_1, t_2) = \mu_X(t_1)\mu_Y^*(t_2)$, for all t_1 and t_2,

(b) *Orthogonal* if $R_{XY}(t_1, t_2) = 0$ for all t_1 and t_2,

(c) *Independent* if for all positive integers n, the nth order PDF of X and Y factors, that is,

$$
\begin{aligned}
&F_{XY}(x_1, y_1, x_2, y_2, \ldots, x_n, y_n; t_1, \ldots, t_n) \\
&\qquad = F_X(x_1, \ldots, x_n; t_1, \ldots, t_n)F_Y(y_1, \ldots, y_n; t_1, \ldots, t_n),
\end{aligned}
$$

for all x_i, y_i and for all $t_1, \ldots, t_n$.

Note that two random processes are orthogonal if they are uncorrelated and at least one of their mean functions is zero. Actually, the orthogonality concept is useful only when the random processes under consideration are zero-mean, in which case it becomes equivalent to the uncorrelated condition. The orthogonality concept was used for estimating random vectors in Section 5.6. This concept will prove useful for estimating random processes and sequences in Chapter 10.

A random process may be uncorrelated, orthogonal, or independent of *itself* at earlier and/or later times. For example, we may have $R_{XX}(t_1, t_2) = 0$ for $t_1 \neq t_2$, in which case we call X an *orthogonal random process*. Similarly $X(t)$ may be independent of $\{X(t_1), \ldots, X(t_n)\}$ for all $t \notin \{t_1, \ldots, t_n\}$ and for all $t_1, \ldots, t_n$ and for all $n \geq 1$. Then we say $X(t)$ is an *independent random process*. Clearly, the sample functions of such processes will be quite rough.

We say a random process is stationary when its statistics do not change with the continuous parameter, often time. The formal definition is:

Definition 7.4–2. A random process $X(t)$ is *stationary* if it has the same nth-order distribution function as $X(t + T)$ for all T and for all positive n.

This definition implies that the mean of a stationary process is a constant. To prove this note that $f(x; t) = f(x; t + T)$ for all T implies $f(x; t) = f(x; 0)$ by taking $T = -t$, which in turn implies that $E[X(t)] = \mu_X$, a constant.

Since the second-order density is also shift invariant, that is,

$$f(x_1, x_2; t_1, t_2) = f(x_1, x_2; t_1 + T, t_2 + T),$$

we have, on choosing $T = -t_2$ that

$$f(x_1, x_2; t_1, t_2) = f(x_1, x_2; t_1 - t_2, 0),$$

which implies $E[X(t_1)X^*(t_2)] = R(t_1 - t_2, 0)$. In the stationary case, therefore, the notation for correlation function can be simplified to a function of just the shift $\tau \triangleq t_1 - t_2$ between the two parameters. Thus we can define the one-parameter correlation function

$$\begin{aligned} R_X(\tau) &\triangleq R_X(\tau, 0) \\ &= E[X(t + \tau)X^*(t)] \end{aligned} \tag{7.4–1}$$

independent of t. An example of this sort of correlation function was seen at the end of Section 7.3.

A weaker form of stationarity does not directly constrain the nth-order PDF's, but rather just the first- and second-order moments. This property, which is easier to verify, is called wide-sense stationarity and will be quite useful in Chapter 9.

Definition 7.4–3. A random process X is *wide-sense stationary* (WSS) if $E[X(t)] = \mu_X$, a constant, and $E[X(t + \tau)X^*(t)] = R_X(\tau)$ for all τ, $-\infty < \tau < +\infty$.

Example 7.4–1: (A random complex exponential.) Let $X(t) = A \exp(j2\pi ft)$ with f = a constant and A a *real* random variable with $E[A] = 0$. Calculating the mean and correlation of $X(t)$, we obtain

$$E[X(t)] = E[A \exp(j2\pi ft)] = E[A]\exp(j2\pi ft) = 0,$$

and

$$\begin{aligned} E[X(t + \tau)X^*(t)] &= E[A \exp(j2\pi f(t + \tau))A \exp(-j2\pi ft)] \\ &= E[A^2]\exp(j2\pi f\tau) = R(\tau). \end{aligned}$$

This example, while shown to be wide-sense stationary, is clearly not stationary. Consider, for example, that $X(0)$ must be pure real while $X(1/4f)$ must always be pure imaginary. We thus conclude that the WSS property is considerably weaker than stationarity.

We could generalize this example to have M complex sinusoids and obtain a rudimentary frequency domain representation for random processes. Consider $X(t) = \sum_{k=1}^{M} A_k \exp(j2\pi f_k t)$ where the generally complex random variables A_k are uncorrelated with mean zero and variances σ_k^2. Then the resulting random process is WSS with mean zero and autocorrelation or autocovariance equal to

$$R(\tau) = \sum_{k=1}^{M} \sigma_k^2 \exp(j2\pi f_k \tau). \tag{7.4–2}$$

For such random processes $X(t)$, the set $\{A_k\}$ constitutes a frequency domain representation. From our experience with Fourier analysis of deterministic functions, we would expect that as M became large and as the f_k became dense, that is, the spacing between the f_k became small, then most random processes would have such an approximate representation.

Wide-Sense Stationary Processes and LSI Systems. If we use the impulse response to represent an LSI system, we can specialize the results of Theorems 7.3–1 and 7.3–2 to the wide-sense stationary case. Rewriting Equation 7.3–2 we have

$$E[Y(t)] = L\{\mu_X\} = \mu_X L\{1\},$$

where we have substituted the constant mean μ_X. Specializing the linear operator L to convolution, we get

$$\begin{aligned} \mu_Y(t) &= \mu_X \int_{-\infty}^{+\infty} h(\tau)\, d\tau \\ &= \mu_X H(0), \quad \text{where } H(\omega) \text{ is the system's frequency response,} \end{aligned}$$

the Fourier transform of $h(t)$. We can do the same thing with Equations 7.3–3a and 7.3–3b resulting in

$$R_{XY}(t_1, t_2) = \int_{-\infty}^{+\infty} h^*(\tau_2) R_{XX}(t_1, t_2 - \tau_2)\, d\tau_2, \tag{7.4–3a}$$

and

$$R_{YY}(t_1, t_2) = \int_{-\infty}^{+\infty} h(\tau_1) R_{XY}(t_1 - \tau_1, t_2)\, d\tau_1, \tag{7.4–3b}$$

which can be written in convolution operator notation as

$$R_{XY}(t_1, t_2) = h^*(t_2) * R_{XX}(t_1, t_2),$$

where the convolution is along the t_2-axis, and

$$R_{YY}(t_1, t_2) = h(t_1) * R_{XY}(t_1, t_2),$$

where the convolution is along the t_1-axis. Combining these two equations, we get

$$R_{YY}(t_1, t_2) = h(t_1) * R_{XX}(t_1, t_2) * h^*(t_2).$$

Equations 7.4–3 can be specialized to the WSS case by noting that R_{XX} is just a function of the shift $\tau \triangleq (t_1 - t_2)$ and using single-parameter functions, as follows:

$$R_{XY}(\tau) = \int_{-\infty}^{+\infty} h^*(\tau_2) R_{XX}(\tau + \tau_2)\, d\tau_2 \qquad (7.4\text{–}4a)$$

and

$$R_{YY}(\tau) = \int_{-\infty}^{+\infty} h(\tau_1) R_{XY}(\tau - \tau_1)\, d\tau_1. \qquad (7.4\text{–}4b)$$

We will come back to this subject in Chapter 9 after we have treated the mean-square calculus of random processes in Chapter 8. Then we will be able to address the questions of interchange of the operators L and E in a more rigorous fashion.

Periodic and Cyclostationary Processes. Besides stationary and its wide-sense version, two other classes of random processes are often encountered. They are periodic and cyclostationary and are here defined.

Definition 7.4–4. A random process $X(t)$ is *wide-sense periodic* if

$$\mu_X(t) = \mu_X(t + T) \qquad \text{for all } t$$

and

$$K_X(t_1, t_2) = K_X(t_1 + T, t_2) = K_X(t_1, t_2 + T) \quad \text{for all } t_1, t_2.$$

Note that $K_X(t_1, t_2)$ is periodic with period T along both axes.

An example of a wide-sense periodic random process is the random complex exponential of Example 7.4–1. In fact, the random Fourier series,

$$X(t) = \sum_{k=0}^{\infty} A_k \exp\left(j\frac{2\pi kt}{T}\right) \qquad (7.4\text{–}5)$$

with r.-v. coefficients A_k, would also be wide-sense periodic. A wide-sense periodic process can also be WSS, in which case we call it *wide-sense periodic stationary*. We will consider these processes in Section 9.6. The covariance function of a wide-sense periodic process is generically indicated in Figure 7.4–1. We see that K_X is doubly periodic with a two-dimensional period of (T, T). In Problem 7.17 the reader is asked to prove that the sample functions of a wide-sense periodic random process are periodic with probability-1, that is,

$$X(t) = X(t + T) \quad \text{(a.s.)} \quad \text{for all } t.$$

A final classification of random processes is cyclostationarity. It is only partially related to periodicity and is often confused with it. The reader should carefully note the difference in the following definition.

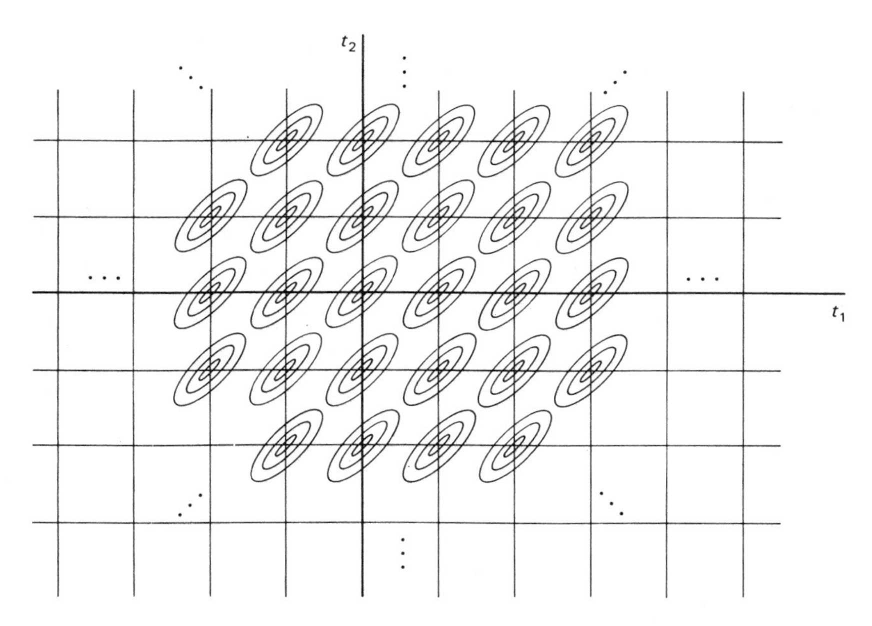

Figure 7.4–1 Wide-sense periodic covariance function.

Definition 7.4–5. A random process $X(t)$ is *wide-sense cyclostationary* if

$$\mu_X(t) = \mu_X(t + T) \quad \text{for all } t$$

and

$$K_X(t_1, t_2) = K_X(t_1 + T, t_2 + T) \quad \text{for all } t_1 \text{ and } t_2.$$

An example of cyclostationarity is the random PSK process of Equation 7.2–8. Its mean function is zero and hence trivially periodic. Its covariance function (Equation 7.2–10) is invariant to a shift by T in *both* its arguments. Note that Equation 7.2–10 is *not* doubly-periodic since $R_X(0, T) = 0 \neq R_X(0, 0)$. Also note that the sample functions of $X(t)$ are not periodic in any sense.

In addition to modulators, scanning sensors tend to produce cyclostationary processes. For example, the line-by-line scanning in television transforms the random image field into a one-dimensional random process that has been modeled as cyclostationary. Cyclostationary processes are not stationary or WSS except in trivial cases. However, it is sometimes appropriate to convert a cyclostationary process into a stationary process as in the following example.

Example 7.4–2: (Wide-sense stationary PSK.) We have seen that the PSK process of Section 7.2 is cyclostationary and hence not wide-sense stationary. This is easily seen with reference to Equation 7.2–10. This cyclostationarity arises from the fact that the analog angle process $\Theta_a(t)$ is step-wise constant and changes only at $t = nT$ for integer n. In many real situations the modulation process starts at an arbitrary time t, which in fact can be modeled as random from the viewpoint of the system designer. Thus in this more practical case, the modulated signal process (Equation 7.2–8) is converted to

$$\tilde{X}(t) = \cos(2\pi f_c t + \Theta_a(t) + 2\pi f_c \mathrm{T}_0), \tag{7.4–6}$$

by the addition of a random variable T_0, which is uniformly distributed on $[0, T]$ and independent of the angle process $\Theta_a(t)$. It is then easy to see that the mean and covariance functions only need to be modified by an ensemble average over T_0, which by the uniformity of T_0, is just an integral over $[0, T]$. We thus obtain

$$\begin{aligned} R_{\tilde{X}}(t_1+\tau, t_1) &= \frac{1}{T}\int_0^T R_X(t_1+\tau + t, t_1+t)\,dt \\ &= \frac{1}{T}\int_{-\infty}^{+\infty} s_Q(t_1+t + \tau)s_Q(t_1+t)\,dt \\ &= \frac{1}{T} s_Q(\tau) * s_Q(-\tau), \end{aligned}$$

which is just a function of the shift τ. Thus $\tilde{X}$ is WSS.

7.5 SUMMARY

In this chapter we introduced the concept of the random process, an ensemble of functions of a continuous parameter. The parameter can be space or time. Most topics in this chapter generalize to two-dimensional parameters. Many modern applications, in fact, require a two-dimensional parameter, for example, the intensity function $I(t_1, t_2)$ of an image. Such random functions are called *random fields* and can be analyzed using extensions of the methods of this chapter. See Reference 7–6 or Chapter 7 of Reference 7–5.

We studied four important processes: the Poisson process, which serves as a basis for many more involved counting processes; the random PSK modulation of a sinusoid, which is basic to digital communications; the Wiener process, a first example of a Gaussian random process; and the Markov process, which is the signal model to be used in the Kalman-Bucy filter of Chapter 10.

We considered the effect of linear systems on the second-order properties of random processes and specialized our results to the useful subcategory of WSS processes and LSI systems. We also briefly considered the classes of wide-sense periodic and cyclostationary processes.

PROBLEMS

7.1. Let $X[n]$ be a real stationary random sequence with mean $E[X[n]] = \mu_X$ and correlation $E[X[n + m]X[n]] = R_X[m]$. If $X[n]$ is the input to a D/A converter, the continuous-time output can be idealized as the random process

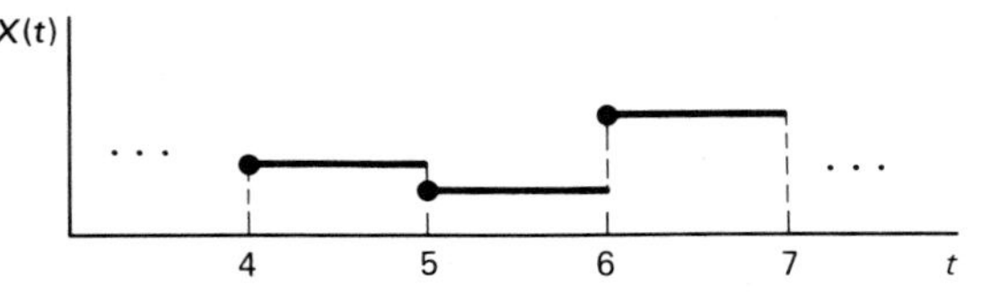

Figure P-7.1

$X(t)$ with

$$X(t) \triangleq X[n] \quad \text{for } n \le t < n+1, \quad \text{for all } n,$$

as shown in Figure P-7.1.

(a) Find the mean $E[X(t)] = \mu_a(t)$ as a function of μ_X.

(b) Find the correlation $E[X(t_1)X(t_2)] = R_a(t_1, t_2)$ in terms of μ_X and $R_X[m]$.

7.2. Let $B[n]$ be a Bernoulli random sequence taking on the values ± 1. Then define the random process,

$$X(t) \triangleq \sqrt{p}\,\sin\!\left(2\pi f_0 t + B[n]\frac{\pi}{2}\right) \quad \text{for } nT \le t < (n+1)T, \quad \text{for all } n,$$

where $\sqrt{p}$ and f_0 are real numbers.

(a) Determine the mean function $\mu_X(t)$.

(b) Determine the covariance function $K_X(t, s)$.

7.3. Let $N(t)$ be a Poisson random process defined on $0 \le t < \infty$ with $N(0) = 0$ with mean arrival rate $\lambda \ge 0$.

(a) Determine the probability $P[N(t) = n]$ as a function of λ and t.

(b) Find the joint probability $P[N(t_1) = n_1, N(t_2) = n_2]$ for $t_2 > t_1$.

(c) Find an expression for the Kth order joint PMF,

$$P_N(n_1, \ldots, n_K; t_1, \ldots, t_K),$$

with $0 \le t_1 < t_2 < \ldots < t_K < \infty$. Be careful to consider the relative values of $n_1, \ldots, n_K$.

7.4. The *nonuniform Poisson counting process* $N(t)$ is defined for $t \ge 0$ as follows:

(1) $N(0) = 0$.

(2) $N(t)$ has independent increments.

(3) For all $t_2 \ge t_1$,

$$P[N(t_2) - N(t_1) = n] = \frac{\left[\int_{t_1}^{t_2} \lambda(\nu)\,d\nu\right]^n}{n!} \exp\!\left(-\int_{t_1}^{t_2} \lambda(\nu)\,d\nu\right), \quad \text{for } n \ge 0.$$

The function $\lambda(t)$ is called the *intensity function* and is everywhere nonnegative, that is, $\lambda(t) \ge 0$ for all t.

(a) Find the mean function $\mu_N(t)$ of the nonuniform Poisson process.

(b) Find the correlation function $R_N(t_1, t_2)$ of $N(t)$.

Define a *warping* of the time axis as follows:

$$\tau(t) \triangleq \int_0^t \lambda(\nu)\,d\nu$$

Now $\tau(t)$ is monotonic increasing if $\lambda(\nu) > 0$ for all ν, so we can then define the inverse mapping:

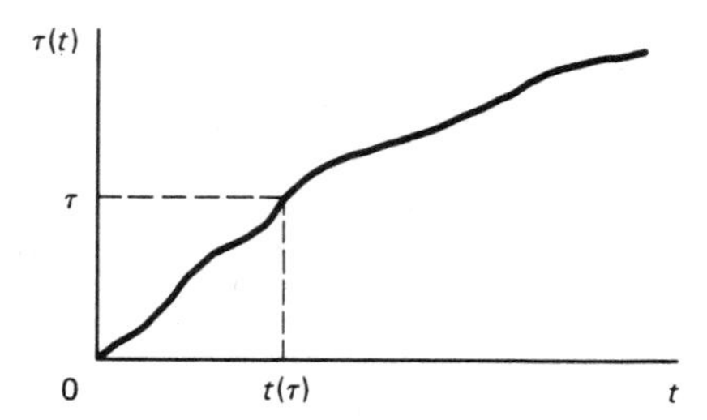

Figure P-7.4

(c) Assume $\lambda(t) > 0$ for all t and define the counting process,

$$N_u(\tau) \triangleq N(t(\tau)).$$

Show that $N_u(\tau)$ is a uniform Poisson counting process with rate $\lambda = 1$; that is, show for $\tau \geq 0$

(1) $N_u(0) = 0$.

(2) $N_u(\tau)$ has independent increments.

(3) For all $\tau_2 \geq \tau_1$,

$$P[N(\tau_2) - N(\tau_1) = n] = \frac{(\tau_2 - \tau_1)^n}{n!} e^{-(\tau_2 - \tau_1)}, \qquad n \geq 0.$$

7.5. This problem concerns the construction of the Poisson counting process as given in Section 7.2.

(a) Show that the density for the nth arrival time $T(n)$ is:

$$f_T(t; n) = \frac{\lambda^n t^{n-1}}{(n-1)!} e^{-\lambda t} u(t), \qquad n > 0.$$

In the derivation of the property that the increments of a Poisson process are Poisson distributed, that is,

$$P[X(t_a) - X(t_b) = n] = \frac{[\lambda(t_a - t_b)]^n}{n!} e^{-\lambda(t_a - t_b)} u(n), \qquad t_a > t_b,$$

we implicitly use the fact that the first interarrival time in $(t_b, t_a]$ is exponentially distributed. Actually, this fact is not clear as the interarrival time in question is only partially in the interval $(t_b, t_a]$. A pictorial diagram is shown in Figure P-7.5.

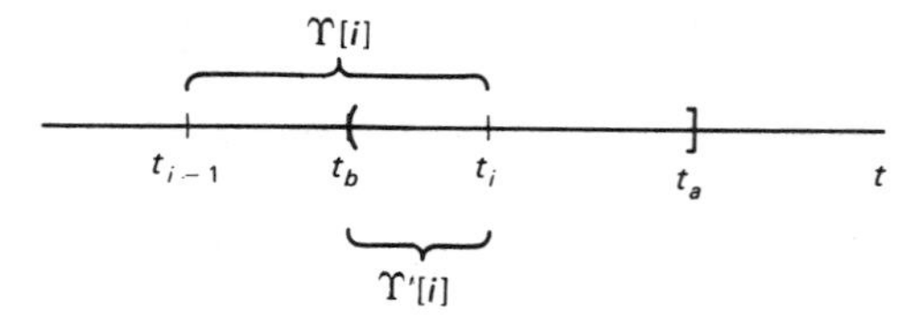

Figure P-7.5 Interarrival time overlapping a time interval

Define $\Upsilon'[i] \triangleq T[i] - t_b$ as the *partial interarrival time*. We note $\Upsilon'[i] = \Upsilon[i] - T$ where the random variable $T \triangleq t_b - T[i-1]$ and $\Upsilon[i]$ denotes the (full) interarrival time.

(b) Fix the random variable $T = t$ and find the PDF

$$F_{\Upsilon'[i]}(\tau' \mid T = t) = P[\Upsilon[i] \leq \tau' + t \mid \Upsilon[i] \geq t].$$

(c) Modify the result of part (b) to account for the fact that T is random variable. (Hint: This part does not involve a lot of calculations.)

Because of the preceding properties, the exponential distribution is called *memoryless*. It is the only continuous distribution with this property.

7.6. Let $N(t)$ be a counting process on $[0, \infty)$ whose average rate $\lambda(t)$ depends on another positive random process $S(t)$, specifically $\lambda(t) = S(t)$. We assume that $N(t)$ given $\{S(t)$ on $[0, \infty)\}$ is a nonuniform Poisson process. We know $\mu_S(t) = \mu_0 > 0$ and also know $K_S(t_1, t_2)$.

(a) Find $\mu_N(t)$ for $t \geq 0$ in terms of $\mu_S(t)$.

(b) Find $\sigma_N^2(t)$ for $t \geq 0$ in terms of $K_S(t_1, t_2)$.

7.7. Let the scan-line of an image be described by the spatial random process $S(x)$, which models the ideal gray-level at the point x. Let us transmit each point independently with an optical channel by modulating the intensity of a photon source:

$$\lambda(t, x) = S(x) + \lambda_0, \qquad 0 \leq t \leq T.$$

In this way we create a family of random processes, indexed by the continuous parameter x,

$$\{N(t, x)\}.$$

For each x, $N(t, x)$ given $S(x)$ is a uniform Poisson process. At the end of the observation interval, we store $N(x) \triangleq N(T, x)$ and inquire about the statistics of this spatial process.

To summarize, $N(x)$ is an integer-valued *spatial* random process that depends on the value of another random process $S(x)$, called the signal process. The spatial random process $S(x)$ is stationary with zero mean and covariance function:

$$K_S(x) = \sigma_S^2 \exp(-\alpha\,|x|),$$

where $\alpha > 0$. The conditional distribution of $N(x)$, given $S(x)$, is Poisson with mean $\lambda(x) \triangleq (S(x) + \lambda_0)T$ where λ_0 is a positive constant; that is,

$$P[N(x) = n \mid S(x)] = \frac{\lambda^n(x)}{n!} e^{-\lambda(x)} u(n).$$

The random variables $N(x)$ are conditionally independent from point to point.

(a) Find the (unconditional) mean and variance

$$E[N(x)] \quad \text{and} \quad E[N(x) - E[N(x)]^2].$$

Hint: First find the conditional mean and mean-square.

(b) Find $R_N(x_1, x_2) \triangleq E[N(x_1)N(x_2)]$.

7.8. Let $W(t)$ be a *standard* Wiener process, defined over $[0, \infty)$ (that is, it's distributed as $N(0, t)$ at time t).

Find the joint density $f_W(a_1, a_2; t_1, t_2)$ for $0 < t_1 < t_2$.

7.9. Let $W(t)$ be a standard Wiener process, and define

$$X(t) \triangleq W^2(t) \quad \text{for} \quad t \geq 0.$$

(a) Find the density $f_X(x; t)$.

(b) Find the conditional density $f_X(x_2 \mid x_1; t_2, t_1), \qquad t_2 > t_1.$

(c) Is $X(t)$ Markov? Justify your answer.
(d) Does $X(t)$ have independent increments?

7.10. Let $X[n]$ be a Markov chain on $n \geq 0$ taking values 1 and 2 with *one-step transition probabilities*,

$$P_{ij} \triangleq P[X[n] = i \mid X[n-1] = j], \qquad 1 \leq i, j \leq 2,$$

given in matrix form as,

$$\mathbf{P} = \begin{bmatrix} 0.9 & 0.2 \\ 0.1 & 0.8 \end{bmatrix} = (P_{ij}).$$

We describe the state probabilities at time n by the vector

$$\mathbf{P}[n] \triangleq \begin{bmatrix} P[X[n] = 1] \\ P[X[n] = 2] \end{bmatrix}.$$

(a) Show that $\mathbf{P}[n] = \mathbf{P}^n \mathbf{P}[0]$.
(b) Draw a two-state transition diagram and label the branches with the one-step transition probabilities P_{ij}. Don't forget the P_{ii} or *self transitions.* (See Figure 7.2–4 for state transition diagram of continuous-time Markov chain.)
(c) Given that $X[0] = 1$, find the probability that the first transition to state 2 occurs at time n. Compare to the exponential distribution of transition times obtained for the continuous-time Markov chain (see Example 7.2–3).

7.11. Let $X(t)$ be a Markov random process on $[0, \infty)$ with initial density $f_X(x; 0) = \delta(x-1)$ and conditional density

$$f_X(x_2 \mid x_1; t_2, t_1) = \frac{1}{\sqrt{2\pi(t_2 - t_1)}} \exp\left(-\frac{1}{2}\frac{(x_2 - x_1)^2}{t_2 - t_1}\right), \qquad \text{for all } t_2 > t_1$$

(a) Find $f_X(x; t)$ for all t.
(b) Repeat part (a) for $f_X(x; 0) \sim N(0, 1)$.

7.12. Let A and B be independent and identically distributed (i.i.d.) random variables with mean 0, variance σ^2, and third moment $\mu \triangleq E[A^3] = E[B^3] \neq 0$. Consider the random process

$$X(t) = A\cos(2\pi ft) + B\sin(2\pi ft), \qquad -\infty < t < +\infty,$$

where f is a given frequency.
(a) Show that the random process $X(t)$ is WSS.
(b) Show that $X(t)$ is not strictly stationary.

7.13. Earlier we proved Theorem 7.3–2 thus deriving Equation 7.3–4. State and prove a corresponding theorem for covariance functions. Do not assume $\mu_X(t) = 0$.

7.14. Let $X(t)$ be a stationary random process with mean μ_X and covariance $K_{XX}(\tau) = \delta(\tau)$. Let the sample functions of $X(t)$ drive the differential equation

$$\dot{Y}(t) + aY(t) = X(t), \qquad -\infty < t < +\infty.$$

$H(\omega) = \dfrac{1}{a + j\omega}$

(a) Find $\mu_Y = E[Y(t)]$.
(b) Find $R_{YY}(\tau)$.
(c) Find σ_Y^2.

7.15. Is the random process $X(t)$ in Problem 7.1 wide-sense stationary?

7.16. Let $X(t)$ be a random process defined by

$$X(t) \triangleq N \cos(2\pi f_0 t + \Theta),$$

where f_0 is a known frequency and N and Θ are independent random variables. The characteristic function for N is

$$\Phi_N(\omega) = E[e^{+j\omega N}] = \exp\{\lambda[e^{j\omega} - 1]\},$$

where λ is a given positive constant (that is, N is a Poisson random variable). The random variable Θ is uniformly distributed on $[-\pi, +\pi]$.

(a) Determine the mean function $\mu_X(t)$.

(b) Determine the covariance function $K_X(t, s)$.

(c) Is $X(t)$ wide-sense stationary? Justify your answer.

(d) Is $X(t)$ stationary? Justify your answer.

7.17. Show that if a random process is wide-sense periodic with period T, then for any t, $E[|X(t + T) - X(t)|^2] = 0$ and hence $X(t) = X(t + T)$ with probability-1.

7.18. Let $X(t)$ be an independent increment random process defined on $t \geq 0$ with initial value $X(0) = X_0$, a random variable. Assume the following characteristic functions exist: $E[e^{j\omega X_0}] \triangleq \Phi_{X_0}(\omega)$ and

$$E[e^{j\omega(X(t)-X(s))}] \triangleq \Phi_{X(t)-X(s)}(\omega) \quad \text{for} \quad t \geq s.$$

(a) On defining $E[e^{j\omega X(t)}] \triangleq \Phi_{X(t)}(\omega)$, show that

$$\Phi_{X(t)}(\omega) = \Phi_{X_0}(\omega)\Phi_{X(t)-X(0)}(\omega).$$

(b) Show that for all $t_2 \geq t_1$, the joint characteristic function of $X(t_2)$ and $X(t_1)$ is given by

$$\Phi_{X(t_2),X(t_1)}(\omega_2, \omega_1) = \Phi_{X_0}(\omega_1 + \omega_2)\Phi_{X(t_1)-X(0)}(\omega_1 + \omega_2)\Phi_{X(t_2)-X(t_1)}(\omega_2).$$

(c) Apply part (a) to Problem 7.11(b) by using Gaussian characteristic functions.

REFERENCES

7–1. S. Karlin and H. M. Taylor, *A First Course in Stochastic Processes*. New York: Academic Press, 1975.

7–2. T. Kailath, *Linear Systems*. Englewood Cliffs, N.J.: Prentice-Hall, 1980.

7–3. L. Kleinrock, *Queueing Systems, Vol. 1: Theory*. New York: John Wiley, 1975.

7–4. H. L. Van Trees, *Detection, Estimation and Modulation Theory: Part I*. New York: John Wiley, 1968, p. 142.

7–5. E. Wong and B. Hajek, *Stochastic Processes in Engineering Systems*, New York: Springer-Verlag, 1985, pp. 62–63.

7–6. P. Whittle, "On Stationary Process in the Plane," *Biometrika*, Vol. 41, 1954, pp. 434–449.

8

MEAN-SQUARE CALCULUS

In the last chapter we introduced random processes, that is, random functions of a continuous parameter. We now go on to develop a calculus for these random functions, complete with concepts of continuity, derivatives, integrals, and differential and integral equations. The definitions of these quantities will involve the limit of a sequence of approximations at each value of the continuous parameter t. These sequences are random sequences in the sense of Chapter 6. The mean-square limit of the approximating sequence will then be used to define the resulting derivative and integral, using and extending the convergence concepts of Section 6.4. These mathematical constructs will allow us to consider the effect of linear circuits and more general linear devices on random processes. Our general approach will relate the mean-square calculus operations on random processes to ordinary-calculus operations on the corresponding correlation and covariance functions.

From our work with limits in Chapter 6, we expect that the mean-square derivative and integral will be weaker concepts than the sample-function derivative and integral that we looked at in Chapter 7. The reason this added abstractness is necessary is that many useful random processes do not have sample-function derivatives. Furthermore, this defect cannot be determined from just examining the mean and correlation or covariance functions. In the first section we begin by looking at the various concepts of continuity for random processes.

8.1 CONTINUITY AND DERIVATIVES [8–1]

We will consider random processes that may be real or complex valued. The concept of continuity for random processes relies on the concept of limit for random processes just the same as in the case of ordinary functions.

However, in the case of random processes, as for random sequences, there are four concepts for limit, which implies that there are four types of stochastic continuity. The strongest continuity would correspond to sure convergence of the sample function limits,

✓ *sample-function continuity,* $\lim_{\varepsilon \to 0} X(t + \varepsilon, \zeta) = X(t, \zeta)$ for all $\zeta \in \Omega$.

The next strongest situation would be to disregard those sample functions in a set of probability zero that are discontinuous at time t. This would yield continuity almost surely (a.s.),

✓ *a.s. continuity,* $$P\left[\left(\lim_{s \to t} X(s)\right) \neq X(t)\right] = 0.$$

Corresponding to the concept of limit in probability we could study the concept of continuity in probability.

✓ *p-continuity,* $\lim_{s \to t} P[|X(s) - X(t)| > \varepsilon] = 0 \quad$ for any $\varepsilon > 0$.

The most useful and tractable concept of continuity turns out to be a mean-square-based definition. This is the concept that we will use almost exclusively.

Definition 8.1–1. A random process $X(t)$ is *continuous in the mean-square sense at the point t* if

$$\text{as } \varepsilon \to 0 \text{ we have } E[|X(t + \varepsilon) - X(t)|^2] \to 0.$$

If the above holds for all t, we say $X(t)$ is *mean-square (m.-s.) continuous.*

One advantage of this definition is that it is readily expressible in terms of correlation functions. By expanding out the expectation of the square of the difference, it is seen that we just require a certain continuity in the correlation function.

Theorem 8.1–1. The random process $X(t)$ is m.-s. continuous at t iff $R_X(t_1, t_2)$ is continuous at the point $t_1 = t_2 = t$.

Proof. Expand the expectation in Definition 8.1–1 to get an expression involving R_X,

$$\begin{aligned} E[|X(t + \varepsilon) - X(t)|^2] &= R_X(t + \varepsilon, t + \varepsilon) - R_X(t, t + \varepsilon) \\ &\quad - R_X(t + \varepsilon, t) + R_X(t, t). \end{aligned}$$

Clearly the right-hand side goes to zero as $\varepsilon \to 0$ iff the deterministic function R_X is continuous at $t_1 = t_2 = t$. ■

Example 8.1–1: (Standard Wiener process.) We investigate the m.-s. continuity of the Wiener process of Section 7.2. By Equation 7.2–16 we have

$$R_X(t_1, t_2) = \min(t_1, t_2), \qquad t_1, t_2 \geq 0.$$

The problem thus reduces to whether the function $\min(t_1, t_2)$ is continuous at the

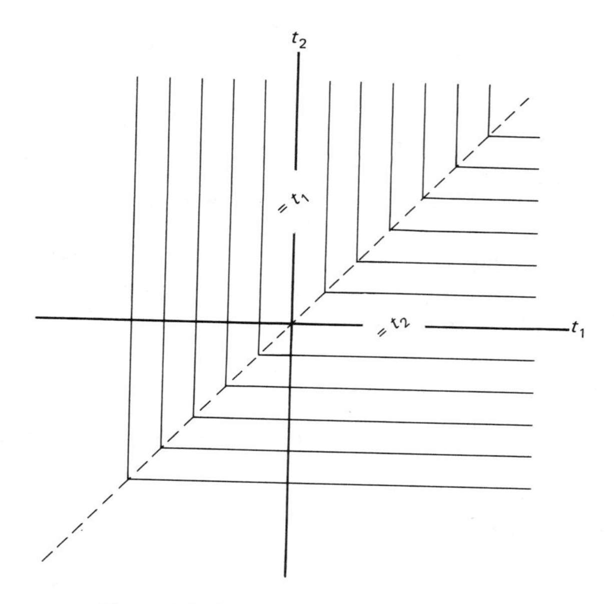

Figure 8.1–1 Contour plot of $\min(t_1, t_2)$.

point (t, t). (See Figure 8.1–1.) The value of the function $\min(t_1, t_2)$ at (t, t) is t so we consider

$$|\min(t_1, t_2) - t|$$

for $t_1 = t + \varepsilon_1$ and $t_2 = t + \varepsilon_2$,

$$|\min(t + \varepsilon_1, t + \varepsilon_2) - t|.$$

But

$$|\min(t + \varepsilon_1, t + \varepsilon_2) - t| \leq \max(\varepsilon_1, \varepsilon_2),$$

so this magnitude can be made arbitrarily small by choice of $\varepsilon_1 > 0$ and $\varepsilon_2 > 0$. Thus the Wiener process is m.-s. continuous.

Lest the reader feel overly confident at this point, note that the Poisson counting process has the same correlation function when centered at its mean, thus the Poisson process is also m.-s. continuous even though its sample functions are not continuous! Evidently m.-s. continuity does not mean that the sample functions are continuous anywhere.

We next look at a special case of Theorem 8.1–1 for wide-sense stationary random processes.

Corollary. A wide-sense stationary random process $X(t)$ is m.-s. continuous for all t iff $R_X(\tau)$ is continuous at $\tau = 0$.

Proof. By Theorem 8.1–1, we need continuity on $t_1 = t_2$, but this is the same as $\tau = 0$. Hence $R_X(\tau)$ must be continuous at $\tau = 0$. ■

We note that in the stationary case we get m.-s. continuity of the random process for all time after verifying only the continuity of a one-dimensional function $R_X(\tau)$ at the origin, $\tau = 0$. Continuity is a necessary condition for the existence of the derivative of an ordinary function. However, considering the case where the difference

$$x(t + \varepsilon) - x(t) = O(\sqrt{\varepsilon}), \tag{8.1–1}$$

we see that it is not a *sufficient* condition in the ordinary calculus. Similarly, in the mean-square calculus, we find that m.-s. continuity is not a sufficient condition for the existence of the m.-s. derivative, which is defined as follows.

Definition 8.1–2. The random process $X(t)$ has a *mean-square derivative at* t if the mean-square limit of $[X(t + \varepsilon) - X(t)]/\varepsilon$ exists as $\varepsilon \to 0$.

If it exists we denote this m.-s. derivative by X', $X^{(1)}$, dX/dt or $\dot{X}$. Generally, we do not know X' when we are trying to determine whether it exists, so we turn to the Cauchy convergence criterion. In this case the test becomes

$$E[|[X(t + \varepsilon_1) - X(t)]/\varepsilon_1 - [X(t + \varepsilon_2) - X(t)]/\varepsilon_2|^2] \\ \to 0 \text{ as } \varepsilon_1 \text{ and } \varepsilon_2 \to 0. \tag{8.1–2}$$

As was the case for continuity, we can express this condition in terms of the correlation function, making it easier to apply. This generally useful condition is stated in the following theorem.

Theorem 8.1–2. A random process $X(t)$ with autocorrelation function $R_X(t_1, t_2)$ has a m.-s. derivative at time t iff $\partial^2 R_X(t_1, t_2)/\partial t_1 \, \partial t_2$ exists at $t_1 = t_2 = t$.

Proof. Expand the square inside the expectation in Equation 8.1–2 to get three terms, the first and last of which look like

$$E[|[X(t + \varepsilon) - X(t)]/\varepsilon|^2] \\ = [R(t + \varepsilon, t + \varepsilon) - R(t, t + \varepsilon) - R(t + \varepsilon, t) + R(t, t)]/\varepsilon^2$$

which converges to

$$\partial^2 R(t_1, t_2)/\partial t_1 \, \partial t_2,$$

if the second mixed partial derivative exists at the point $(t_1, t_2) = (t, t)$. The second or cross term is

$$-2E[\,[X(t + \varepsilon_1) - X(t)]/\varepsilon_1 \cdot [X(t + \varepsilon_2) - X(t)^*/\varepsilon_2\,]$$
$$= -2[R(t + \varepsilon_1, t + \varepsilon_2) - R(t, t + \varepsilon_2) - R(t + \varepsilon_1, t) + R(t, t)]/\varepsilon_1\varepsilon_2$$
$$= -2\{[R(t + \varepsilon_1, t + \varepsilon_2) - R(t + \varepsilon_1, t)]/\varepsilon_2 - [R(t, t + \varepsilon_2) - R(t, t)]/\varepsilon_2\}/\varepsilon_1$$
$$\to -2 \frac{\partial}{\partial t_1}\left(\frac{\partial R(t_1, t_2)}{\partial t_2}\right)\Bigg|_{(t_1,t_2)=(t,t)}$$
$$= -2\partial^2 R(t_1, t_2)/\partial t_1 \, \partial t_2 \,|_{(t_1,t_2)=(t,t)}$$

iff this second mixed partial derivative exists at the point $(t_1, t_2) = (t, t)$.

Combining all three of these terms, we get convergence to

$$2\,\partial^2 R/\partial t_1\,\partial t_2 - 2\,\partial^2 R/\partial t_1\,\partial t_2 = 0. \quad \blacksquare$$

In the preceding theorem the reader should note that we are talking about a two-dimensional function $R(t_1, t_2)$ and its second mixed-partial derivative evaluated on the diagonal points $(t_1, t_2) = (t, t)$. This is clearly not the same as the second derivative of the one-dimensional function $R(t, t)$, which is the restriction of $R(t_1, t_2)$ to the diagonal line $t_1 = t_2$. In some cases the derivative of $R(t, t)$ will exist while the partial derivative of $R(t_1, t_2)$ will not.

Example 8.1–2: Let $X(t)$ be a random process with correlation function $R(t_1, t_2) = \sigma^2 \exp(-\alpha\,|t_1 - t_2|)$. To test for the existence of a m.-s. derivative X', we attempt to compute the second mixed partial derivative of R. We first compute

$$\partial R/\partial t_2 = \begin{cases} \dfrac{\partial}{\partial t_2}[\sigma^2 \exp -\alpha(t_2 - t_1)], & t_1 < t_2 \\[2ex] \dfrac{\partial}{\partial t_2}[\sigma^2 \exp -\alpha(t_1 - t_2)], & t_1 \geq t_2 \end{cases}$$

$$= \begin{cases} -\alpha\sigma^2 \exp -\alpha(t_2 - t_1), & t_1 < t_2 \\ +\alpha\sigma^2 \exp -\alpha(t_1 - t_2), & t_1 \geq t_2. \end{cases} \qquad (8.1\text{–}3)$$

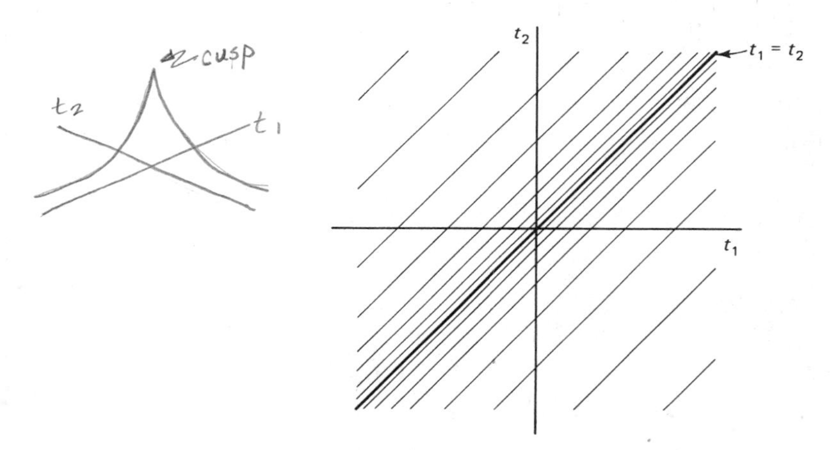

Figure 8.1–2 Contour plot of R of Example 8.1–2.

Graphing the function $R(t_1, t_2)$ as shown in Figure 8.1–2, we see that there is a cusp on the diagonal line $t_1 = t_2$. Thus there is no partial derivative there for any t. So the second mixed-partial cannot exist there either, and we conclude that no m.-s. derivative exists for an $X(t)$ with this correlation function. Evidently it is not smooth enough.

Example 8.1–3: We look at another random process $X(t)$ with mean function $\mu_X = 5$ and correlation function,

$$R(t_1, t_2) = \sigma^2 \exp(-\alpha(t_1 - t_2)^2) + 25,$$

which is smoother on the diagonal line. The first partial with respect to t_2 is

$$\frac{\partial R}{\partial t_2} = 2\alpha(t_1 - t_2)\sigma^2 \exp(-\alpha(t_1 - t_2)^2).$$

Then the second mixed partial becomes

$$\frac{\partial^2 R}{\partial t_1\, \partial t_2} = 2\alpha\sigma^2[1 - 2\alpha(t_1 - t_2)^2]\exp(-\alpha(t_1 - t_2)^2),$$

which evaluated at $t_1 = t_2 = t$ becomes

$$\left.\frac{\partial^2 R}{\partial t_1\, \partial t_2}\right|_{(t_1,t_2)=(t,t)} = 2\alpha\sigma^2,$$

so that in this case the m.-s. derivative $X'(t)$ exists for all t.

Given the existence of the m.-s. derivative, the next question we might be interested in is: What is its probability law? Or more simply: (a) What is its mean and correlation function; and (b) How is X' correlated with X? To answer (a) and (b), we start by considering the expectation

$$E[X'(t)] = E[dX(t)/dt].$$

Now assuming that the m.-s. derivative is a linear operator, we would expect to be able to interchange the derivative and the expectation under certain conditions to obtain,

$$\checkmark\quad E[X'(t)] = dE[X(t)]/dt = d\mu_X(t)/dt. \tag{8.1–4}$$

We can show that this equation is indeed true for the m.-s. derivative by making use of the inequality

$$|E[Z]|^2 \leq E[|Z|^2],$$

for a complex random variable Z, a consequence of the nonnegativity of $\text{Var}[Z]$. First we set $A_n \triangleq n[X(t + 1/n) - X(t)]$ and note that the right-hand side of Equation 8.1–4 is just $\lim_{n\to\infty} E[A_n]$. Thus Equation 8.1–4 will be true if $\lim_{n\to\infty} E[X'(t) - A_n] = 0$. Then making use of the above inequality with $Z \triangleq X'(t) - A_n$, we get

$$|E[X'(t) - A_n]|^2 \leq E[|X'(t) - A_n|^2],$$

where the right-hand side goes to zero by the definition of m.-s. derivative. Thus the left-hand side must also go to zero and hence we are free to interchange the order of m.-s. differentiation and mathematical expectation, that is, Equation 8.1–4 is correct.

To calculate the correlation function of the mean-square derivative process we first define,

$$\checkmark\quad R_{X'}(t_1, t_2) = E[X'(t_1)X'^*(t_2)], \tag{8.1–5}$$

and formally compute, interchanging the order of expectation and differentiation,

$$R_{X'}(t_1, t_2) = \lim_{m,n\to\infty} E\left[n\left(X\left(t_1 + \frac{1}{n}\right) - X(t_1)\right) \cdot m\left(X\left(t_2 + \frac{1}{m}\right) - X(t_2)\right)^*\right]$$

$$= \lim_{n\to\infty} n\left[\frac{\partial R_X\left(t_1 + \frac{1}{n}, t_2\right)}{\partial t_2} - \frac{\partial R_X(t_1, t_2)}{\partial t_2}\right]$$

$$= \partial^2 R_X(t_1, t_2)/\partial t_1\, \partial t_2.$$

This is really just the first step in the proof of Theorem 8.1–2 generalized to allow $t_1 \neq t_2$. To justify this interchange of the m.-s. derivative and expectation operators, we make use of the Schwarz inequality (Equation 3.4–13), which has been derived for real-valued r.v.'s in Section 3.4. (It also holds for the complex case as will be shown in the next section.)

We first define

$$R_{XX'}(t_1, t_2) \triangleq E[X(t_1)X'^*(t_2)]. \tag{8.1–6}$$

We use the Schwarz inequality $|E[AB_n^*]| \leq \sqrt{E[|A|^2]E[|B_n|^2]}$ with $A \triangleq X(t_1)$ and

$$B_n \triangleq X'(t_2) - [X(t_2 + \varepsilon_n) - X(t_2)]/\varepsilon_n,$$

to obtain

$$R_{XX'}(t_1, t_2) = \lim_{n\to\infty} E[X(t_1)(X(t_2 + \varepsilon_n) - X(t_2))^*/\varepsilon_n]$$

$$= \lim_{n\to\infty} ([R_X(t_1, t_2 + \varepsilon_n) - R_X(t_1, t_2)]/\varepsilon_n)$$

$$= \frac{\partial R_X(t_1, t_2)}{\partial t_2}.$$

Then $E[X'(t_1)X'^*(t_2)]$ is obtained similarly as,

$$\lim_{n\to\infty} E\left[\left(\frac{X(t_1 + \varepsilon_n) - X(t_1)}{\varepsilon_n}\right)X'^*(t_2)\right]$$

$$= \lim_{n\to\infty} ([R_{XX'}(t_1 + \varepsilon_n, t_2) - R_{XX'}(t_1, t_2)]/\varepsilon_n)$$

$$= \partial R_{XX'}(t_1, t_2)/\partial t_1.$$

Thus we have finally obtained the following theorem.

Theorem 8.1–3. If a random process $X(t)$ with mean function $\mu_X(t)$ and correlation function $R_X(t_1, t_2)$ has an m.-s. derivative $X'(t)$, then the mean and correlation function of $X'(t)$ are given by

$$\mu_{X'}(t) = d\mu_X(t)/dt$$

and

$$R_{X'}(t_1, t_2) = \partial^2 R_X(t_1, t_2)/\partial t_1\, \partial t_2.$$

Example 8.1–3 (cont'd.): We now continue to study the m.-s. derivative of the process $X(t)$ of Example 8.1–3 by calculating its mean function and its correlation function. We obtain

$$\mu_{X'}(t) = d\mu_X(t)/dt = 0,$$

and

$$R_{X'}(t_1, t_2) = \partial^2 R_X(t_1, t_2)/\partial t_1 \, \partial t_2$$
$$= 2\alpha\sigma^2[1 - 2\alpha(t_1 - t_2)^2]\exp(-\alpha(t_1 - t_2)^2).$$

We note that in the course of calculating $R_{X'}$, we are effectively checking existence of the m.-s. derivative by noting whether this derivative exists at $t_1 = t_2$.

In Example 8.1–2 we found that the correlation function had a cusp at $t_1 = t_2$ which precluded the existence of a m.-s. derivative. However, this second mixed-partial derivative does exist in the sense of singularity functions. In earlier experience with linear systems we have worked with singularity functions and have found them operationally very elegant and simple to use. They are properly treated mathematically through the rather abstract theory of *generalized functions.* If we proceed to take the partial derivative, in this generalized sense, with respect to t_1 of $\partial R/\partial t_2$ in Equation 8.1–3, the step discontinuity on the diagonal gives rise to an impulse in t_1. We obtain

$$\frac{\partial^2 R_X(t_1, t_2)}{\partial t_1 \, \partial t_2} = \begin{Bmatrix} -\alpha^2\sigma^2 \exp(-\alpha(t_2 - t_1)), & t_1 < t_2 \\ -\alpha^2\sigma^2 \exp(-\alpha(t_1 - t_2)), & t_1 \geq t_2 \end{Bmatrix} + 2\alpha\sigma^2\delta(t_1 - t_2)$$
$$= 2\alpha\sigma^2\delta(t_1 - t_2) - \alpha^2\sigma^2 \exp(-\sigma\,|t_1 - t_2|). \quad (8.1\text{–}7)$$

We can call the random process with this autocorrelation function a *generalized random process* and say it is the *generalized m.-s. derivative* of the conventional process X. When we say this we mean that the defining mean-square limit (Equation 8.1–2) is zero in the sense of generalized functions. A more detailed justification of generalized random processes may be found in [8–2]. In this text we will be content to use it formally as a notation for a limiting behavior of conventional random processes.

The surprising thing about the autocorrelation function in Equation 8.1–7 is the term $2\alpha\sigma^2\delta(t_1 - t_2)$. In fact, if we single out this term and consider the autocorrelation function

$$R(t_1, t_2) = \sigma^2\delta(t_1 - t_2), \quad (8.1\text{–}8)$$

we can show that it corresponds to the generalized m.-s. derivative of the Wiener process defined in Section 7.2. By definition $\mu_X(t) = 0$ and $R_X(t_1, t_2) = \alpha \min(t_1, t_2)$ for the Wiener process. We proceed by calculating,

$$\partial R_X/\partial t_2 = \frac{\partial}{\partial t_2}\begin{cases} \alpha t_2, & t_2 < t_1 \\ \alpha t_1, & t_2 \geq t_1, \end{cases}$$
$$= \begin{cases} \alpha, & t_2 < t_1 \\ 0, & t_2 > t_1. \end{cases}$$

Then

$$\frac{\partial^2 R_X}{\partial t_1\,\partial t_2} = \frac{\partial}{\partial t_1}\begin{cases} 0, & t_1 < t_2 \\ \alpha, & t_1 > t_2, \end{cases}$$

$$= \frac{\partial}{\partial t_1}[\alpha u(t_1 - t_2)]$$

$$= \alpha\delta(t_1 - t_2),$$

which is the same as Equation 8.1–8 if we set $\sigma^2 = \alpha$.

The generalized m.-s. derivative of the Wiener process is called *white Gaussian noise.* It is not a random process in the conventional sense since, for example, its mean-square value $R_{X'}(t, t)$ at time t is infinite. Nevertheless, it is the formal limit of approximating conventional processes whose correlation function is a narrow pulse of area α. These approximating processes will often yield almost the same system outputs (up to mean-square equivalence), thus the white noise can be used to simplify the analysis in these cases in essentially the same way that singularity functions are used in deterministic system analysis. We will return to this matter later when we study the concept of power spectral density in Chapter 9.

Many random processes are stationary or approximately so. In the stationary case we have seen that we can write the correlation function as a 1-D function $R(\tau)$. Then we can express the conditions and results of Theorems 8.1–2 and 8.1–3 in terms of this function. Unfortunately, the resulting formulas are not as intuitive and tend to be somewhat confusing. For the special case of a stationary or wide-sense stationary random process we get the following:

Theorem 8.1–4. The m.s. derivative of a WSS random process $X(t)$ exists at time t iff the autocorrelation $R_X(\tau)$ has derivatives up to order two at $\tau = 0$.

Proof. By the previous result we need $\partial^2 R_X(t_1, t_2)/\partial t_1\,\partial t_2 \,|_{(t_1,t_2)=(t,t)}$. Now

$$R_X(\tau) = R_X(t + \tau, t) \quad \text{independent of } t,$$

so

$$\left.\frac{\partial R_X(t_1, t_2)}{\partial t_1}\right|_{(t_1,t_2)=(t+\tau,t)} = dR_X(\tau)/d\tau$$

since $t_2 = t$ is held constant, and

$$\left.\frac{\partial R_X(t_1, t_2)}{\partial t_2}\right|_{(t_1,t_2)=(t,t-\tau)} = \partial R_X(t, t - \tau)/\partial(-\tau)$$

$$= -dR_X(\tau)/d\tau,$$

since $t_1 = t$ is held constant here, thus

$$\left.\frac{\partial^2 R_X(t_1, t_2)}{\partial t_1\,\partial t_2}\right|_{(t_1,t_2)=(t+\tau,t)} = -d^2 R_X(\tau)/d\tau^2. \quad \blacksquare$$

Calculating the second-order properties of X', we have:

$$E[X'(t)] = \mu_{X'}(t) = 0,$$

$$E[X'(t+\tau)X^*(t)] = R_{X'X}(\tau) = +dR_X(\tau)/d\tau, \quad (8.1\text{–}9a)$$

$$E[X(t+\tau)X'^*(t)] = R_{XX'}(\tau) = -dR_X(\tau)/d\tau, \quad (8.1\text{–}9b)$$

and

$$E[X'(t+\tau)X'^*(t)] = R_{X'X'}(\tau) = -d^2R_X(\tau)/d\tau^2, \quad (8.1\text{–}9c)$$

which follow from the formulas used in the above proof.

One can also derive Equations 8.1–9 directly, for example,

$$E[X(t+\tau)X'^*(t)] = \lim_{\varepsilon\to 0}\left(\frac{E[X(t+\tau)X^*(t+\varepsilon)] - E[X(t+\tau)X^*(t)]}{\varepsilon}\right)$$

$$= \lim_{\varepsilon\to 0}\left(\frac{R_{XX}(\tau-\varepsilon) - R_X(\tau)}{\varepsilon}\right)$$

$$= -dR_X(\tau)/d\tau = R_{XX'}(\tau)$$

and similarly for $R_{X'}(\tau) \equiv R_{X'X'}(\tau)$.

Example 8.1–4: Let $X(t)$ have zero mean and correlation function

$$R_X(\tau) = \sigma^2 \exp(-\alpha^2\tau^2).$$

Here the m.-s. derivative exists because $R(\tau)$ is infinitely differentiable at $\tau = 0$. Computing the first and second derivatives, we get

$$dR_X/d\tau = -2\alpha^2\tau\sigma^2\exp(-\alpha^2\tau^2).$$

Then

$$R_{X'}(\tau) = -d^2R_X/d\tau^2 = 2\alpha^2\sigma^2(1 - 2\alpha^2\tau^2)\exp(-\alpha^2\tau^2).$$

8.1–1 Further Results on Mean-Square Convergence [8–1]

We now consolidate and present further results on mean-square convergence that are helpful in this chapter. We will have use for two inequalities for the moments of complex random variables: the Schwarz inequality and another inequality called the triangle inequality.

Complex Schwarz Inequality. Let X and Y be second-order complex random variables, that is,

$$E[|X|^2] < \infty \quad \text{and} \quad E[|Y|^2] < \infty;$$

then we have the inequality

$$|E[XY^*]| \le \sqrt{E[|X|^2]E[|Y|^2]}.$$

Proof. Consider $W \triangleq aX + Y$. Then minimize $E[|W|^2]$ as a function of a, where a may be complex. Then, since the minimum must be nonnegative, the preceding inequality is obtained. We leave this as an exercise for the student. ■

Another inequality that will be useful is the triangle inequality, which shows that the rms value $\sqrt{E[|X|^2]}$ can be considered as a "length" for random variables!

Triangle Inequality. Let X and Y be second-order complex random variables; then

$$\sqrt{E[|X+Y|^2]} \le \sqrt{E[|X|^2]} + \sqrt{E[|Y|^2]}$$

Proof.

$$\begin{aligned} E[|X+Y|^2] &= E[(X+Y)(X^*+Y^*)] \\ &= E[|X|^2] + E[XY^*] + E[X^*Y] + E[|Y|^2] \\ &\le E[|X|^2] + 2\,|E[XY^*]| + E[|Y|^2] \\ &\le E[|X|^2] + 2\sqrt{E[|X|^2]E[|Y|^2]} + E[|Y|^2] \end{aligned}$$

(by the Schwarz inequality)

$$= \left(\sqrt{E[|X|^2]} + \sqrt{E[|Y|^2]}\right)^2. \quad \blacksquare$$

Note that the quantity $\sqrt{E[|X|^2]}$ obeys the equation for a distance or norm,

$$\|X+Y\| \le \|X\| + \|Y\|,$$

and hence the name "triangle inequality." In fact, one can define the linear space of second-order complex random variables with *norm*

$$\|X\| \triangleq \sqrt{E[|X|^2]}.$$

This space is then a Hilbert Space with *inner product*

$$(X, Y) \triangleq E[XY^*].$$

Figure 8.1–3 illustrates this geometric viewpoint.

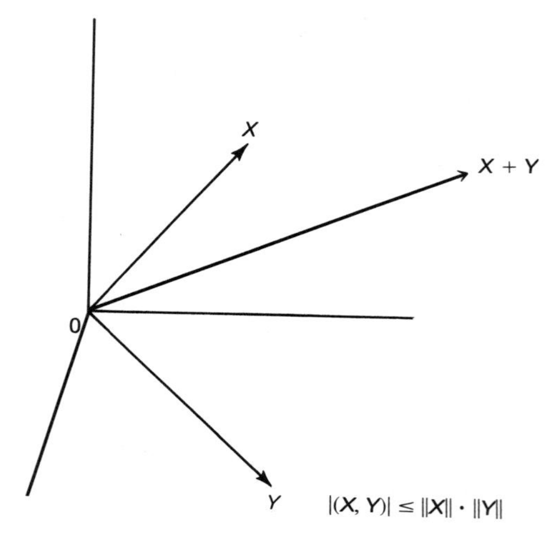

Figure 8.1–3 Illustration of linear space of random variables.

Later in this chapter we will be working with m.-s. derivatives and integrals and also m.-s. differential equations. As such, we will have need of the following general results concerning the moments of mean-square limits.

Theorem 8.1–5. Let $X[n] \to X$ and $Y[n] \to Y$ in the m.-s. sense with $E[|X|^2] < \infty$ and $E[|Y|^2] < \infty$. Then we have the properties:

(1) $\lim_{n\to\infty} E[X[n]] = E[X]$,

(2) $\lim_{n\to\infty} E[X[n]Y^*] = E[XY^*]$,

(3) $\lim_{n\to\infty} E[|X[n]|^2] = E[|X|^2]$,

(4) $\lim_{n\to\infty} E[X[n]Y^*[n]] = E[XY^*]$,

and

(5) if $X = X_1$ (m.s.) then $P[X \neq X_1] = 0$, i.e., $X = X_1$ (a.s.).

Proof.

(1) Let $A_n = X - X[n]$; then we need $\lim_{n\to\infty} E[A_n] = 0$. But $|E[A_n]|^2 \leq \|A_n\|^2$ and $\|A_n\| \to 0$ by the definition of m.-s. convergence, thus $\lim_{n\to\infty} E[A_n] = 0$.

(2) We note $XY^* - X[n]Y^* = A_n Y^*$. Then by the Schwarz inequality,

$$|E[A_n Y^*]|^2 \leq \|A_n\|^2 \cdot \|Y\|^2 \to 0 \quad \text{as} \quad n \to \infty,$$

since Y is second order.

(3) This is the same as $\|X[n]\| \to \|X\|$ in terms of the norm notation. Now by the triangle inequality,

$$\|X[n]\| = \|X[n] - X + X\| \leq \|X[n] - X\| + \|X\|.$$

This tells us that $\lim_{n\to\infty} \|X[n]\| \leq \|X\|$. But also $\|X\| = \|X - X[n] + X[n]\| \leq \|X - X[n]\| + \|X[n]\|$, which implies $\lim_{n\to\infty} \|X[n]\| \geq \|X\|$. Therefore, $\lim_{n\to\infty} \|X[n]\| = \|X\|$.

(4) Consider $XY^* - X[n]Y^*[n]$, and add and subtract $X[n]Y^*$ to obtain

$$(X - X[n])Y^* + X[n](Y - Y[n])^*.$$

Next use the triangle inequality followed by the Schwarz inequality to obtain,

$$|E[XY^* - X[n]Y^*[n]]| \leq \|X - X[n]\| \cdot \|Y\| + \|X[n]\| \cdot \|Y - Y[n]\|$$
$$\to 0 \text{ by definition of m.-s. convergence and by property (3).}$$

(5) Let $\varepsilon > 0$; then by the Chebyshev inequality we have,

$$P[|X - X_1| > \varepsilon] \leq E[|X - X_1|^2]/\varepsilon^2 = 0.$$

Since ε is arbitrary,

$$P[X \neq X_1] = 0. \quad \blacksquare$$

Another useful property for our work later in this chapter will be the fact that the m.-s. limit is a linear operator.

Theorem 8.1–6. Let $X[n] \rightarrow X$ (m.s.) and $Y[n] \rightarrow Y$ (m.s.) with both X and Y second-order random variables. Then

$$\lim_{n \to \infty} \{aX[n] + bY[n]\} = aX + bY. \qquad \text{(m.s.)}$$

Proof. We have to show that

$$\|aX[n] + bY[n] - (aX + bY)\| \rightarrow 0.$$

By the triangle inequality

$$\begin{aligned} \|a(X[n] - X) + b(Y[n] - Y)\| &\leq \|a(X[n] - X)\| + \|b(Y[n] - Y)\| \\ &= |a| \cdot \|X[n] - X\| + |b| \cdot \|Y[n] - Y\| \\ &\rightarrow 0 \text{ as } n \rightarrow \infty, \end{aligned}$$

by m.-s. convergence of $X[n]$ and $Y[n]$. $\blacksquare$

Corollary. The m.-s. derivative is a linear operator, that is,

$$\frac{d}{dt}[aX(t) + bY(t)] = aX'(t) + bY'(t) \qquad \text{(m.s.)}$$

Proof. We leave this as an exercise to the reader.

8.2. STOCHASTIC INTEGRALS

In this section we continue our study of the calculus of random processes by considering the stochastic integral. Stochastic integrals are important in applications for representing linear operators such as convolution, which arise when random processes are passed through linear systems. Our discussion will be followed in the next section by a look at stochastic differential equations, which of course will involve the stochastic integral in their solution. Another application of the stochastic integral is in forming averages for estimating the moments and probability functions of stationary processes. This topic is related to ergodic theory and will be introduced in Section 8.4.

We will be interested in the mean-square stochastic integral. It is defined as the mean-square limit of its defining sum as the partition of the integration interval gets finer and finer. We first look at the integration of a random process $X(t)$ over a finite interval (T_1, T_2). The operation of the integral is then just a simple averager. First we create a partition of (T_1, T_2) consisting of n points using $(t_1, t_2, \ldots, t_n)$. Then the approximate integral is the sum

$$I_n \triangleq \sum_{i=1}^{n} X(t_i)\Delta t_i. \tag{8.2-1}$$

On defining the m.-s. limit random variable as I, we have the following definition of the mean-square integral.

Definition 8.2–1. The *mean-square integral* of $X(t)$ over the interval (T_1, T_2) is denoted I. It exists when

$$\lim_{n\to\infty} E\left[\left|I - \sum_{i=1}^{n} X(t_i)\Delta t_i\right|^2\right] = 0. \tag{8.2-2}$$

We give I the following symbol:

$$I \triangleq \int_{T_1}^{T_2} X(t)\, dt.$$

Because the m.-s. limit is a linear operator by Theorem 8.1–6, it follows that the integral just defined is linear, that is, it obeys

$$\int_{T_1}^{T_2} \{aX_1(t) + bX_2(t)\}\, dt = a\int_{T_1}^{T_2} X_1(t)\, dt + b\int_{T_1}^{T_2} X_2(t)\, dt,$$

whenever the integrals on the right exist.

To study the existence of the mean-square integral, as before we make use of the Cauchy criterion for the existence of a limit. Thus the integral I exists iff $\lim_{m,n\to\infty} E[|I_n - I_m|^2] = 0$. If we expand this expression into three terms, we get $E[|I_n|^2] - 2Re\{E[I_n I_m^*]\} + E[|I_m|^2]$. Without loss of generality we concentrate on the cross term and evaluate

$$E[I_n I_m^*] = \sum_{i,j} R_X(t_i, t_j)\, \Delta t_i\, \Delta t_j,$$

where the sums range over $1 \le i \le n$ and $1 \le j \le m$. As $m, n \to +\infty$, this converges to the deterministic integral

$$\int_{T_1}^{T_2}\int_{T_1}^{T_2} R_X(t_1, t_2)\, dt_1\, dt_2, \tag{8.2-3}$$

if this deterministic integral exists. Clearly, the other two terms $E[|I_n|^2]$ and $E[|I_m|^2]$ converge to the same double integral. Thus we see that the m.-s. integral exists whenever the double integral of Equation 8.2–3 exists in the sense of the ordinary calculus.

The mean and mean-square (power) of I are directly computed as

$$E[I] = E\left[\int_{T_1}^{T_2} X(t)\, dt\right] = \int_{T_1}^{T_2} E[X(t)]\, dt, \tag{8.2-4a}$$

and

$$E[|I|^2] = E\left[\int_{T_1}^{T_2}\int_{T_1}^{T_2} X(t_1)X^*(t_2)\,dt_1\,dt_2\right]$$

$$= \int_{T_1}^{T_2}\int_{T_1}^{T_2} R_X(t_1, t_2)\,dt_1\,dt_2. \tag{8.2–4b}$$

The variance of I is given as

$$\sigma_I^2 = \int_{T_1}^{T_2}\int_{T_1}^{T_2} K_X(t_1, t_2)\,dt_1\,dt_2. \tag{8.2–4c}$$

Example 8.2–1: (Integral of white noise over $(0, t)$.) Let the random process $X(t)$ be zero mean with covariance function $K_X(\tau) = \sigma^2\delta(\tau)$ and define the running m.-s. integral as

$$Y(t) \triangleq \int_0^t X(s)\,ds, \qquad t \geq 0.$$

For fixed t, $Y(t)$ is a random variable; therefore $Y(t)$ is a stochastic process. Its mean is given as $E[Y(t)] = \int_0^t \mu_X(s)\,ds$ and equals 0 since $\mu_X = 0$. The covariance is calculated for $t_1 \geq 0, t_2 \geq 0$ as

$$\begin{aligned}
K_Y(t_1, t_2) &= \int_0^{t_1}\int_0^{t_2} K_X(s_1, s_2)\,ds_1\,ds_2,\\
&= \sigma^2\int_0^{t_1}\int_0^{t_2} \delta(s_1 - s_2)\,ds_1\,ds_2,\\
&= \sigma^2\int_0^{t_2} u(t_1 - s_2)\,ds_2,\\
&= \sigma^2\int_0^{\min(t_1,t_2)} ds_2,\\
&= \sigma^2\min(t_1, t_2),
\end{aligned}$$

which we recognize as the covariance of the Wiener process (Section 7.2). Thus the m.-s. integral of white Gaussian noise is the Wiener process. Note that $Y(t)$ must be Gaussian if $X(t)$ is Gaussian, since $Y(t)$ is the m.-s. limit of a weighted sum of samples of $X(t)$ (see Problem 8.6).

We can generalize this integral by including a weighting function $h(t)$ multiplying the random process $X(t)$. It would thus be the mean-square limit of the following generalization of Equation 8.2–1:

$$I_n \triangleq \sum_{i=1}^{n} h(t_i)X(t_i)\,\Delta t_i.$$

This amounts to the following definition for the weighted integral.

Definition 8.2–2. The weighted mean-square integral of $X(t)$ over the interval (T_1, T_2) is defined by

$$\lim_{n\to\infty} E\left[\left|I - \sum_{i=1}^{n} h(t_i)X(t_i)\,\Delta t_i\right|^2\right] = 0,$$

when the limit exists. We give it the following symbol:

$$I \triangleq \int_{T_1}^{T_2} h(t)X(t)\,dt.$$

Example 8.2–2: (Application to linear systems.) A linear system L has response $h(t, s)$ at time t to an impulse applied at time s. Then for a deterministic function $x(t)$ as input, we have the output $y(t)$ given as,

$$y(t) = L\{x(t)\} = \int_{-\infty}^{+\infty} h(t, s)x(s)\,ds, \tag{8.2–5}$$

whenever the foregoing integral exists. If $x(t)$ is bounded and integrable, one condition for the existence of Equation 8.2–5 would be

$$\int_{-\infty}^{+\infty} |h(t, s)|\,ds < +\infty \qquad \text{for all } -\infty < t < +\infty. \tag{8.2–6}$$

We can generalize this integral to an m.-s. stochastic integral if the following double integral exists:

$$\iint_{-\infty}^{+\infty} h(t_1, s_1)h^*(t_2, s_2)R_X(s_1, s_2)\,ds_1\,ds_2,$$

in the ordinary calculus. A condition for this, in the case where R_X is bounded and integrable, is Equation 8.2–6. Given the existence of such a condition, $Y(t)$ exists as an m.-s. limit and defines a random process,

$$Y(t) \triangleq \int_{-\infty}^{+\infty} h(t, s)X(s)\,ds,$$

whose mean and covariance functions are:

$$\mu_Y(t) = \int_{-\infty}^{+\infty} h(t, s)\mu_X(s)\,ds, \tag{8.2–7a}$$

and

$$R_Y(t_1, t_2) = \iint_{-\infty}^{+\infty} h(t_1, s_1)h^*(t_2, s_2)R_X(s_1, s_2)\,ds_1\,ds_2 \tag{8.2–7b}.$$

8.3 STOCHASTIC DIFFERENTIAL EQUATIONS

Having investigated stochastic derivatives and integrals, we now turn to the subject of stochastic differential equations. The simplest stochastic differential equation is

$$dY(t)/dt = X(t),$$

where the derivative on the left is an m.-s. derivative. The solution turns out to be

$$Y(t) = \int_{t_0}^{t} X(s)\,ds + Y(t_0), \qquad t \geq t_0,$$

where the integral is an m.-s. integral.

Using the general linear, constant coefficient, differential equation (LCCDE) as a model, we form the general stochastic LCCDE,

$$a_n Y^{(n)}(t) + a_{n-1} Y^{(n-1)}(t) + \ldots + a_0 Y(t) = X(t), \tag{8.3–1}$$

for $t \geq 0$ with prescribed initial conditions,

$$Y(0),\ Y^{(1)}(0), \ldots,\ Y^{(n-1)}(0),$$

where the equality is in the m.-s. sense.

To appreciate the meaning of Equation 8.3–1 more fully, we point out that it is a mean-square equality at each t separately. Thus Equation 8.3–1 is an equality at each t with probability-1 and hence at any countable collection of t's by Property 5 of Theorem 8.1–5. However, this does not say that the sample functions of $Y(t)$, which in fact may not even be differentiable, satisfy the differential equation driven by the sample functions of $X(t)$. The sample function interpretation of Equation 8.3–1 would require that it hold with probability-1 for *all* $t \geq 0$, which is an uncountable collection of times.

We thus must think of the m.-s. differential equation as an idealization approached in the limit, much as the impulse and the ideal lowpass filter. Also, the m.-s. differential equation must be treated with care because it is not quite what it seems. Just as we would not put an impulse into a linear circuit in reality, similarly if we simulate a stochastic differential equation we would not use a random process lacking sample function derivatives as the input. Instead, we would use a smoother approximation to the process and if this smoothing is slight, we would expect that the idealized solution obtained from our m.-s. differential equation would have similar properties. This, of course, needs further justification, but for the present we will assume that it is correct.

One may raise the question: Why work with such extreme processes, that is, processes without sample-function derivatives? The answer is that the analysis of these m.-s. differential equations can proceed very well using the basic methods of linear system analysis. If we had instead included the extra "smoothing" to guarantee sample-function derivatives, then the analysis would be more complicated. For comparison, imagine trying to find the response of a linear system to a very narrow Gaussian-shaped pulse of area one versus finding the ideal impulse response.

We proceed by finding the mean function $\mu_Y(t)$ and correlation function $R_{YY}(t_1, t_2)$ of the m.-s. Equation 8.3–1. Since we have equality with probability-1 for each t, we can compute the expectations of both sides of Equation 8.3–1 and use

$$d^i E[Y(t)]/dt^i = E[Y^{(i)}(t)],$$

to obtain

$$a_n \mu_Y^{(n)}(t) + a_{n-1}\mu_Y^{(n-1)}(t) + \ldots + a_0 \mu_Y(t) = \mu_X(t) \tag{8.3–2}$$

with prescribed initial conditions at $t = 0$,

$$\mu_Y^{(i)}(0) = E[Y^{(i)}(0)] \quad \text{for } i = 0, 1, \ldots, n-1.$$

Thus the mean function of $Y(t)$ is the solution to this linear differential equation, whose input is the mean function of $X(t)$. So knowledge of μ_X is sufficient to determine μ_Y, that is, we do not have to know any of the higher moment functions of $X(t)$. Note that this could not be true if we were considering nonlinear differential equations. However, essentially no change would be necessary to accommodate time-varying coefficients, but of course the resulting equations would be much harder to solve. Thus we will stay with the constant-coefficient case. Parenthetically, we note that if $\mu_X = 0$ for all $t > 0$ and the initial conditions are zero, then clearly $\mu_Y = 0$ for all $t > 0$.

Next we determine the cross-correlation function $R_{XY}(t_1, t_2)$, which is the correlation between the input to Equation 8.3–1 at time t_1 and the output at time t_2. This quantity can be useful in system identification studies. We will assume for simplicity that the coefficients a_i are real. Then we substitute t_2 for t in Equation 8.3–1, conjugate, and multiply both sides by $X(t_1)$ to obtain

$$X(t_1)\left[\sum_{i=0}^{n} a_i Y^{(i)}(t_2)^*\right] = X(t_1)X^*(t_2), \qquad t_1 \geq 0, \qquad t_2 \geq 0,$$

which holds with probability-1 since only a countable number of times (two !) is being considered. Taking expectations we obtain

$$\sum_{i=0}^{n} a_i E[X(t_1)Y^{(i)*}(t_2)] = E[X(t_1)X^*(t_2)],$$

or

$$\sum_{i=0}^{n} a_i R_{XY^{(i)}}(t_1, t_2) = R_{XX}(t_1, t_2),$$

which, using $R_{XY^{(i)}} = \partial^{(i)} R_{XY}/\partial t_2^i$, is the same as

$$\checkmark \quad \sum_{i=0}^{n} a_i \partial^{(i)} R_{XY}(t_1, t_2)/\partial t_2^i = R_{XX}(t_1, t_2), \qquad t_2 \geq 0, \tag{8.3–3}$$

for each $t_1 \geq 0$, subject to the initial conditions

$$\checkmark \quad \partial^{(i)} R_{XY}(t_1, 0)/\partial t_2^i \quad \text{for} \quad i = 0, 1, \ldots, n-1.$$

To obtain a differential equation for R_{YY}, we multiply both sides of Equation 8.3–1 by $Y^*(t_2)$ and similarly obtain, for each $t_2 \geq 0$,

$$\checkmark \quad \sum_{i=0}^{n} a_i \, \partial^{(i)} R_{YY}(t_1, t_2)/\partial t_1^i = R_{XY}(t_1, t_2) \quad \text{for } t_1 \geq 0, \tag{8.3–4}$$

with initial conditions $\partial^{(i)} R_{YY}(0, t_2)/\partial t_1^i$, for $i = 0, 1, \ldots, n-1$.

One can obtain equations identical to Equations 8.3–3 and 8.3–4 for the covariance functions K_{XY} and K_{YY} by noting that Equation 8.3–2 can be used to center Equation 8.3–1 at the means of X and Y. This follows from the linearity of Equation 8.3–1 and converts it to a m.-s. differential equation in Y_c and X_c. This then yields Equations 8.3–3 and 8.3–4 for the above covariance functions.

11/19

We now turn to the solution of these partial differential equations. We will solve Equation 8.3–3 first followed by Equation 8.3–4. We also note that Equation 8.3–3 is not really a partial differential equation since t_1 just plays the role of a constant parameter. Thus we must first solve the LCCDE Equation 8.3–3 for each t_1, thereby obtaining the cross-correlation R_{XY}. Then we use this function as input to Equation 8.3–4, which in turn is solved for each value of the parameter t_2. What remains is the problem of obtaining the appropriate initial conditions for these two deterministic LCCDEs from the given stochastic LCCDE Equation 8.3–1. This is illustrated by the following example, which also shows the formal advantages of working with the idealization called white noise.

Example 8.3–1: Let $X(t)$ be a stationary random process with mean μ_X and covariance function $K_{XX}(\tau) = \sigma^2\delta(\tau)$. Let $Y(t)$ be the solution to the m.-s. differential equation,

$$dY(t)/dt + \alpha Y(t) = X(t), \qquad t \geq 0, \tag{8.3–5}$$

subject to the initial condition $Y(0) = 0$. We assume $\alpha > 0$. Then the mean μ_Y is the solution to the first order differential equation,

$$\mu_Y' + \alpha\mu_Y(t) = \mu_X, \qquad t \geq 0,$$

subject to $\mu_Y(0) = 0$. This initial condition coming from the fact that the initial random variable $Y(0)$ equals the constant 0 and therefore $E[Y(0)] = 0$. The solution is then easily obtained as

$$\mu_Y(t) = (\mu_X/\alpha)(1 - \exp(-\alpha t)) \quad \text{for } t \geq 0.$$

Next we use the covariance version of Equation 8.3–3 specialized to the first order m.-s. differential Equation 8.3–5 to obtain

$$\partial K_{XY}(t_1, t_2)/\partial t_2 + \alpha K_{XY}(t_1, t_2) = \sigma^2\delta(t_1 - t_2),$$

to be solved for $t_2 \geq 0$ subject to the initial condition: $K_{XY}(t_1, 0) = 0$, which follows from $Y(0) = 0$. For $0 \leq t_2 < t_1$, the solution is just 0 since the input is zero for $t_2 < t_1$. For the interval $t_2 \geq t_1$, we get the delayed impulse response $\sigma^2 \exp(-\alpha(t_2 - t_1))$ since the input is a delayed impulse occurring at $t_2 = t_1$. Thus the overall solution is

$$K_{XY}(t_1, t_2) = \begin{cases} 0, & 0 \leq t_2 < t_1, \\ \sigma^2 \exp(-\alpha(t_2 - t_1)), & t_2 \geq t_1. \end{cases}$$

This cross-covariance function is plotted in Figure 8.3–1.

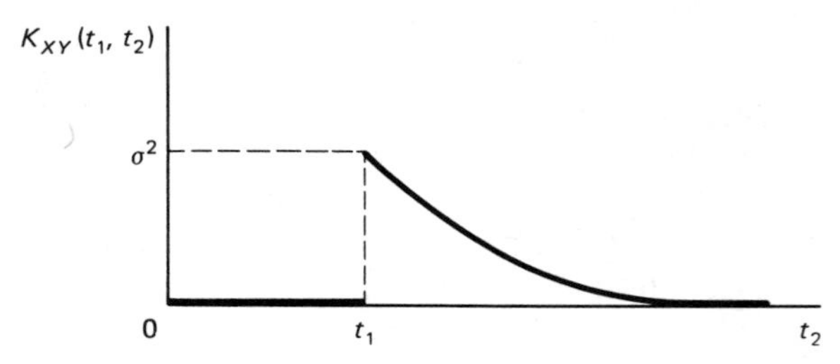

8.3–1 Cross-covariance function of input and output (t_1 fixed).

Next we obtain the differential equation for K_{YY} by specializing the covariance version of Equation 8.3–4 to the first order m.-s. differential Equation 8.3–5:

$$\frac{\partial K_{YY}(t_1, t_2)}{\partial t_1} + \alpha K_{YY}(t_1, t_2) = K_{XY}(t_1, t_2),$$

subject to the initial condition at $t_1 = 0$ (for each t_2):

$$K_{YY}(0, t_2) = 0.$$

for the interval $0 < t_1 \leq t_2$,

$$K_{XY}(t_1, t_2) = \sigma^2 \exp(-\alpha(t_2 - t_1)),$$

so

$$\partial K_{YY}(t_1, t_2)/\partial t_1 + \alpha K_{YY}(t_1, t_2) = \sigma^2 \exp(-\alpha(t_2 - t_1)),$$

which has solution $K_{YY}(t_1, t_2) = \sigma^2/2\alpha e^{-\alpha t_2}(e^{\alpha t_1} - e^{-\alpha t_1})$.

For $t_1 \geq t_2$, $K_{XY}(t_1, t_2) = 0$, we then have to solve

$$\partial K_{YY}(t_1, t_2)/\partial t_1 + \alpha K_{YY}(t_1, t_2) = 0,$$

subject to $K_{YY}(t_2, t_2) = (\sigma^2/2\alpha)(1 - \exp(-2\alpha t_2))$. We obtain,

$$K_{YY}(t_1, t_2) = \frac{\sigma^2}{2\alpha}(1 - \exp(-2\alpha t_2)) \exp(-\alpha(t_1 - t_2)), \qquad t_1 \geq t_2.$$

The overall function is plotted versus t_1 in Figure 8.3–2.

We note that as t_1 and $t_2 \to \infty$, the variance of Y tends to a constant, that is, $K_{YY}(t, t) \to \sigma^2/2\alpha$. In fact with $t_1 = t + \tau$ and $t_2 = t$, the covariance of Y becomes

$$K_{YY}(t + \tau, t) = \begin{cases} (\sigma^2/2\alpha)(1 - e^{-2\alpha t})\exp(-\alpha\tau), & \tau \geq 0 \\ (\sigma^2/2\alpha)(\exp \alpha\tau - e^{-2\alpha t} \exp(-\alpha\tau)), & \tau < 0. \end{cases}$$

Now if we let $t \to +\infty$ for any fixed value of τ, we obtain

$$K_{YY}(\tau) = (\sigma^2/2\alpha)\exp(-\alpha\,|\tau|).$$

This is an example of what is called *asymptotic wide-sense stationarity.* It happens here because the input random process is WSS and the LCCDE is stable. In fact the only thing creating the nonstationarity is the zero initial conditions, the effect of which decays away with time due to the stability assumption.

You may wonder at this point whether the random process $Y(t)$ with correlation R_{YY} and cross-correlation R_{XY} given by Equations 8.3–3 and

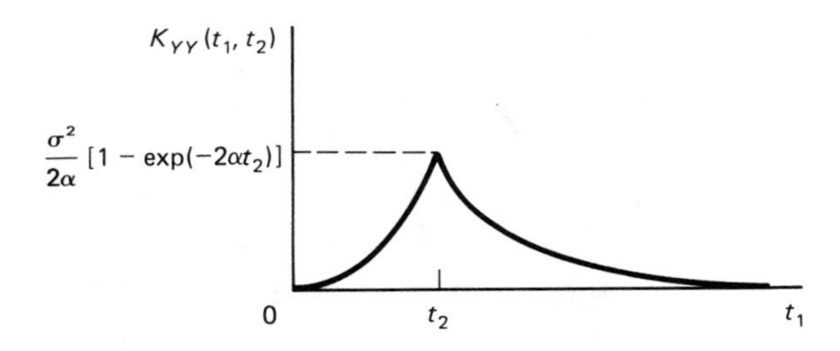

Figure 8.3–2 Plot of output covariance function for t_2 fixed.

8.3–4 actually satisfies Equation 8.3–1 in the m.-s. sense. This necessary topic is taken up in Problem 8.8 at the end of this chapter.

8.4 ERGODICITY [8–3]

Until now we have generally assumed that a statistical description of the random process is available. Of course, this is seldom true in practice; thus we must develop some way of learning the needed statistical quantities from the observed sample functions of the random processes of interest. Fortunately, for many stationary random processes, we can substitute time-averages for the unknown ensemble averages. We can use the stochastic integral defined in Section 8.2 to form this time-average of the random process $X(t)$ over the interval $[-T, T]$,

$$\frac{1}{2T}\int_{-T}^{+T} X(t)\,dt.$$

In many cases this average will tend to the ensemble mean as T goes to infinity. When this happens for a random process, we say the random process is ergodic in the mean. Other types of ergodicity can be defined, such as ergodic in mean square or ergodic in correlation function or ergodic in the probability distribution function. Each type of ergodicity means that the corresponding time-average converges to the named ensemble average.

This gives us a way of learning the unknown functions, which can then be used to statistically characterize the random processes of interest. To study convergence of the respective time-averages and hence determine whether the required ergodicity holds, we need to decide on a type of convergence for the random variables in question. For most of our work we will adopt the relatively simple concept of mean-square convergence. For example, we might say that the process $X(t)$ is mean-square ergodic in both the mean function and the covariance function.

The property of mean-square ergodicity occurs when the random process decorrelates sufficiently rapidly with time shift so that the time-average in question looks like the average of many almost uncorrelated random variables, which in turn—by appropriate forms of the weak law of large numbers—will converge to the appropriate ensemble average. That is, we can write upon setting $\Delta T = T/N$,

$$\begin{aligned}\frac{1}{2T}\int_{-T}^{+T} X(t)\,dt &= \frac{1}{2N\,\Delta T}\int_{-N\Delta T}^{+N\Delta T} X(t)\,dt \\ &= \frac{1}{2N}\sum_{n=-N}^{+(N-1)}\left(\frac{1}{\Delta T}\int_{n\Delta T}^{(n+1)\Delta T} X(t)\,dt\right),\end{aligned}$$

where the terms in the sum are approximately uncorrelated if ΔT is large enough. If the random process stays correlated over infinitely long time intervals, then we would not expect such behavior, and indeed ergodicity may not hold; for example:

(1) $X(t) = A$ where A is a random variable,

(2) $X(t) = A \cos 2\pi ft + B \sin 2\pi ft$, where A and B are random variables with $E[A] = E[B] = 0$, $E[A^2] = E[B^2] > 0$, and $E[AB] = 0$.

Example 1 is clearly not ergodic because any time average of $X(t)$ is just A, an r.v.; thus there is no convergence to the ensemble mean $E[A]$. Example 2, while WSS and ergodic in the mean, is not ergodic in power.

Definition 8.4–1. A wide-sense stationary random process is *ergodic in the mean* if the time-average of $X(t)$ converges to the ensemble average $E[X(t)] = \mu_X$, that is,

$$\hat{M} \triangleq \frac{1}{2T} \int_{-T}^{+T} X(t)\, dt \rightarrow \mu_X \qquad \text{(m.s.)} \qquad \text{as } T \rightarrow \infty$$

In the above equation we observe that the time average $\hat{M}$ is a r.v. Hence we can compute its mean and variance using the theory of m.-s. integrals obtaining

$$E[\hat{M}] = \frac{1}{2T} \int_{-T}^{+T} E[X(t)]\, dt = \mu_X, \tag{8.4–1}$$

and

$$\sigma_{\hat{M}}^2 = \frac{1}{(2T)^2} \iint_{-T}^{+T} K_{XX}(t_1 - t_2)\, dt_1\, dt_2. \tag{8.4–2}$$

The mean of $\hat{M}$ is thus the ensemble mean of $X(t)$. So if the variance is small the estimate will be good. Estimates that have the correct mean value are said to be *unbiased* (cf. Chapter 5). Noting the mean value from Equation 8.4–1 we see that

$$\sigma_{\hat{M}}^2 = E[|\hat{M} - \mu_X|^2];$$

thus convergence of the integral in Equation 8.4–2 to zero is the same as the convergence of $\hat{M}$ to μ_X in the m.-s. sense. Since we will mostly deal with m.-s. ergodicity, we can omit its mention with the understanding that unless otherwise stated time-averages are computed in the m.-s. sense.

To evaluate the integral in Equation 8.4–2 we look at Figure 8.4–1, which shows the area over which the integration is performed. Realizing that the random process $X(t)$ is WSS and the covariance function is thus only a function of the difference between t_1 and t_2, we can make the following change of variables that simplifies the integral:

$$s \triangleq t_1 + t_2$$

$$\tau \triangleq t_1 - t_2 \text{ with Jacobian } |J| = \left| \begin{vmatrix} \frac{\partial s}{\partial t_1} & \frac{\partial s}{\partial t_2} \\ \frac{\partial \tau}{\partial t_1} & \frac{\partial \tau}{\partial t_2} \end{vmatrix} \right| = \left| \begin{vmatrix} 1 & 1 \\ 1 & -1 \end{vmatrix} \right| = |-2| = 2,$$

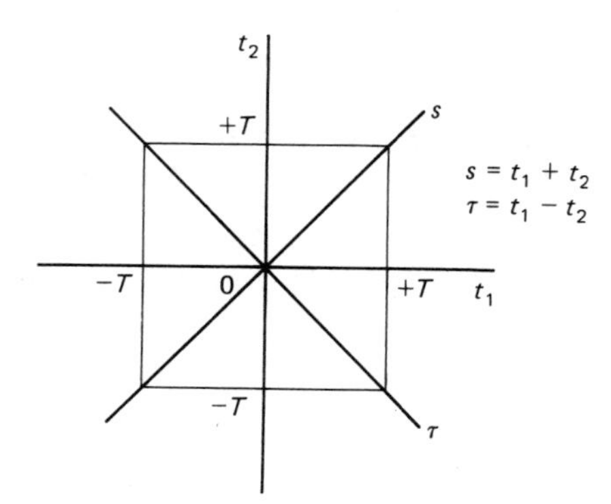

Figure 8.4–1 Square region in (t_1, t_2) plane.

so $dt_1\, dt_2 = |J|^{-1}\, ds\, d\tau = \frac{1}{2}\, ds\, d\tau$, and Equation 8.4–2 becomes

$$\frac{1}{(2T)^2}\int_{-2T}^{+2T}\left[\int_{-(2T-|\tau|)}^{+(2T-|\tau|)} \tfrac{1}{2}K_X(\tau)\, ds\right] d\tau = \frac{1}{(2T)^2}\int_{-2T}^{+2T} K_X(\tau)(2T - |\tau|)\, d\tau.$$

Thus we arrive at the following condition for the ergodicity in the mean of a WSS random process,

$$\frac{1}{2T}\int_{-2T}^{+2T}\left(1 - \frac{|\tau|}{2T}\right)K_X(\tau)\, d\tau \to 0 \qquad T \to \infty.$$

Note: A sufficient condition for ergodicity in the mean is thus

$$\int_{-\infty}^{+\infty} |K_X(\tau)|\, d\tau < \infty,$$

that is, that the covariance function is absolutely integrable.

Theorem 8.4–1. A WSS random process $X(t)$ is ergodic in the mean iff its covariance function $K_X(\tau)$ satisfies

$$\lim_{T\to\infty}\left[\frac{1}{2T}\int_{-2T}^{+2T}\left(1 - \frac{|\tau|}{2T}\right)K_X(\tau)\, d\tau\right] = 0.$$

We note that this is the same as saying that the estimate $\hat{M}$ converges to μ_X in the mean-square sense, hence the name m.-s. ergodicity. Since m.-s. convergence also implies convergence in probability, we have that $\hat{M}$ also converges to the ensemble mean μ_X in probability. This is analogous to the weak law of large numbers for uncorrelated random sequences as studied in Section 6.5.

We can also define ergodicity in the higher moments, for example, mean square or power.

Definition 8.4–2. A WSS random process $X(t)$ is *ergodic in mean square* if

$$\lim_{T\to\infty}\left[\frac{1}{2T}\int_{-T}^{+T} X^2(t)\, dt\right] = R_X(0) \qquad \text{(m.s.)}.$$

Similarly, we can define ergodicity in correlation and covariance.

Definition 8.4–3. A WSS random process $X(t)$ is *ergodic in correlation at the shift* λ iff

$$\lim_{T\to\infty}\left[\frac{1}{2T}\int_{-T}^{+T} X(t+\lambda)X^*(t)\,dt\right] = R_X(\lambda). \qquad \text{(m.s.)}$$

If this condition is true for all λ, we say X is *ergodic in correlation*. The conditions for the preceding two types of ergodicities are covered by the following theorem on ergodicity in correlation, where we have defined the random process for each λ,

$$\Phi_\lambda(t) \triangleq X(t+\lambda)X^*(t).$$

Theorem 8.4–2. The WSS random process $X(t)$ is ergodic in correlation at the shift λ iff

$$\lim_{T\to\infty}\left[\frac{1}{2T}\int_{-T}^{+T}\left(1-\frac{|\tau|}{2T}\right)K_{\Phi_\lambda}(\tau)\,d\tau\right] = 0. \qquad (8.4\text{–}3)$$

Proof. The time-average estimate of the correlation function found in Definition 8.4–3 is just the time-average of the random process $\Phi_\lambda(t)$ for each fixed λ. Thus we can apply Theorem 8.4–1 to the ergodicity in the mean problem for $\Phi_\lambda(t)$ so long as this process is WSS, that is, $X(t)$ has stationary second and fourth moments. The preceding condition is then seen to be the same as in Theorem 8.4–1 with the substitution of the covariance function $K_{\Phi_\lambda}(\tau)$ for $K_X(\tau)$. ■

Note that

$$K_{\Phi_\lambda}(\tau) = R_{\Phi_\lambda}(\tau) - |R_X(\tau)|^2,$$

where

$$\begin{aligned} R_{\Phi_\lambda}(\tau) &= E[\Phi_\lambda(t+\tau)\Phi_\lambda^*(t)] \\ &= E[X(t+\tau+\lambda)X^*(t+\tau)X^*(t+\lambda)X(t)], \end{aligned}$$

which shows explicitly that X must have the fourth-moment stationarity here mentioned for this theorem to apply. Some examples of WSS random processes that are ergodic in various senses follow.

Example 8.4–1: Consider the WSS process of Example 7.1–2, which is a random amplitude cosine at frequency f_0 with random phase,

$$X(t) = A\cos(2\pi f_0 t + \Theta), \quad -\infty < t < +\infty.$$

Here A is $N(0, 1)$, Θ is uniformly distributed over $[-\pi, +\pi]$ and both A and Θ are independent. Then

$$E[X(t)] = E[A]E[\cos(2\pi f_0 t + \Theta)] = 0$$

and

$$\begin{aligned} E[X(t+\tau)X(t)] &= \frac{1}{2\pi}\int_{-\pi}^{+\pi}\cos(2\pi f_0 t + 2\pi f_0\tau + \theta)\cos(2\pi f_0 t + \theta)\,d\theta \\ &= \frac{1}{2\pi}\cos(2\pi f_0\tau), \qquad -\infty < \tau < +\infty, \end{aligned}$$

so that $X(t)$ is indeed wide-sense stationary. (In fact it can be shown that the process is also strict-sense stationary.) We first inquire whether $X(t)$ is ergodic in the mean; hence we consider

$$\sigma_{\hat{M}}^2 = \frac{1}{2T}\int_{-2T}^{+2T}\left[1 - \frac{|\tau|}{2T}\right]\frac{1}{2}\cos(2\pi f_0 \tau)\, d\tau.$$

If we realize that the triangular pulse can be expressed as the convolution of two rectangular pulses, as shown in Figure 8.4-2, then we can write for the Fourier transform of the triangular pulse:

$$\left(\sqrt{2T}\frac{\sin 2\pi f T}{2\pi f T}\right)^2 = 2T\left(\frac{\sin 2\pi f T}{2\pi f T}\right)^2$$

and then use Parsevals's theorem [8–4] to evaluate the above variance as

$$\frac{1}{2T}\int_{-2T}^{+2T}\left[I - \frac{|\tau|}{2T}\right]\frac{1}{2}\cos 2\pi f_0 \tau\, d\tau$$

$$= \frac{1}{2T}\int_{-\infty}^{+\infty} 2T\left(\frac{\sin 2\pi f T}{2\pi f T}\right)^2 \pi[\delta(f + f_0) + \delta(f - f_0)]\, df.$$

Thus

$$\sigma_{\hat{M}}^2 = \frac{1}{2}\left(\frac{\sin 2\pi f_0 T}{2\pi f_0 T}\right) \to 0 \quad \text{as } T \to \infty \text{ for } f_0 \neq 0.$$

Hence this random process is ergodic in the mean for any $f_0 \neq 0$.

To determine whether it's ergodic in power, we could use the condition of Equation 8.4–3. However, in this simple case we can obtain the result by examining the time average directly. Thus

$$\frac{1}{2T}\int_{-T}^{+T} X^2(t)\, dt = A^2 \frac{1}{2T}\int_{-T}^{+T}\cos^2(2\pi f_0 t + \Theta)\, dt.$$

Clearly, this time average will converge to $A^2/2$ not to $E[A^2]/2$ since for any Θ the time average of the cosine squared will converge to 1/2 for $f_0 \neq 0$. Thus this random cosine is not ergodic in power and hence not ergodic in correlation either. This is not unexpected since $K(\tau)$ does not tend to zero as $|\tau|$ tends to infinity. ($K(\tau)$ is in fact periodic!)

Another useful type of ergodicity is ergodicity in distribution function. Here we consider using time-averages to estimate the distribution function of some random process X, which is at least stationary to first order; that is, the first-order PDF is shift-invariant. We can form such a time-average by first forming the *indicator process* $I_x(t)$,

$$I_x(t) \triangleq \begin{cases} 1 & \text{if } X(t) \leq x, \\ 0 & \text{else,} \end{cases}$$

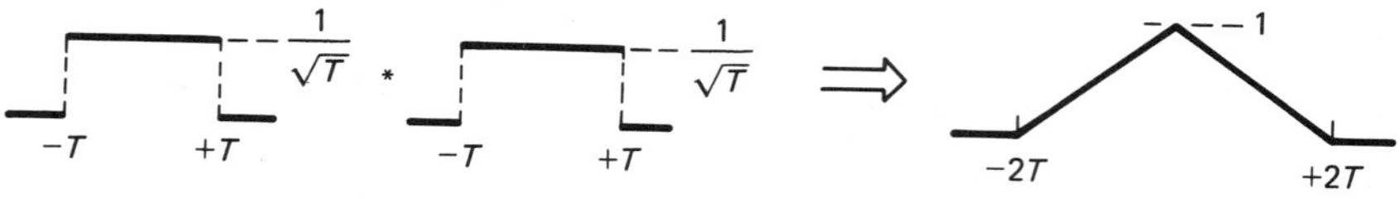

Figure 8.4–2 Convolution of rectangular pulses.

thus the random process I_x for each fixed x is one if the event $\{X(t) \le x\}$ occurs and zero if the event $\{X(t) \le x\}$ does not occur. The function I_x thus "indicates" the event in the sense of Boolean logic. Since $I_x(t)$ is a function of the random process $X(t)$, it in turn is a random process. The time-average value of I_x will be used to estimate the distribution function $F_X(x; t) = P[X(t) \le x]$,

$$\hat{F}_X(x) \triangleq \frac{1}{2T} \int_{-T}^{+T} I_x(t)\, dt. \tag{8.4–4}$$

First we consider the mean of this estimate. It is directly seen that

$$E[\hat{F}_X(x)] = E[I_x(t)] = 1 \cdot P[X(t) \le x] = F_X(x; t).$$

Next we consider the correlation function of I_x,

$$\begin{aligned} E[I_x(t_1)I_x(t_2)] &= P[X(t_1) \le x, \quad X(t_2) \le x] \\ &= F_X(x, x; t_1, t_2) \\ &= F_X(x, x; t_1 - t_2), \end{aligned}$$

where the last line follows if X is stationary of order 2. Thus we can say that I_x will be a WSS random process iff $X(t)$ is stationary of order 2. In this case we can apply Theorem 8.4–1 to I_x to get the following result.

Theorem 8.4–3. The random process $X(t)$, which is stationary up to order 2, is ergodic in distribution iff

$$\lim_{T\to\infty} \left[\frac{1}{2T} \int_{-T}^{+T} \left[1 - \frac{|\tau|}{2T} \right] K_{I_x}(\tau)\, d\tau \right] = 0,$$

where

$$\begin{aligned} K_{I_x}(\tau) &= E[I_x(t + \tau)I_x(t)] - E^2[I_x(t)] \\ &= F_X(x, x; \tau) - F_X^2(x; 0) \\ &= F_X(x, x; \tau) - F_X^2(x). \end{aligned}$$

Thus $K_{I_x}(\tau)$ must generally decay to zero as $|\tau| \to +\infty$ for the foregoing condition to be met; that is, we generally need

$$F_X(x, x; \tau) \to F_X^2(x) \qquad \text{as } |\tau| \to +\infty. \tag{8.4–5}$$

This would be true if $X(t + \tau)$ and $X(t)$ are asymptotically independent as $|\tau| \to +\infty$.

Example 8.4–2: Let $X(t)$ be a random process with covariance function

$$K_X(\tau) = \sigma_X^2 \exp(-\alpha\, |\tau|).$$

Then $X(t)$ is ergodic in the mean since $K_X(\tau)$ is absolutely integrable. If we further assume that $X(t)$ is a Gaussian random process, then using the Gaussian fourth-order moment property (cf. Problem 6.6), we can show ergodicity in correlation and hence in power and covariance. Also, again invoking the Gaussian property, we can conclude ergodicity in distribution since the distribution function is a continuous function of the mean and covariance so that Equation 8.4–5 is satisfied.

We note that the three theorems of this section give the same kind of condition for the three types of ergodicity considered. The difference between them lies in the three different covariance functions. The general form of the condition is

$$\lim_{T\to\infty} \frac{1}{2T} \int_{-2T}^{+2T} \left(1 - \frac{|\tau|}{2T}\right) K(\tau)\, d\tau = 0. \tag{8.4-6}$$

We now present an equivalent simpler condition for the case when $K(\tau)$ has a limit as $|\tau| \to +\infty$.

Theorem 8.4–4. If $K(\tau)$ has a limit as $|\tau| \to +\infty$, then the condition 8.4–6 is equivalent to $\lim_{|\tau|\to\infty} K(\tau) = 0$.

Proof. If K tends to a nonzero value, then clearly Equation 8.4–6 will not hold. So assume $\lim_{|\tau|\to\infty} K(\tau) = 0$, then for T large enough, $|K| < \varepsilon$ so that

$$\frac{1}{2T} \left| \int_{-2T}^{+2T} \left(1 - \frac{|\tau|}{2T}\right) K(\tau)\, d\tau \right| \le \frac{1}{2T}[4T\varepsilon + M], \quad \text{where } M < \infty,$$

$$\to 0 \text{ as } T \to \infty$$

which was to be shown.

8.5. KARHUNEN-LOÈVE EXPANSION [8–5]

Another application of the stochastic integral is to the Karhunen-Loève expansion. The idea is to decompose a general second-order random process into an orthonormal expansion whose coefficients are uncorrelated random variables. The expansion functions are just deterministic functions that have been orthonormalized to serve as a basis set for this decomposition. The Karhunen-Loeve (K-L) expansion has proved to be a very useful theoretical tool in such diverse areas as detection theory, pattern recognition, and image coding. It is often used as an intermediate step in deriving general results in these areas. We will present an example to show how the K-L expansion is used in optimal detection of a known signal in white Gaussian noise (Example 8.5–4), thus extending the detection decision theory of Section 5.7 to random processes.

Theorem 8.5–1. Let $X(t)$ be a zero-mean, second-order random process defined over $[-T/2, +T/2]$ with continuous covariance function $K_X(t_1, t_2)$. Then we can write

$$X(t) = \sum_{n=1}^{\infty} X_n \phi_n(t) \quad \text{(m.s.)} \quad \text{for } |t| \le T/2 \tag{8.5-1}$$

with

$$X_n \triangleq \int_{-T/2}^{+T/2} X(t)\phi_n^*(t)\, dt, \tag{8.5-2}$$

where the set of functions $\{\phi_n(t)\}$ is a complete, orthonormal set of solutions to the integral equation

$$\int_{-T/2}^{+T/2} K_X(t_1, t_2)\phi_n(t_2)\, dt_2 = \lambda_n \phi_n(t_1), \qquad |t_1| \leq T/2, \tag{8.5–3}$$

and the coefficients X_n are orthogonal; that is,

$$E[X_n X_m^*] = \lambda_n \delta_{mn}, \tag{8.5–4}$$

with δ_{mn} the Kronecker delta function.

The functions $\phi_n(t)$ are orthonormal in the sense that

$$\int_{-T/2}^{+T/2} \phi_n(t)\phi_m^*(t)\, dt = \delta_{mn}. \tag{8.5–5}$$

In fact it is easy to show that any two normalized solutions $\phi_n(t)$ and $\phi_m(t)$ to the integral Equation 8.5–3 must be orthonormal if $(\lambda_n \neq \lambda_m)$ and both λ's are not zero. See Problem 8.12.

The interesting thing about this expansion is that the coefficients are uncorrelated or orthogonal. Otherwise any expansion such as the Fourier series expansion would suffice†. The point of this theorem is that there exists at least one set of orthormal functions with the special property that the coefficients in its expansion are uncorrelated random variables. We break up the proof of this important theorem into two steps or lemmas. We also need a result from the theory of integral equations known as Mercer's Theorem, which states that

$$K(t_1, t_2) = \sum_{n=1}^{\infty} \lambda_n \phi_n(t_1)\phi_n^*(t_2). \tag{8.5–6}$$

This result is derived in Appendix 8.APP at the end of this chapter. Also, a constructive method is shown in Facts 1 to 3 of this appendix, to find $\{\phi_n(t)\}$ a set of orthonormal solutions to Equation 8.5–3.

Lemma 8.5–1. If $X(t) = \sum_{n=1}^{\infty} X_n \phi_n(t)$ (m.s.) and the X_n's are orthogonal, then the $\phi_n(t)$ must satisfy integral Equation 8.5–3.

Proof. We compute $X(t)X_n^* = \sum_m X_m X_n^* \phi_m(t)$, thus

$$E[X(t)X_n^*] = \sum_m E[X_m X_n^*]\phi_m(t) = E[|X_n|^2]\phi_n(t),$$

but also

$$E[X(t)X_n^*] = E\left[X(t)\int_{-T/2}^{+T/2} X^*(s)\phi_n(s)\, ds\right]$$

$$= \int_{-T/2}^{+T/2} K(t, s)\phi_n(s)\, ds,$$

† Incidently, it can be shown that the m.-s. Fourier series coefficients of $X(t)$ become asymptotically uncorrelated as $T \to \infty$ and further that the K-L eigenfunctions approach complex exponentials as $T \to \infty$. [8–6]

hence

$$\int_{-T/2}^{+T/2} K(t, s)\phi_n(s)\,ds = \lambda_n\phi_n(t) \quad \text{with} \quad \lambda_n \triangleq E[|X_n|^2]. \quad \blacksquare$$

We next show a partial converse to this lemma.

Lemma 8.5–2. If the orthonormal set $\{\phi_n(t)\}$ satisfies integral Equation 8.5–3, then the random-variable coefficients X_n given in Equation 8.5–2 must be orthogonal.

Proof. By Equation 8.5–2

$$X_m = \int_{-T/2}^{+T/2} X(t)\phi_m^*(t)\,dt,$$

so

$$E[X(t)X_m^*] = \int_{-T/2}^{+T/2} E[X(t)X^*(s)]\phi_m(s)\,ds = \lambda_m\phi_m(t),$$

because the ϕ_n's satisfy Equation 8.5–3. Thus

$$\begin{aligned} E[X_nX_m^*] &= E\left[\int_{-T/2}^{+T/2} X(t)\phi_n^*(t)\,dt\,X_m^*\right] \\ &= \int_{-T/2}^{+T/2} E[X(t)X_m^*]\phi_n^*(t)\,dt \\ &= \int_{-T/2}^{+T/2} \lambda_m\phi_m(t)\phi_n^*(t)\,dt = \lambda_m\delta_{mn}. \quad \blacksquare \end{aligned}$$

By combining the results in the above two lemmas we can see that the K-L coefficients will be orthogonal if and only if the orthonormal basis functions are chosen as solutions to Equation 8.5–3. What remains to show is that the K-L expansion does in fact produce a mean-square equality, which we show below with the help of Mercer's theorem.

Proof of Theorem 8.5–1. Define $\hat{X}(t) \triangleq \sum_{n=1}^{\infty} X_n\phi_n(t)$ and consider $E[|X(t) - \hat{X}(t)|^2] = E[X(t)(X(t) - \hat{X}(t))^*] - E[\hat{X}(t)(X(t) - \hat{X}(t))^*]$. The second term is zero since

$$\begin{aligned} E[\hat{X}(t)(X(t) - \hat{X}(t))^*] &= E\left[\sum_m X_mX^*(t)\phi_m(t) - \sum_{m,n} X_mX_n^*\phi_m(t)\phi_n^*(t)\right] \\ &= \sum_m E[X_mX^*(t)]\phi_m(t) - \sum_{m,n} E[X_mX_n^*]\phi_m(t)\phi_n^*(t) \\ &= \sum_m E[X_mX^*(t)]\phi_m(t) - \sum_{m,n} E[X_mX_n^*]\phi_m(t)\phi_n^*(t) \\ &= \sum_m \lambda_m\,|\phi_m(t)|^2 - \sum_m \lambda_m\,|\phi_m(t)|^2 = 0, \end{aligned}$$

by the first step in proof of Lemma 8.5–2; so evaluating the first term, we get

$$E[|X(t) - \hat{X}(t)|^2] = R(t, t) - \sum_{n=1}^{\infty} \lambda_n \phi_n(t) \phi_n^*(t),$$

which is zero by Equation 8.5–6, Mercer's Theorem. ■

We now present some examples of the calculation of the K-L expansion.

Example 8.5–1: (K-L expansion of white noise.) Let $K(t_1, t_2) = \sigma^2 \delta(t_1 - t_2)$. Then we must find the ϕ_n's to satisfy the K-L integral equation,

$$\sigma^2 \int_{-T/2}^{+T/2} \delta(t_1 - t_2)\phi(t_2)\, dt_2 = \lambda \phi(t_1), \qquad -T/2 \le t_1 \le +T/2,$$

or

$$\sigma^2 \phi(t) = \lambda \phi(t).$$

Thus in this case the $\phi(t)$ functions are arbitrary and all the λ_n's equal σ^2. So the expansion functions $\phi_n(t)$ can be any complete orthonormal set with corresponding eigenvalues taken to be $\lambda_n = \sigma^2$.

Note that this example, though easy, violates the second-order constraint on the covariance function in the K-L Theorem 8.5–1. Nevertheless, the resulting expansion can be shown to be valid in the sense of generalized random processes (cf. definition on p. 304).

Example 8.5–2: (Random process plus white noise.) Here we look at what happens if we add a random process to the white noise of the previous example and then want to know the K-L expansion for the noisy process,

$$Y(t) = X(t) + W(t),$$

where $W(t)$ is white noise and X and W are orthogonal (Definition 7.4–1), which we denote as $X \perp W$. Plugging into the K-L equation as before, we obtain

$$\int_{-T/2}^{+T/2} [R_X(t, s) + \sigma^2 \delta(t - s)]\phi^{(y)}(s)\, ds = \lambda^{(y)} \phi^{(y)}(t)$$

$$= \int_{-T/2}^{+T/2} R_X(t, s)\phi^{(y)}(s)\, ds + \sigma^2 \phi^{(y)}(t),$$

so

$$\int_{-T/2}^{+T/2} R_X(t, s)\phi^{(y)}(s)\, ds = (\lambda^{(y)} - \sigma^2)\phi^{(y)}(t),$$

where we use the superscripts x and y to denote the respective eigenvalues and eigenfunctions. But since

$$\int_{-T/2}^{+T/2} R_X(t, s)\phi^{(x)}(s)\, ds = \lambda^{(x)} \phi^{(x)}(t),$$

we immediately obtain

$$\phi^{(y)}(t) = \phi^{(x)}(t)$$

and

$$\lambda^{(t)} - \sigma^2 = \lambda^{(x)}.$$

We see that the eigenfunctions, that is, the K-L basis functions, are the same for both the X and Y processes. The K-L coefficients $Y_n = X_n + W_n$, then have variance $\lambda_n^{(y)} = \lambda_n^{(x)} + \sigma^2$.

Example 8.5–3: (K-L expansion for Wiener process.) For this example let the time interval be $(0, T)$ to match our definition of the Wiener process. Using Equation 7.2–10 in the K-L integral Equation 8.5–3, we obtain,

$$\sigma^2 \int_0^T \min(t, s)\phi(s)\, ds = \lambda\phi(t), \quad 0 < t < T. \tag{8.5–7}$$

or

$$\sigma^2\left[\int_0^t s\phi(s)\, ds + t\int_t^T \phi(s)\, ds\right] = \lambda\phi(t).$$

We temporarily agree to set $\sigma^2 = 1$ to simplify the equations. The standard method of solution of Equation 8.5–7 is to differentiate it as many times as necessary to convert it over to a differential equation. We then evaluate the boundary conditions needed to solve the differential equation by plugging the general solution back into the integral equation. Here we take one derivative with respect to t and obtain

$$\int_t^T \phi(s)\, ds = \lambda\dot{\phi}(t). \tag{8.5–8}$$

Taking a second derivative, we obtain a differential equation,

$$-\phi(t) = \lambda\ddot{\phi}(t),$$

with general solution,

$$\phi(t) = A \sin\frac{t}{\sqrt{\lambda}} + B \cos\frac{t}{\sqrt{\lambda}}.$$

Next we use the boundary conditions at 0 and T to determine the coefficients A, B, and λ. From Equation 8.5–7 at $t = 0+$ we get $\phi(0+) = 0$, which implies $B = 0$. From Equation 8.5–8 at $t = T-$ we get $\dot{\phi}(T-) = 0$, which implies that

$$\cos\left(T/\sqrt{\lambda}\right) = 0, \text{ i.e. } \lambda = \lambda_n = \left(\frac{T}{(n - \frac{1}{2})\pi}\right)^2 \text{ for } n \geq 1.$$

Finally, A is chosen such that ϕ is normalized, that is,

$$\int_0^T \phi_n^2(t)\, dt = 1, \text{ which implies } A = \sqrt{2/T}.$$

Thus we get the following solution for the K-L basis functions,

$$\phi_n(t) = \sqrt{2/T}\sin[(n - \tfrac{1}{2})\pi t/T],\ n \geq 1.$$

Note that by Problem 8.12, the $\phi_n(t)$ must satisfy $\phi_n \perp \phi_m$ for $n \neq m$ since the eigenvalues $\lambda_n = [T/(n - \frac{1}{2})\pi]^2$ are distinct. This is the K-L expansion for the standard Wiener process, that is, $\sigma = 1$. If $\sigma \neq 1$, then just replace λ_n with $\sigma^2\lambda_n$.

We now present an application of the Karhunen–Loève expansion to a simple problem from the area of communications known as Detection Theory [8–6].

Example 8.5–4: (Application to Detection Theory.) Assume we observe a waveform $X(t)$ over the interval $[-T/2, +T/2]$ and wish to decide whether it contains a signal buried in noise or just noise alone. To be more precise we define two hypotheses H_1 and H_0, and consider the decision theory problem: (cf. Section 5.7):

$$X(t) = \begin{cases} m(t) + W(t): & H_1 \\ W(t) \quad : & H_0, \end{cases}$$

where $m(t)$ is a deterministic function, that is, the *signal*, and $W(t)$ is the *noise* modeled by a zero-mean, white Gaussian process. Using the K–L expansion, we can simplify the preceding decision problem by replacing this waveform problem by a sequence of simpler scalar problems,

$$X_n = \begin{cases} m_n + W_n : & H_1 \\ W_n & : H_0 \end{cases}$$

where m_n and W_n are the respective K–L coefficients.

Effectively we take the K–L transform of the original received signal X. The transform space is then just the space of sequences of K–L coefficients. Using the fact that the noise is zero-mean Gaussian and that the expansion coefficients are orthogonal, we conclude that the r.-v.'s W_n are jointly independent, that is, W_n is an independent random sequence. The problem can be simplified even further by observing that $K_W(t_1, t_2) = \sigma_W^2 \delta(t_1 - t_2)$ permits the ϕ_n's to be any complete set of orthonormal solutions to the K–L integral Equation 8.5–3. It is convenient to take $\phi_1(t) = km(t)$ where k is the normalizing constant

$$k \triangleq \left[\int_{-T/2}^{+T/2} m^2(t)\, dt \right]^{-1/2},$$

and then to complete the orthonormal set in any valid way. We then notice that all the m_k will be zero except for $k = 1$, thus only X_1 is affected by the presence or absence of the signal. One can then show that this detection problem can finally be reduced to the scalar problem,

$$X_1 = \begin{cases} k^{-1} + W_1 : & H_1 \\ W_1 & : H_0. \end{cases}$$

To compute X_1 we note that it is just the stochastic integral

$$X_1 = k \int_{-T/2}^{+T/2} X(t) m(t)\, dt$$

that is often referred to as a matching operation. In fact, it can be performed by sampling the output of a filter whose impulse response is $h(t) = km(T - t)$, where T is chosen large enough to make this impulse response causal. The filter output at time T is then X_1. This filter is called the *matched filter* and is widely used in communications and pattern recognition.

Another application of the K–L expansion is in the derivation of important results in linear estimation theory, the topic of Chapter 10. Analogously to the preceding example, the approach will be to reduce the waveform estimation problem to the simpler one of estimating the individual K–L coefficients. Scalar linear estimation problems were discussed earlier in Chapter 5.

8.6 SUMMARY

We have studied an extension of many of the concepts of calculus to random processes by studying the mean-square calculus. This enabled us to define useful differential-equation and integral operators for a wide class of second-order random processes. In so doing, we derived further results on m.-s. convergence, including the notion that random variables with finite mean-square value, that is, second-order random variables, can be viewed as vectors in a Hilbert space with the expectation of their conjugate product serving as an inner product. This viewpoint will be important for applications to linear estimation in Chapter 10.

We defined the stochastic integral and applied it to two problems: ergodicity, the problem of estimating the parameter functions of random processes, and the Karhunen-Loève expansion, an important theoretical tool that decomposes a possibly nonstationary process into a sum of products of orthogonal random variables and orthonormal basis functions. This countable-basis representation will be used in Chapter 10 to derive linear estimates for finite-interval observations.

8.APP Integral Equations

In this appendix we look at some of the properties of the solution of integral equations that are needed to appreciate the Karhunen-Loève expansion presented in Section 8.5. In Facts 1 to 3 following, we develop a method to solve for a complete set of orthonormal solutions $\{\lambda_n, \phi_n(t)\}$. Consider the integral equation,

$$\int_a^b R(t, s)\phi(s)\, ds = \lambda\phi(t) \quad \text{on } a \leq t \leq b, \tag{8.APP–1}$$

where the kernel $R(t, s)$ is continuous, Hermitian, and positive semidefinite. Thus R fulfills the conditions to be a correlation function, although in the integral equation setting there may be no such interpretation. The solution to Equation 8.APP–1 consists of the function $\phi(t)$, called an *eigenfunction*, and the scalar λ, called the corresponding *eigenvalue*. To avoid the trivial case, we rule out the solution $\phi(t) = 0$ as a valid eigenfunction.

A fundamental theorem concerns the existence of solutions to Equation 8.APP–1. Its proof can be found in the book on functional analysis by F. Reisz and B. Sz–Nagy [8–7].

Existence Theorem. If the continuous and Hermitian kernel $R(t, s)$ in the integral Equation 8.APP–1 is nonzero and positive semidefinite, then there exists at least one eigenfunction with nonzero eigenvalue.

While the proof of this theorem is omitted, it should seem reasonable based on our experience with computing the eigenvalues and eigenvectors of correlation matrices in Section 4.3. To see the connection, note that the integral in Equation 8.APP–1 is the limit of a sum involving samples in s. If we require its solution only at samples in t, then we have equivalently the vector-matrix eigenvector problem of Chapter 4. So with some form of continuity in R and ϕ, the properties of the eigenvectors and eigenvalues should carry over to the present eigenfunctions and eigenvalues.

The existence theorem allows us to conclude several useful results concerning the solutions to Equation 8.APP–1, culminating in Mercer's theorem, which is applied in Section 8.5 to prove the Karhunen-Loeve representation. This method is adapted from Reference 8–2.

Fact 1. All the eigenvalues must be real and nonnegative. Additionally, when R is positive definite, they must all be positive.

Proof. Since R is positive semidefinite,

$$\iint \phi^*(t)R(t, s)\phi(s)dt\, ds \geq 0,$$

but

$$\int \phi^*(t)\left[\int R(t, s)\phi(s)\, ds\right] dt$$

$$= \int \phi^*(t)\lambda\phi(t)\, dt$$

$$= \lambda\int |\phi(t)|^2\, dt,$$

thus λ is real and nonnegative since $\int |\phi(t)|^2\, dt \neq 0$. Also if R is positive definite, then λ cannot be zero. ■

Fact 2. Let $R(t, s)$ be Hermitian and positive semidefinite. Let ϕ_1, λ_1 be a corresponding eigenfunction and eigenvalue pair. Then

$$R_1(t, s) \triangleq R(t, s) - \lambda_1\phi_1(t)\phi_1^*(s)$$

is also positive semidefinite.

Fact 3. Either $R_1(t, s) = 0$ or else there is another eigenfunction and eigenvalue ϕ_2, λ_2 with $\phi_2 \perp \phi_1$†, such that

$$R_2(t, s) \triangleq R_1(t, s) - \lambda_2\phi_2(t)\phi_2^*(s)$$

is positive semidefinite.

Continuing with this procedure, we eventually obtain

$$R_N(t, s) \triangleq R(t, s) - \sum_{n=1}^{N} \lambda_n\phi_n(t)\phi_n^*(s).$$

Now since

$$R_N(t, t) = R(t, t) - \sum_{n=1}^{N} \lambda_n\, |\phi_n(t)|^2 \geq 0,$$

we have that the increasing sum $\sum_{n=1}^{N} \lambda_n\, |\phi_n(t)|^2$ is bounded from above so that it must tend to a limit. Thus $R_\infty(t, s)$ exists for $s = t$. For $s \neq t$ consider

† By $\phi_2 \perp \phi_1$ we mean $\int_a^b \phi_1(t)\phi_2^*(t)\, dt = 0$, so that upon normalization ϕ_1 and ϕ_2 become orthonormal (Equation 8.5–5).

the partial sum for $m > n$,

$$\Delta R_{m,n}(t, s) \triangleq \sum_{k=n}^{m} \lambda_k \phi_k(t)\phi_k^*(s).$$

Then $|\Delta R_{m,n}(t, t)| \to 0$ as $m, n \to \infty$. Using the Schwarz inequality† we conclude

$$|\Delta R_{m,n}(t, s)| \leq \sqrt{\Delta R_{m,n}(t, t) \cdot \Delta R_{m,n}(s, s)}.$$

Thus $|\Delta R_{m,n}(t, s)| \to 0$ as $m, n \to \infty$. Therefore we can write

$$R_\infty(t, s) \triangleq R(t, s) - \sum_{n=1}^{\infty} \lambda_n \phi_n(t)\phi_n^*(s). \tag{8.APP–2}$$

Mercer's theorem now can be seen as an affirmative answer to the question of whether or not $R_\infty(t, s) = 0$. We now turn to the proofs of Facts 2 and 3 prior to proving Mercer's theorem.

Proof of Fact 2. We must show that R_1 is positive semidefinite. We do this by defining the random process

$$Y(t) \triangleq X(t) - \phi_1(t)\int_a^b X(\tau)\phi_1^*(\tau)\, d\tau$$

and showing that R_1 is its correlation function. We have

$$\begin{aligned} E[Y(t)Y^*(s)] &= E[X(t)X^*(s)] - \phi_1^*(s)\int_a^b E[X^*(\tau)X(t)]\phi_1(\tau)\, d\tau \\ &\quad -\phi_1(t)\int_a^b E[X(\tau)X^*(s)]\phi_1^*(\tau)\, d\tau \\ &\quad + \phi_1(t)\phi_1^*(s)\int_a^b\int_a^b E[X(\tau_1)X^*(\tau_2)]\phi_1^*(\tau_1)\phi_1(\tau_2)\, d\tau_1\, d\tau_2 \\ &= R_X(t, s) - \phi_1^*(s)\int_a^b R_X(t, \tau)\phi_1(\tau)\, d\tau \\ &\quad - \phi_1(t)\int_a^b R_X(\tau, s)\phi_1^*(\tau)\, d\tau \\ &\quad +\phi_1(t)\phi_1^*(s)\int_a^b\int_a^b R_X(\tau_1, \tau_2)\phi_1^*(\tau_1)\phi_2(\tau_2)\, d\tau_1\, d\tau_2, \end{aligned}$$

but $\lambda_1\phi_1(t) = \int_a^b R_X(t, \tau)\phi_1(\tau)\, d\tau$ and using $R_X^*(t, \tau) = R_X(\tau, t)$ we get

$$\lambda_1\phi_1^*(s) = \int_a^b R_X(\tau, s)\phi_1^*(\tau)\, d\tau,$$

† We can regard $\Delta R_{m,n}(t, s)$ as the correlation of two r.v.'s with variance $\Delta R_{m,n}(t, t)$ and $\Delta R_{m,n}(s, s)$. Alternatively we can use the Schwarz inequality for complex numbers: $|\sum a_i b_i|^2 \leq (\sum |a_i|^2)(\sum |b_i|^2)$.

so that the two cross terms are each equal to $-\lambda_1\phi_1(t)\phi_1^*(s)$. Evaluating

$$\int_a^b \int_a^b R_X(\tau_1, \tau_2)\phi_1^*(\tau_1)\phi_1(\tau_2)\, d\tau_1\, d\tau_2$$

$$= \int_a^b \phi_1^*(\tau_1)\left[\int_a^b R_X(\tau_1, \tau_2)\phi_1(\tau_2)\, d\tau_2\right] d\tau_1$$

$$= \int_a^b \phi_1^*(\tau_1)\lambda_1\phi_1(\tau_1)\, d\tau_1$$

$$= \lambda_1 \int_a^b |\phi_1(t)|^2\, dt$$

$$= \lambda_1,$$

and combining, we get

$$R_Y(t, s) = R_X(t, s) - \lambda_1\phi_1(t)\phi_1^*(s),$$

which agrees with the definition of R_1 in Fact 2. ■

Proof of Fact 3. Just repeat the proof of Fact 2, with R_1 in place of R to conclude that R_2 is positive semidefinite. To show that $\phi_2 \perp \phi_1$, we proceed as follows. We first note that

$$\lambda_2\phi_2(t) = \int_a^b R_1(t, s)\phi_2(s)\, ds$$

and then plug in the definition of R_1 to obtain

$$\lambda_2\phi_2(t) = \int_a^b R(t, s)\phi_2(s)\, ds - \lambda_1\phi_1(t)\int_a^b \phi_1^*(s)\phi_2(s)\, ds.$$

Then we multiply by $\phi_1^*(t)$ and integrate over (a, b) to obtain

$$\lambda_2\int_a^b \phi_1^*(t)\phi_2(t)\, dt = \int_a^b \int_a^b \phi_1^*(t)\phi_2(s)R(t, s)\, dt\, ds$$

$$- \lambda_1\int_a^b |\phi_1(t)|^2\, dt \cdot \int_a^b \phi_1^*(s)\phi_2(s)\, ds$$

$$= \int_a^b \phi_2(s)\left[\int_a^b R(t, s)\phi_1^*(t)\, dt\right] ds - \lambda_1\int_a^b \phi_1^*(s)\phi_2(s)\, ds$$

$$= \int_a^b \phi_2(s)\lambda_1^*\phi_1^*(s)\, ds - \lambda_1\int_a^b \phi_1^*(s)\phi_2(s)\, ds,$$

by the Hermitian symmetry of R, that is, $R(t, s) = R^*(s, t)$, we thus obtain

$$\lambda_2\int_a^b \phi_1^*(t)\phi_2(t)\, dt = (\lambda_1^* - \lambda_1)\int_a^b \phi_1^*(s)\phi_2(s)\, ds.$$

Now λ_1 is real so

$$\lambda_2 \int_a^b \phi_1^*(t)\phi_2(t)\,dt = 0.$$

Thus either $\lambda_2 = 0$ or $\phi_1 \perp \phi_2$. We can reject the first possibility by the existence theorem and hence we are done.

Mercer's theorem. Let the kernel $R(t, s)$ be continuous, Hermitian, and positive semidefinite. Let $\{\lambda_n, \phi_n(t)\}$ be the possibly infinite complete set of orthonormal solutions to the integral Equation 8.APP–1. Then the following expansion holds for all $t, s \in [a, b]$,

$$R(t, s) = \sum_{n=1}^{\infty} \lambda_n \phi_n(t)\phi_n^*(s).$$

Proof. By Equation 8.APP–2 we know that the question reduces to whether the positive semidefinite kernel R_∞ is equal to zero. If it is not zero then by the existence theorem there is an eigenfunction and nonzero eigenvalue λ for R_∞. Since $\lambda > 0$, adding this new eigenfunction-eigenvalue pair to the right-hand side of Equation 8.APP–2, we get a change in the value of R_∞, which contradicts the assumed convergence. Thus $R_\infty = 0$ and the theorem is proved.

PROBLEMS

8.1. Use the theorems on mean-square limits to show the following properties of the mean-square derivative

(a) $\frac{d}{dt}(aX_1(t) + bX_2(t)) = a\frac{dX_1(t)}{dt} + b\frac{dX_2(t)}{dt}$

(b) $E[X_1(t_1)\dot{X}_2^*(t_2)] = \frac{\partial}{\partial t_2} R_{X_1X_2}(t_1, t_2)$.

8.2. Let $X(t)$ be a stationary random process with mean μ_X and covariance function

$$K_X(\tau) = \frac{\sigma_X^2}{1 + \alpha^2\tau^2}, \qquad -\infty < \tau < +\infty.$$

(a) Show that a mean-square derivative exists for all t.

(b) Find $\mu_{\dot{X}}(t)$ and $K_{\dot{X}}(\tau)$ for all t and τ.

$K_{\dot{X}}(\tau) = \frac{-d^2}{d\tau^2} K_{XX}(\tau)$

8.3. To estimate the mean of a stationary random process, we often consider an integral average

$$I(T) \triangleq \frac{1}{T}\int_0^T X(t)\,dt, \qquad T > 0.$$

(a) Find the mean of $I(T)$ as a function of T, that is, $\mu_I(T)$ for $T > 0$.

(b) Find the variance of $I(T)$, that is, $\sigma_I^2(T)$ for $T > 0$.

8.4. Let $X(t)$ be a WSS Gaussian random process. Show that the m.s. derivative of $Y(t) \triangleq X^2(t)$ is $\dot{Y}(t) = 2X(t)\dot{X}(t)$ and find the correlation function of $\dot{Y}$ in terms of R_X and its derivatives.

8.5. (a) If $U(t)$ is an independent increments random process, with zero mean, show that $R_{UU}(t_1, t_2) = f(\min(t_1, t_2))$ where $f(t) \triangleq E[U^2(t)]$.

(b) Using the definition of an independent process (see Section 7.4), show that the generalized m.-s. derivative $U'(t)$ is an independent process.

(c) What is the condition on the function $f(t)$ such that the random process $U'(t)$ be wide-sense stationary?

8.6. (a) Let $X[n]$ be a sequence of Gaussian random variables. Let Y be the limit of this sequence where we assume Y exists in the mean-square sense. Use the fact that convergence in mean square implies convergence in distribution to conclude that Y is also Gaussian.

(b) Repeat the above argument for a sequence of Gaussian random vectors $\mathbf{X}[n] = [X_1[n], \ldots, X_K[n]]^T$.

Note: Mean-square convergence for random vectors means

$$E[|\mathbf{X}[n] - \mathbf{X}|^2] \to 0$$

where $|\mathbf{X}|^2 \triangleq \sum_{i=1}^{K} X_i^2$. And Chebyshev's inequality for random vectors is

$$P[|\mathbf{X}| > \varepsilon] \leq \int \cdots \int \frac{|\mathbf{x}|^2}{\varepsilon^2} f(\mathbf{x})\, d\mathbf{x} = \frac{E\{|\mathbf{X}|^2\}}{\varepsilon^2}.$$

(c) Let $\mathbf{X}[n] \triangleq [X_n(t_1), \ldots, X_n(t_K)]^T$ and use the result of part (b) to conclude that a m.-s. limit of Gaussian random processes is another Gaussian random process.

Note: After you finish this problem you might note that one application of the result of part (c) is that the m.-s. derivative $X'(t)$ is Gaussian whenever $X(t)$ is Gaussian.

8.7. Consider the m.s. differential equation

$$\frac{dY(t)}{dt} + 2Y(t) = X(t)$$

for $t > 0$ subject to the initial condition $Y(0) = 0$. The input is

$$X(t) = 5 \cos 2t + W(t),$$

where $W(t)$ is a Gaussian white noise with mean zero and covariance $K_W(\tau) = \sigma^2 \delta(\tau)$.

(a) Find $\mu_Y(t)$ for $t > 0$.

(b) Find the covariance $K_Y(t_1, t_2)$ for t_1 and $t_2 > 0$.

(c) What is the maximum value of σ such that

$$P[|Y(t) - \mu_Y(t)| < 0.1] > 0.99 \quad \text{for all } t > 0?$$

Use Table 2.4–1 for erf($\cdot$) in Chapter 2.

$$\left(\text{erf}(x) \triangleq \frac{1}{\sqrt{2\pi}} \int_0^x e^{-\frac{1}{2}u^2}\, du, \qquad x \geq 0\right)$$

8.8. Show that the (m.-s.) solution to

$$\dot{Y}(t) + \alpha Y(t) = X(t),$$

that is, the random process $Y(t)$ having $R_{YY}(t_1, t_2)$ and $R_{XY}(t_1, t_2)$ of Equations 8.3–3 and 8.3–4 actually satisfies the differential equation in the mean-square sense.

More specifically, show that

$$E[(\dot{Y}(t) + \alpha Y(t) - X(t))^2] = 0.$$

8.9. To detect a constant signal of amplitude A in white Gaussian noise of variance σ^2 and mean zero, we consider two hypotheses (that is, events):

$$\left.\begin{aligned} H_0 &: R(t) = W(t) \\ H_1 &: R(t) = A + W(t) \end{aligned}\right\} \quad \text{for } t \in [0, T].$$

It can be shown that the optimal detector to decide on the correct hypothesis, first computes the integral

$$\Lambda \triangleq \int_0^T R(t)\, dt,$$

and then does a threshold test.

(a) Find the mean value of Λ under each hypothesis.
(b) Find the variance of Λ under each hypothesis.
(c) An optimal detector would compare Λ to a threshold $\Lambda_0 \triangleq AT/2$ in the case when each hypothesis is equally likely, that is, $P[H_0] = P[H_1] = 1/2$. Under these conditions, find

$$P[\Lambda \geq \Lambda_0 \mid H_0]$$

expressing your result in terms of the error function erf defined in Chapter 2. Note: By Problem 8.6(c), Λ is a Gaussian random variable.

8.10. (a) State the general definition of "ergodic in the mean" for a wide-sense stationary process $X(t)$.
(b) Let $X(t)$ be a wide-sense stationary Gaussian random process with zero-mean and correlation function

$$R_X(\tau) = \sigma^2 e^{-\alpha|\tau|} \cos 2\pi f \tau,$$

where σ, α, and f are all positive constants.
Show that $X(t)$ is ergodic in the mean.
Hint: You may want to use a sufficient condition.

8.11. Let the random sequence $X[n]$ be stationary over the range $0 \leq n < \infty$. Define the time average

$$\hat{M} \triangleq \frac{1}{N} \sum_{n=1}^{N} X[n].$$

Analogously to the concept of "ergodic in the mean" for random processes, we have the following:

Definition. A stationary random sequence $X[n]$ is ergodic in the mean if $\hat{M}$ converges to the ensemble mean μ_X in the mean-square sense as $N \to \infty$.

(a) Find a suitable condition on the covariance function $K_X[m]$ for $X[n]$ to be ergodic in the mean.
(b) Show that this condition can be put in the form

$$\lim_{N\to\infty} \left\{ \frac{1}{N} \sum_{n=-N}^{+N} \left(1 - \frac{|n|}{N}\right) K_X[n] \right\} = 0.$$

(c) Is the stationary random sequence with covariance

$$K_X[m] = 5(0.9)^{|m|} + 15(0.8)^{|m|}$$

ergodic in the mean?

8.12. If the K–L integral Equation 8.5–3 has two solutions $\phi_1(t)$ and $\phi_2(t)$ corresponding to the eigenvalues λ_1 and λ_2, then show that if $\lambda_1 \neq 0$ and $\lambda_2 \neq \lambda_1$ we must have

$$\int_{-T/2}^{+T/2} \phi_1(t)\phi_2^*(t)\, dt = 0.$$

Hint: Substitute for $\phi_1(t)$ in the above expression and use the Hermitian symmetry of $K(t, s)$, that is, $K(t, s) = K^*(s, t)$.

8.13. In this problem we use the Loève K-L expansion to get LMMSE estimates of a Gaussian signal in white Gaussian noise. We observe $X(t)$ on $[0, T]$ where

$$X(t) = S(t) + W(t), \quad \text{with } S \perp W.$$

$W(t)$ is zero-mean and has covariance $K_W(\tau) = \sigma_W^2\delta(\tau)$.
$S(t)$ is a zero-mean with covariance $K_S(t_1, t_2)$.

(a) Show that any set of orthonormal functions $\{\phi_n(t)\}$ satisfies the K-L integral equation for the random process $W(t)$.

(b) Using part (a), show that the same set of orthonormal functions may be used for the K-L expansion of $X(t)$ and $S(t)$.

(c) Show that for $X_n = S_n + W_n$, the LMMSE estimate of S_n is given as

$$\hat{E}[S_n \mid X_n] = \frac{\sigma_{S_n}^2}{\sigma_{S_n}^2 + \sigma_W^2} X_n,$$

where X_n, S_n, and W_n are the K-L coefficients of the respective random processes.

(d) Using the above, argue that

$$\hat{S}(t) = \sum_{n=1}^{\infty} \frac{\sigma_{S_n}^2}{\sigma_{S_n}^2 + \sigma_W^2} X_n\phi_n(t).$$

Hint: Expand $S(t)$ in a K-L expansion,

$$S(t) = \sum_{n=1}^{\infty} S_n\phi_n(t)$$

8.14. Derive the Karhunen-Loève expansion for random sequences: If $X[n]$ is a second-order random sequence with correlation $R[n_1, n_2]$. Then

$$X[n] = \sum_{k=1}^{N+1} X_k\phi_k[n] \quad \text{with } |n| \leq N/2 \text{ and } N \text{ even}$$

$$\text{where } X_k \triangleq \sum_{n=-N/2}^{+N/2} X[n]\phi_k^*[n]$$

$$\text{and } E[X_k X_l^*] = \lambda_k\delta[k - l]$$

$$\text{and } \sum_{n=-N/2}^{+N/2} \phi_k[n]\phi_l^*[n] = \delta[k - l].$$

You may assume Mercer's theorem holds in the form

$$R[n_1, n_2] = \sum_{k=1}^{N+1} \lambda_k \phi_k[n_1] \phi_k^*[n_2],$$

which is just the eigenvalue-eigenvector decomposition of the matrix **R** with entries $R[i, j]$ for $i, j = -N/2, \ldots, +N/2$. (Note: It may be helpful to rewrite the above in matrix-vector form.)

REFERENCES

8–1. W. B. Davenport, Jr., *Probability and Random Processes.* New York: McGraw-Hill, 1970.

8–2. E. Wong and B. Hajek, *Stochastic Processes in Engineering Systems.* New York: Springer-Verlag, 1985.

8–3. A. Papoulis, *Probability, Random Variables, and Stochastic Processes.* New York: McGraw-Hill, 1984.

8–4. A. V. Oppenheim, A. S. Willsky, and I. T. Young, *Signals and Systems.* Englewood Cliffs, N. J.: Prentice-Hall, 1983, p. 224.

8–5. M. Loève, *Probability Theory.* New York: Van Nostrand, 1963.

8–6. H. L. Van Trees, *Detection Estimation and Modulation Theory: Part I.* New York: John Wiley, 1968.

8–7. F. Riesz and B. Sz.-Nagy, *Functional Analysis.* New York: Ungar, 1955.

9

STATIONARY PROCESSES AND SEQUENCES

In this chapter we will concentrate on the study of random processes and sequences that are *stationary*, meaning that their statistical properties do not change with time. This class of processes is important because stationarity provides the possibility of learning the statistical properties under various ergodicity hypotheses (cf. Section 8.4). Also the amount of information required to statistically describe stationary processes is greatly reduced. Furthermore, the various systems that are designed to process signals modeled as stationary processes can be constant-coefficient systems, which greatly simplifies both their design and implementation. Finally, frequency-domain methods can be used for the design and analysis of LSI systems with stationary input processes.

In reality, many stochastic processes have statistics that slowly vary with time. In such cases it is often appropriate to treat the problem as a succession of stationary problems for which the material developed in this chapter can be easily extended. Various "adaptive" schemes often use this approach.

The first section introduces the Fourier transform of the autocorrelation function, which is termed the power spectral density. We show that the power spectral density has a precise interpretation as a density for average power versus frequency. We then reexamine and study the white noise process and LSI systems with stationary random inputs, this time in the frequency domain. We will treat the corresponding concepts for random sequences and consider discrete-time simulation. We discuss the case of vector random processes and state equations as generators of Markov random processes. We also treat state equations for random sequences and introduce the vector Markov random sequence. Finally, we consider stationary band-limited and periodic processes.

Many of the results in this chapter require only the WSS condition in place of the relatively stronger stationarity condition. In fact, the chapter could almost have been entitled "WSS Processes and Sequences." The WSS condition is clearly all that is required if we only consider moments of order 2 or less. Otherwise, stationarity may be required.

9.1 POWER SPECTRAL DENSITY

In this section we will assume that the random processes of interest are all jointly stationary and of *second order*, that is,

$$E[|X(t)|^2] < \infty.$$

Some important properties of the auto- and cross-correlation functions of stationary processes are summarized as follows. They, of course, also hold for the respective covariance functions.

1. $|R_{XX}(\tau)| \leq R_{XX}(0)$,

which directly follows from $E[|X(t + \tau) - X(t)|^2] \geq 0$.

2. $|R_{XY}(\tau)| \leq \sqrt{R_{XX}(0)R_{YY}(0)}$,

which is derived using the Schwarz inequality. (See Sections 3.3 and 7.2.)

3. $R_{XX}(\tau) = R^*_{XX}(-\tau)$

(since $E[X(t + \tau)X^*(t)] = E[X(t)X^*(t - \tau)] = E^*[X(t - \tau)X^*(t)]$ for WSS random processes), which is called the *conjugate symmetry property*. In the special case of a real valued process, this property becomes that of even symmetry, that is,

3a. $R_{XX}(\tau) = R_{XX}(-\tau)$.

Another property of the autocorrelation function of a complex valued, stationary random process is that it be positive semidefinite,

4. For all $N > 0$, all $t_1 < t_2 < \ldots < t_N$ and all complex $a_1, a_2, \ldots, a_N$,

$$\sum_{k=1}^{N} \sum_{l=1}^{N} a_k a_l^* R_{XX}(t_k - t_l) \geq 0.$$

This was shown in Chapter 7 to be a necessary condition for an arbitrary function $g(t, s) = g(t - s)$ to be an autocorrelation function. We will show that this property is also a sufficient condition. In general, however, it is very difficult to directly check property 4. We next define the power spectral density.

Definition 9.1–1. Let $R_{XX}(\tau)$ be an autocorrelation function. Then we define the *power spectral density* $S_{XX}(\omega)$ to be its Fourier transform (if it exists), that is,

$$S_{XX}(\omega) \triangleq \int_{-\infty}^{+\infty} R_{XX}(\tau)e^{-j\omega\tau}\, d\tau. \tag{9.1–1}$$

We often abbreviate S_{XX} as S_X when no confusion can result.

Under quite general conditions one can define the inverse Fourier transform, and it equals $R_{XX}(\tau)$ at all points of continuity,

$$R_{XX}(\tau) = \frac{1}{2\pi}\int_{-\infty}^{+\infty} S_{XX}(\omega)e^{+j\omega\tau}\,d\omega \tag{9.1–2}$$

In operator notation we have,

$$S_{XX} = FT\{R_{XX}\}$$

and

$$R_{XX} = \text{IFT}\{S_{XX}\}$$

where FT and IFT stand for the respective Fourier transform operators.

The name power spectral density (psd) will be justified later. All that we have done thus far is define it as the Fourier transform of $R_{XX}(\tau)$. We can also define the Fourier transform of the cross-correlation function R_{XY} to obtain a frequency function called the *cross-power spectral density*,

$$S_{XY}(\omega) \triangleq \int_{-\infty}^{+\infty} R_{XY}(\tau)e^{-j\omega\tau}\,d\tau. \tag{9.1–3}$$

We will see later that the power spectral density or psd is real and everywhere nonnegative and in fact, as the name implies, has the interpretation of a density function for average power versus frequency. By contrast, the cross-power spectral density has no such interpretation and is generally complex valued.

We next list some properties of the psd S_{XX}:

1. $S_{XX}(\omega)$ is real valued since R_{XX} is conjugate symmetric.
2. If $X(t)$ is real valued, then $S_{XX}(\omega)$ is an even function since $R_{XX}(\tau)$ is real and even. Otherwise S_{XX} is not even.
3. $S_{XX}(\omega) \geq 0$ (to be shown in Section 9.3)

Additional properties of the psd are shown in the table below.

Table of Transforms

Random Process	Correlation Function	psd
$X(t)$	$R_X(\tau)$	$S_X(\omega)$
$aX(t)$	$\lvert a\rvert^2 R_X(\tau)$	$\lvert a\rvert^2 S_X(\omega)$
$X_1(t) + X_2(t)$ *with* X_1 and X_2 *orthogonal*	$R_{X_1}(\tau) + R_{X_2}(\tau)$	$S_{X_1}(\omega) + S_{X_2}(\omega)$
$X'(t)$	$-d^2R_X(\tau)/d\tau^2$	$\omega^2 S_X(\omega)$
$X^{(n)}(t)$	$(-1)^n d^{2n}R_X(\tau)/d\tau^{2n}$	$\omega^{2n}S_X(\omega)$
$X(t)\exp(+j\omega_0 t)$	$\exp(+j\omega_0\tau)R_X(\tau)$	$S_X(\omega - \omega_0)$
$X(t)\cos(\omega_0 t + \theta)$ with θ uniform on $[-\pi, +\pi]$	$R_X(\tau)\cos(\omega_0\tau)$	$\frac{1}{2}[S_X(\omega + \omega_0) + S_X(\omega - \omega_0)]$
$X(t) + b$ (if $\mu_X = 0$)	$R_X(\tau) + \lvert b\rvert^2$	$S_X(\omega) + 2\pi\lvert b\rvert^2\delta(\omega)$

One could continue with such a table, but it will suit our purposes to stop at this point. One comment is in order: We note the simplicity of the operations in the frequency domain. This suggests that for LSI systems and stationary random processes, we should solve for output correlation functions by first transforming the input correlation function into the frequency domain, carry out the indicated operations, and then transform back to the correlation domain. This would be completely analogous to the situation in deterministic linear system theory for shift-invariant systems.

Another comment would be that, if the interpretations of $S_X(\omega)$ as a density of average power is correct, then the constant or mean component has all its average power concentrated at $\omega = 0$ by the last entry in the table. Also by the next-to-last two entries in the table, modulation by the frequency ω_0 shifts the distribution of average power in frequency by $+\omega_0$. Both of these results seem quite reasonable.

An Interpretation of the psd. Given a WSS process $X(t)$, consider the finite support piece,

$$X_T(t) \triangleq X(t) I_{[-T,+T]}(t),$$

where $I_{[-T,+T]}$ is an indicator function, and $T > 0$. We can compute the Fourier transform of X_T by the m.-s. integral

$$FT\{X_T(t)\} = \int_{-T}^{+T} X(t) e^{-j\omega t}\, dt. \qquad \text{(m.s.)}$$

The magnitude squared of this random variable is

$$|FT\{X_T(t)\}|^2 = \int_{-T}^{+T}\int_{-T}^{+T} X(t_1) X^*(t_2) e^{-j\omega(t_1 - t_2)}\, dt_1\, dt_2.$$

Dividing by $2T$ and taking the expectation we get

$$\frac{1}{2T} E[|FT\{X_T(t)\}|^2] = \frac{1}{2T}\int_{-T}^{+T}\int_{-T}^{+T} R_X(t_1 - t_2) e^{-j\omega(t_1 - t_2)}\, dt_1\, dt_2,$$

which becomes (using the transformation of Figure 8.4–1)

$$\int_{-T}^{+T}\left[1 - \frac{|\tau|}{2T}\right] R_X(\tau) e^{-j\omega\tau}\, d\tau.$$

In the limit as $T \to +\infty$, this integral tends to Equation 9.1–1 for an integrable R_X, thus

$$S_X(\omega) = \lim_{T\to\infty} \frac{1}{2T} E[|FT\{X_T(t)\}|^2] \qquad (9.1\text{–}4)$$

so that S_X is real and nonnegative and is related to average power at frequency ω.†

† In practice the psd is often estimated by Equation 9.1–4 with the expectation replaced by a time-average (cf. Section 8.4).

We next look at two examples of the computation of power spectral densities corresponding to correlation functions we have seen earlier.

Example 9.1–1: Find the power spectral density for the following exponential autocorrelation function with parameter $\alpha > 0$,

$$R(\tau) = \exp(-\alpha\,|\tau|), \quad -\infty < \tau < +\infty.$$

$$S(\omega) = \int_{-\infty}^{+\infty} R(\tau)e^{-j\omega\tau}\,d\tau = \int_{-\infty}^{+\infty} e^{-\alpha|\tau|}e^{-j\omega\tau}\,d\tau$$

$$= \int_{-\infty}^{0} e^{(\alpha-j\omega)\tau}\,d\tau + \int_{0}^{\infty} e^{-(\alpha+j\omega)\tau}\,d\tau$$

$$= 2\alpha/[\alpha^2 + \omega^2], \quad -\infty < \omega < +\infty.$$

This function is plotted in Figure 9.1–1. We see that the peak value is at the origin and equal to $2/\alpha$. The "bandwidth" of the process is seen to be α on a 3dB basis if S is indeed a power density (to be shown). We note that, while there is a cusp at the origin of the correlation function R, there is no cusp in its spectral density S. In fact S is continuous and differentiable everywhere. (S will always be continuous if R is absolutely integrable.)

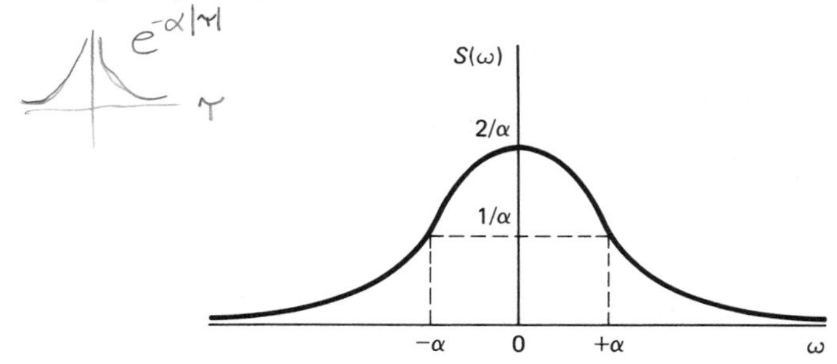

Figure 9.1–1 Plot of psd for exponential autocorrelation.

Example 9.1–2. Consider an autocorrelation function that is triangular in shape such that the correlation goes to zero at shift T,

$$R(\tau) = \max\left[1 - \frac{|\tau|}{T}, 0\right].$$

This function is plotted as Figure 9.1–2. If we realize that this triangle can be written as the convolution of two rectangular pulses, each of width T and height $1/\sqrt{T}$, then we see that the psd of the triangular correlation function is just the square of the

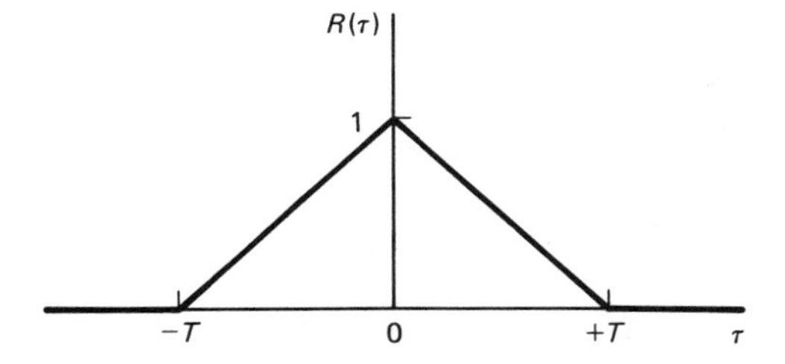

Figure 9.1–2 Triangular autocorrelation function.

Fourier transform of the rectangular pulse, that is, the sinc function. The transform of this rectangular pulse is

$$\sqrt{T}\frac{\sin(\omega T/2)}{(\omega T/2)},$$

and the power spectral density S is thus

$$S(\omega) = T\left(\frac{\sin(\omega T/2)}{\omega T/2}\right)^2. \tag{9.1–5}$$

As a check we note that $S(0)$ is just the area under the correlation function that in this case is easily seen to be T. Thus checking,

$$S(0) = \int_{-\infty}^{+\infty} R(\tau)\, d\tau = 2 \cdot \tfrac{1}{2} \cdot 1 \cdot T.$$

One question that arises at this point is how to generate these correlation functions. The correlation function in Example 9.1–1 was seen in Chapter 8 as the result of a white noise input to a first order differential equation. The correlation function in Example 9.1–2 can be generated with a running integral averager operating also on white noise. Consider

$$X(t) \triangleq \frac{1}{\sqrt{T}}\int_{t-T}^{t} W(\tau)\, d\tau,$$

with $W(t)$ a white noise with zero mean and correlation function $R_W(\tau) = \delta(\tau)$. Then $\mu_X(t) = 0$ and $E[X(t_1)X(t_2)]$ can be computed as

$$\begin{aligned} R_X(t_1, t_2) &= \frac{1}{T}\int_{t_1-T}^{t_1}\int_{t_2-T}^{t_2} R_W(s_1 - s_2)\, ds_1\, ds_2 \\ &= \frac{1}{T}\int_{t_1-T}^{t_1}\left[\int_{t_2-T}^{t_2} \delta(s_2 - s_1)\, ds_2\right] ds_1. \end{aligned}$$

Now defining the inner integral as

$$g_{t_2}(s_1) \triangleq \int_{t_2-T}^{t_2} \delta(s_2 - s_1)\, ds_2 = \begin{cases} 1, & t_2 - T \le s_1 \le t_2 \\ 0, & \text{else} \end{cases}$$

which as a function of s_1 looks as shown in Figure 9.1–3, so

$$\begin{aligned} R_X(t_1, t_2) &= \frac{1}{T}\int_{t_1-T}^{t_1} g_{t_2}(s_1)\, ds_1 \\ &= \max\left[1 - \frac{|t_1 - t_2|}{T}, 0\right]. \end{aligned}$$

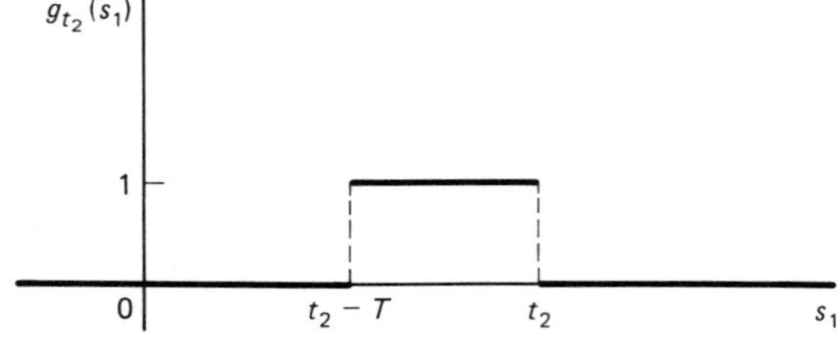

Figure 9.1–3 Plot of equation in s_1 for $t_2 > T$.

9.2 MORE ON WHITE NOISE

The correlation function of white noise is an impulse (Equation 8.1–8), so its psd is a constant

$$S(\omega) = \sigma^2, \qquad -\infty < \omega < +\infty.$$

The name white noise thus arises out of the fact that the power spectral density is constant at all frequencies just as in white light, which contains all wavelengths in equal amounts.† Here we look at the white noise process as a mean-square limit approached by a sequence of second-order processes. To this end consider an independent increment process (cf. Definition 7.2–1) with zero mean such as the Wiener process ($R_X(t_1, t_2) = \sigma^2 \min(t_1, t_2)$) or a centered Poisson process, that is, $N_c(t) = N(t) - \lambda t$, with correlation $R_{N_c}(t_1, t_2) = \lambda \min(t_1, t_2)$. Actually we only need uncorrelated increments here; thus $X(t)$ will be taken to have just uncorrelated increments. For such processes we have by Equation 7.2–15,

$$E[(X(t + \Delta) - X(t))^2] = \alpha \Delta,\ddagger$$

where the variance parameter is α.

Thus upon letting $X_\Delta(t)$ denote the first-order difference divided by Δ,

$$X_\Delta(t) \triangleq [X(t + \Delta) - X(t)]/\Delta,$$

we have

$$E[X_\Delta^2(t)] = \alpha/\Delta$$

and

$$E[X_\Delta(t_1)X_\Delta(t_2)] = 0 \quad \text{for} \quad (t_1, t_1 + \Delta] \cap (t_2, t_2 + \Delta] = \phi.$$

If we consider t_1 and t_2 closer than Δ, we can do the following calculation, which shows that the resulting correlation function is triangular, just as in Example 9.1–2. Since $X(t_1 + \Delta) - X(t_1)$ is $N(0, \Delta)$, taking $t_1 < t_2$ and shifting t_1 to 0, the expectation becomes

$$\frac{1}{\Delta^2} E[X(\Delta)(X(t_2 - t_1 + \Delta) - X(t_2 - t_1))]$$

$$= \frac{1}{\Delta^2} E[X(\Delta)(X(\Delta) - X(t_2 - t_1))] \quad \text{since } (\Delta, t_2 - t_1 + \Delta] \cap (0, \Delta] = \phi,$$

$$= \frac{1}{\Delta^2}[\alpha\Delta - \alpha(t_2 - t_1)] = \frac{\alpha}{\Delta}[1 - (t_2 - t_1)/\Delta].$$

Thus the process generated by the first-order difference is wide-sense

† Obviously an idealization. Modern physics tells us that the power density eventually goes to zero as $\omega \to \infty$.

‡ As an aside (cf. introduction to Chapter 8), we note that this equation precludes the existence of a *sample-function derivative* for the Wiener process. In the limit as $\Delta \to dt$, we get $E[(dX)^2] = \alpha dt$ or $E[(dX/dt)^2] \sim \alpha(dt)^{-1} = \infty$! One can consider *sample-function differentials* though, that is, $dX(t)$. A thorough treatment of this approach is contained in Reference 9–1.

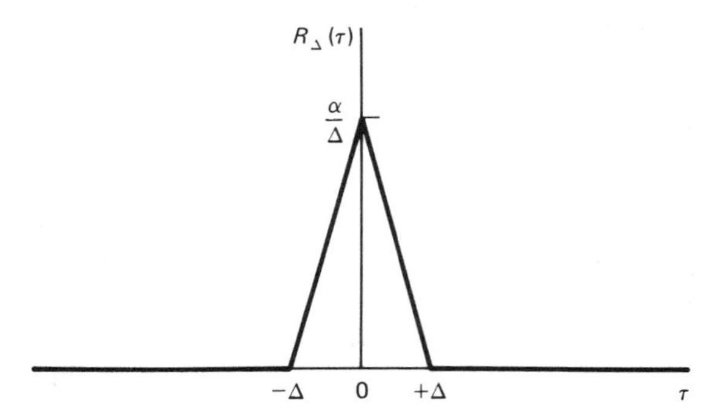

Figure 9.2–1 Correlation function of $X_\Delta(t)$.

stationary (the mean is zero) and has correlation function $R_\Delta(\tau)$ given as

$$R_\Delta(\tau) = \frac{\alpha}{\Delta}\max\left[1 - \frac{|\tau|}{\Delta}, 0\right].$$

We note from Figure 9.2–1 that as Δ goes to zero this correlation function tends to a delta function.

Since we just computed the Fourier transform of a triangular function in Example 9.1–2, we can write the psd by inspection as

$$S_\Delta(\omega) = \alpha\left(\frac{\sin(\omega\Delta/2)}{\omega\Delta/2}\right)^2.$$

The psd will be approximately flat for arguments out to $\pi/6$, which is $|\omega| = \pi/(3\Delta)$. As $\Delta \to 0$, S_Δ approaches the constant α everywhere. Thus as $\Delta \to 0$, $X_\Delta(t)$ converges (m.s.) to white noise, the m.-s. derivative of an uncorrelated increments process,

$$\begin{aligned} R_{\dot{X}}(t_1, t_2) &= \frac{\partial^2}{\partial t_1\,\partial t_2}[\sigma^2 \min(t_1, t_2)] \\ &= \frac{\partial}{\partial t_1}[\sigma^2 u(t_1 - t_2)] \\ &= \sigma^2\delta(t_1 - t_2). \end{aligned}$$

If one has a system that is continuous in its response to stimuli, then we say that the *system is continuous*; that is, the system operator is a continuous operator. This would mean, for example, that the output would change only slightly if the input changed slightly. A stable differential or difference equation is an example of such a continuous operator. We will see that for linear shift-invariant systems that are described by system functions, the response to the random process $X_\Delta(t)$ will change only slightly when Δ changes, if Δ is small and if the systems are lowpass in the sense that the system function tends to zero as $|\omega| \to +\infty$. Thus the white noise will be seen as a convenient artifice for more easily constructing this limiting output. (See Problem 9.5.)

9.3 STATIONARY PROCESS AND LINEAR SYSTEMS

If we input the stationary random process $X(t)$ to a linear shift-invariant(LSI) system with impulse response $h(t)$, then the output random process can be expressed as the m.-s. convolution integral,

$$Y(t) = \int_{-\infty}^{+\infty} h(\tau)X(t - \tau)\, d\tau, \tag{9.3–1}$$

when this integral exists (see Example 8.2–2). Computing the mean of the output process $Y(t)$, we get

$$E[Y(t)] = \int_{-\infty}^{+\infty} h(\tau)E[X(t - \tau)]\, d\tau \quad \text{by Theorem 8.1–5,}$$

$$= \int_{-\infty}^{+\infty} h(\tau)\mu_X\, d\tau = \mu_X \int_{-\infty}^{+\infty} h(\tau)\, d\tau,$$

$$= \mu_X H(0). \tag{9.3–2}$$

We thus see that the mean of the output is constant and equals the mean of the input times the system function evaluated at $\omega = 0$, that is, the d.c. gain of the system. If we compute the cross-correlation function between the input process and the output process we find,

$$R_{YX}(\tau) = E[Y(t + \tau)X^*(t)]$$

$$= E[Y(t)X^*(t - \tau)] \quad \text{by substituting } t \text{ for } t - \tau,$$

$$= \int_{-\infty}^{+\infty} h(\alpha)E[X(t - \alpha)X^*(t - \tau)]\, d\alpha$$

by Theorem 8.1–5, property 2

$$= \int_{-\infty}^{+\infty} h(\alpha)R_{XX}(\tau - \alpha)\, d\alpha$$

which can be rewritten as

$$R_{YX}(\tau) = h(\tau) * R_{XX}(\tau). \tag{9.3–3}$$

Thus the cross-correlation R_{YX} equals h convolved with the autocorrelation R_{XX}. This fact can be used to identify unknown systems (see Problem 9.2). If we take Fourier transforms of both sides of Equation 9.3–3 we obtain the relation,

$$S_{YX}(\omega) = H(\omega)S_{XX}(\omega). \tag{9.3–4}$$

Since S_{YX} is a frequency-domain representation of the cross-correlation function R_{YX}, Equation 9.3–4 tells us that $Y(t)$ and $X(t)$ will have high cross-correlation at those frequencies ω where $H(\omega)$ and $S_{XX}(\omega)$ are large.

Proceeding to calculate the autocorrelation function of the output, we write

$$R_{YY}(\tau) = E[Y(t+\tau)Y^*(t)] = E\left[Y(t+\tau)\int_{-\infty}^{+\infty} X^*(t-\alpha)h^*(\alpha)\,d\alpha\right]$$

$$= \int_{-\infty}^{+\infty} E[Y(t+\tau)X^*(t-\alpha)]h^*(\alpha)\,d\alpha$$

by Theorem 8.1–5, property 2

$$= \int_{-\infty}^{+\infty} R_{YX}(\tau+\alpha)h^*(\alpha)\,d\alpha.$$

Letting $\beta \triangleq -\alpha$, we obtain

$$= \int_{-\infty}^{+\infty} R_{YX}(\tau-\beta)h^*(-\beta)\,d\beta$$

so

$$R_{YY}(\tau) = R_{YX}(\tau) * h^*(-\tau), \tag{9.3–5}$$

which upon Fourier transformation becomes

$$S_{YY}(\omega) = S_{YX}(\omega)H^*(\omega). \tag{9.3–6}$$

Combining the preceding results, we obtain a fundamental equation relating the autocorrelation function of the output to the autocorrelation function of the input

$$R_{YY}(\tau) = h(\tau) * R_{XX}(\tau) * h^*(-\tau), \tag{9.3–7}$$

which becomes, in the spectral domain,

$$S_{YY}(\omega) = |H(\omega)|^2\, S_{XX}(\omega). \tag{9.3–8}$$

These two equations are among the most important in the theory of stationary random processes. In particular, Equation 9.3–8 shows how the average power in the output process is composed solely of the average input power at that frequency multiplied by $|H(\omega)|^2$ the power gain of the LSI system.

The following comment on Equation 9.3–3 through Equation 9.3–8 is related to keeping track of the conjugates and minus signs. Notice that the conjugate and negative argument on the impulse response, which becomes simply a conjugate in the frequency domain, arises in connection with the second factor in the correlation. The $h(\tau)$ without the conjugate or negative time argument comes from the linear operation implied by the first subscript, that is, the first factor in the correlation.

With reference to Equation 9.3–4 we see that the cross-spectral density function can be complex and hence has no positivity or conjugate symmetry properties, since those that S_{XX} has will be lost upon multiplication with an arbitrary, generally complex H. On the other hand, as shown in Equation 9.3–8, the psd of the output will share the real and nonnegative aspects of the psd of the input, since multiplication with $|H|^2$ will not change these properties.

We are now in a position to show that the psd S has a precise interpretation as a density for average power versus frequency. We will show directly that $S(\omega) \geq 0$ for all ω and that the average power in the frequency band $(\omega_1, \omega_2]$ is given by the integral of S over that band.

Theorem 9.3–1. Let $X(t)$ be a stationary, second-order, random process with correlation function $R(\tau)$ and power spectral density $S(\omega)$. Then $S(\omega) \geq 0$ and for all $\omega_2 \geq \omega_1$,

$$\frac{1}{2\pi}\int_{\omega_1}^{\omega_2} S(\omega)\, d\omega$$

is the average power in the frequency band $(\omega_1, \omega_2]$.

Proof. Let $\omega_2 \geq \omega_1$ both be real numbers. Define a filter transfer function as follows:

$$H(\omega) \triangleq \begin{cases} 1 & \text{if } \omega \in (\omega_1 \omega_2] \\ 0 & \text{elsewhere} \end{cases}$$

and note that it passes signals only in the band $(\omega_1, \omega_2]$. If $X(t)$ is input to this filter, the psd of the output Y is (by Equation 9.3–8)

$$S_Y(\omega) = \begin{cases} S_X(\omega), & \omega \in (\omega_1 \omega_2] \\ 0, & \text{elsewhere.} \end{cases}$$

Now the output power in $Y(t)$ has average value $R_Y(0)$,

$$R_Y(0) = \frac{1}{2\pi}\int_{-\infty}^{+\infty} S_Y(\omega)\, d\omega = \frac{1}{2\pi}\int_{\omega_1}^{\omega_2} S_X(\omega)\, d\omega \geq 0,$$

and this holds for all $\omega_2 > \omega_1$. So by choosing $\omega_2 \simeq \omega_1$ we can conclude that $S_X(\omega) \geq 0$ for all ω and that the function S_X thus has the interpretation of a power density in the sense that if one integrates this function across a frequency band, one gets the average power in that band. ■

We have seen earlier that the conditions that a function must meet to be a valid correlation or covariance function are rather strong. In fact, we have seen that the function must be positive semidefinite, although we have not in fact shown that this condition is sufficient. It turns out that one more advantage of working in the frequency domain is the ease with which we can specify when a given frequency function qualifies as a power spectral density. The function simply must be real and nonnegative, that is, $S(\omega) \geq 0$. We can see this for a given function $F(\omega) \geq 0$ by taking a filter with transfer function $H(\omega) = \sqrt{F(\omega)}$ and letting the input be white noise with $S_W = 1$. Then by Equation 9.3–8 the output psd is $S_X(\omega) = F(\omega)$, thus showing that F is a valid psd. If the random process is real, as it frequently is, then we also need $F(\omega)$ to be an even function to satisfy property 2 of the psd (cf. page 339.) All this can be formalized as follows.

Theorem 9.3–2. Let $F(\omega)$ be an integrable function that is real and nonnegative; that is, $F(\omega) \geq 0$ for all ω. Then there exists a stationary random process with power spectral density $S(\omega) = F(\omega)$. In particular, if $F(\omega)$ is even then the random process is real-valued.

We now see that the test for a valid spectral density function is much easier than the condition of positive semidefiniteness for the correlation function. In fact, it is relatively easy to show that the positive semidefinite condition on a function is equivalent to the nonnegativity of its transform and hence that positive semidefiniteness is a sufficient condition for a function to be a valid correlation or covariance function. First, by Theorem 9.3–2 we know that the positive semidefinite condition is implied by the nonnegativity of $S(\omega)$. To show equivalence, what remains is to show that the positive semidefinite condition on a function $f(\tau)$ implies that its Fourier transform $F(\omega)$ is nonnegative. We proceed as follows: Since $f(\tau)$ is positive semidefinite we have,

$$\sum_{n=1}^{N} \sum_{m=1}^{N} a_n a_m^* f(\tau_n - \tau_m) \geq 0.$$

Also since

$$f(\tau) = \frac{1}{2\pi} \int_{-\infty}^{+\infty} F(\omega) e^{+j\omega\tau} \, d\omega,$$

we have

$$\frac{1}{2\pi} \sum_n \sum_m \left(a_n a_m^* \int_{-\infty}^{+\infty} F(\omega) e^{+j\omega(\tau_n - \tau_m)} \right) d\omega \geq 0.$$

which can be rewritten as

$$\frac{1}{2\pi} \int_{-\infty}^{+\infty} F(\omega) \left[\sum_n \sum_m a_n a_m^* e^{+j\omega(\tau_n - \tau_m)} \right] d\omega$$

$$= \frac{1}{2\pi} \int_{-\infty}^{+\infty} F(\omega) \left| \sum_{n=1}^{N} a_n e^{+j\omega\tau_n} \right|^2 d\omega \geq 0,$$

where we recognize the term inside the magnitude square sign as a so-called transversal or tapped delay-line filter. Thus by choosing N large enough, with the τ_n equally spaced, we can select the a_n's to arbitrarily approximate any ideal filter transfer function $H(\omega)$. Then by choosing H to be very narrow bandpass filters centered at each value of ω, we can eventually conclude that $F(\omega) \geq 0$ for all ω, $-\infty < \omega < +\infty$. We have thereby established the following theorem.

Theorem 9.3–3. A necessary and sufficient condition for $f(\tau)$ to be a correlation function is that it be positive semidefinite.†

Incidentally, there is an analogy here for probability density functions, which can be regarded as the Fourier transforms of their characteristic functions.

† Additionally, if $f(\tau)$ is even, then the associated random process is real.

As we know, nonnegativity is the sufficient condition for a function to be a valid probability density function (assuming that it is normalized to integrate to one); thus the probability density is analogous to the power spectral density; and in fact one can consider a spectral distribution function [9–2] analogous to the probability distribution function. Thus the characteristic function and the correlation function are analogous and so both must be positive semidefinite to be valid for their respective roles. Also for the characteristic function the normalization of the density to integrate to one imposes the condition $\Phi(0) = 1$, which is easily met by scaling an arbitrary positive semidefinite function that is not identically zero.

Stochastic Differential Equations Revisited. We shall now reexamine the stochastic differential equation, this time with a stationary or at least WSS input, and also with the linear constant coefficient differential equation (LCCDE) holding for all time. We assume that the equation is stable in the bounded-input bounded-output (BIBO) sense, so that the resulting output process is also stationary (or WSS if that is the condition on the input process).

Thus consider the following general LCCDE, which we interpret in the mean-square sense,

$$a_N Y^{(N)}(t) + a_{N-1} Y^{(N-1)}(t) + \ldots + a_0 Y(t)$$
$$= b_M X^{(M)}(t) + b_{M-1} X^{(M-1)}(t) + \ldots + b_0 X(t), \qquad -\infty < t < +\infty.$$

This represents the relationship between output $Y(t)$ and input $X(t)$ in a linear system with frequency response

$$H(\omega) = B(\omega)/A(\omega),$$

where

$$B(\omega) \triangleq \sum_{m=0}^{M} b_m (j\omega)^m$$

and

$$A(\omega) \triangleq \sum_{n=0}^{N} a_n (j\omega)^n,$$

which is a *rational function* with *numerator polynomial B* and *denominator polynomial A*. Because the system is stable, we can apply the results of the previous section to obtain

$$\mu_Y = \mu_x H(0) \qquad (a_0 \neq 0),$$
$$S_{YX}(\omega) = H(\omega) S_{XX}(\omega),$$

and

$$S_{YY}(\omega) = |H(\omega)|^2 S_{XX}(\omega),$$

where

$$H(0) = b_0/a_0 \quad \text{and} \quad |H(\omega)|^2 = |B(\omega)|^2/|A(\omega)|^2.$$

So

$$\mu_Y = (b_0/a_0)\mu_X \quad \text{and} \quad S_{YY}(\omega) = (|B(\omega)|^2/|A(\omega)|^2) S_{XX}(\omega).$$

This frequency-domain analysis method is generally preferable to the time-domain approach but is restricted to the case where both the input and output processes are at least WSS. After one obtains the various spectral densities, then one can use the IFT to obtain the correlation and covariance functions if they are desired. The calculation of the required IFT's is often easier if viewed as an inverse *Laplace transform*. The Laplace transform of Equation 9.3–3 is (cf. Section 9.APP).

$$S_{YX}(s) = H(s)S_{XX}(s) \tag{9.3–9}$$

while the Laplace transform of Equation 9.3–7 is written

$$S_{YY}(s) = H(s)H(-s)S_{XX}(s) \tag{9.3–10}$$

in light of $h^*(-\tau) \leftrightarrow H(-s)$. Recalling the definition of the Laplace transform [9–4]

$$S_{..}(s) \triangleq \int_{-\infty}^{+\infty} R_{..}(\tau)e^{-s\tau}\,d\tau,$$

we note that such functions of the complex variable s, that is, $S_{..}(s)$, may be obtained from the functions of the real variable ω, that is the Fourier transform $S_{..}(\omega)$, by a two-step procedure. First set

$$S_{..}(j\omega) \triangleq S_{..}(\omega)$$

and then replace $j\omega$ by s. Two examples follow.

Example 9.3–1: Consider the first order m.-s. differential equation

$$Y'(t) + \alpha Y(t) = X(t), \qquad \alpha > 0,$$

with stationary input $X(t)$ with mean $\mu_X = 0$ and covariance function $K_{XX}(\tau) = \delta(\tau)$. The system function is easily seen to be

$$H(\omega) = \frac{1}{\alpha + j\omega},$$

and the psd of the input process is

$$S_{XX}(\omega) = 1,$$

so we have the following cross- and output-power spectral densities:

$$S_{YX}(\omega) = H(\omega)S_{XX}(\omega) = \frac{1}{\alpha + j\omega},$$

$$S_{YY}(\omega) = |H(\omega)|^2\, S_{XX}(\omega) = \frac{1}{|\alpha + j\omega|^2} = \frac{1}{\alpha^2 + \omega^2}.$$

Using the residue method of Section 9.APP, one can then directly obtain the following output correlation function

$$R_Y(\tau) = \frac{1}{2\alpha}\exp(-\alpha\,|\tau|), \qquad -\infty < \tau < +\infty,$$

which is also the output covariance function since $\mu_Y = 0$. By the above equation for S_{YX} we also obtain the cross-correlation function $R_{YX}(\tau) = \exp(-\alpha\tau)u(\tau)$.

In Example 9.3–1 it is interesting that $R_{YX}(\tau)$ is 0 for $\tau < 0$. This means that the output Y is orthogonal to all future values of the input X, which is a white noise. This occurs because of two reasons: The system is causal and the input is a white noise. The system causality requires that the output not depend *directly* on (that is, not be a function of) future inputs but only depend directly on present and past inputs. The whiteness of the input X guarantees that the past and present inputs will be uncorrelated with future inputs. Combining both conditions we see that there will be no cross-correlation between the present output and the future inputs. If we assume additionally that the input is Gaussian, then the input process is an independent process and the output becomes independent of all future inputs. Then we can say that the causality of the system prevents the *direct dependence* of the present output on future inputs, and the independent process input prevents any *indirect dependence.* These concepts are important to the theory of Markov processes in Section 9.5.

Example 9.3–2: Consider the following second order LCCDE

$$\frac{d^2Y(t)}{dt^2} + 3\frac{dY(t)}{dt} + 2Y(t) = 5X(t),$$

again with white noise input as in the previous example. Here the system function is

$$H(\omega) = \frac{5}{(j\omega)^2 + 3j\omega + 2} = \frac{5}{(2 - \omega^2) + j3\omega},$$

thus analogously to Example 9.3–1 the output psd becomes

$$S_Y(\omega) = \frac{25}{(2 - \omega^2)^2 + (3\omega)^2} = \frac{25}{\omega^4 + 5\omega^2 + 4}.$$

Applying the residue method of Section 9.APP to evaluate the IFT, we define the function of a complex variable $S_Y(j\omega) \triangleq S_Y(\omega)$ and rewrite the right-hand side in terms of the complex variable $j\omega$ to obtain

$$S_Y(j\omega) = \frac{25}{(j\omega)^4 - 5(j\omega)^2 + 4}.$$

Substituting $s = j\omega$, we get

$$S_Y(s) = \frac{25}{s^4 - 5s^2 + 4},$$

which factors as

$$= \frac{5}{(s + 2)(s + 1)} \cdot \frac{5}{(-s + 2)(-s + 1)} = H(s)H(-s),$$

where $H(s)$ is the Laplace transform system function. Then the inverse Laplace transform yields the output correlation function

$$R_{YY}(\tau) = 25[\tfrac{1}{6}\exp(-|\tau|) - \tfrac{1}{12}\exp(-2|\tau|)], \qquad -\infty < \tau < +\infty.$$

9.4 WSS RANDOM SEQUENCES

We can also consider WSS and stationary random sequences as defined in Chapter 6. From Equation 6.3–6 specialized to the LSI convolution case, we have

$$R_Y[m] = h[m] * R_X[m] * h^*[-m]. \tag{9.4–1}$$

Defining the power spectral density as the FT of the discrete-time correlation function, we obtain

$$S_Y[\omega] = |H[\omega]|^2 \, S_X[\omega], \tag{9.4–2}$$

which looks similar to Equation 9.3–8. However, in Equation 9.4–2 the various frequency-domain quantities are discrete-time Fourier transforms (cf. Def. 6.2–3). If the LSI system is a linear constant coefficient difference equation (LCCDE) as in Equation 6.2–2, then the Z-transform system function $H[z]$ is

$$H[z] = B[z]/A[z],$$

where the polynomials B and A have the coefficients of the right side and left side of the LCCDE (Equation 6.2–2) respectively. If this linear system is excited by a white random sequence $W[n]$, then the output psd is given as

$$S_Y[\omega] = \frac{|B[\omega]|^2}{|A[\omega]|^2}\sigma_W^2. \tag{9.4–3}$$

We can write this as a Z-transform when the coefficients of the LCCDE are real as

$$S_Y[z] = H[z]H[z^{-1}]\sigma_W^2, \tag{9.4–4}$$

since $H^*[e^{j\omega}] \leftrightarrow h[-m] \leftrightarrow H[z^{-1}]$ in agreement with Equation 9.4–1. Thus $S_Y[z]$ is rational with poles and zeros distributed in the z-plane in reciprocal pairs. In going from the functions of a *real variable* in Equation 9.4–2 to the functions of a *complex variable* in Equation 9.4–4, we first establish $S_{\cdot}[z]$ on the unit circle by writing

$$S_{\cdot}[e^{j\omega}] \triangleq S_{\cdot}[\omega].$$

Then we substitute z for $e^{j\omega}$.

Given any rational $S_X[z]$, that is, one with a finite number of poles and zeros in the finite z-plane, then one can find such a *spectral factorization* as Equation 9.4–3 by defining $H[z]$ to have all the poles and zeros that are inside the unit circle, $\{|z| < 1\}$, and then $H[z^{-1}]$ will necessarily have all the poles and zeros outside the unit circle, $\{|z| > 1\}$.

If a zero occurs on the unit circle, then it must be of even order, since otherwise one can easily show that $S_X[e^{j\omega}]$ must go through zero and hence be negative in its vicinity. Thus we can assign half the zeros to $H[z]$ and the other half to $H[z^{-1}]$. Since $H[z]$ contains only poles inside the unit circle, it will be BIBO stable [9–3]. Except in the case of a zero on the unit circle, its inverse will also be stable. (This last point is important in the development of the Wiener filter in Chapter 10.) The other factor $H[z^{-1}]$ has all its poles outside the unit circle, so it is stable in the anticausal sense. Denoting the largest pole magnitude inside the unit circle by $p_{\max}$, we thus have that $S_X[z]$ is analytic, that is, free of singularities, in the annular region of convergence, $p_{\max} < |z| < 1/p_{\max}$.

Discrete-Time Simulation. We thus obtain the system function H of an LCCDE that, when driven by a white noise $W[n]$, will generate the random sequence $X[n]$ with *psd* $S_X[\omega]$. This can be the basis for a discrete-time simulation on a computer. The white random sequence $W[n]$ is easily obtained by using the computer's random number generator. Then one specifies appropriate initial conditions and proceeds to recursively calculate $X[n]$ using the LCCDE of the system function $H[z]$.

To achieve a Gaussian distribution for X, one could transform the output of the random number generator to achieve a Gaussian distribution for W, which would carry across to X. Another method that is often used is to average 6 to 10 calls to the random number generator to obtain an approximate Gaussian distribution for W via the Central Limit Theorem. When simulating a non-Gaussian r.v., the distribution for X and W is not the same. Thus the preceding method will not work. One approach is to use the LCCDE to generate samples of $W[n]$ from some real data and then use the resulting distribution for $W[n]$ in the simulation.

Example 9.4–1: In order to simulate a zero-mean Markov random sequence with average power $R_X[0] = \sigma^2$ correlation at shift one $R_X[1] = \rho\sigma^2$, we want to find the parameters of a first-order stochastic difference equation to achieve these values. Thus consider

$$X[n] = aX[n-1] + bW[n], \tag{9.4–5}$$

where $W[n]$ is a zero-mean white noise source with unit power. Computing the impulse response, we get

$$h[n] = ba^n u[n]$$

and the corresponding system function

$$H[z] = \frac{b}{1 - az^{-1}}.$$

Since the mean is zero we calculate the covariance of the output of Equation 9.4–5

$$\begin{aligned}
K_X[m] &= h[m] * h[-m] * K_W[m] \\
&= h[m] * h[-m] \\
&= b^2(a^m u[m]) * (a^{-m} u[-m]) \\
&= b^2 \sum_{k=-\infty}^{+\infty} a^k u[k] a^{m+k} u[m+k] \\
&= b^2 a^m \sum_{k=\max(0,-m)}^{+\infty} a^{2k} \\
&= \frac{b^2}{1-a^2} a^{|m|}, \qquad -\infty < m < +\infty.
\end{aligned}$$

From the specifications at $m = 0$ and $m = 1$, we need

$$K_X[0] = \sigma^2 = b^2/(1 - a^2),$$
$$K_X[1] = \rho\sigma^2 = a \cdot b^2/(1 - a^2),$$

Thus

$$a = \rho \text{ and } b^2 = \sigma^2(1 - \rho^2).$$

To compute the resulting psd we use Equation 9.4–2 to get

$$S_X[\omega] = \frac{b^2}{|1 - ae^{-j\omega}|^2} = \frac{\sigma^2(1 - \rho^2)}{1 - 2\rho \cos \omega + \rho^2},$$

which can be compared to the psd of Example 9.APP–2.

9.5 VECTOR PROCESSES AND STATE EQUATIONS

In this section we will generalize some of the results of Section 9.3 to the important class of vector random processes. This will lead into a brief discussion of state equations and vector Markov processes. Vector random processes occur in two-channel systems that are used in communications to model the in-phase and quadrature components of bandpass signals. Vector processes are used extensively in control systems to model plants with many inputs and outputs. Also, vector models are created artificially from high-order scalar models in order to employ the useful concept of *state* in control theory.

Let $X_1(t)$ and $X_2(t)$ be two jointly stationary random processes that are input to the systems H_1 and H_2 respectively. Call the outputs Y_1 and Y_2 as shown in Figure 9.5–1.

From earlier discussions we know how to calculate $R_{X_1Y_1}$, $R_{X_2Y_2}$, $R_{Y_1Y_1}$, $R_{Y_2Y_2}$. We now look at how to calculate the correlations across the systems, that is, $R_{X_1Y_2}$, $R_{X_2Y_1}$ and $R_{Y_1Y_2}$. Given $R_{X_1X_2}$ we first calculate,

$$\begin{aligned} R_{X_1Y_2}(\tau) &= E[X_1(t + \tau)Y_2^*(t)] \\ &= \int_{-\infty}^{+\infty} E[X_1(t + \tau)X_2^*(t - \beta)]h_2^*(\beta)\, d\beta \\ &= \int_{-\infty}^{+\infty} R_{X_1X_2}(\tau + \beta)h_2^*(\beta)\, d\beta \\ &= \int_{-\infty}^{+\infty} R_{X_1X_2}(\tau - \beta')h_2^*(-\beta')\, d\beta', \qquad (\beta' = -\beta), \end{aligned}$$

so

$$R_{X_1Y_2}(\tau) = R_{X_1X_2}(\tau) * h_2^*(-\tau),$$

and by symmetry

$$R_{X_2Y_1}(\tau) = R_{X_2X_1}(\tau) * h_1^*(-\tau).$$

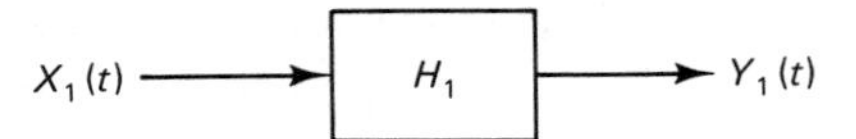

Figure 9.5–1 A two-channel system, uncoupled.

The cross correlation at the output is

$$R_{Y_1Y_2}(\tau) = h_1(\tau) * R_{X_1X_2}(\tau) * h_2^*(-\tau).$$

Expressing these results in the spectral domain, we have

$$S_{X_1Y_2}(\omega) = S_{X_1X_2}(\omega)H_2^*(\omega)$$

and

$$S_{Y_1Y_2}(\omega) = H_1(\omega)H_2^*(\omega)S_{X_1X_2}(\omega).$$

In passing, we note the following important fact: If the supports† of the two system functions H_1 and H_2 do not overlap, then Y_1 and Y_2 are orthogonal random processes independent of the correlation of the input processes. We can generalize the above to a two-channel system with internal coupling as seen in Figure 9.5–2. Here two additional system functions have been added to cross-couple the inputs and outputs. They are denoted by H_{12} and H_{21}.

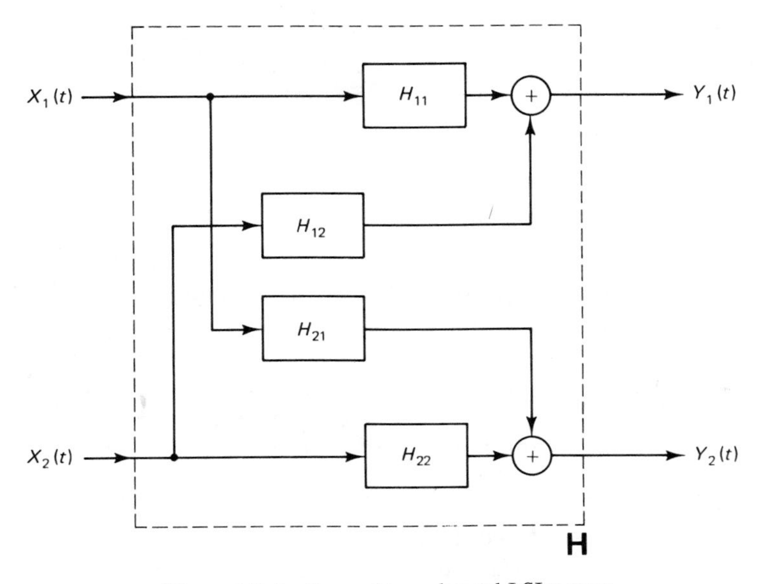

Figure 9.5–2 General two-channel LSI system.

† We recall that the support of a function g is defined as

$$\text{supp}(g) \triangleq \{x \mid g(x) \neq 0\}.$$

This case is best treated with vector notation; thus we define

$$\mathbf{X}(t) \triangleq [X_1(t), X_2(t)]^T, \qquad \mathbf{Y}(t) \triangleq [Y_1(t), Y_2(t)]^T,$$

and

$$\mathbf{h}(t) \triangleq \begin{bmatrix} h_{11}(t) & h_{12}(t) \\ h_{21}(t) & h_{22}(t) \end{bmatrix},$$

where h_{ij} is the impulse response of the subsystem with system function H_{ij}. We then have

$$\mathbf{Y}(t) = \mathbf{h}(t) * \mathbf{X}(t), \tag{9.5–1}$$

where the vector convolution is defined by

$$(\mathbf{h}(t) * \mathbf{X}(t))_i \triangleq \sum_{j=1}^{N} h_{ij}(t) * X_j(t).$$

If we define the following relevant input and output correlation matrices

$$\mathbf{R}_{\mathbf{XX}}(\tau) \triangleq \begin{bmatrix} R_{X_1X_1}(\tau) & R_{X_1X_2}(\tau) \\ R_{X_2X_1}(\tau) & R_{X_2X_2}(\tau) \end{bmatrix} \tag{9.5–2}$$

$$\mathbf{R}_{\mathbf{YY}}(\tau) \triangleq \begin{bmatrix} R_{Y_1Y_1}(\tau) & R_{Y_1Y_2}(\tau) \\ R_{Y_2Y_1}(\tau) & R_{Y_2Y_2}(\tau) \end{bmatrix},$$

one can show that (Problem 9.10)

$$\mathbf{R}_{\mathbf{YY}}(\tau) = \mathbf{h}(\tau) * R_{XX}(\tau) * \mathbf{h}^{\dagger}(-\tau), \tag{9.5–3}$$

where the dagger † indicates the Hermitian (or conjugate) transpose.

Taking the matrix Fourier transformation, we obtain

$$\mathbf{S}_{\mathbf{YY}}(\omega) = \mathbf{H}(\omega)\mathbf{S}_{\mathbf{XX}}(\omega)\mathbf{H}^{\dagger}(\omega) \tag{9.5–4}$$

with

$$\mathbf{H}(\omega) = FT\{\mathbf{h}(t)\},$$

and

$$\mathbf{S}(\omega) = FT\{\mathbf{R}(\tau)\},$$

where this notation indicates a term-by-term Fourier transform. This multichannel generalization clearly extends to the M input and N output case by just enlarging the matrix dimensions accordingly.

State Equations. As shown in Problem 9.9, it is possible to rewrite an Nth order LCCDE in the form of a first-order vector differential equation where the dimension of the output vector is equal to N,

$$\dot{\mathbf{Y}}(t) = \mathbf{A}\mathbf{Y}(t) + \mathbf{B}\mathbf{X}(t), \qquad -\infty < t < +\infty. \tag{9.5–5}$$

This is just a multi-channel system as seen in Equation 9.5–1. We can take the vector Fourier transform and calculate the system function

$$\mathbf{H}(\omega) = (j\omega\mathbf{I} - \mathbf{A})^{-1}\mathbf{B} \tag{9.5–6}$$

to specify the LSI operation in the frequency domain. Here $\mathbf{I}$ is the identity matrix. Alternately, we can express the operation in terms of a matrix convolution

$$\mathbf{Y}(t) = \mathbf{h}(t) * \mathbf{X}(t),$$

where we assume the multi-channel system is stable; that is, all the impulse responses h_{ij} are BIBO stable. The solution proceeds much the same as in the scalar case for the first-order equation; in fact, it can be shown that

$$\mathbf{h}(t) = (\exp \mathbf{A}t)u(t). \tag{9.5–7}$$

The matrix exponential function $\exp(\mathbf{A}t)$ was encountered in Chapter 7 in the solution of the probability vector for a continuous-time Markov chain. This function is widely used in linear system theory, where its properties have been studied extensively [9–4].

If we compute the cross-correlation matrices in the WSS case, we obtain

$$\mathbf{R}_{\mathbf{YX}}(\tau) = \mathbf{h}(\tau) * \mathbf{R}_{\mathbf{XX}}(\tau)$$

and

$$\mathbf{R}_{\mathbf{XY}}(\tau) = \mathbf{R}_{\mathbf{XX}}(\tau) * \mathbf{h}^{\dagger}(-\tau),$$

with output correlation matrix,

$$\mathbf{R}_{\mathbf{YY}}(\tau) = \mathbf{h}(\tau) * \mathbf{R}_{\mathbf{XX}}(\tau) * \mathbf{h}^{\dagger}(-\tau). \tag{9.5–8}$$

Upon vector Fourier transformation this becomes

$$\mathbf{S}_{\mathbf{YY}}(\omega) = \mathbf{H}(\omega)\mathbf{S}_{\mathbf{XX}}(\omega)\mathbf{H}^{\dagger}(\omega). \tag{9.5–9}$$

If $\mathbf{R}_{\mathbf{XX}}(\tau) = \mathbf{Q}\delta(\tau)$, then since the system $\mathbf{H}$ is assumed causal, that is, $\mathbf{h}(t) = \mathbf{0}$ for $t < 0$, we have that the cross-correlation matrix $\mathbf{R}_{\mathbf{YX}}(\tau) = \mathbf{0}$ for $\tau < 0$; that is, $E[\mathbf{Y}(t + \tau)\mathbf{X}^{\dagger}(t)] = \mathbf{0}$ for $\tau < 0$. In words we say that $\mathbf{Y}(t + \tau)$ is orthogonal to $\mathbf{X}(t)$ for $\tau < 0$. Said another way: The past of $\mathbf{Y}(t)$ is orthogonal to the present and future of $\mathbf{X}(t)$. If we additionally assume that the input process $\mathbf{X}(t)$ is a Gaussian process, then the uncorrelatedness condition becomes an independence condition. Under the Gaussian assumption the output $\mathbf{Y}(t)$ is independent of the present and future of $\mathbf{X}(t)$. A similar result was obtained in Section 9.3 for the scalar-valued case. We can use this result to show that the m.-s. solution to a first-order vector LCCDE is a vector Markov random process with the following definition.

Definition 9.5–1. A random process $\mathbf{Y}(t)$ is *vector Markov* if for all $n > 0$ and for all $t_n > t_{n-1} > \ldots > t_1$, and for all values $\mathbf{y}(t_{n-1}), \ldots, \mathbf{y}(t_1)$, we have

$$P[\mathbf{Y}(t_n) \leq \mathbf{y}_n \,|\, \mathbf{y}(t_{n-1}), \ldots, \mathbf{y}(t_1)] = P[\mathbf{Y}(t_n) \leq \mathbf{y}_n \,|\, \mathbf{y}(t_{n-1})]$$

for all values of the real vector $\mathbf{y}_n$. Here $\mathbf{A} < \mathbf{B}$ means

$$\{A_N \leq B_N, A_{N-1} \leq B_{N-1}, \ldots, A_1 \leq B_1\}.$$

Before discussing vector m.-s. differential equations we briefly recall a result for *deterministic* vector LCCDE's. The first-order vector equation,

$$\dot{\mathbf{y}}(t) = \mathbf{A}\mathbf{y}(t) + \mathbf{B}\mathbf{x}(t), \qquad t \geq t_0,$$

subject to the initial condition $\mathbf{y}(t_0)$, can be shown to have solution, employing the matrix exponential,

$$\mathbf{y}(t) = [\exp \mathbf{A}(t - t_0)]\mathbf{y}(t_0) + \int_{t_0}^{t} \mathbf{h}(t - \nu)\mathbf{x}(\nu)\, d\nu, \qquad t \geq t_0,$$

thus generalizing the scalar case. This deterministic solution can be found in any graduate text on linear systems theory, for example in Reference 9–4. The first term is called the *zero-input solution* and the second term is called the *zero-state solution* analogously to the solution for scalar LCCDEs.

We can extend this theory to the stochastic case by considering the m.-s. differential Equation 9.5–5 over the semi-infinite domain $t_0 \leq t < \infty$, and replacing the above deterministic solution with the following stochastic solution, expressed with the help of an m.-s. integral,

$$\mathbf{Y}(t) = [\exp \mathbf{A}(t - t_0)]\mathbf{Y}(t_0) + \int_{t_0}^{t} \mathbf{h}(t - \nu)\mathbf{X}(\nu)\, d\nu. \qquad \text{(m.s.)} \tag{9.5–10}$$

If the LCCDE 9.5–5 is BIBO stable, in the limit as $t_0 \to -\infty$, we get the solution for all time that is, $t_0 = -\infty$,

$$\mathbf{Y}(t) = \int_{-\infty}^{t} \mathbf{h}(t - \nu)\mathbf{X}(\nu)\, d\nu = \mathbf{h}(t) * \mathbf{X}(t), \tag{9.5–11}$$

which is the same as already derived for the stationary infinite time-interval case. In effect, we use the stability of the system to conclude that the resulting zero-input part of the solution is zero at any finite time.

We can now prove the following.

Theorem 9.5–1. Let the input to the state equation

$$\dot{\mathbf{X}}(t) = \mathbf{A}\mathbf{X}(t) + \mathbf{B}\mathbf{W}(t)$$

be the white Gaussian process $\mathbf{W}(t)$. Then the output $\mathbf{X}(t)$ is a vector Markov random process.

Proof. We write the solution at t_n in terms of the solution at an earlier time t_{n-1} as

$$\mathbf{X}(t_n) = [\exp \mathbf{A}(t_n - t_{n-1})]\mathbf{X}(t_{n-1}) + \int_{t_{n-1}}^{t_n} \mathbf{h}(t_n - \nu)\mathbf{W}(\nu)\, d\nu.$$

Then we denote the integral term as $\mathbf{I}(t_n)$ and note that it is independent of $\mathbf{X}(t_{n-1})$. Thus we can deduce that

$$\begin{aligned} P[\mathbf{X}(t_n) \leq \mathbf{x}_n \mid \mathbf{x}(t_{n-1}), \ldots, \mathbf{x}(t_1)] \\ &= P[\mathbf{I}(t_n) \leq \mathbf{x}_n - e^{\mathbf{A}(t_n - t_{n-1})}\mathbf{x}(t_{n-1}) \mid \mathbf{x}(t_{n-1}), \ldots, \mathbf{x}(t_1)] \\ &= P[\mathbf{I}(t_n) \leq \mathbf{x}_n - e^{\mathbf{A}(t_n - t_{n-1})}\mathbf{x}(t_{n-1}) \mid \mathbf{x}(t_{n-1})] \\ &= P[\mathbf{X}(t_n) \leq \mathbf{x}_n \mid \mathbf{x}(t_{n-1})] \end{aligned}$$

and hence that $\mathbf{X}(t)$ is a vector Markov process. ■

If in Theorem 9.5–1 we did not have the Gaussian condition on the input $\mathbf{W}(t)$ but just the white noise condition, then we could not conclude that the output was Markov. This is because we would not have the independence condition required in the proof but only the weaker uncorrelatedness condition.

Similar results hold for random sequences. The state equation becomes

$$\mathbf{Y}[n] = \mathbf{AY}[n-1] + \mathbf{BX}[n], \qquad (9.5\text{–}12)$$

which is a first-order vector difference equation in the m.-s. sense. The impulse response is

$$\mathbf{h}[n] = \mathbf{A}^n\mathbf{B}u[n],$$

and the matrix system function is

$$\mathbf{H}[z] = (\mathbf{I} - \mathbf{A}z^{-1})\mathbf{B},$$

as can be easily verified. The cross-correlation matrices between input and output become

$$\mathbf{R_{YX}}[m] = \mathbf{h}[m] * \mathbf{R_{XX}}[m],$$

$$\mathbf{R_{XY}}[m] = \mathbf{R_{XX}}[m] * \mathbf{h}^\dagger[-m],$$

where we note that for a causal $\mathbf{h}$, we have $\mathbf{Y}[n]$ uncorrelated with the future of the input $\mathbf{X}[n]$ when the input $\mathbf{X}$ is a white noise vector sequence. The output correlation matrix is

$$\mathbf{R_{YY}}[m] = \mathbf{h}[m] * \mathbf{R_{XX}}[m] * \mathbf{h}^\dagger[-m]$$

and the output psd matrix becomes upon Fourier transformation

$$\mathbf{S_{YY}}[\omega] = \mathbf{H}[\omega]\mathbf{S_{XX}}[\omega]\mathbf{H}^\dagger[\omega].$$

The total solution of Equation 9.5–12 can be written as

$$\mathbf{Y}[n] = \mathbf{A}^{n-n_0}\mathbf{Y}[n_0] + \sum_{k=n_0}^{n} \mathbf{h}[n-k]\mathbf{X}[k], \qquad n \geq n_0$$

in terms of an initial condition at n_0. In the limit as $n_0 \to -\infty$, this then becomes the convolution summation

$$\mathbf{Y}[n] = \mathbf{h}[n] * \mathbf{X}[n], \qquad -\infty < n < +\infty.$$

Definition 9.5–2. A random sequence $\mathbf{Y}[n]$ is *vector Markov* if for all $K > 0$ and for all $n_K > n_{K-1} > \ldots > n_1$, we have

$$P[\mathbf{Y}[n_K] \leq \mathbf{y}_K \mid \mathbf{y}[n_{K-1}], \ldots, \mathbf{y}[n_1]] = P[\mathbf{Y}[n_K] \leq \mathbf{y}_K \mid \mathbf{y}[n_{K-1}]]$$

for all real values of the vector $\mathbf{y}_K$, and all conditioning vectors $\mathbf{y}[n_{K-1}], \ldots, \mathbf{y}[n_1]$.

We can now prove the following theorem for random sequences, which is analogous to Theorem 9.5–1 for random processes.

Theorem 9.5–2. In the state equation

$$\mathbf{X}[n] = \mathbf{AX}[n-1] + \mathbf{BW}[n],$$

let $\mathbf{W}[n]$ be a white Gaussian random sequence. Then the output $\mathbf{X}[n]$ is a vector Markov random sequence.

Proof. The proof is analogous to that of Theorem 9.5–1 and is left to the reader as an exercise. See Problem 9.11.

9.6 REPRESENTATION OF BANDLIMITED AND PERIODIC PROCESSES

Here we return to the scalar valued case and consider expansions of random processes in terms of sets of random variables. An example that we have already seen would be the Karhunen-Loève expansion of Section 8.5. In general, the sets of random variables will contain an infinite number of elements; thus we are equivalently representing a random process by a random sequence. This representation is essential for digital processing of waveforms. Also, when the coefficients in the representation or expansion are uncorrelated or independent, then important additional simplifications result. We start out by considering WSS processes whose power spectral densities have finite support; that is, the respective correlation functions are bandlimited. We then develop an m.-s. sampling theorem for random processes.

Bandlimited Processes

Definition 9.6–1. A random process $X(t)$ that is WSS is said to be *bandlimited* to $[\omega_1, \omega_2]$ if $S_X(\omega) = 0$ for $|\omega| \notin [\omega_1, \omega_2]$. When $\omega_1 = 0$ we say the process is *lowpass*, and we set $\omega_2 = \omega_c$, called the *cutoff* frequency.

In the case of a lowpass process we can use the ordinary sampling theorem for deterministic signals [9–3] to write the following representation for the lowpass function $R(\tau)$ in terms of the infinite set of samples $R(nT)$ taken at spacing $T = \pi/\omega_c$

$$R(\tau) = \sum_{n=-\infty}^{+\infty} R(nT)\frac{\sin \omega_c(\tau - nT)}{\omega_c(\tau - nT)}. \qquad (9.6\text{–}1)$$

It turns out that one can define a mean-square sampling theorem for WSS random processes, which we next state and prove.

Theorem 9.6–1. If a second-order WSS random process $X(t)$ is lowpass with cutoff frequency ω_c, then

$$X(t) = \sum_{n=-\infty}^{+\infty} X(nT)\frac{\sin \omega_c(t - nT)}{\omega_c(t - nT)} \qquad \text{(m.s.)}$$

where $T \triangleq \pi/\omega_c$.

We point out that the foregoing equality is in the sense of a m.-s. limit, that is, with

$$X_N(t) \triangleq \sum_{n=-N}^{+N} X(nT)\frac{\sin \omega_c(t - nT)}{\omega_c(t - nT)},$$

then $\lim_{N\to\infty} E[|X(t) - X_N(t)|^2] = 0$ for each t.

Proof. First we observe that

$$E[|X(t) - X_N(t)|^2] = E[(X(t) - X_N(t))X^*(t)] - E[(X(t) - X_N(t))X_N^*(t)]. \tag{9.6–2}$$

Since $X_N^*(t)$ is just a weighted sum of the $X^*(mT)$, we begin by obtaining the preliminary result that $E[(X(t) - X_N(t))X^*(mT)] = 0$:

$$E\left[\left(X(t) - \sum_n X(nT)\frac{\sin \omega_c(t - nT)}{\omega_c(t - nT)}\right)X^*(mT)\right]$$

$$= R(t - mT) - \sum_{n=-N}^{+N} R(nT - mT)\frac{\sin \omega_c(t - nT)}{\omega_c(t - nT)}$$

$$\xrightarrow[N\to\infty]{} R(t - mT) - R(t - mT) = 0,$$

where the last equality follows by replacing τ with $t - mT$ in $R(\tau)$ and writing the sampling expansion for this bandlimited function of t. Setting $\hat{X}(t) \triangleq \lim X_N(t)$ in the m.-s. sense, we get

$$E[(X(t) - \hat{X}(t))X^*(mT)] = 0,$$

that is, the error $X(t) - \hat{X}(t)$ is orthogonal to $X(mT)$ for all m. We write this symbolically as

$$X(t) - \hat{X}(t) \perp X(mT), \qquad \forall_m.$$

But then we also have that $X(t) - \hat{X}(t) \perp X_N(t)$ because $X_N(t)$ is just a weighted sum of $X(mT)$. Then letting $N \to +\infty$, we get

$$X(t) - \hat{X}(t) \perp \hat{X}(t), \qquad \forall_t,$$

which just means $E[(X(t) - \hat{X}(t))\hat{X}^*(t)] = 0$; thus the second term in Equation 9.6–2 is asymptotically zero. Considering the first term in Equation 9.6–2 we get

$$E[(X(t) - X_N(t))X^*(t)] = R(0) - \sum_{n=-N}^{+N} R(nT - t)\frac{\sin \omega_c(t - nT)}{\omega_c(t - nT)},$$

which tends to zero as $n \to +\infty$ by virtue of the representation

$$R(0) = \sum_{n=-\infty}^{+\infty} R(nT - t)\frac{\sin \omega_c(t - nT)}{\omega_c(t - nT)}$$

obtained by right-shifting the bandlimited $R(\tau)$ in Equation 9.6–1 by the shift t, thereby obtaining

$$R(\tau - t) = \sum_{n=-\infty}^{+\infty} R(nT - t)\frac{\sin \omega_c(\tau - nT)}{\omega_c(\tau - nT)}$$

and then setting the free parameter $\tau = t$. Thus $E[|X(t) - X_N(t)|^2] \to 0$ as $N \to +\infty$. ■

In words the m.-s. sampling theorem tells us that knowledge of the sampled sequence is sufficient for determining the random process at time t up to an event of probability zero. This since two random variables equal in the m.-s. sense are equal with probability-1. (See Theorem 8.1–5.)

X_a random process
X random sequence

To consider digital processing of the resulting random sequence, we change the notation slightly by writing X_a for the random process and use X to denote a corresponding random sequence. Then the mean of the random sequence $X[n] \triangleq X_a(nT)$ is $\mu_X = E[X_a(t)]$, and the correlation function is $R[m] = R_a(mT)$. This then gives the following power spectral density for the random sequence:

$$S_X[\omega] = S_{X_a}\left(\frac{\omega}{T}\right), \qquad |\omega| \leq \pi,$$

if $X(t)$ is lowpass with cutoff $\omega_c = \pi/T$. After digital processing we can restore the continuous-time random process using the m.-s. sampling expansion. Note that we assume perfect sampling and reconstruction. Here the WSS random process becomes a stationary random sequence and the reconstructed process is again wide-sense stationary. If sample-and-hold type suboptimal reconstruction is used (as in Problem 7.1), then the reconstructed random process may not even be WSS.

We see that the coefficients in this expansion, that is, the samples of X_a at spacing T, most often will be correlated. However, there is one case where they will be uncorrelated. That case is when the process has a flat power spectral density and is lowpass.

Example 9.6–1: Let $X_a(t)$ be WSS and bandlimited to $(-\omega_c, +\omega_c)$ with flat psd

$$S_{X_a}(\omega) = S_{X_a}(0) I_{(-\omega_c, +\omega_c)}(\omega)$$

as seen in Figure 9.6–1. Then $R_{X_a}(\tau)$ is given as

$$R_{X_a}(\tau) = R_{X_a}(0)\frac{\sin \omega_c \tau}{\omega_c \tau} \quad \text{with} \quad R_{X_a}(0) = \frac{\omega_c}{\pi} S_{X_a}(0).$$

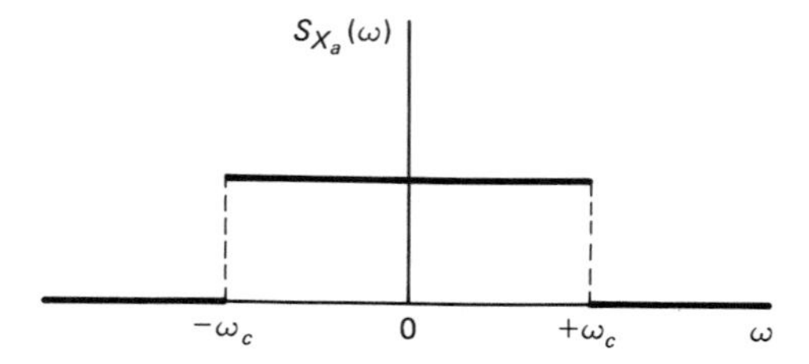

Figure 9.6–1 The psd of a bandlimited white process.

Since $R_{X_a}(mT) = 0$ for $m \neq 0$ we see that the samples are orthogonal, all with the same average power $R_{X_a}(0)$. Thus the random sequence $X[n]$ is a white noise and its psd is flat:

$$S_X[\omega] = S_{X_a}(0), \qquad |\omega| < \pi$$

with correlation function a discrete-time impulse

$$R_X[m] = R_{X_a}(0)\delta[m].$$

Bandpass Random Processes. Next we consider the treatment of bandpass random processes. Such processes are good models for random signals or noise that have been modulated by a carrier wave to enable long-distance transmission. Also, sometimes the frequency selective nature of the transmission medium converts a wideband signal into an approximately narrow band signal. Common examples are radio waves in the atmosphere and pressure waves in the Earth or underwater.

First we show that we can construct a WSS bandpass random process using two lowpass WSS processes. Thus consider a bandpass random process whose psd, for positive frequencies, is centered at ω_0:

$$U(t) = X(t)\cos(\omega_0 t) + Y(t)\sin(\omega_0 t), \tag{9.6-3}$$

where the lowpass processes X and Y are real valued, have zero means, are wide-sense stationary, and satisfy

$$K_{XX}(\tau) = K_{YY}(\tau),$$
$$K_{XY}(\tau) = -K_{YX}(\tau).$$

In a representation such as Equation 9.6–3, X is called the *in-phase component* of U, and Y is called the *quadrature component*.

Then it follows that

$$K_{UU}(\tau) = K_{XX}(\tau)\cos(\omega_0\tau) + K_{YX}(\tau)\sin(\omega_0\tau)$$

or

$$S_{UU}(\omega) = \tfrac{1}{2}[S_{XX}(\omega - \omega_0) + S_{XX}(\omega + \omega_0)] + \frac{1}{2j}[S_{YX}(\omega - \omega_0) - S_{YX}(\omega + \omega_0)].$$

Now $K_{XY}(\tau) = -K_{YX}(\tau)$. Also, $K_{YX}(\tau) = K^*_{XY}(-\tau) = K_{XY}(-\tau)$ since X and Y are real valued, so

$$S_{YX}(\omega) = S^*_{XY}(\omega) = S_{XY}(-\omega) = -S_{XY}(+\omega).$$

From $S^*_{XY}(\omega) = -S_{XY}(\omega)$ we get that S_{XY} is pure imaginary. From $S_{XY}(-\omega) = -S_{XY}(\omega)$ we get that S_{XY} is an odd function of ω. The same holds for S_{YX} since $K_{XY}(\tau) = K^*_{YX}(-\tau) = K_{YX}(-\tau)$; thus $\tfrac{1}{2}S_{XX}(\omega - \omega_0)$ is the in-phase part of the psd at ω_0, and $(1/2j)S_{YX}(\omega - \omega_0)$ is the quadrature part. These properties are shown in Figure 9.6–2.

If we want to represent a general WSS bandpass process as in Equation 9.6–3 we could decompose the process $U(t)$ as shown in Figure 9.6–3; however, the random processes obtained after multiplication by cos and sin

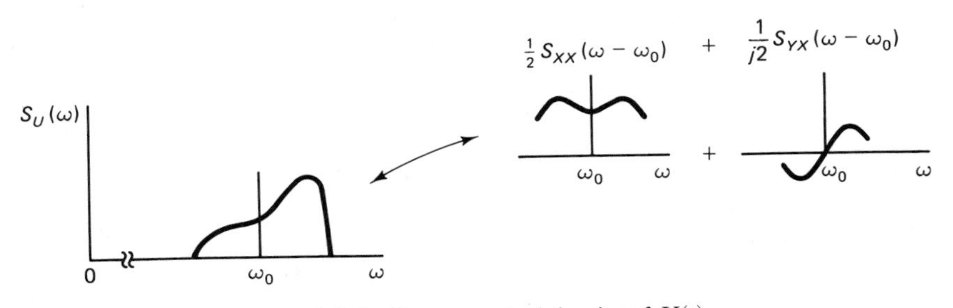

Figure 9.6–2 Power spectral density of $U(t)$.

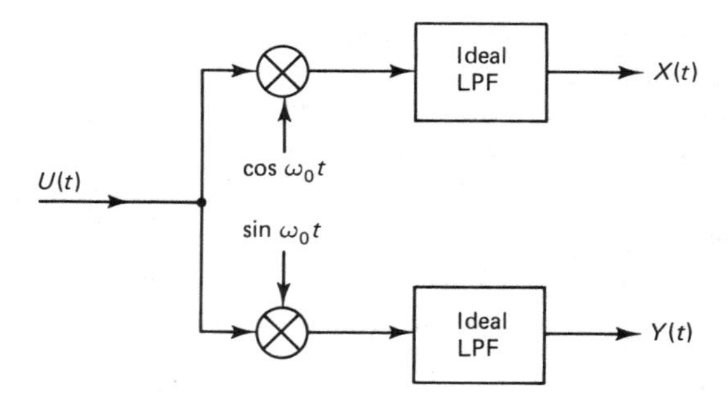

Figure 9.6–3 Decomposition of bandpass process.

are not WSS so that the system of Figure 9.6–3 cannot be analyzed in the frequency domain. An alternative approach is through the *Hilbert transform*, defined as filtering with the system function

$$H(\omega) = -j\,\mathrm{sgn}(\omega).$$

Using this operator we can define an *analytic signal process Z*,

$$Z(t) \triangleq U(t) + j\check{U}(t), \tag{9.6–4}$$

where the superscript $\vee$ indicates a Hilbert transformation. Then it turns out that we can take

$$X(t) = Re[Z(t)e^{-j\omega_0 t}]$$

and

$$Y(t) = -Im[Z(t)e^{-j\omega_0 t}],$$

to achieve the desired representation (Equation 9.6–3). These X and Y are the same as in Figure 9.6–3. The psd of $Z(t)$ is

$$S_Z(\omega) = 4S_U(\omega)u(\omega), \tag{9.6–5}$$

where $u(\omega)$ is the unit-step function. This psd is sketched in Figure 9.6–4, where we note that its support is restricted to $\omega \geq 0$.

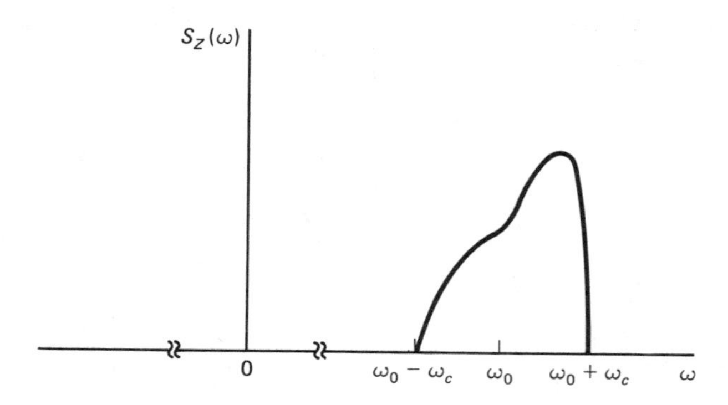

Figure 9.6–4 A psd of analytic signal process $Z(t)$.

One can use this theory for computer simulation of bandpass processes by first decomposing the process into its X and Y lowpass components and then representing these two lowpass processes by their sampled equivalents. Thus one can simulate bandpass processes and bandpass systems by discrete-time processing of coupled pairs of random sequences, that is, vector random sequences of dimension two.

WSS Periodic Processes. A stationary random process may have a correlation function that is periodic; that is $R(\tau) = R(\tau + T)$ for all τ. This is a special case of the general periodic process introduced in Chapter 7.

Definition 9.6–2. A WSS random process is *mean-square periodic* if for some $T > 0$ we have $R_X(\tau) = R_X(\tau + T)$ for all τ. We call the smallest such T the *period*.

In Problem 7.15 the reader asked to show that a m.-s. periodic process was also periodc in the stronger sense:

$$E[|X(t) - X(t + T)|^2] = 0.$$

We now show this directly for the WSS periodic case. Evaluating we get

$$E[|X(t) - X(t + T)|^2] = 2[R(0) - Re(R(T))].$$

Now since $R(0) = R(T)$, it follows that $R(T)$ is real and $Re[R(T)] = R(T) = R(0)$. Hence $X(t) = X(t + T)$ (m.s.) and hence also with probability-1.

Turning to the psd of a WSS periodic process, we know that R has a Fourier series

$$R(\tau) = \sum_{n=-\infty}^{+\infty} \alpha_n e^{+j\omega_0 n\tau}, \qquad \left(\omega_0 = \frac{2\pi}{T}\right)$$

with coefficients

$$\alpha_n = \frac{1}{T}\int_{-T/2}^{+T/2} R(\tau)e^{-j\omega_0 n\tau}\, d\tau.$$

Thus the psd of a WSS periodic process is a line spectra of the form

✓ $$S(\omega) = 2\pi \sum_{n=-\infty}^{+\infty} \alpha_n \delta(\omega - n\omega_0), \tag{9.6–6}$$

which can be summarized as follows. We note that the α_n's are necessarily nonnegative.

Theorem 9.6–2. If a WSS random process $X(t)$ is m.-s. periodic, then its psd is a line spectra with impulses at multiples of the fundamental frequency ω_0. The impulse areas are given by the Fourier coefficients of the periodic correlation function R.

Example 9.6–2: Let the input to the LSI system H be the periodic random process $X(t)$ as indicated in Figure 9.6–5. In general we have

$$S_Y(\omega) = |H(\omega)|^2 S_X(\omega). \qquad (9.3\text{-}8)$$

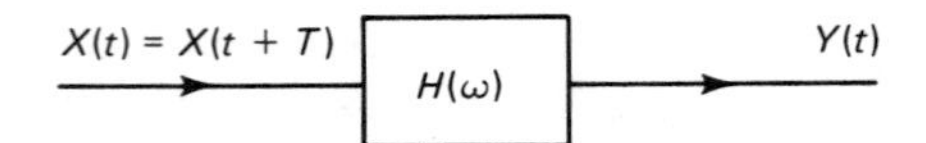

Figure 9.6–5 Mean-square periodic process as input to LSI system.

Using Equation 9.6–6, we get

$$S_Y(\omega) = 2\pi \sum_{n=-\infty}^{+\infty} \alpha_n \, |H(\omega_0 n)|^2 \, \delta(\omega - n\omega_0).$$

Hence the output $Y(t)$ is m.-s. periodic with the same period T and has correlation function given by

$$R_Y(\tau) = \sum_{n=-\infty}^{+\infty} \alpha_n \, |H(\omega_0 n)|^2 \, e^{+j\omega_0 n\tau}.$$

Deterministic functions that are periodic can be represented by Fourier series as we have just done for the periodic correlation function. We now show that the WSS periodic process itself may be represented as a Fourier series in the mean-square sense.

Theorem 9.6–3. Let the WSS random process $X(t)$ be m.-s. periodic with period T. Then we can write

$$X(t) = \sum_{n=-\infty}^{+\infty} A_n e^{+j\omega_0 nt}, \qquad \text{(m.s.)}$$

where $\omega_0 = 2\pi/T$, with random Fourier coefficients

$$A_n \triangleq \frac{1}{T} \int_{-T/2}^{+T/2} X(\tau) e^{-j\omega_0 n\tau} \, d\tau, \qquad \text{(m.s.)}$$

with mean

$$E[A_n] = \mu_X \delta[n],$$

and correlation

$$E[A_n A_m^*] = \alpha_n \delta[m - n],$$

and mean-square value

$$\alpha_n \triangleq \frac{1}{T} \int_{-T/2}^{+T/2} R_X(\tau) e^{-j\omega_0 n\tau} \, d\tau. \qquad (9.6\text{–}7)$$

Thus the periodic random process can be expanded in a Fourier series whose coefficients are orthogonal. Thus the Fourier series is the Karhunen-Loève expansion (see Section 8.5) for an WSS periodic process. You are asked to show this in Problem 9.14.

Proof. First we show that the A_n are orthogonal. We readily see that

$$E[A_n] = \frac{1}{T} \int_{-T/2}^{+T/2} E[X(u)] e^{-j\omega_0 nu} \, du = \mu_X \delta[n],$$

then

$$E[A_n^* X(t)] = \frac{1}{T}\int_{-T/2}^{+T/2} E[X^*(u)X(t)]e^{+j\omega_0 nu}\,du$$
$$= \frac{1}{T}\int_{-T/2}^{+T/2} R_X(t-u)e^{+j\omega_0 nu}\,du$$
$$= \left[\frac{1}{T}\int_{-T/2+t}^{T/2+t} R_X(\tau)e^{-j\omega_0 n\tau}\,d\tau\right]e^{+j\omega_0 nt}$$
$$= \alpha_n e^{+j\omega_0 nt},$$

since the integrand is periodic in τ with period T and also by Equation 9.6–7. Next we consider

$$E[A_k A_n^*] = \frac{1}{T}\int_{-T/2}^{+T/2} E[A_n^* X(u)]e^{-j\omega_0 ku}\,du.$$
$$= \alpha_n\left[\frac{1}{T}\int_{-T/2}^{+T/2} e^{+j\omega_0 nu}e^{-j\omega_0 ku}\,du\right]$$
$$= \alpha_n \delta[n-k].$$

It remains to show that

$$E\left[\left|X(t) - \sum_{n=-\infty}^{+\infty} A_n e^{+j\omega_0 nt}\right|^2\right] = 0.$$

Expanding the left-hand side, we get

$$E[|X(t)|^2] - \sum_n E[A_n^* X(t)]e^{-j\omega_0 nt}$$
$$-\sum_n E[A_n X^*(t)]e^{+j\omega_0 nt} + \sum_{nk}\sum E[A_k A_n^*]e^{+j\omega_0(k-n)t}$$
$$= R_X(0) - \sum_n \alpha_n - \sum_n \alpha_n^* + \sum_n \alpha_n.$$

But the α_n are real since $\alpha_n = E[|A_n|^2]$, and also $R_X(0) = \sum \alpha_n$ so that we finally have

$$\sum \alpha_n - \sum \alpha_n - \sum \alpha_n + \sum \alpha_n = 0. \quad \blacksquare$$

As shown in Problem 9.18 analogous results hold for WSS periodic random sequences; that is, $R[m] = R[m+T]$ or all m where T = integer.

9.7 SUMMARY

This chapter was devoted to the study of stationary random processes and sequences. Most of the results also remain valid if the WSS condition is substituted for the stationary condition. The WSS condition may safely be substituted whenever the result depends only on second-order properties.

We introduced the broadly useful concept of power spectral density for both random processes and random sequences and showed its interpretation as a density for average power in a stationary process. We especially studied the interaction of LSI systems with WSS random processes.

We extended previous results on scalar processes to the vector-valued case and introduced state equations for both random processes and sequences. We learned that the solution to the state equations with a white Gaussian input was Markov. This result will be used in Chapter 10 in the derivation of the Kalman filter for random sequences.

Finally, we introduced some representations for WSS periodic and band-limited processes, the latter being quite useful in communications and computer simulation of continuous-time systems.

9.APP RESIDUE METHOD FOR INVERSE FOURIER TRANSFORMATION

In Section 9.1 we defined the power spectral density (psd) $S(\omega)$ and showed that it is useful for analyzing LSI systems with random process inputs. We often want to take an inverse transform to find the correlation function corresponding to a given psd to obtain a time-domain characterization. This appendix presents the residue method for accomplishing the necessary inverse Fourier transformation.

We start by recalling the relation between the psd and correlation function for a WSS random process,

$$S(\omega) = \int_{-\infty}^{+\infty} R(\tau)e^{-j\omega\tau}\, d\tau,$$

$$R(\tau) = \frac{1}{2\pi}\int_{-\infty}^{+\infty} S(\omega)e^{+j\omega\tau}\, d\tau.$$

To apply the powerful residue method of complex variable theory [9–5] to the evaluation of the above IFT, we must first express this integral as an integral along a contour in the *complex s-plane*. We define a new function S of the *complex variable* $s = \sigma + j\omega$ as follows.

First we define $S(s)$ on the imaginary axis in terms of the function of a *real variable* $S(\omega)$ as

$$S(j\omega) \triangleq S(\omega).\dagger$$

Then we replace $j\omega$ by s to extend the function $S(j\omega)$ to the entire complex plane. Thus

$$S(j\omega) = S(\omega) = \int_{-\infty}^{+\infty} R(\tau)e^{-j\omega\tau}\, d\tau$$

† It is important to realize that two *different* functions here have the same symbol S. We distinguish them by noting whether the argument is real or complex.

so

$$S(s) = \int_{-\infty}^{+\infty} R(\tau)e^{-s\tau}\, d\tau, \qquad (9.\text{APP–1})$$

which is just the two-sided Laplace transform of the correlation function R. Also

$$\begin{aligned} R(\tau) &= \frac{1}{2\pi j}\int_{-\infty}^{+\infty} S(j\omega)e^{j\omega\tau}\, d(j\omega) \\ &= \frac{1}{2\pi j}\int_{-j\infty}^{+j\infty} S(s)e^{s\tau}\, ds. \qquad (9.\text{APP–2}) \end{aligned}$$

This is an integral along the imaginary axis of the s-plane.

The integral in Equation 9.APP–2 is called a *contour integral* in the theory of functions of a complex variable [9–5], where it is shown that one can evaluate such an integral over a closed contour by the *method of residues*. This method is particularly easy to apply when the functions are rational; that is, the function is the ratio of two polynomials in s. Since this situation often occurs in linear systems whose behavior is modeled by differential equations, this method of evaluation can be very useful. We state the main result as a fact from the theory of complex variables.

Fact. Let $F(s)$ be a function of the complex variable s, which is analytic inside and on a closed *counterclockwise* contour C encircling the origin except at P poles located inside C at $s = p_i$, $i = 1, \ldots, P$. Then

$$\frac{1}{2\pi j}\oint_C F(s)\, ds = \sum_{\substack{p_i \text{ inside} \\ C}} \text{Res}[F(s); s = p_i], \qquad (9.\text{APP–3})$$

where:

1. at a first-order pole, $\text{Res}[F(s); s = p] = [F(s)(s - p)]|_{s=p}$,
2. at a second-order pole, $\text{Res}[F(s); s = p] = \dfrac{d}{ds}[F(s)(s - p)^2]|_{s=p}$,

and at an nth order pole

3. $\text{Res}[F(s); s = p] = \dfrac{1}{(n-1)!}\left(\dfrac{d^{(n-1)}}{ds^{(n-1)}}[F(s)(s-p)^n]\right)\Big|_{s=p}$.

In applying this result to our problem we first have to close the contour in some fashion. If we close the contour with a half-circle of infinite radius C_L as shown in Figure 9.APP–1, then provided that the function being integrated, $S(s)e^{s\tau}$, tends to zero fast enough as $|s| \to +\infty$, the value of the integral will not be changed by this closing of the contour. In other words, the integral over the semicircular part of the contour will be zero.

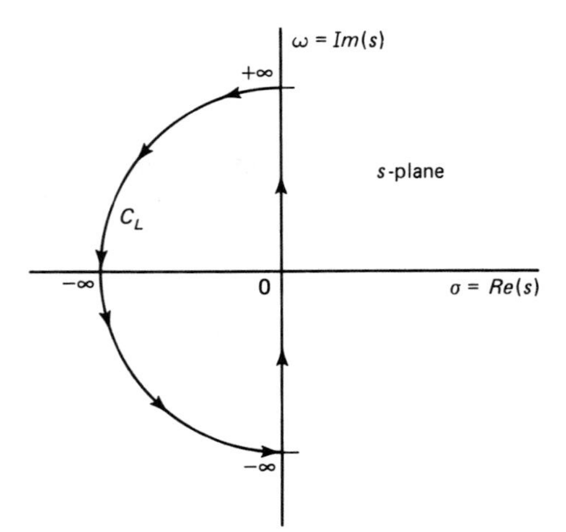

Figure 9.APP–1 Closed contour in left half of s-plane.

The conditions for this are $|S(s)| \to 0$ as $|s| \to +\infty$, and

$$|e^{s\tau}| \to 0 \quad \text{as} \quad Re(s) \to -\infty,$$

the latter of which is satisfied for $\tau > 0$. Thus for positive τ we have

$$R(\tau) = \frac{1}{2\pi j} \oint_{C_L} S(s)e^{s\tau}\, ds = \sum_{\substack{p_i \\ \text{inside } C_L}} \text{Res}[S(s)e^{s\tau}; s = p_i]. \tag{9.APP–4a}$$

Similarly, for $\tau < 0$ one can show that it is permissible to close the contour to the right as shown in Figure 9.APP–2, in which case we get

$$|e^{s\tau}| \to 0 \quad \text{as} \quad Re(s) \to +\infty,$$

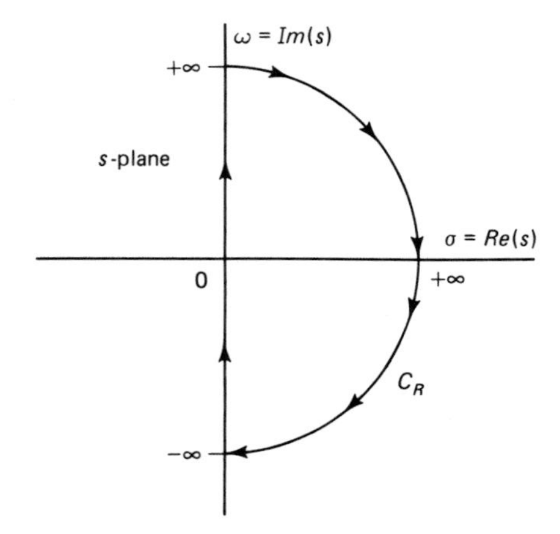

Figure 9.APP–2 Closed contour in right half of s-plane.

so that we have

$$R(\tau) = \frac{1}{2\pi j}\oint_{C_R} S(s)e^{s\tau}\,ds = -\sum_{\substack{p_i \\ \text{inside } C_R}} \text{Res}[S(s)e^{s\tau}; s = p_i], \tag{9.APP–4b}$$

for $\tau < 0$, the minus sign arising from the clockwise traversal of the contour.

Example 9.APP–1: (First-order psd.) Let

$$S(\omega) = 2\alpha/(\alpha^2 + \omega^2), \qquad 0 < \alpha < 1,$$

then

$$S(j\omega) = S(\omega) = 2\alpha/(\alpha^2 + \omega^2) = 2\alpha/(j\omega + \alpha)(-j\omega + \alpha),$$

so

$$S(s) = \frac{2\alpha}{(s + \alpha)(-s + \alpha)},$$

where the configuration of the poles in the s-plane is shown in Figure 9.APP–3.

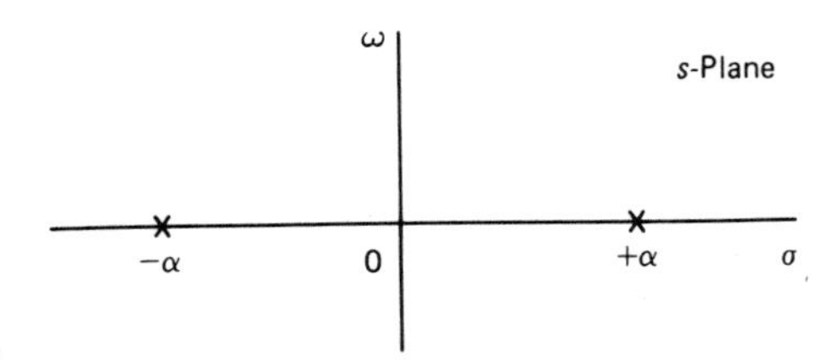

Figure 9.APP–3 Pole-zero diagram.

Evaluating the residues for $\tau > 0$, we get

$$R(\tau) = \text{Res}[S(s)e^{s\tau}; s = -\alpha] = \left.\frac{2\alpha e^{s\tau}}{(-s + \alpha)}\right|_{s=-\alpha}$$

$$= \frac{2\alpha}{2\alpha}e^{-\alpha\tau},$$

while for $\tau < 0$ we get

$$R(\tau) = -\text{Res}[S(s)e^{s\tau}; s = +\alpha]$$

$$= -\left.\frac{2\alpha e^{s\tau}(s - \alpha)}{(s + \alpha)(-s + \alpha)}\right|_{s=+\alpha}$$

$$= \left.\frac{-2\alpha e^{s\tau}}{(s + \alpha)(-1)}\right|_{s=\alpha} = \frac{2\alpha}{2\alpha}e^{\alpha\tau}.$$

Combining the results into a single formula, we get

$$R(\tau) = \exp(-\alpha\,|\tau|), \qquad -\infty < \tau < +\infty.$$

Inverse Fourier Transform for psd of Random Sequence. In the case of a random sequence one can do a similar contour integral evaluation in the complex z-plane. We recall the transform and inverse transform for a sequence:

$$S[\omega] = \sum_{m=-\infty}^{+\infty} R[m]e^{-j\omega m},$$

$$R[m] = \frac{1}{2\pi}\int_{-\pi}^{+\pi} S[\omega]e^{+j\omega m}\, d\omega.$$

We rewrite the latter integral as a contour integral around the unit circle in a complex plane by defining the function of a complex variable, $S[e^{j\omega}] \triangleq S[\omega]$, and then substituting $z = e^{j\omega}$ into this new function to obtain

$$S[z] = \sum_{m=-\infty}^{+\infty} R[m]z^{-m} \quad \text{and}$$

$$R[m] = \frac{1}{2\pi j}\oint_C S[z]z^{m-1}\, dz \quad \text{where } C = \{|z| = 1\}. \tag{9.APP–5}$$

In this case the contour is already closed and it encircles the origin in a counterclockwise direction, so we can apply Equation 9.APP–3 directly to obtain

$$R[m] = \sum_{\substack{p_i \\ \text{inside } C}} \text{Res}[S[z]z^{m-1};\ z = p_i],$$

where the sum is over the residues at the poles inside the unit circle. This formula is valid for all values of the integer m; however, it is awkward to evaluate for negative m due to the variable-order pole contributed by z^{m-1} at $z = 0$. Fortunately, a transformation mapping z to $1/z$ conveniently solves this problem, and we have [9–3],

$$R[m] = \frac{1}{2\pi j}\oint_C S\left[\frac{1}{z}\right]z^{-m+1}(-z^{-2}\, dz),$$

$$= \frac{1}{2\pi j}\oint_C S[z^{-1}]z^{-m-1}\, dz,$$

avoiding the variable-order pole for $m < 0$. We thus arrive at the prescription:

For $m \geq 0$

$$R[m] = \sum_{\substack{i:\text{poles} \\ \text{inside unit} \\ \text{circle}}} \text{Res}[S[z]z^{m-1};\ z = p_i]. \tag{9.APP–6a}$$

and for $m < 0$

$$R[m] = \sum_{\substack{i:\text{poles} \\ \text{outside unit} \\ \text{circle}}} \text{Res}[S[z^{-1}]z^{-m-1};\ z = p_i^{-1}]. \tag{9.APP–6b}$$

Example 9.APP–2: (First-order psd of random sequence.) We consider a psd given as

$$S[\omega] = \frac{2(1-\rho^2)}{(1+\rho^2) - 2\rho\cos\omega}, \qquad |\omega| \leq \pi, \tag{9.APP–7}$$

which is plotted in Figure 9.APP–4.

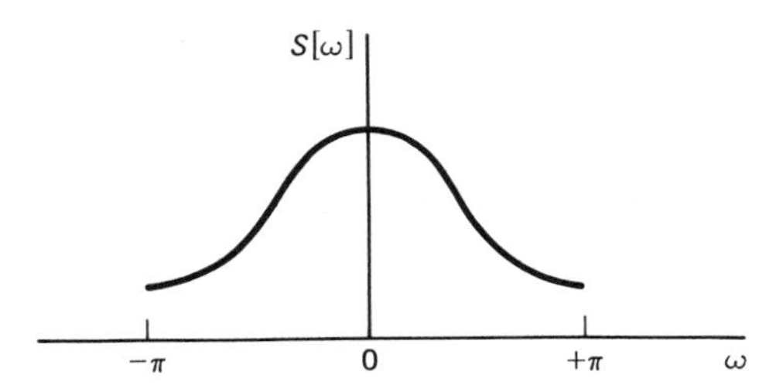

Figure 9.APP–4 Plot of psd for $0<\rho<1$.

Using the identity $\cos\omega = 0.5(\exp j\omega + \exp -j\omega)$, we can make this substitution in Equation 9.APP–7 to obtain the function of a complex variable,

$$S[e^{jw}] = S[\omega] = \frac{2(1-\rho^2)}{(1+\rho^2) - 2\rho\cos\omega}$$

$$= \frac{2(1-\rho^2)}{(1+\rho^2) - \rho(e^{+j\omega} + e^{-j\omega})}.$$

Then we replace $e^{j\omega}$ by z to obtain the function of z,

$$S[z] = \frac{2(1-\rho^2)}{(1+\rho^2) - \rho(z+z^{-1})}$$

$$= -2(\rho^{-1} - \rho)\frac{z}{(z-\rho)(z-\rho^{-1})}.$$

The z-plane pole-zero configuration of this function is shown in Figure 9.APP–5. The overall transformation from $S[\omega]$ to $S[z]$ is given by

$$\cos\omega \leftarrow \tfrac{1}{2}(z + z^{-1}). \tag{9.APP–8}$$

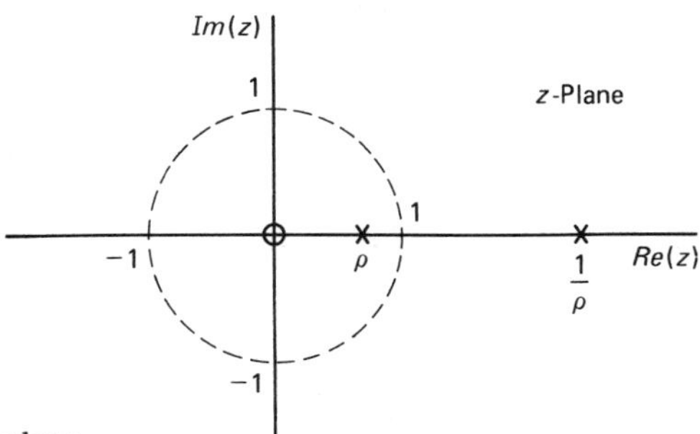

Figure 9.APP–5 Pole-zero diagram in z-plane.

For $m \geq 0$ we get

$$R[m] = \text{Res}[S[z]z^{m-1}; z = \rho]$$

$$= -2(\rho^{-1} - \rho)\frac{\rho\rho^{m-1}}{(\rho - \rho^{-1})}$$

$$= 2\rho^m.$$

For $m < 0$ we have

$$R[m] = \text{Res}[S[z^{-1}]z^{-m-1}; z = \rho].$$

Now

$$S[1/z] = -2(\rho^{-1} - \rho)\frac{z^{-1}}{(z^{-1} - \rho)(z^{-1} - \rho^{-1})}$$

$$= -2(\rho^{-1} - \rho)\frac{z}{(z - \rho^{-1})(z - \rho)},$$

which could easily have been foretold from the symmetry evident in Equation 9.APP–8. Hence

$$\text{Res}[S[z^{-1}]z^{-m-1}; z = \rho] = -2(\rho^{-1} - \rho)\frac{z^{-m}(z - \rho)}{(z - \rho^{-1})(z - \rho)}\Bigg|_{z=\rho}$$

$$= -2\frac{(\rho^{-1} - \rho)\rho^{-m}}{(\rho - \rho^{-1})}$$

$$= 2\rho^{-m}.$$

Combining, we get

$$R[m] = 2\rho^{|m|}, \qquad -\infty < m < +\infty.$$

PROBLEMS

9.1. The psd of a random process is given as $S(\omega) = 1/(1 + \omega^2)^2$ for $-\infty < \omega < +\infty$. Find its autocorrelation function $R(\tau)$.

9.2. Consider the LSI system shown in Figure P9.2, whose input is the zero-mean random process $W(t)$ and whose output is the random process $X(t)$. The frequency response of the system is $H(\omega)$.

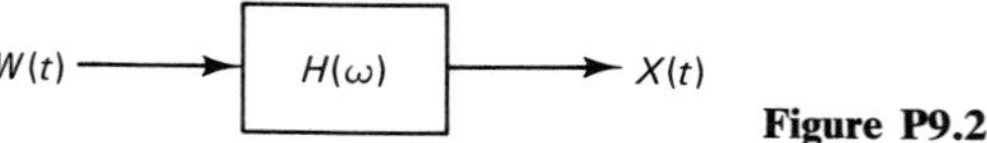

Figure P9.2

Given that $K_{WW}(\tau) = \delta(\tau)$, find $H(\omega)$ in terms of the cross-covariance $K_{XW}(\tau)$ or its Fourier transform.

9.3. Consider the first-order stochastic differential equation:

$$\frac{dX(t)}{dt} + X(t) = W(t) \qquad \text{(m.s.)}$$

driven by the zero-mean white noise $W(t)$ with correlation function $R_W(t, s) = \delta(t - s)$.

(a) If this m.-s. differential equation is valid for all time, $-\infty < t < +\infty$, find the psd of the resulting wide-sense stationary process $X(t)$.
(b) Using residue theory (or any other method), find the inverse Fourier transform of S_X, the autocorrelation function $R_X(\tau)$, $-\infty < \tau < +\infty$.
(c) If the above m.-s. differential equation is run only for $t > 0$, is it possible to choose an initial condition random variable $X(0)$ such that $X(t)$ is wide-sense stationary for all $t \geq 0$? If such an r.v. exists, find its mean and variance. *Justify your answer.* You may assume that the r.v. $X(0)$ is orthogonal to $W(t)$ on $t \geq 0$; that is, $X(0) \perp W(t)$. Hint: Express $X(t)$ for $t > 0$ in terms of the initial condition and a stochastic integral involving $W(t)$.

9.4. Let the random process $X(t)$ with mean value 128 and covariance function

$$K_X(\tau) = 1000 \exp(-10\,|\tau|)$$

be filtered by the lowpass filter

$$H(\omega) = \frac{1}{1 + j\omega}$$

to produce the output process $Y(t)$.
(a) Find $\mu_Y(t)$.
(b) Find $K_Y(\tau)$.

9.5. In this problem we consider using white noise as an approximation to a smoother process (cf. Section 9.2), which is the input to a lowpass filter. The output process from the filter is then investigated to determine the error resulting from the white noise approximation. Let the stationary random process $X(t)$ have zero mean and autocovariance function:

$$K_X(\tau) = \frac{1}{2\tau_0} \exp(-|\tau|/\tau_0),$$

which can be written as $h(\tau) * h(-\tau)$ with $h(\tau) = \frac{1}{\tau_0} e^{-t/\tau_0} u(t)$.

(a) Let $X(t)$ be input to the lowpass filter shown in Figure P9.5–1, with output $Y(t)$. Find the output psd $S_{YY}(\omega)$,

for $G(\omega) \triangleq \begin{cases} 1, & |\omega| \leq \omega_0 \\ 0, & |\omega| > \omega_0. \end{cases}$

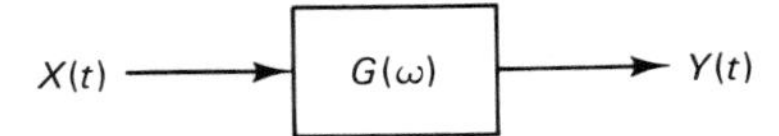

Figure P9.5–1

(b) Alternatively we may, at least formally, excite the system directly with a standard white noise $W(t)$, with mean zero and $K_W(\tau) = \delta(\tau)$. Call the output $V(t)$ as shown in Figure P9.5–2. Find the output psd $S_{VV}(\omega)$

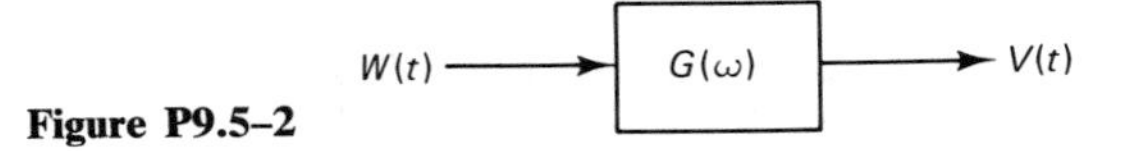

Figure P9.5–2

(c) Show that for $|\omega_0\tau_0| \ll 1$, $S_{YY} \simeq S_{VV}$ and find an upper bound on the *power error*

$$|R_{VV}(0) - R_{YY}(0)|.$$

9.6. Consider the LSI system shown below

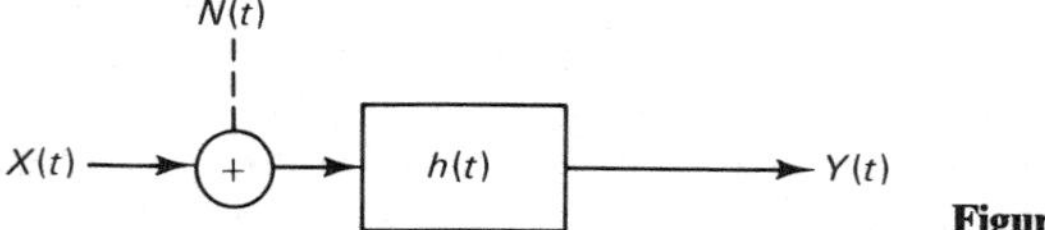

Figure P9.6

Let $X(t)$ and $N(t)$ be WSS and mutually uncorrelated with power spectral densities $S_{XX}(\omega)$ and $S_{NN}(\omega)$ and zero means.

(a) Find the psd of the output $Y(t)$.

(b) Find the cross-power spectral density of X and Y.

(c) Define the error $\varepsilon(t) \triangleq Y(t) - X(t)$ and evaluate the psd of $\varepsilon(t)$.

(d) Assume that $h(t) = a\delta(t)$ and choose the value of a which minimizes $E[\varepsilon^2(t)] = R_{\varepsilon\varepsilon}(0)$.

9.7. The power spectral density of a random sequence is given as $S[\omega] = 1/[(1 + \alpha^2) - 2\alpha \cos \omega]^2$ for $-\pi \leq +\pi$. Find its correlation function $R[m]$.

9.8. Derive a similar result to that of Problem 9.2 in the discrete-time case, that is, $W[n]$ a white-noise random sequence with $\mu_W[n] = 0$ and $K_W[m] = \delta[m]$.

√ **9.9.** Let the WSS random process $X(t)$ be the input to the third-order m.-s. differential equation,

$$\frac{d^3Y}{dt^3} + a_2\frac{d^2Y}{dt^2} + a_1\frac{dY}{dt} + a_0Y(t) = X(t),$$

with WSS output random process $Y(t)$.

(a) Put this equation into the form of a first-order vector m.s. differential equation,

$$\frac{d\mathbf{Y}}{dt} = \mathbf{A}\mathbf{Y}(t) + \mathbf{B}\mathbf{X}(t).$$

by defining $\mathbf{Y}(t) \triangleq \begin{bmatrix} Y(t) \\ \dot{Y}(t) \\ \ddot{Y}(t) \end{bmatrix}$ and $\mathbf{X}(t) \triangleq [X(t)]$ and evaluating the matrices $\mathbf{A}$ and $\mathbf{B}$.

(b) Find a first order matrix-differential equation for $\mathbf{R}_{XY}(\tau)$ with input $\mathbf{R}_{XX}(\tau)$.

(c) Find a first-order matrix-differential equation for $\mathbf{R}_{YY}(\tau)$ with input $\mathbf{R}_{XY}(\tau)$.

(d) Using matrix Fourier transforms, show that the output psd matrix $\mathbf{S}_{YY}$ is given as

$$\mathbf{S}_{YY}(\omega) = (\omega\mathbf{I} - \mathbf{A})^{-1}\mathbf{B}\mathbf{S}_{XX}(\omega)\mathbf{B}^{\dagger}(-j\omega\mathbf{I} - \mathbf{A}^{\dagger})^{-1}.$$

9.10. Let $\mathbf{X}(t)$ be a WSS vector random process, which is input to the LSI system with impulse response matrix $\mathbf{h}(t)$.

(a) Show that the correlation matrix of the output $\mathbf{Y}(t)$ is given by Equation 9.5–3.

(b) Derive the corresponding equation for matrix covariance functions.

9.11. Prove Theorem 9.5–2 by an argument analogous to that used in the proof of Theorem 9.5–1.

9.12. Derive Equation 9.6–5 directly from the definition of the Hilbert-transform system function,

$$H(\omega) \triangleq -jsgn(\omega).$$

where $j = \sqrt{-1}$ and $\text{sgn}(\omega) = 1$ for $\omega > 0$ and -1 for $\omega < 0$.

9.13. If a stationary random process is periodic, then we can represent it by a Fourier series with orthogonal coefficients. This is not true in general when the random process though stationary is not periodic. Thus point out the fallacy in the following proposition, which purports to show that the Fourier series coefficients are always orthogonal: First take a segment of length T from a stationary random process $X(t)$. Repeat the corresponding segment of the correlation function periodically. This then corresponds to a periodic random process. If we expand this process in a Fourier series, its coefficients will be orthogonal. Furthermore, the periodic process and the original process will agree over the original time interval.

9.14. Prove Theorem 9.6–3 by showing that

$$\phi_n(t) = \frac{1}{\sqrt{T}}\exp(j2\pi f_0 nt)$$

are the Karhunen-Loève basis functions for a WSS periodic random process. (Refer to Section 8.5.)

9.15. In geophysical signal processing one often has to simulate a multichannel random process. The following problem brings out an important constraint on the power spectral density matrix of such a vector random process. Let the N-dimensional vector random process $\mathbf{X}(t)$ be wide-sense stationary (WSS) with correlation matrix

$$\mathbf{R}_X(t) \triangleq E[\mathbf{X}(t+\tau)\mathbf{X}^{\dagger}(t)]$$

and power spectral density matrix

$$\mathbf{S}_X(\omega) \triangleq FT\{\mathbf{R}_X(\tau)\}.$$

Here $FT\{\cdot\}$ denotes the matrix Fourier transform, that is, the (i, j)th component of $\mathbf{S}_X$ is the Fourier transform of the (i, j)th component of $\mathbf{R}_X$, which is $E[X_i(t+\tau)X_j^*(t)]$ where $X_i(t)$ is the ith component of $\mathbf{X}(t)$.

(a) For constants $a_1, \ldots, a_N$ define the WSS scalar process,

$$Y(t) \triangleq \sum_{i=1}^{N} a_i X_i(t).$$

Find the power spectral density of $Y(t)$ in terms of the components of the matrix $\mathbf{S}_X(\omega)$.

(b) Show that the power spectral density matrix $\mathbf{S}_X(\omega)$ must be a positive semidefinite matrix for each fixed ω; that is, we must have $\mathbf{a}^T\mathbf{S}_X(\omega)\mathbf{a}^* \geq 0$ for all complex column vectors $\mathbf{a}$.

9.16. Consider the linear system shown in Figure P9.16 excited by the two *orthogonal*, zero-mean, jointly wide-sense stationary random processes $X(t)$, "the signal," and $U(t)$, "the noise." Then the input to the system G is

$$Y(t) = h(t) * X(t) + U(t),$$

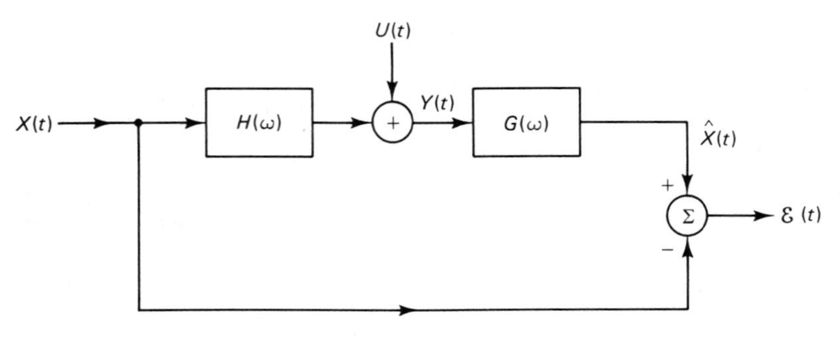

Figure P9.16

which models a distorted-signal-in-noise estimation problem. If we pass this $Y(t)$, "the received signal" through the filter G, we get an estimate $\hat{X}(t)$. Finally $\varepsilon(t)$ can be thought of as the "estimation error"

$$\varepsilon(t) = \hat{X}(t) - X(t).$$

In this problem we will calculate some relevant power spectral densities and cross-power spectral density for this problem.

(a) Find $S_{YY}(\omega)$.

(b) Find $S_{\hat{X}X}(\omega) = S^*_{X\hat{X}}(\omega)$. In terms of H, G, S_{XX} and S_{UU}.

(c) Find $S_{\varepsilon\varepsilon}(\omega)$.

(d) Use your answer to part (c) to show that to minimize $S_{\varepsilon\varepsilon}(\omega)$ at those frequencies where

$$S_{XX}(\omega) \gg S_{UU}(\omega),$$

we should have $G \approx H^{-1}$ and where

$$S_{XX}(\omega) \ll S_{UU}(\omega)$$

we should have $G \approx 0$.

9.17. Let $X(t)$, the input to the system in Figure P9.17 be a stationary Gaussian random process.

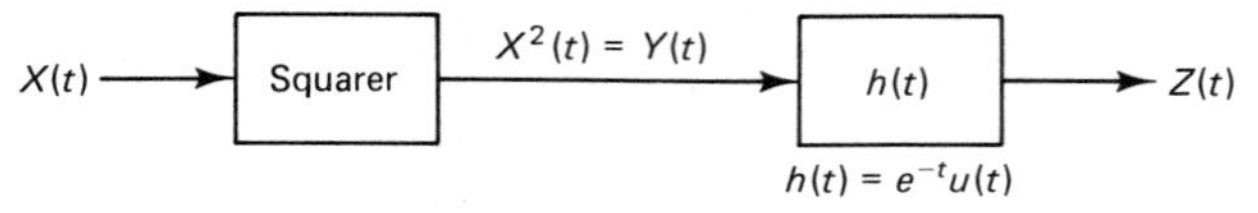

Figure P9.17

The power spectral density of $Z(t)$ is measured experimentally and found to be

$$S_Z(\omega) = \pi\delta(\omega) + \frac{2\beta}{(\omega^2 + \beta^2)(\omega^2 + 1)}$$

(a) Find the correlation function of $Y(t)$ in terms of β.
(b) Find the correlation function of $X(t)$.

9.18. Prove an analogous result to Theorem 9.6–3 for WSS periodic random sequences. Hint: Perform the expansion for $X[n]$ over $0 \leq n \leq T - 1$ where T = integer. Only use T Fourier coefficient $A_0, \ldots, A_{T-1}$.

REFERENCES

9–1. E. Wong & B. Hajek, *Stochastic Processes in Engineering Systems.* New York: Springer-Verlag, 1985, pp. 53–54.

9–2. J. L. Doob, *Stochastic Processes.* New York; John Wiley, 1953, p. 474.

9–3. A. V. Oppenheim and R. W. Schafer, *Digital Signal Processing,* Englewood Cliffs, N. J.: Prentice-Hall, 1975, Section 2.4.

9–4. T. Kailath, *Linear Systems.* Englewood Cliffs, N. J.: Prentice-Hall, 1980, pp. 161–166.

9–5. E. Hille, *Analytic Function Theory, Vol. I.* New York: Blaisdell, 1965, Chapters 7 and 9.

10

ESTIMATION THEORY II

In this chapter we will continue the study of estimation that was begun in Chapter 5. Here we will focus on the estimation and prediction of random sequences and random processes. We begin with a review of the conditional mean and its fundamental role in estimation problems that use the mean-square error criteria. We extend the estimates of Section 5.6 to complex random vectors. Then we look at linear estimation and the orthogonality principle and introduce the optimal linear estimation operator. We also will consider innovations sequences for prediction problems and then generalize our results to obtain equations for noisy prediction and estimation. After this we turn to continuous time and study the Wiener filter for random processes.

10.1 THE CONDITIONAL MEAN REVISITED

In Section 5.6 we introduced the concept of the minimum mean-square error (MMSE) estimate of one random vector from another. Theorem 5.6–1 showed that this optimal estimate was equal to the conditional mean. Specifically the MMSE estimate of the random vector $\mathbf{X}$ based on observation of the random vector $\mathbf{Y}$ is $E[\mathbf{X} \mid \mathbf{Y}]$, which is Equation 5.6–6. Note that the roles of $\mathbf{X}$ and $\mathbf{Y}$ are interchanged as compared to Chapter 5 because of the different emphasis there.

To extend this result to complex random vectors, we define the MSE as

$$\begin{aligned}\varepsilon_{\min} &= E[\|\mathbf{X} - \mathbf{g}(\mathbf{Y})\|^2] \\ &= \sum_{i=1}^{N} E[|X_i - g_i(\mathbf{Y})|^2], \qquad (10.1\text{–}1)\end{aligned}$$

generalizing Equation 5.6–5, and we must now show that $g_i(\mathbf{Y}) = E[X_i \mid \mathbf{Y}]$ minimizes each term in this sum of square values. Thus we need a version of Theorem 5.6–1 for complex random variables. This can be established by first modifying the MSE in the proof of Theorem 5.6–1 to

$$\begin{aligned} \varepsilon &= E[|X - E[X \mid \mathbf{Y}] - \delta g(\mathbf{Y})|^2] \\ &= E[|X - E[X \mid \mathbf{Y}]|^2] - 2\,\mathrm{Re}[(X - E[X \mid \mathbf{Y}])\delta g^*] + E[|\delta g|^2] \end{aligned}$$

and then proceeding to show that the cross-term is zero, using the smoothing property of conditional expectation, thus obtaining

$$\varepsilon = E[|X - E[X \mid \mathbf{Y}]|^2] + E[|\delta g(\mathbf{Y})|^2]. \tag{10.1–2}$$

From this equation, we conclude that δg is zero just as before. Thus $g_i(\mathbf{Y}) = E[X_i \mid \mathbf{Y}]$ will minimize each term in Equation 10.1–1, thus establishing the conditional mean as the MSE optimal estimate for complex random vectors.

If we observe a complex random sequence starting at $n \geq 0$, we can consider the problem of estimating $X[n]$ based on the present and past of $Y[n]$. Based on the preceding results, we have immediately upon definition of the random vector,

$$\mathbf{Y}_n \triangleq [Y[n], Y[n-1], \ldots, Y[0]]^T,$$

that the MMSE estimate of $X[n]$, denoted $\hat{X}[n]$, is

$$\hat{X}[n] = E[X[n] \mid \mathbf{Y}_n]$$

at each n. Such an estimate is called *causal* or sometimes a *filter* estimate since it does not involve the input sequence Y at future times $m > n$. For infinite-length observations Y, one can define the estimate of $X[n]$ based on the vector $\mathbf{Y}_N^{(n)} \triangleq [Y[n], Y[n-1], \ldots, Y[n-N]]^T$ and then let N go to infinity to define the mean-square limit under appropriate conditions,

$$\lim_{N \to \infty} E[X[n] \mid \mathbf{Y}_N^{(n)}], \tag{10.1–3}$$

thus defining the expectation conditioned on infinite length sequences. One can also show that this limit exists with probability-1 by using Martingale theory. (cf. Section 6.5.) We can show (see Problem 6.21) that for each fixed n, the random sequence with time parameter N,

$$G[N] \triangleq E[X[n] \mid \mathbf{Y}_N^{(n)}] \tag{10.1–4}$$

is a Martingale. We thus can use the Martingale Convergence Theorem 6.5–4 to conclude the probability-1 existence of the limit (Equation 10.1–2) if the variance of the random sequence $G[N]$ is uniformly bounded in the sense of the theorem; that is, for all $N \geq 1$,

$$\sigma_G^2(N) \leq C < \infty \quad \text{for some } C. \tag{10.1–5}$$

In fact, this variance is bounded by the variance of $X[n]$. We leave the demonstration of this result to the reader (Problem 10.2).

Similar expressions can be obtained for random processes under appropriate continuity conditions. For instance, by conditioning on ever more dense samplings of a random process $Y(t)$ over ever larger intervals of the past, one can define the conditional mean of the random process $X(t)$ based on causal observation of the random process $Y(t)$,

$$E[X(t) \mid Y(\tau), \tau \leq t]. \tag{10.1–6}$$

We will consider the corresponding linear estimation problem for random processes in Section 10.6.

We also learned in Section 5.6 that when the observations Y_i are Gaussian and zero-mean, then the conditional mean is linear in these conditioning random variables (compare to Theorem 5.6–2). For the case of estimating the random vector $\mathbf{X}$ from the random vector $\mathbf{Y}$, this becomes

$$E[\mathbf{X} \mid \mathbf{Y}] = \mathbf{AY}, \tag{10.1–7}$$

where the coefficients A_{ij} are determined by the orthogonality conditions

$$(\mathbf{X} - \mathbf{AY})_i \perp Y_k \qquad 1 \leq i, k \leq N, \tag{10.1–8}$$

which is just the orthogonality condition of Theorem 5.6–2 applied to the estimation of each component X_i of $\mathbf{X}$. The proof of this theorem goes through just the same with the proper definition of the complex Gaussian random vector. However, the definition is somewhat restrictive and the development would take us too far afield. The interested reader may consult [10–1] for the complex Gaussian theory, which is often applied to model narrowband data, such as in Section 9.6.

10.2 ORTHOGONALITY AND LINEAR ESTIMATION

In the previous section we saw that the MMSE estimate is given by the conditional mean and that it is linear in the observations when the data is jointly Gaussian. Unfortunately, in the non-Gaussian case the conditional mean estimate can be nonlinear and is often very difficult to obtain. In general, the derivation of an optimal nonlinear estimate will depend on higher-order moment functions that may not be available. For these reasons, in this section we concentrate on the best linear estimate for minimizing the MSE. We denote this estimate as LMMSE, standing for *linear* minimum mean-square error. Of course, for Gaussian data the LMMSE and the MMSE estimators are the same. Sometimes we will use the phrase *optimal linear* to describe the LMMSE estimate or estimator.

Consider the random sequence $Y[n]$ observed for $n \geq 0$. We wish to linearly estimate the random signal sequence $X[n]$. We assume both sequences are zero mean to simplify the discussion, since the student should be able to extend the argument to the nonzero mean case with no difficulty. We denote the *LMMSE estimate* by

$$\hat{E}[X[n] \mid Y[n], \ldots, Y[0]], \tag{10.2–1}$$

where the hat distinguishes this linear estimate from the nonlinear conditional mean estimate $E[X[n] \mid Y[n], \ldots, Y[0]]$. For the moment, we will treat Equation 10.2–1 as just a notation for the LMMSE estimate, but at the end of the section we will introduce the $\hat{E}$ operator. The following theorem establishes that the LMMSE estimate for a complex random sequence is determined by the orthogonality principle.

Theorem 10.2–1. The LMMSE estimate of the zero-mean random sequence $X[n]$ based on the zero-mean random sequence $Y[n]$, is given as

$$\hat{E}[X[n] \mid Y[n], \ldots, Y[0]] = \sum_{i=0}^{n} a_i^{(n)} Y[i],$$

where the $a_i^{(n)}$'s satisfy the orthogonality condition,

$$\left[X[n] - \sum_{i=0}^{n} a_i^{(n)} Y[i]\right] \perp Y[k], \qquad 0 \leq k \leq n.$$

Furthermore, the LMMSE is given by

$$\varepsilon_{\min} = E[|X[n]|^2] - \sum_{i=0}^{n} a_i^{(n)} E[Y[i]X^*[n]]. \tag{10.2–2}$$

Proof. Let the $a_i^{(n)}$'s be the coefficients determined by the orthogonality principle and let $b_i^{(n)}$ be some other set of coefficients. Then we can write the error using this other set of coefficients as

$$X[n] - \sum_{i=0}^{n} b_i^{(n)} Y[i] = \left[X[n] - \sum_{i=0}^{n} a_i^{(n)} Y[i]\right] + \sum_{i=0}^{n} (a_i^{(n)} - b_i^{(n)}) Y[i],$$

where we have both added and subtracted $\sum a_i^{(n)} Y[i]$. Because the first term on the right of the equal sign is orthogonal to $Y[i]$ for $i = 0, \ldots, n$ we have

$$\left[X[n] - \sum_{i=0}^{n} a_i^{(n)} Y[i]\right] \perp \sum_{i=0}^{n} (a_i^{(n)} - b_i^{(n)}) Y[i],$$

which implies

$$\begin{aligned} E\left[\left|X[n] - \sum_{i=0}^{n} b_i^{(n)} Y[i]\right|^2\right] &= E\left[\left|X[n] - \sum_{i=0}^{n} a_i^{(n)} Y[i]\right|^2\right] \\ &\quad + E\left[\left|\sum_{i=0}^{n} (a_i^{(n)} - b_i^{(n)}) Y[i]\right|^2\right] \\ &\geq E\left[\left|X[n] - \sum_{i=0}^{n} a_i^{(n)} Y[i]\right|^2\right] \end{aligned}$$

with equality if $a_i^{(n)} = b_i^{(n)}$ for $i = 0, \ldots, n$. This result is just the same as in Equation 10.1–2 only using the notation of random sequences.

To evaluate the MSE, we compute

$$\varepsilon = E\left[\left|X[n] - \sum a_i^{(n)} Y[i]\right|^2\right]$$

$$= E\left[\left(X[n] - \sum a_i^{(n)} Y[i]\right) X^*[n]\right]$$

$$- E\left[\left(X[n] - \sum a_i^{(n)} Y[i]\right)\left(\sum a_i^{(n)} Y[i]\right)^*\right].$$

We note also that

$$\left[X[n] - \sum_{i=0}^{n} a_i Y[i]\right] \perp \sum_{i=0}^{n} a_i^{(n)} Y[i],$$

and thus by orthogonality we have

$$\varepsilon = E[|X[n]|^2] - \sum_{i=0}^{n} a_i^{(n)} E[Y[i] X^*[n]]$$

$$= \sigma_X^2[n] - \sum_{i=0}^{n} a_i^{(n)} K_{YX}[i, n]. \quad \blacksquare$$

Solving for $a_i^{(n)}$, we suppress the superscript (n) for notational simplicity and write out the orthogonality condition of the preceding theorem:

$$E[X[n] Y^*[k]] = \sum_{i=0}^{n} a_i E[Y[i] Y^*[k]], \qquad 0 \leq k \leq n. \quad (10.2\text{–}3)$$

We define the column vector,

$$\mathbf{a} \triangleq [a_0, a_1, \ldots, a_n]^T$$

and row vector,

$$\mathbf{k_{XY}} \triangleq E[X[n] Y^*[0], X[n] Y^*[1], \ldots, X[n] Y^*[n]]$$

and covariance matrix

$$\mathbf{K_{YY}} \triangleq E[\mathbf{YY}^\dagger], \qquad \mathbf{Y} \triangleq (Y[0], \ldots, Y[n])^T$$

Then Equation 10.2–3 becomes

$$\mathbf{k_{XY}} = \mathbf{a}^T \mathbf{K_{YY}},$$

with solution

$$\mathbf{a}^T = \mathbf{k_{XY}} \mathbf{K_{YY}^{-1}}. \quad (10.2\text{–}4)$$

The MSE of Equation 10.2–2 then becomes

$$\varepsilon_{\min} = \sigma_X^2[n] - \mathbf{k_{XY}} \mathbf{K_{YY}^{-1}} \mathbf{k_{XY}^\dagger}. \quad (10.2\text{–}5)$$

One comment on the error expression is that the maximum error output from an LMMSE estimator is $\sigma_X^2[n]$, which is obtained by setting the $a_i^{(n)}$'s equal to zero. This is optimal when there is no cross-covariance between the observations and the signal $X[n]$. Any nonzero cross-convariance causes the

MSE to decrease from $\sigma_X^2[n]$ down to Equation 10.2–5, the amount of *decrease* being

$$\mathbf{k}_{XY}\mathbf{K}_{YY}^{-1}\mathbf{k}_{XY}^{\dagger}.$$

We next look at an example of the above linear estimation procedure applied to the problem of estimating a signal contaminated by white noise.

Example 10.2–1: (Estimation of a signal in noise.) Assume we have a random signal sequence $X[n]$, which is immersed in white noise $V[n]$ of variance σ_V^2, where the signal and noise are uncorrelated and zero mean. Let the observations be for $n \geq 0$,

$$Y[n] = X[n] + V[n], \qquad X \perp V.$$

We want to determine the causal LMMSE estimate $\hat{X}[n]$,

$$\begin{aligned}\hat{X}[n] &= \hat{E}[X[n] \mid Y[n], \ldots, Y[0]] \\ &= \sum_{i=0}^{n} a_i^{(n)} Y[i],\end{aligned}$$

in terms of the covariance function of the signal X and the variance of the white noise V. From Theorem 10.2–1 the coefficients of the optimal linear estimator $a_i^{(n)}$ will be determined by the orthogonality conditions (Equation 10.2–3) specialized to this example. The solution is thus Equation 10.2–4, which must be solved for each value of $n \geq 0$. It remains to determine the covariance matrix $\mathbf{K}_{YY}$ and cross-covariance vector $\mathbf{k}_{XY}$. Looking at the ijth component of $\mathbf{K}_{YY}$ we compute

$$\begin{aligned}(\mathbf{K}_{YY})_{ij} &= E[Y[i]Y^*[j]] \\ &= E[X[i]X^*[j]] + E[V[i]V^*[j]] \\ &= (\mathbf{K}_{XX})_{ij} + \sigma_V^2\delta_{ij}.\end{aligned}$$

For the $\mathbf{k}_{XY}$ we obtain

$$\begin{aligned}(\mathbf{k}_{XY})_i &= E[X[n]Y^*[i]] = E[X[n](X[i] + V[i])^*] \\ &= E[X[n]X^*[i]] \triangleq (\mathbf{k}_X)_i, \qquad 0 \leq i \leq n,\end{aligned}$$

since the signal is orthogonal to the noise. Thus we obtain the estimator coefficient vector

$$\mathbf{a}^T = \mathbf{k}_X[\mathbf{K}_{XX} + \sigma_V^2\mathbf{I}]^{-1}.$$

A special case that allows considerable simplification of the preceding example is when the signal covariance is also diagonal; that is, $X[n]$ is also a white noise,

$$\mathbf{K}_{XX} \triangleq \sigma_X^2\mathbf{I} \quad \text{and} \quad \mathbf{k}_X \triangleq \sigma_X^2[0, 0, \ldots, 0, 1].$$

Then the solution for the coefficient vector is

$$\mathbf{a}^T = \frac{\sigma_X^2}{\sigma_X^2 + \sigma_V^2}[0, 0, \ldots, 0, 1],$$

which means that $a_i^{(n)} = 0$ except for $i = n$, so that the LMMSE estimate is

$$\hat{X}[n] = [\sigma_X^2/(\sigma_X^2 + \sigma_V^2)]Y[n].$$

A situation where this special case would arise is when we estimate the coefficients of a Karhunen-Loeve expansion of a signal process from the corresponding expansion coefficients of the random signal process with additive white noise (see Example 8.5–2).

In general, the signal random sequence is not white and the estimate takes the *growing-memory* form:

$$\hat{X}[n] = \sum_{i=0}^{n} a_i^{(n)} Y[i].$$

It is called growing memory because all past data must be used in the current estimate. The MSE is given by the formula

$$E[|X[n] - \hat{X}[n]|^2] = \sigma_X^2[n] - \sum_{i=0}^{n} a_i^{(n)} K_{YX}[i, n].$$

We next turn to the application of these results to the prediction problem for random sequences.

Example 10.2–2: If we set $X[n] \triangleq Y[n+1]$ for all n, we can use the result of Theorem 10.2–1 to evaluate the LMMSE estimate

$$\hat{E}[X[n] \mid X[n-1], \ldots, X[0]],$$

which is called the LMMSE *one-step predictor* for $X[n]$. Specializing our results, we replace n by $n-1$ and obtain

$$\hat{X}[n] = \sum_{i=0}^{n-1} a_i^{(n)} X[i],$$

where

$$\begin{aligned} \mathbf{a}^{(n)T} &= \mathbf{k}_{\mathbf{X}} \mathbf{K}_{\mathbf{X}}^{-1} \\ &= \{K_{\mathbf{X}}[n, j] K_{\mathbf{X}}[i, j]\}^{-1}, \end{aligned}$$

where $0 \leq i, j \leq n-1$.

In the wide-sense stationary case, the matrix $\mathbf{K}_{\mathbf{X}}$ is Toeplitz; that is,

$$(\mathbf{K}_{\mathbf{X}})_{ij} = g[i-j]$$

for some g since the covariance depends only on the difference of the two time parameters. Efficient algorithms exist for computing the inverse of Toeplitz matrices, which allow the recursive calculation of the coefficient vectors $\mathbf{a}^{(n)}$ for increasing n. Such an algorithm is the Levinson algorithm, which is described in [10–2]. Linear prediction, as the foregoing is called, is widely used in speech analysis, synthesis, and coding [10–3].

One difficulty in the preceding approach is that the resulting predictors and estimators, though linear, are nevertheless growing memory except in the simplest cases. We will overcome this problem in Section 10.3 by incorporating a Markov signal model. We first pause briefly to discuss the properties of the LMMSE operator.

Some Properties of the Operator Ê. We have introduced the symbol $\hat{E}$ in Equation 10.2–1 for the LMMSE linear estimate. Here we regard $\hat{E}$ as an operator and establish certain linearity properties of this operator that will be useful in the next section. We will see that this operator can be used to simplify the derivation of important results in linear estimation.

Theorem 10.2–2. The operator $\hat{E}$ has the following additional linearity properties:

$$\text{(a)}\quad \hat{E}[X_1 + X_2 \mid \mathbf{Y}] = \hat{E}[X_1 \mid \mathbf{Y}] + \hat{E}[X_2 \mid \mathbf{Y}]$$

and

(b) when $\mathbf{Y}_1$ and $\mathbf{Y}_2$ are orthogonal,

$$\hat{E}[X \mid \mathbf{Y}_1, \mathbf{Y}_2] = \hat{E}[X \mid \mathbf{Y}_1] + \hat{E}[X \mid \mathbf{Y}_2].$$

Proof. To prove (a) we note that

$$\hat{E}[X \mid \mathbf{Y}] = \mathbf{c}^T\mathbf{Y},$$

where the vector $\mathbf{c}$ is given as

$$\mathbf{c}^T = \mathbf{k}_{X\mathbf{Y}}\mathbf{K}_{\mathbf{YY}}^{-1}.$$

Clearly $\mathbf{c} = \mathbf{c}_1 + \mathbf{c}_2$ since $\mathbf{k}_{X\mathbf{Y}} = \mathbf{k}_{X_1\mathbf{Y}} + \mathbf{k}_{X_2\mathbf{Y}}$; thus

$$\mathbf{c}^T\mathbf{Y} = \mathbf{c}_1^T\mathbf{Y} + \mathbf{c}_2^T\mathbf{Y}.$$

To show (b) we note that since $\mathbf{Y}_1$ and $\mathbf{Y}_2$ are orthogonal; that is, $E[\mathbf{Y}_1\mathbf{Y}_2^\dagger] = \mathbf{0}$, which we write as $\mathbf{Y}_1 \perp \mathbf{Y}_2$, then the orthogonalities of the individual estimates

$$(X - \mathbf{c}_1^T\mathbf{Y}_1) \perp \mathbf{Y}_1 \quad \text{and} \quad (X - \mathbf{c}_2^T\mathbf{Y}_2) \perp \mathbf{Y}_2$$

imply that

$$(X - \mathbf{c}_1^T\mathbf{Y}_1 - \mathbf{c}_2^T\mathbf{Y}_2) \perp \mathbf{Y}_1 \quad \text{and} \quad \mathbf{Y}_2,$$

which can be seen from Figure 10.2–1. Then (b) follows. ■

With reference to Figure 10.2–1, we see that the operator $\hat{E}$ projects the signal vector X onto the linear subspace spanned by the observation vectors $\mathbf{Y}_i$. Thus $\hat{E}$ is sometimes referred to as an *orthogonal projection*

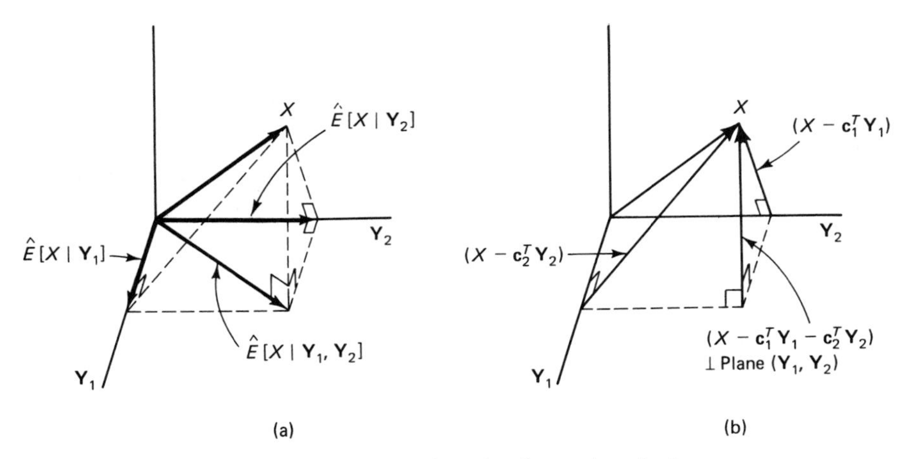

Figure 10.2–1 Illustration of orthogonal projection.

operator. Geometrically, it is clear that such an orthogonal projection will minimize the error in an estimate, which by linearity is constrained to lie in the subspace spanned by the observation vectors. Property (a) then says that the orthogonal projection of the sum of two vectors is the sum of their orthogonal projections, a result that is geometrically satisfying. Property (b) says that the orthogonal projection onto a linear subspace can be computed by summing the orthogonal projections onto each of its orthogonal basis vectors.

All this reinforces the Hilbert or linear-space concept of random variables introduced in Section 8.1–1, where we defined the norm

$$\|X\|^2 \triangleq E[|X|^2]$$

and inner product

$$(X, Y) = E[XY^*].$$

It is important to note that the word "vector" above does not refer to the number of components of $\mathbf{X}$ but rather to the vector space of random variables.

10.3 INNOVATIONS SEQUENCES AND PREDICTION

In this section we look at the use of signal models to avoid the growing memory aspect of the prediction solution found in the last section. We first consider the vector difference equation or state equation driven by a white random input sequence $\mathbf{W}[n]$,

$$\mathbf{X}[n] = \mathbf{AX}[n-1] + \mathbf{BW}[n], \qquad n \geq 0, \tag{10.3–1}$$

with $\mathbf{X}[-1] = \mathbf{X}_{-1}$ given and

$$\mathbf{R_W}[m] = E[\mathbf{W}[m+n]\mathbf{W}^\dagger[n]] = \boldsymbol{\sigma}^2_{\mathbf{W}}\delta[m],$$

and $\mathbf{W}[n]$ is orthogonal to the past of $\mathbf{X}[n]$, including $\mathbf{X}_{-1}$. In symbols this becomes

$$\mathbf{W}[n] \perp \mathbf{X}[m] \quad \text{for} \quad m < n \quad \text{and} \quad \mathbf{W}[n] \perp \mathbf{X}_{-1} \quad \text{for} \quad n \geq 0.$$

By Theorem 9.5–2, $\mathbf{X}[n]$ of Equation 10.3–1 is vector Markov. Using a technique analogous to that used in Problem 9.9, any scalar LCCDE driven by white noise can be put into the form Equation 10.3–1. The resulting matrix $\mathbf{A}$ will always be nonsingular if the dimension of the state vector is equal to the order of the scalar LCCDE. Thus we will assume $\mathbf{A}$ is a nonsingular matrix, (cf. Example 10.3–1 below).

Starting from the initial condition $\mathbf{X}_{-1}$ we can recursively compute the following forwards-in-time or causal solution,

$$\begin{aligned}
\mathbf{X}[0] &= \mathbf{AX}_{-1} + \mathbf{BW}[0], \\
\mathbf{X}[1] &= \mathbf{AX}[0] + \mathbf{BW}[1] \\
&= \mathbf{A}^2\mathbf{X}_{-1} + \mathbf{ABW}[0] + \mathbf{BW}[1], \\
\mathbf{X}[2] &= \mathbf{A}^3\mathbf{X}_{-1} + \mathbf{A}^2\mathbf{BW}[0] + \mathbf{ABW}[1] + \mathbf{BW}[2],
\end{aligned}$$

and so forth.

We thus infer the general solution

$$\mathbf{X}[n] = \sum_{k=0}^{n} \mathbf{A}^{n-k}\mathbf{B}\mathbf{W}[k] + \mathbf{A}^{n+1}\mathbf{X}_{-1}, \qquad (10.3\text{–}2)$$

the first term of which is just a convolution of the vector input sequence with the matrix impulse response,

$$\mathbf{H}[n] \triangleq \mathbf{A}^{n}\mathbf{B}u[n].$$

It is important to note that Equation 10.3–2 is a causal and linear transformation from the space of input random sequences (including $\mathbf{X}_{-1}$) to the space of output random sequences. If we use Equation 10.3–1 and assume that $\mathbf{B}^{-1}$ exists, we can also write the input sequence as a causal linear transformation on the output random sequence,

$$\mathbf{W}[n] = \mathbf{B}^{-1}[\mathbf{X}[n] - \mathbf{A}\mathbf{X}[n-1]], \qquad n \geq 0.$$

We say the white sequence $\mathbf{W}[n]$ is *causally equivalent* to $\mathbf{X}[n]$ and call it the *innovations sequence* associated with $\mathbf{X}$. The name comes from the fact that $\mathbf{W}$ contains the new information obtained when we observe $\mathbf{X}[n]$ given the past $[\mathbf{X}[n-1], \mathbf{X}[n-2], \ldots, \mathbf{X}[0]]$. This statement will subsequently become more clear. In general, we make the following definition.

Definition 10.3–1. The *innovations sequence* of a random sequence $\mathbf{X}[n]$ is defined to be a white random sequence, which is a causal and causally invertible linear transformation of the sequence $\mathbf{X}[n]$.

It follows immediately that we can write

$$\hat{E}[\mathbf{X}[n] \,|\, \mathbf{X}[n-1], \ldots, \mathbf{X}[0], \mathbf{X}_{-1}] = \hat{E}[\mathbf{X}[n] \,|\, \mathbf{W}[n-1], \ldots, \mathbf{W}[0], \mathbf{X}_{-1}],$$

because the LMMSE estimate can always undo a causally invertible linear transformation, that is, the required inverse can, if needed, be part of the general causal linear transformation $\hat{E}$. To see the benefit of the innovations concept consider evaluating

$$\hat{E}[\mathbf{X}[n] \,|\, \mathbf{W}[n-1], \ldots, \mathbf{W}[0], \mathbf{X}_{-1}].$$

Expressing $\mathbf{X}[n]$ with Equation 10.3–2, we see

$$\mathbf{X}[n] = \sum_{k=0}^{n-1} \mathbf{A}^{n-k}\mathbf{B}\mathbf{W}[k] + \mathbf{B}\mathbf{W}[n] + \mathbf{A}^{n+1}\mathbf{X}_{-1}.$$

Applying the $\hat{E}$ operator, and using linearity property (a) of Theorem 10.2–2, we obtain

$$\begin{aligned}
&\hat{E}[\mathbf{X}[n] \,|\, \mathbf{W}[n-1], \ldots, \mathbf{W}[0], \mathbf{X}_{-1}] \\
&\quad = \sum_{k=0}^{n-1} \mathbf{A}^{n-k}\mathbf{B}\hat{E}[\mathbf{W}[k] \,|\, \mathbf{W}[n-1], \ldots, \mathbf{W}[0], \mathbf{X}_{-1}] \\
&\qquad + \mathbf{B}\hat{E}[\mathbf{W}[n] \,|\, \mathbf{W}[n-1], \ldots, \mathbf{W}[0], \mathbf{X}_{-1}] \\
&\qquad + \mathbf{A}^{n+1}\hat{E}[\mathbf{X}_{-1} \,|\, \mathbf{W}[n-1], \ldots, \mathbf{W}[0], \mathbf{X}_{-1}].
\end{aligned}$$

Then repeatedly using linearity property (b) of the same Theorem, since the innovations $[\mathbf{W}[n-1], \ldots, \mathbf{W}[0], \mathbf{X}_{-1}]$ are orthogonal, we have

$$\hat{E}[\mathbf{X}[n] \mid \mathbf{W}[n-1], \ldots, \mathbf{W}[0], \mathbf{X}_{-1}] = \sum_{k=0}^{n-1} \mathbf{A}^{n-k}\mathbf{B}\mathbf{W}[k] + \mathbf{A}^{n+1}\mathbf{X}_{-1}. \tag{10.3–3}$$

From Equation 10.3–2 we also have

$$\mathbf{X}[n-1] = \sum_{k=0}^{n-1} \mathbf{A}^{n-1-k}\mathbf{B}\mathbf{W}[k] + \mathbf{A}^{n}\mathbf{X}_{-1},$$

so combining, we get

$$\hat{\mathbf{X}}[n] = \mathbf{A}\mathbf{X}[n-1]. \tag{10.3–4}$$

This is the final form of the LMMSE predictor for the state equation model with a white random input sequence. (Note: We are assuming that the mean of the input sequence is zero. This is incorporated in the white noise definition.) The overall operation is shown in Figure 10.3–1.

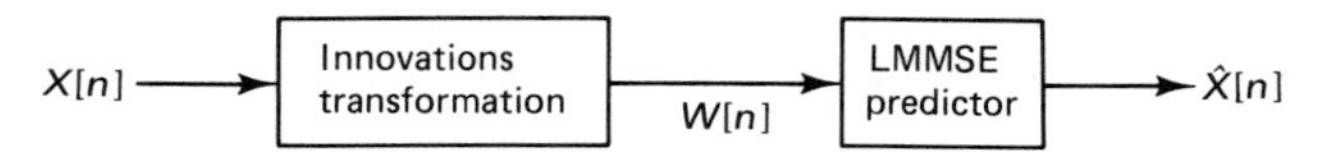

Figure 10.3–1 Innovations decomposition of LMMSE predictor.

Equation 10.3–4 can also be derived by applying the $\hat{E}$ operator directly to Equation 10.3–1 and using the fact that $\mathbf{X}[n-1] \perp \mathbf{W}[n]$. We can thus view Equation 10.3–1 as a one-step innovations representation in the same sense that Equation 10.3–2 is an n-step innovations representation. The innovations method is of quite general use in linear estimation theory, the basic underlying concept being a representation of the observed data as an orthogonal decomposition. We will make better use of the innovations method in Section 10.4 to derive the Kalman-Bucy filter.

With reference to our state equation model (Equation 10.3–1) we note that if we added the condition that the driving noise $\mathbf{W}[n]$ be Gaussian, then $\mathbf{X}[n]$ would be both Gaussian and Markov. In this case the preceding LMMSE estimate would also be the MMSE estimate; that is, $\hat{E}$ would be E. This motivates the following weakening of the Markov property.

Definition 10.3–2. A random sequence $\mathbf{X}[n]$ is called *wide-sense Markov* if for all n

$$\hat{E}[\mathbf{X}[n] \mid \mathbf{X}[n-1], \mathbf{X}[n-2], \ldots] = \hat{E}[\mathbf{X}[n] \mid \mathbf{X}[n-1]].$$

We note that this definition has built in the concept of limited memory in that the LMMSE prediction cannot be changed by the incorporation of earlier data. From Equation 10.3–4, it follows immediately that the solution to Equation 10.3–1 is a wide-sense Markov random sequence. One can prove that any wide-sense Markov random sequence would satisfy a first-order vector state equation with a white input that is uncorrelated with the past of $\mathbf{X}[n]$.

Theorem 10.3–1. Let $\mathbf{X}[n]$ be a wide-sense Markov zero-mean sequence. Then there exists an innovations sequence $\mathbf{W}[n]$ such that Equation 10.3–1 is satisfied and the sequence $\mathbf{W}$ is orthogonal to the past of $\mathbf{X}$.

Proof. Since $\mathbf{X}$ is wide-sense Markov and zero-mean,

$$\hat{E}[\mathbf{X}[n] \mid \mathbf{X}[n-1], \mathbf{X}[n-2], \ldots] = \hat{E}[\mathbf{X}[n] \mid \mathbf{X}[n-1]] \triangleq \mathbf{A}\mathbf{X}[n-1],$$

where $\mathbf{A}$ is defined as the matrix of the indicated LMMSE one-step prediction. Next we define $\mathbf{W}$ as

$$\mathbf{W}[n] \triangleq \mathbf{X}[n] - \mathbf{A}\mathbf{X}[n-1].$$

It then follows that $E[\mathbf{W}[n]] = \mathbf{0}$ and

$$\mathbf{W}[n] \perp (\mathbf{X}[n-1], \mathbf{X}[n-2], \ldots)$$

because of the fact that $\mathbf{X}$ is wide-sense Markov and the prediction error must be orthogonal to the past data used in the prediction. Furthermore,

$$E[\mathbf{W}[n]\mathbf{W}^{\dagger}[m]] = \mathbf{0} \quad \text{for} \quad m < n,$$

since $\mathbf{W}[m]$ is a linear function of the past and present of $\mathbf{X}[m]$, which is in the past of $\mathbf{X}[n]$ for $m < n$. Thus

$$E[\mathbf{W}[n]\mathbf{W}^{\dagger}[m]] = \boldsymbol{\sigma}_{\mathbf{W}}^2[n]\,\delta[m-n],$$

where

$$\boldsymbol{\sigma}_{\mathbf{W}}^2[n] \triangleq E[\mathbf{W}[n]\mathbf{W}^{\dagger}[n]]. \quad \blacksquare$$

Example 10.3–1: Consider the second-order *scalar* difference equation driven by white noise,

$$X[n] = \alpha X[n-1] + \beta X[n-2] + W[n],$$

where $X[-1] = X[-2] = 0$ and where $W[n]$ has zero mean and variance $\text{Var}[W[n]] = \sigma_W^2$. To apply the vector wide-sense Markov prediction results, we construct the random vector

$$\mathbf{X}[n] \triangleq [X[n], X[n-1]]^{\mathrm{T}}$$

and obtain the first-order vector equation

$$\mathbf{X}[n] = \mathbf{A}\mathbf{X}[n-1] + \mathbf{b}W[n]$$

on setting

$$\mathbf{A} \triangleq \begin{pmatrix} \alpha & \beta \\ 1 & 0 \end{pmatrix} \quad \text{and} \quad \mathbf{b} \triangleq \begin{pmatrix} 1 \\ 0 \end{pmatrix}.$$

Using Equation 10.3–4 we then have

$$\hat{E}[\mathbf{X}[n] \mid \mathbf{X}[n-1], \ldots, \mathbf{X}[0]] = \mathbf{A}\mathbf{X}[n-1],$$

so that

$$\hat{E}\left[\begin{pmatrix} X[n] \\ X[n-1] \end{pmatrix} \middle| X[n-1], \ldots, X[0]\right] = \begin{pmatrix} \alpha & \beta \\ 1 & 0 \end{pmatrix}\begin{pmatrix} X[n-1] \\ X[n-2] \end{pmatrix}$$

or

$$\hat{X}[n] = \alpha X[n-1] + \beta X[n-2].$$

Sometimes such a scalar random sequence is called wide-sense *Markov of order* 2. More generally a p-th order, scalar LCCDE driven by white noise generates a scalar wide-sense Markov random sequence of order p.

Predicting Gaussian Random Sequences. Here we look at the special case where the random sequence is Gaussian so that the results of the last section having to do with wide-sense Markov become strict-sense Markov and the orthogonal random sequence $\mathbf{W}[n]$ becomes an independent random sequence. We start with a theorem, which is in part a restatement and in part a strengthening of some of the previous results.

Theorem 10.3–2. Let $\mathbf{X}[n]$ be a zero-mean, Gauss-Markov random sequence. Then $\mathbf{X}[n]$ satisfies the difference equation

$$\mathbf{X}[n] = \mathbf{A}_n\mathbf{X}[n-1] + \mathbf{B}_n\mathbf{W}[n]$$

for some $\mathbf{A}_n$, $\mathbf{B}_n$ and white Gaussian, zero-mean sequence $\mathbf{W}[n]$.

Proof. Since $\mathbf{X}[n]$ is Gaussian, the conditional mean $E[\mathbf{X}[n] \mid \mathbf{X}[n-1], \mathbf{X}[n-2], \ldots]$ is a linear function of $[\mathbf{X}[n-1], \mathbf{X}[n-2], \ldots]$ because in the Gaussian case the MMSE estimator is linear. Since the random sequence is also Markov we have that

$$\begin{aligned} E[\mathbf{X}[n] \mid \mathbf{X}[n-1], \ldots] &= E[\mathbf{X}[n] \mid \mathbf{X}[n-1]] \\ &\triangleq \mathbf{A}_n\mathbf{X}[n-1], \end{aligned}$$

for some $\mathbf{A}_n$ that may depend on n. In fact, we know the matrix $\mathbf{A}_n$ can be determined by the orthogonality relation,

$$(\mathbf{X}[n] - \mathbf{A}_n\mathbf{X}[n-1]) \perp \mathbf{X}[n-1].$$

What remains to be shown is that the prediction-error sequence $\mathbf{X}[n] - \mathbf{A}_n\mathbf{X}[n-1]$ is a white Gaussian random sequence. First, we know it is Gaussian because it is a linear operation on a Gaussian random sequence. Second, we know that the prediction error is orthogonal to all previous $\mathbf{X}[k]$ for all $k < n$. Thus

$$(\mathbf{X}[n] - \mathbf{A}_n\mathbf{X}[n-1]) \perp (\mathbf{X}[k] - \mathbf{A}_k\mathbf{X}[k-1])$$

for all $k < n$. Hence it is an orthogonal random sequence, but this is the same as saying it is white and zero mean. Thus the proof is completed by setting

$$\mathbf{B}_n\boldsymbol{\sigma}_{\mathbf{W}}^2[n]\mathbf{B}_n^\dagger \triangleq E[(\mathbf{X}[n] - \mathbf{A}_n\mathbf{X}[n-1])(\mathbf{X}[n] - \mathbf{A}_n\mathbf{X}[n-1])^\dagger].$$

In fact, we can just as well take $\mathbf{B}_n = \mathbf{I}$† and then

$$\begin{aligned}\boldsymbol{\sigma}_{\mathbf{W}}^2[n] &= E[(\mathbf{X}[n] - \mathbf{A}_n\mathbf{X}[n-1])(\mathbf{X}[n] - \mathbf{A}_n\mathbf{X}[n-1])^\dagger] \\ &= E[\mathbf{X}[n]\mathbf{X}^\dagger[n]] - \mathbf{A}_n E[\mathbf{X}[n-1]\mathbf{X}^\dagger[n]] \\ &= \mathbf{K}_{\mathbf{X}}[n, n] - \mathbf{A}_n\mathbf{K}_{\mathbf{X}}[n-1, n],\end{aligned}$$

but

$$\mathbf{A}_n = \mathbf{K}_{\mathbf{X}}[n, n-1]\mathbf{K}_{\mathbf{X}}^{-1}[n-1, n-1]$$

so

$$\boldsymbol{\sigma}_{\mathbf{W}}^2[n] = \mathbf{K}_{\mathbf{X}}[n, n] - \mathbf{K}_{\mathbf{X}}[n, n-1]\mathbf{K}_{\mathbf{X}}^{-1}[n-1, n-1]\mathbf{K}_{\mathbf{X}}[n-1, n]. \quad \blacksquare$$

Note that the difference between Theorem 10.3–1 and Theorem 10.3–2 is that in the latter we are assuming that the vector random sequence $\mathbf{X}[n]$ is Gaussian while in the former theorem this condition is not assumed. The sequence $\mathbf{W}[n]$ then is independent and Gaussian in the case where the output sequence $\mathbf{X}$ is Gaussian as well as Markov.

We also can note that if $\mathbf{X}[n]$ were also stationary in Theorem 10.3–2, then the coefficient matrices $\mathbf{A}_n$ and $\mathbf{B}_n$ would be constants, and the innovations variance matrix would be constant $\boldsymbol{\sigma}_{\mathbf{W}}^2$. Finally note that since we use the expectation operator E in Theorem 10.3–2 but only the LLMSE operator $\hat{E}$ in Theorem 10.3–1, the representation in the former theorem is really much stronger than that of the latter theorem.

10.4 KALMAN PREDICTOR AND FILTER

Here we extend the prediction results developed in the last section by enlarging the class of problems to allow prediction from noisy data. This generalization when combined with the Gauss-Markov signal model will result in the celebrated Kalman-Bucy prediction filter. In 1960 R. E. Kalman published the discrete-time theory [10–4]. A year later the continuous-time theory was published by R. E. Kalman and R. S. Bucy [10–5]. Actually, we do not need to make the Gauss-Markov assumption. We could just assume the signal is wide-sense Markov and derive the LMMSE filter. The result would be the same as what we will derive here essentially because the MMSE filter is linear for Gaussian data. We will assume that the Gauss-Markov sequence to be predicted is stationary with zero-mean and is defined for $n \geq 0$ with *known* initial condition $\mathbf{X}[-1] = \mathbf{0}$. We will also restrict attention in this section to *real-valued* random vector sequences. By Theorem 10.3–2 we have that the Gauss-Markov signal can be represented as

$$\mathbf{X}[n] = \mathbf{A}\mathbf{X}[n-1] + \mathbf{B}\mathbf{W}[n], \qquad n \geq 0, \tag{10.4–1}$$

† Thus insuring that $\mathbf{B}^{-1}$ exists where it is useful such as for insuring the innovations property of $\mathbf{W}[n]$.

subject to $\mathbf{X}[-1] = \mathbf{0}$, where $\mathbf{W}[n]$ is white Gaussian noise with zero mean and variance matrix, $\boldsymbol{\sigma}_{\mathbf{W}}^2$. As earlier, the matrix $\mathbf{A}$ is taken to be nonsingular.

The observations will no longer be assumed noiseless. Instead, we will assume the more practical case where there is some noise added to the signal prior to our observation,

$$\mathbf{Y}[n] = \mathbf{X}[n] + \mathbf{V}[n]\dagger, \qquad n \geq 0 \tag{10.4–2}$$

where the random sequence $\mathbf{V}[n]$, called the *observation noise*, is white, Gaussian, and zero-mean. We take the observation noise $\mathbf{V}[n]$ to be stationary with variance matrix $\boldsymbol{\sigma}_{\mathbf{V}}^2 \triangleq E[\mathbf{V}[n]\mathbf{V}[n]^T]$, where we remember that all random sequences are assumed real valued. Additionally, we assume that $\mathbf{V}$ and $\mathbf{W}$ are orthogonal at all pairs of times, that is,

$$\mathbf{V}[n] \perp \mathbf{W}[k] \qquad \text{for all } n, k.$$

Since the two noises are zero-mean, this amounts to saying that $\mathbf{V}$ and $\mathbf{W}$ are uncorrelated at all times n and k. We can write this more compactly as $\mathbf{V} \perp \mathbf{W}$. Furthermore we take $\mathbf{V}$ and $\mathbf{W}$ as jointly Gaussian so that they are *jointly independent* random sequences.

Our method of solution will be to first find the innovation sequence for the noisy observations (Equation 10.4–2) and then to base our estimate on it. The Kalman predictor and filter are then derived. We will see that they have a convenient predictor–corrector structure. Finally, we will solve for certain error-covariance functions necessary to determine so-called gain matrices in the filter.

Now we know that the MMSE estimate of the signal sequence $\mathbf{X}[n]$ based on the observation set $\{\mathbf{Y}[k], k < n\}$ is the corresponding conditional mean. Thus we look for $\hat{\mathbf{X}}[n] \triangleq E[\mathbf{X}[n] \mid \mathbf{Y}[n-1], \mathbf{Y}[n-2], \ldots, \mathbf{Y}[0]]$. We first define an innovations sequence for $\mathbf{Y}[n]$ analogously to the method of Section 10.3 for the $\mathbf{X}[n]$, which were the observations there. Motivated by the requirements of the innovations Definition 10.3–1, we define the sequence $\mathbf{Y}$ as follows:

$$\tilde{\mathbf{Y}}[0] \triangleq \mathbf{Y}[0] \qquad \text{and} \qquad \text{for } n \geq 1$$

$$\tilde{\mathbf{Y}}[n] \triangleq \mathbf{Y}[n] - E[\mathbf{Y}[n] \mid \mathbf{Y}[n-1], \mathbf{Y}[n-2], \ldots, \mathbf{Y}[0]]. \tag{10.4–3}$$

We now must show that $\tilde{\mathbf{Y}}[n]$ thus defined is an innovations sequence for $\mathbf{Y}[n]$. To do this we must prove that $\tilde{\mathbf{Y}}[n]$ satisfies the three defining properties (see Definition 10.3–1):

(1) $\tilde{\mathbf{Y}}[n]$ is a causal, linear transformation on $\mathbf{Y}[n]$,

(2) $\mathbf{Y}[n]$ is a causal, linear transformation on $\tilde{\mathbf{Y}}[n]$, and

(3) $\tilde{\mathbf{Y}}[n]$ is an orthogonal random sequence.

Now (1) is immediate by the definition $\tilde{\mathbf{Y}}[n]$ since $\mathbf{Y}[n]$ is a Gaussian

† A more general observation model $\mathbf{Y}[n] = \mathbf{C}_n\mathbf{X}[n] + \mathbf{V}[n]$ is treated in Problem 10.7.

random sequence. To show (2) we note that we can recursively solve Equation 10.4–3 for $\mathbf{Y}[n]$ as

$$\mathbf{Y}[0] = \tilde{\mathbf{Y}}[0] \qquad \text{and}$$

$$\mathbf{Y}[n] = \tilde{\mathbf{Y}}[n] + E[\mathbf{Y}[n] \,|\, \mathbf{Y}[n-1], \mathbf{Y}[n-2], \ldots, \mathbf{Y}[0]]$$

$$= \tilde{\mathbf{Y}}[n] + \sum_{k=0}^{n-1} \mathbf{D}_k^{(n)} \tilde{\mathbf{Y}}[k], \qquad n \geq 1,$$

where the $\mathbf{D}_k^{(n)}$ are a known sequence of matrices, thus establishing (2). As an additional piece of terminology, when (1) and (2) hold simultaneously we say that $\mathbf{Y}$ and $\tilde{\mathbf{Y}}$ are *causally linearly equivalent.*

To establish (3) we note that, for $k < n$,

$$E[\tilde{\mathbf{Y}}[n]\tilde{\mathbf{Y}}[k]^T] = E[\tilde{\mathbf{Y}}[n](\mathbf{Y}[k] - \hat{\mathbf{Y}}[k])^T] = \mathbf{0},$$

since $\tilde{\mathbf{Y}}[n] \perp (\mathbf{Y}[n-1], \mathbf{Y}[n-2], \ldots, \mathbf{Y}[0])$ and hence is orthogonal to any linear combination of the $\mathbf{Y}[k]$ for $k < n$. Similarly, we have $E[\tilde{\mathbf{Y}}[n]\tilde{\mathbf{Y}}[k]^T] = \mathbf{0}$ for $n > k$, thus combining we have

$$E[\tilde{\mathbf{Y}}[n]\tilde{\mathbf{Y}}[k]^T] = \boldsymbol{\sigma}_{\tilde{\mathbf{Y}}}^2[n]\,\delta[n-k]$$

for some variance matrix $\boldsymbol{\sigma}_{\tilde{\mathbf{Y}}}^2[n]$.† Combining (1), (2), and (3) we see that $\tilde{\mathbf{Y}}[n]$ is the desired innovations sequence for the noisy observations $\mathbf{Y}[n]$.

Since $\tilde{\mathbf{Y}}[n]$ and $\mathbf{Y}[n]$ are causally linearly equivalent, we can base our estimate on $\tilde{\mathbf{Y}}[n]$ instead of $\mathbf{Y}[n]$ with expected simplifications due to the orthogonality of the innovations sequence $\tilde{\mathbf{Y}}[n]$. Since the data are Gaussian and the estimate must be linear, we can thus write

$$\hat{\mathbf{X}}[n] = \sum_{k=0}^{n-1} \mathbf{C}_k^{(n)} \tilde{\mathbf{Y}}[k],$$

where the $\mathbf{C}_k^{(n)}$ are to be determined. To this end we write,

$$E[\hat{\mathbf{X}}[n]\tilde{\mathbf{Y}}[k]^T] = \mathbf{C}_k^{(n)} \boldsymbol{\sigma}_{\tilde{\mathbf{Y}}}^2[k]. \qquad (10.4\text{–}4)$$

In order to solve Equation 10.4–4 for $\mathbf{C}_k^{(n)}$ we must eliminate the $\hat{\mathbf{X}}[n]$ under the expectation sign on the left-hand side. In fact, the equation remains valid with the hat over $\mathbf{X}[n]$ removed, a result that can be seen from the following:

$$\hat{\mathbf{X}}[n]\tilde{\mathbf{Y}}[k]^T = E[\mathbf{X}[n] \,|\, \tilde{\mathbf{Y}}[n-1], \tilde{\mathbf{Y}}[n-2], \ldots, \tilde{\mathbf{Y}}[0]]\tilde{\mathbf{Y}}[k]^T,$$

which for $k < n$ equals,

$$E[\mathbf{X}[n]\tilde{\mathbf{Y}}[k]^T \,|\, \tilde{\mathbf{Y}}[n-1], \tilde{\mathbf{Y}}[n-2], \ldots, \tilde{\mathbf{Y}}[0]],$$

so that upon taking the unconditional expectation of both sides we get

$$E[\hat{\mathbf{X}}[n]\tilde{\mathbf{Y}}[k]^T] = E[\mathbf{X}[n]\tilde{\mathbf{Y}}[k]^T] \qquad \text{for } k < n,$$

† The reader should understand that $\sigma_{\tilde{\mathbf{Y}}}^2[n]$ will not be constant since observations start at $n = 0$.

so that Equation 10.4–4 becomes

$$E[\mathbf{X}[n]\tilde{\mathbf{Y}}[k]^T] = \mathbf{C}_k^{(n)}\boldsymbol{\sigma}_{\tilde{\mathbf{Y}}}^2[k], \qquad 0 \le k < n.$$

Assuming that the variance matrix $\boldsymbol{\sigma}_{\tilde{\mathbf{Y}}}^2[k]$ is nonsingular, this equation can be solved to yield

$$\mathbf{C}_k^{(n)} = E[\mathbf{X}[n]\tilde{\mathbf{Y}}[k]^T][\boldsymbol{\sigma}_{\tilde{\mathbf{Y}}}^2[k]]^{-1}. \tag{10.4–5}$$

Then we have the prediction estimate

$$\hat{\mathbf{X}}[n] = \sum_{k=0}^{n-1} E[\mathbf{X}[n]\tilde{\mathbf{Y}}[k]^T][\boldsymbol{\sigma}_{\tilde{\mathbf{Y}}}^2[k]]^{-1}\tilde{\mathbf{Y}}[k]. \tag{10.4–6}$$

Since the signal model Equation 10.4–1 is recursive, we suspect there is a way around the growing memory estimate that appears in Equation 10.4–6. Using Equation 10.4–1, we can write

$$E[\mathbf{X}[n]\tilde{\mathbf{Y}}[k]^T] = \mathbf{A}E[\mathbf{X}[n-1]\tilde{\mathbf{Y}}[k]^T] + \mathbf{B}E[\mathbf{W}[n]\tilde{\mathbf{Y}}[k]^T],$$

but for $k < n$, $\mathbf{W}[n] \perp \mathbf{X}[k]$ and $\mathbf{W}[n] \perp \mathbf{V}[k]$, which implies $\mathbf{W}[n] \perp \mathbf{Y}[k]$ and hence also $\mathbf{W}[n] \perp \tilde{\mathbf{Y}}[k]$, so that we have

$$E[\mathbf{X}[n]\tilde{\mathbf{Y}}[k]^T] = \mathbf{A}E[\mathbf{X}[n-1]\tilde{\mathbf{Y}}[k]^T] \quad \text{for all } k < n.$$

Hence we can also express the prediction estimate $\hat{\mathbf{X}}[n]$ as

$$\hat{\mathbf{X}}[n] = \mathbf{A}\sum_{k=0}^{n-1} E[\mathbf{X}[n-1]\tilde{\mathbf{Y}}[k]^T][\boldsymbol{\sigma}_{\tilde{\mathbf{Y}}}^2[k]]^{-1}\tilde{\mathbf{Y}}[k]. \tag{10.4–7}$$

But Equation 10.4–6 must hold at $n - 1$ as well as n, thus also

$$\hat{\mathbf{X}}[n-1] = \sum_{k=0}^{n-2} E[\mathbf{X}[n-1]\tilde{\mathbf{Y}}[k]^T][\boldsymbol{\sigma}_{\tilde{\mathbf{Y}}}^2[k]]^{-1}\tilde{\mathbf{Y}}[k].$$

Combining this equation with Equation 10.4–7, we finally obtain

$$\hat{\mathbf{X}}[n] = \mathbf{A}\hat{\mathbf{X}}[n-1] + \mathbf{A}E[\mathbf{X}[n-1]\tilde{\mathbf{Y}}[n-1]^T][\boldsymbol{\sigma}_{\tilde{\mathbf{Y}}}^2[n-1]]^{-1}\tilde{\mathbf{Y}}[n-1],$$

which is an efficient way of calculating the prediction estimate of Equation 10.4–7. If we define the *Kalman gain* matrix,

$$\mathbf{G}_{n-1} \triangleq E[\mathbf{X}[n-1]\tilde{\mathbf{Y}}[n-1]^T][\boldsymbol{\sigma}_{\tilde{\mathbf{Y}}}^2[n-1]]^{-1}, \tag{10.4–8}$$

we can rewrite the preceding result as

$$\hat{\mathbf{X}}[n] = \mathbf{A}(\hat{\mathbf{X}}[n-1] + \mathbf{G}_{n-1}\tilde{\mathbf{Y}}[n-1]), \qquad n \ge 0 \tag{10.4–9}$$

with initial condition $\hat{\mathbf{X}}[0] = \mathbf{0}$.

We can eliminate the innovations sequence $\tilde{\mathbf{Y}}(n)$ in Equation 10.4–9 as follows

$$\tilde{\mathbf{Y}}[n] = \mathbf{Y}[n] - E[\mathbf{Y}[n] \mid \mathbf{Y}[n-1], \ldots, \mathbf{Y}[0]],$$

so using Equation 10.4–2, we have

$$\begin{aligned} E[\mathbf{Y}[n] \mid \mathbf{Y}[n-1], \ldots, \mathbf{Y}[0]] &= E[\mathbf{X}[n] \mid \mathbf{Y}[n-1], \ldots, \mathbf{Y}[0]] \\ &\quad + E[\mathbf{V}[n] \mid \mathbf{Y}[n-1], \ldots, \mathbf{Y}[0]], \\ &= \hat{\mathbf{X}}[n] + \mathbf{0}. \end{aligned}$$

This last step is justified by noting that $\mathbf{V}[n] \perp \mathbf{X}[k]$ and $\mathbf{V}[n] \perp \mathbf{V}[k]$ for $k < n$, so that we have $\mathbf{V}[n] \perp \mathbf{Y}[k]$ for all $k < n$. Thus we obtain $\tilde{\mathbf{Y}}[n] = \mathbf{Y}[n] - \hat{\mathbf{X}}[n]$. So inserting this into Equation 10.4–9, we finally have

$$\hat{\mathbf{X}}[n] = \mathbf{A}(\hat{\mathbf{X}}[n-1] + \mathbf{G}_{n-1}(\mathbf{Y}[n-1] - \hat{\mathbf{X}}[n-1])), \quad (10.4\text{–}10)$$

which is the most well-known form of the *Kalman predictor*, whose system diagram is shown in Figure 10.4–1.

We can denote the prediction estimate in Equation 10.4–10 more explicitly as

$$\hat{\mathbf{X}}[n \mid n-1] \triangleq E[\mathbf{X}[n] \mid \mathbf{Y}[n-1], \mathbf{Y}[n-2], \ldots, \mathbf{Y}[0]].$$

On the other hand, we may be interested in calculating the causal estimate

$$\hat{\mathbf{X}}[n \mid n] \triangleq E[\mathbf{X}[n] \mid \mathbf{Y}[n], \mathbf{Y}[n-1], \ldots, \mathbf{Y}[0]].$$

which uses all the data up to the present time n. The Kalman predictor can be modified to provide this causal estimate. The resulting recursive formula is called the *Kalman filter*. One method to derive it from Equation 10.4–10 is the following. Consider the prediction

$$\hat{\mathbf{X}}[n \mid n-1] = E[\mathbf{X}[n] \mid \mathbf{Y}[n-1], \ldots, \mathbf{Y}[0]],$$

and use the signal model Equation 10.4–1 to obtain

$$\hat{\mathbf{X}}[n \mid n-1] = \mathbf{A}\hat{\mathbf{X}}[n-1 \mid n-1] + \mathbf{0},$$

since $\mathbf{W}[n] \perp \mathbf{Y}[k]$ for $k < n$. So

$$\hat{\mathbf{X}}[n-1 \mid n-1] = \mathbf{A}^{-1}\hat{\mathbf{X}}[n \mid n-1]$$

since $\mathbf{A}$ is nonsingular. Using this result we pre-multiply Equation 10.4–10 by $\mathbf{A}^{-1}$ to get

$$\hat{\mathbf{X}}[n-1 \mid n-1] = \hat{\mathbf{X}}[n-1 \mid n-2] + \mathbf{G}_{n-1}[\mathbf{Y}[n-1] - \hat{\mathbf{X}}[n-1 \mid n-2]],$$

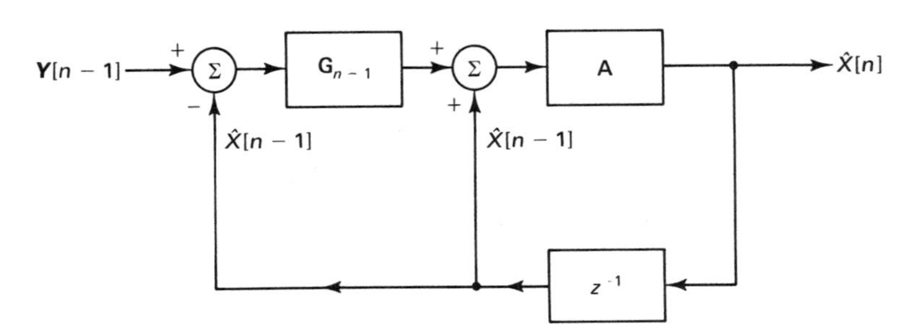

Figure 10.4–1 System diagram of Kalman predictor.

which can be written equivalently for $n \geq 0$ as

$$\hat{\mathbf{X}}[n \mid n] = \mathbf{A}\hat{\mathbf{X}}[n-1 \mid n-1] + \mathbf{G}_n[\mathbf{Y}[n] - \mathbf{A}\hat{\mathbf{X}}[n-1 \mid n-1]], \tag{10.4–11}$$

which is known as the Kalman filter equation. Here $\hat{\mathbf{X}}[-1 \mid -1] \triangleq \mathbf{0}$.

By examining either Equation 10.4–10 or Equation 10.4–11 we can see a predictor-corrector structure. The first term is the MMSE prediction of $\mathbf{X}[n]$ based on the past $\mathbf{Y}[k]$, $k < n$. The second term involving the current data is called the *update.* It is the product of a gain matrix $\mathbf{G}_n$ (which can be precomputed and stored) and the prediction error on the noisy observations, which we have called the innovations, that is, the new information contained in the current data point.

One can go on to derive *Kalman smoothers*, which are fixed-delay estimators of the form

$$\hat{\mathbf{X}}[n \mid n+k] \triangleq E[\mathbf{X}[n] \mid \mathbf{Y}[n+k], \mathbf{Y}[n+k-1], \ldots, \mathbf{Y}[0]], \qquad k > 0.$$

These delayed estimators are of importance in the study of various communication and control systems. We have not yet found efficient algorithms to calculate the sequence of gain matrices $\mathbf{G}_n$. This is the subject of the next section.

Error-Covariance equations. We have to find a method for recursively calculating the gain matrix sequence $\mathbf{G}_n$ of Equation 10.4–8. We are also interested in evaluating the error of the estimate. The *prediction-error variance* matrix is the covariance matrix of $\tilde{\mathbf{X}}[n] \triangleq \hat{\mathbf{X}}[n \mid n-1] - \mathbf{X}[n]$. We write it as

$$\boldsymbol{\varepsilon}[n] \triangleq E[\tilde{\mathbf{X}}[n]\tilde{\mathbf{X}}^T[n]].$$

We start by inserting the observation Equation 10.4–2 into the innovations Equation 10.4–3 to obtain

$$\begin{aligned} \tilde{\mathbf{Y}}[n] &= \mathbf{X}[n] + \mathbf{V}[n] - \hat{\mathbf{X}}[n]^{\dagger} \\ &= -\tilde{\mathbf{X}}[n] + \mathbf{V}[n]. \end{aligned} \tag{10.4–12}$$

But $\mathbf{X}[n] \perp \mathbf{V}[n]$, so

$$\begin{aligned} E[\mathbf{X}[n]\tilde{\mathbf{Y}}^T[n]] &= -E[\mathbf{X}[n]\tilde{\mathbf{X}}^T[n]] \\ &= E[(\hat{\mathbf{X}}[n] - \mathbf{X}[n])\tilde{\mathbf{X}}^T[n]] \quad \text{since } \tilde{\mathbf{X}} \perp \hat{\mathbf{X}} \\ &= E[\tilde{\mathbf{X}}[n]\tilde{\mathbf{X}}^T[n]] \\ &= \boldsymbol{\varepsilon}[n]. \end{aligned}$$

Also $\tilde{\mathbf{Y}}[n] \perp \mathbf{V}[n]$, so we have from Equation 10.4–12,

$$E[\tilde{\mathbf{Y}}[n]\tilde{\mathbf{Y}}[n]^T] = E[\tilde{\mathbf{X}}[n]\tilde{\mathbf{X}}[n]^T] + E[\mathbf{V}[n]\mathbf{V}[n]^T]$$

or

$$\boldsymbol{\sigma}_{\tilde{\mathbf{Y}}}^2[n] = \boldsymbol{\varepsilon}[n] + \boldsymbol{\sigma}_{\mathbf{V}}^2[n];$$

† Remember $\hat{\mathbf{X}}[n]$ is the prediction estimate.

thus by Equation 10.4–8 we have

$$\mathbf{G}_n = \boldsymbol{\varepsilon}[n][\boldsymbol{\varepsilon}[n] + \boldsymbol{\sigma}_{\mathbf{V}}^2[n]]^{-1}, \qquad n \geq 0. \qquad (10.4\text{–}13)$$

The problem is now reduced to calculating $\boldsymbol{\varepsilon}[n]$. From Problem 10.1 we find we can write

$$\boldsymbol{\varepsilon}[n] = E[\mathbf{X}[n]\mathbf{X}^T[n]] - E[\hat{\mathbf{X}}[n]\hat{\mathbf{X}}^T[n]]. \qquad (10.4\text{–}14)$$

To evaluate the right side of Equation 10.4–14 we use Equation 10.4–1 and $\mathbf{X}[n-1] \perp \mathbf{W}[n]$ to get

$$E[\mathbf{X}[n]\mathbf{X}^T[n]] = \mathbf{A}E[\mathbf{X}[n-1]\mathbf{X}^T[n-1]]\mathbf{A}^T + \mathbf{B}\boldsymbol{\sigma}_{\mathbf{W}}^2\mathbf{B}^T. \quad (10.4\text{–}15)$$

Likewise using $\hat{\mathbf{X}}[n] = \mathbf{A}(\hat{\mathbf{X}}[n-1] + \mathbf{G}_{n-1}\tilde{\mathbf{Y}}[n-1])$ and $\hat{\mathbf{X}}[n-1] \perp \tilde{\mathbf{Y}}[n-1]$, we get

$$\begin{aligned} E[\hat{\mathbf{X}}[n]\hat{\mathbf{X}}^T[n]] &= \mathbf{A}E[\hat{\mathbf{X}}[n-1]\hat{\mathbf{X}}^T[n-1]]\mathbf{A}^T + \mathbf{A}\mathbf{G}_{n-1}\boldsymbol{\sigma}_{\tilde{\mathbf{Y}}}^2[n-1]\mathbf{G}_{n-1}^T\mathbf{A}^T \\ &= \mathbf{A}E[\hat{\mathbf{X}}[n-1]\hat{\mathbf{X}}^T[n-1]]\mathbf{A}^T + \mathbf{A}\boldsymbol{\varepsilon}[n-1]\mathbf{G}_{n-1}^T\mathbf{A}^T \end{aligned} \qquad (10.4\text{–}16)$$

where we have used $\mathbf{G}_{n-1}\boldsymbol{\sigma}_{\tilde{\mathbf{Y}}}^2[n-1] = \boldsymbol{\varepsilon}[n-1]$.

Substituting Equations 10.4–15 and 10.4–16 into Equation 10.4–14 and simplifying then yields

$$\boldsymbol{\varepsilon}[n] = \mathbf{A}\boldsymbol{\varepsilon}[n-1][\mathbf{I} - \mathbf{G}_{n-1}^T]\mathbf{A}^T + \mathbf{B}\boldsymbol{\sigma}_{\mathbf{W}}^2\mathbf{B}^T \qquad (10.4\text{–}17)$$

for $n \geq 0$, where

$$\boldsymbol{\varepsilon}[-1] \triangleq E[\tilde{\mathbf{X}}[-1]\tilde{\mathbf{X}}^T[-1]]$$

and

$$E[\tilde{\mathbf{X}}[-1]\tilde{\mathbf{X}}^T[-1]] = \begin{cases} \mathbf{0} & \text{if } \mathbf{X}[-1] = \mathbf{0} \text{ (known)} \\ \boldsymbol{\sigma}_{\mathbf{X}}^2 & \text{if } \mathbf{X} \text{ is WSS.}\dagger \end{cases}$$

In summary, the Kalman filter for the state equation (10.4–1) and the observation model (10.4–2) thus consists of the filtering equation (10.4–11), the gain equation (10.4–13), and the prediction-error covariance equation (10.4–17). The *filtering-error covariance* equation can also be calculated as

$$\begin{aligned} \mathbf{e}[n] &\triangleq E[(\mathbf{X}[n] - \hat{\mathbf{X}}[n \mid n])(\mathbf{X}[n] - \hat{\mathbf{X}}[n \mid n])^T] \\ &= \boldsymbol{\varepsilon}[n][\mathbf{I} - \mathbf{G}_n^T] \\ &= \mathbf{A}\boldsymbol{\varepsilon}[n-1]\mathbf{A}^T + \mathbf{B}\boldsymbol{\sigma}_{\mathbf{W}}^2\mathbf{B}, \qquad n \geq 1. \end{aligned}$$

The proof of this is left as an exercise to the reader.

Example 10.4–1: (A Kalman filter application.) Consider the Gauss-Markov signal model

$$X[n] = 0.9X[n-1] + W[n], \qquad n \geq 0,$$

with means equal to zero and $\sigma_W^2 = 0.19$. Also $X[-1] = 0$. Let the observations be

† This would require a minor modification of our development, since we have assumed $\mathbf{X}[-1] = \mathbf{0}$.

given by the scalar equation

$$Y[n] = X[n] + V[n], \qquad n \geq 0,$$

with $\sigma_V^2 = 1$. We have $A = 0.9$ and $B = 1$. The Kalman filter (Equation 10.4–11) then becomes

$$\hat{X}[n \mid n] = 0.9\hat{X}[n-1 \mid n-1] + G_n(Y[n] - 0.9\hat{X}[n-1 \mid n-1]),$$

with initial condition $\hat{X}[-1 \mid -1] = X[-1] = 0$. The Kalman gain (Equation 10.4–13) is

$$G_n = \varepsilon[n]/(1 + \varepsilon[n]),$$

and prediction-error variance (Equation 10.4–17)

$$\begin{aligned}\varepsilon[n] &= 0.81\varepsilon[n-1][1 - G_{n-1}] + 0.19\\ &= 0.81\varepsilon[n-1]\left(1 - \frac{\varepsilon[n-1]}{1 + \varepsilon[n-1]}\right) + 0.19\\ &= \frac{0.19 + \varepsilon[n-1]}{1 + \varepsilon[n-1]}, \qquad n \geq 0,\end{aligned}$$

with initial condition $\varepsilon[-1] = 0$.

We can solve this equation for the steady-state solution $\varepsilon[\infty]$

$$\varepsilon[\infty] = \frac{0.19 + \varepsilon[\infty]}{1 + \varepsilon[\infty]}.$$

and discarding the negative root we obtain

$$\varepsilon[\infty] = 0.436.$$

Thus $\varepsilon[n] \rightarrow 0.436$ and hence $G_n \rightarrow 0.304$ so that the Kalman filter is asymptotically given by

$$\begin{aligned}\hat{X}[n \mid n] &= 0.9\hat{X}[n-1 \mid n-1] + 0.304(Y[n] - 0.9\hat{X}[n-1 \mid n-1])\\ &= 0.626\hat{X}[n-1 \mid n-1] + 0.304Y[n]\end{aligned}$$

and the steady-state filtering error is $e[\infty] = 0.304$.

We have developed the Kalman filter as a computationally attractive solution for estimating a Gauss-Markov signal (Equation 10.4–1) observed in additive white Gaussian noise (Equation 10.4–2) over the *semi-infinite time* interval $0 \leq n < \infty$. The filter equations are time-variant because the initial estimates near $n = 0$ effectively have a truncated past. For our constant parameter and constant noise variance assumptions, the estimation error should thus decrease toward an asymptotic or steady-state value as we move away from $n = 0$.

The Kalman filter derivation can easily be generalized to allow time-varying parameters $\mathbf{A}_n$ and $\mathbf{B}_n$ in Equation 10.4–1. We can also permit time-varying noise variances $\boldsymbol{\sigma}_{\mathbf{W}}^2[n]$ and $\boldsymbol{\sigma}_{\mathbf{V}}^2[n]$. In fact, the present derivation will also serve for this time-variant case, by just inserting the new quantities as required.

The observation Equation 10.4–2 is also overly restrictive. See Problem 10.7 for a generalization to allow the observation vectors to be of different dimension than the signal vectors,

$$\mathbf{Y}[n] = \mathbf{C}_n\mathbf{X}[n] + \mathbf{V}[n] \tag{10.4–18}$$

In this equation, the rectangular matrix $\mathbf{C}_n$ permits linear combinations of the state vector $\mathbf{X}$ to appear in the observations.

10.5 WIENER FILTERS FOR RANDOM SEQUENCES

In this section we will investigate optimal linear estimates for signals that are not necessarily Gaussian-Markov of any order. This theory predates the Kalman-Bucy filter of the last section. The optimum linear filter is associated with Norbert Wiener and Andrei N. Kolmogorov, who performed this work in the 1940s. The discrete-time theory that is presented in this section is in fact the work of Kolmogorov [10–6], while Wiener developed the continuous-time theory [10–7] to be presented in Section 10.6 on random processes. Nevertheless, it has become conventional to refer to both types of filters as Wiener filters. These filters are mainly appropriate for WSS random processes (sequences) observed over *infinite time intervals.*

We start with the general problem of finding the LMMSE estimate of the random sequence $X[n]$ from observations of the random sequence $Y[n]$ for all time. We assume both random sequences are WSS and for convenience are zero-mean and real-valued. Our approach to find the LMMSE estimate will be based on the innovations sequence. For these infinite time-interval observations, the innovations may be obtained using spectral factorization (cf. Section 9.4).

First we spectrally factor the psd of the observations $S_{YY}[z]$ into its causal and anticausal factors,

$$S_Y[z] = \sigma^2 B[z]B[z^{-1}], \tag{10.5–1}$$

where $B[z]$ contains all the poles and zeros of S_Y that are inside the unit circle in the Z-plane. Hence B and B^{-1} are stable and causal. We can thus operate on $Y[n]$ with the LSI operator B^{-1} to produce the random sequence $\tilde{Y}[n]$ as shown in Figure 10.5–1. The psd of $\tilde{Y}[n]$ is easily seen to be

$$S_{\tilde{Y}}[z] = \sigma^2.$$

Thus $\tilde{Y}[n]$ satisfies the three defining properties of the innovations sequence as listed in Section 10.3. We can then base our estimate on $\tilde{Y}[n]$.

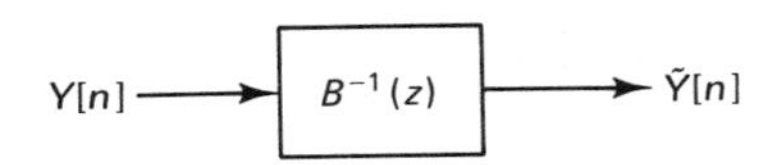

Figure 10.5–1 Whitening filter.

Unrealizable Case (Smoothing). Consider an LSI operator G with convolution representation

$$\hat{X}[n] = \sum_{k=-\infty}^{+\infty} g[k]\tilde{Y}[n-k]. \tag{10.5–2}$$

We want to choose the $g[k]$ to minimize the MSE,

$$E[(X(n) - \hat{X}(n))^2].$$

Expanding this expression, we obtain

$$\begin{aligned} E[X^2[n]] - 2E\Big[\sum_k g[k]X[n]\tilde{Y}[n-k]\Big] &+ E\Big[\Big(\sum_k g[k]\tilde{Y}[n-k]\Big)^2\Big] \\ &= K_X[0] - \sum_k g[k]K_{X\tilde{Y}}[k] + \sigma^2 \sum_k g^2[k] \\ &= K_X[0] + \sum_{k=-\infty}^{+\infty}\Big[\sigma g[k] - \frac{K_{X\tilde{Y}}[k]}{\sigma}\Big]^2 - \frac{1}{\sigma^2}\sum_{k=-\infty}^{+\infty} K^2_{X\tilde{Y}}[k], \end{aligned} \tag{10.5–3}$$

where the last line is obtained by completing the square. Examining this equation we see that only the middle term depends on the choice of $g[k]$. The minimum of this term is obviously zero for the choice,

$$g[k] = \frac{1}{\sigma^2}K_{X\tilde{Y}}[k], \qquad -\infty < k < +\infty \tag{10.5–4}$$

the LMMSE then being given as

$$\varepsilon_u = K_X[0] - \frac{1}{\sigma^2}\sum_{k=-\infty}^{+\infty} K^2_{X\tilde{Y}}[k]. \tag{10.5–5}$$

This result can also be derived directly by using the $\hat{E}$ operator. (See Problem 10.8.)

In the Z-transform domain the operator G is expressed as

$$G[z] = \frac{1}{\sigma^2}S_{X\tilde{Y}}[z] = \frac{1}{\sigma^2}S_{XY}[z]B^{-1}[z^{-1}].$$

The overall transfer function, including the whitening filter, then becomes

$$\begin{aligned} H_u[z] &= G[z]B^{-1}[z] \\ &= S_{XY}[z]/(\sigma^2 B[z]B[z^{-1}]) \\ &= S_{XY}[z]/S_{YY}[z], \end{aligned} \tag{10.5–6}$$

where the subscript "u" on H_u denotes the unrealizable estimator. The MSE is given from Equation 10.5–5 as

$$\frac{1}{2\pi j}\oint_{|z|=1}\Big[S_X[z] - \frac{1}{\sigma^2}S_{X\tilde{Y}}[z]S_{X\tilde{Y}}[z^{-1}]\Big]\frac{dz}{z}$$

using Parseval's Theorem [10–8], and then simplifies to

$$\frac{1}{2\pi}\int_{-\pi}^{+\pi}(S_X[\omega] - |S_{XY}[\omega]|^2/S_{YY}[\omega])\,d\omega.$$

Example 10.5–1: (Additive noise.) Here we take the special case where the observations consist of a signal plus noise $V[n]$ with $X \perp V$,

$$Y[n] = X[n] + V[n], \qquad -\infty < n +\infty.$$

The psd S_Y and cross-psd S_{XY} then become,

$$S_Y[z] = S_X[z] + S_V[z],$$
$$S_{XY}[z] = S_X[z].$$

Thus the optimal unrealizable filter is

$$H_u[z] = \frac{S_X[z]}{S_X[z] + S_V[z]}.$$

The MSE expression becomes

$$\frac{1}{2\pi}\int_{-\pi}^{+\pi}\frac{S_X[\omega]S_V[\omega]}{S_X[\omega] + S_V[\omega]}\,d\omega.$$

Examining the frequency response

$$H_u[\omega] = \frac{S_X[\omega]}{S_X[\omega] + S_V[\omega]},$$

we see the interpretation of H_u as a frequency-domain weighting function. When the signal-to-noise (SNR) ratio is high at a given frequency, $H_u[\omega]$ is close to 1, that is,

$$H_u[\omega] \simeq 1 \quad \text{when } S_X[\omega]/S_V[\omega] \gg 1.$$

Similarly, when the SNR is low at ω, then reasonably enough $H_u[\omega]$ is near zero.

Causal Filter. If we add the constraint that the LMMSE estimate must be causal, we get a filter that can be approximated more easily in the time-domain. We can proceed as before to Equation 10.5–3 at which point we must apply the constraint that $g[k] = 0$ for $k < 0$. Thus the optimal solution is to set

$$g[n] = \frac{1}{\sigma^2}K_{X\tilde{Y}}[n]u[n]. \tag{10.5–7}$$

The overall error then becomes

$$\varepsilon_c = K_X[0] - \frac{1}{\sigma^2}\sum_{k=0}^{\infty}K_{X\tilde{Y}}^2[k], \tag{10.5–8}$$

which is necessarily larger than the unrealizable error (Equation 10.5–5).

To express the optimal filter in the transform domain, we introduce the notation

$$[F[z]]_+ \triangleq \sum_{n=0}^{\infty} f[n]z^{-n} \tag{10.5–9}$$

and then write the Z-transform of Equation 10.5–7 as

$$G[z] = \frac{1}{\sigma^2}[S_{X\tilde{Y}}[z]]_+.$$

The overall LMMSE causal filter then becomes

$$H_c[z] = \frac{1}{\sigma^2 B[z]}\left[\frac{S_{XY}[z]}{B[z^{-1}]}\right]_+, \tag{10.5–10}$$

where the subscript c denotes a causal estimator. The error is expressed as

$$\frac{1}{2\pi}\int_{-\pi}^{+\pi}(S_X[\omega] - H_c[\omega]S^*_{XY}[\omega])\,d\omega.$$

Example 10.5–2: (First-order signal in white noise.) Let the signal psd be given as

$$S_X[z] = \frac{0.19}{(1-0.9z^{-1})(1-0.9z)}.$$

Let the observations be given as

$$Y[n] = X[n] + V[n],$$

where $V[n]$ is white noise with variance $\sigma_V^2 = 1$. Then the psd of the observations is

$$\begin{aligned} S_Y[z] &= S_X[z] + S_V[z] \\ &= 1.436\left[\frac{1-0.627z^{-1}}{1-0.9z^{-1}}\right]\left[\frac{1-0.627z}{1-0.9z}\right] \\ &= \sigma^2 \cdot B[z]\cdot B[z^{-1}], \end{aligned}$$

so that

$$\begin{aligned} \left[\frac{S_X[z]}{B[z^{-1}]}\right]_+ &= \left[\frac{0.19}{(1-0.9z^{-1})(1-0.627z)}\right]_+ \\ &= \left[\frac{0.436}{1-0.9z^{-1}} + \frac{0.273}{z^{-1}-0.627}\right]_+ \\ &= \frac{0.436}{1-0.9z^{-1}}, \end{aligned}$$

where we have used the partial fraction expansion to recover the causal part as required by Equation 10.5–9. Thus by Equation 10.5–10 the optimal filter is

$$\begin{aligned} H_c[z] &= \frac{1}{\sigma^2 B[z]}\,\frac{0.436}{1-0.9z^{-1}} \\ &= \frac{0.304}{1-0.627z^{-1}}. \end{aligned}$$

This result should be compared to that of Example 10.4–1, where an asymptotically equivalent problem is addressed.

10.6 LINEAR ESTIMATION FOR RANDOM PROCESSES

In the previous two sections we looked at estimation problems for random sequences. In Section 10.4 we considered the Kalman filter for optimal estimation in a nonstationary Gaussian environment. Then in the last section we looked at the Wiener filter for optimal linear estimation, having its main application in the WSS case. Now we consider continuous-time and study Wiener filters for random processes.

The general linear estimation problem for random processes can be stated as follows:

> We observe a possibly nonstationary stochastic process $Y(t)$ over an interval I and find the LMMSE estimate of a related process $X(t)$. The process $X(t)$ is assumed in turn related to a signal process $S(t)$. These processes may be complex valued.

Of course, we assume that all three random processes are defined on the same probability space $(\Omega, \mathcal{F}, P)$ so that we may calculate cross-correlations and covariances as needed. This, of course, restricts us to the second-order case where the required covariances exist.

Within this general problem we will be interested in several sub-problems, which are indicated below:

1. *Prediction*: $X(t) = S(t + \lambda)$ with $\lambda > 0$ and $Y(t) = S(t)$ and the interval $I = \{\tau \mid \tau \leq t\}$.
2. *Filtering*: $X(t) = S(t)$, $Y(t) = S(t) + N(t)$, and $I = \{\tau \mid \tau \leq t\}$.
3. *Smoothing*: $X(t) = S(t)$, $Y(t) = S(t) + N(t)$, and $I = \{-\infty < \tau < +\infty\}$.
4. *Linear Operation*: $X(t) = g(t) * S(t)$, $Y(t) = S(t) + N(t)$, and various observation intervals I.
5. *Deconvolution*: $X(t) = S(t)$, $Y(t) = g(t) * S(t) + N(t)$, and various observation intervals I.

A special case of 4. would involve estimating the derivative of a random process from noisy observations, a problem of considerable importance, for example FM demodulation. In many information-processing applications one must compute the derivative of a signal contaminated by a small amount of noise, for example estimating velocity from the record of a displacement sensor. However, the conventional first-order difference approximation to the derivative will greatly exaggerate this noise and in many practical cases overpower the signal. The present theory offers an optimal solution to this dilemma, optimal that is, within the confines of its assumptions of linearity and (in most cases) stationarity. We start out by showing that the orthogonality principle holds for random processes.

Theorem 10.6–1. The optimal linear operator to solve the general LMMSE estimation problem is the operator L, which satisfies

$$(X(t) - L[Y(t)]) \perp Y(\tau), \quad \text{for all } \tau \in I$$

or

$$E[(X(t) - L[Y(t)])Y^*(\tau)] = 0, \tag{10.6–1}$$

for all $\tau \in I$ and for each t.

Proof. Denote by L_o the optimal LMMSE operator and define L to be the linear operator that satisfies the orthogonality relation (Equation 10.6–1) analogously to Theorem 5.6–1. Let $\delta L \triangleq L - L_o$ be the difference operator; then

$$\begin{aligned}\varepsilon_{\min} &= E[|X(t) - L_o[Y(t)]|^2] \\ &= E[|X(t) - L[Y(t)] + \delta L[Y(t)]|^2].\end{aligned}$$

By orthogonality we have,

$$(X(t) - L[Y(t)]) \perp \delta L[Y(t)]$$

since δL is a linear operator on the data, so

$$\varepsilon_{\min} = E[|X(t) - L[Y(t)]|^2] + E[|\delta L[Y(t)]|^2].$$

Since $\varepsilon_{\min}$ is minimal and $E[|\delta L[Y(t)]|^2] \geq 0$, it must be that $\delta L = 0$, that is, $L = L_o$. ■

The preceeding result is not quite rigorous because we are tacitly assuming the existence of the operator L, which satisfies Equation 10.6–1. If the operator is continuous, then it can be represented as

$$L[Y(t)] = \int_I h(t, s) Y(s)\, ds,$$

where the kernel $h(t, s)$ is the response at t to an impulse at s. Then we can use the theory of m.-s. integrals developed in Section 8.2 to facilitate the solution of Equation 10.6–1 by allowing the interchange of E and L to obtain the integral equation

$$E[X(t) Y^*(\tau)] = \int_I h(t, s) E[Y(s) Y^*(\tau)]\, ds \qquad \text{for all } t, \tau \in I.$$

If this integral equation has a solution $h(t, s)$ that permits the existence of the m.-s. integral $L[Y(t)] = \int_I h(t, s) Y(s)\, ds$, then we are done. Otherwise, we would have to look further for a proper interpretation of the operation $L[Y(t)]$. One such condition on the kernal $h(t, s)$ was obtained in Equation 8.2–6. Fortunately, as we will see, this operator will exist in many practical situations.

One should note at this point that in the case where the random processes X and S are jointly Gaussian, then the above estimate is overall optimal, that is, MMSE. Thus in this case the operator L gives the conditional mean, the optimal estimator unconstrained by any linearity conditions:

$$L[Y(t)] = E[X(t) \mid Y(\tau), \tau \in I].$$

Wiener–Kolmogorov Theory. Assume that we observe the random process $Y(t)$ over the interval $I = [a, b]$ and wish to estimate the related process $X(t)$. We assume that both processes are zero-mean. We wish to use the linear estimate

$$\hat{X}(t) = \lim_{N\to\infty} \sum_{i=1}^{N} h(t, s_i)Y(s_i)\,\Delta s_i, \qquad \text{with } s_N \equiv b, \ldots, s_1 \equiv a$$

$$= \int_a^b h(t, s)Y(s)\,ds, \qquad \text{(m.s.)} \tag{10.6–2}$$

with error

$$E\left[\left|X(t) - \int_a^b h(t, s)Y(s)\,ds\right|^2\right].$$

We need to determine the kernel $h(t, s)$. Using the orthogonality principle (Equation 10.6–1), we know that the LMMSE estimate must satisfy

$$\left[X(t) - \int_a^b h(t, s)Y(s)\,ds\right] \perp Y(\tau)$$

for all $t, \tau \in I$. This is equivalent to

$$E[X(t)Y^*(\tau)] = \int_a^b E[Y(s)h(t, s)Y^*(\tau)]\,ds$$

if the indicated m.-s. integral exists. Writing this in terms of the relevant covariance functions we have

$$K_{XY}(t, \tau) = \int_a^b h(t, s)K_{YY}(s, \tau)\,ds \tag{10.6–3a}$$

for $t, \tau \in I$, and also

$$\varepsilon_{\min}(t) = E\left[X(t)\left(X(t) - \int_a^b h(t, s)Y(s)\,ds\right)^*\right],$$

which becomes, in terms of covariance functions,

$$\varepsilon_{\min}(t) = K_{XX}(t, t) - \int_a^b h^*(t, s)K_{XY}(t, s)\,ds. \tag{10.6–3b}$$

In the WSS case using the one-parameter covariance functions, we have

$$K_{XY}(t - \tau) = \int_a^b h(t, s)K_{YY}(s - \tau)\,ds, \tag{10.6–4a}$$

and

$$\varepsilon_{\min}(t) = K_{XX}(0) - \int_a^b h^*(t, s)K_{XY}(t - s)\,ds. \tag{10.6–4b}$$

Example 10.6–1: (Predicting the future of a random process.) Consider a zero-mean, WSS random process $S(t)$ with covariance K_S observed over the interval $I \triangleq (t - T, t)$ with $T > 0$. We want to find the best linear estimate of $S(t + \lambda)$ where $\lambda > 0$. We have

$$\hat{S}(t + \lambda) = \int_{t-T}^{t} h(t, \xi)S(\xi)\,d\xi.$$

Setting $X(t) \triangleq S(t + \lambda)$ and $Y(t) \triangleq S(t)$ in Equation 10.6–4a we obtain,

$$K_S(t + \lambda - \tau) = \int_{t-T}^{t} h(t, s)K_S(s - \tau)\,ds; \qquad t - T \le \tau \le t.$$

Setting $\alpha = t - \tau$ and $\beta = t - s$ in this equation,

$$K_S(\lambda + \alpha) = \int_0^T h(t, t - \beta)K_S(\alpha - \beta)\,d\beta; \qquad 0 \le \alpha \le T,$$

we see that we can take $h(t, s) \triangleq h(t - s) = h(\beta)$ since t does not appear anywhere else in the equation. Thus we obtain

$$K_S(\lambda + \alpha) = \int_0^T K_S(\alpha - \beta)h(\beta)\,d\beta \tag{10.6–5}$$

for $\alpha \in [0, T]$.

The solution to this integral equation yields the LMMSE predictor h. While the integral in Equation 10.6–5 looks like a convolution, the finite limits on this integral in fact preclude an elementary transform-domain solution. One method to solve Equation 10.6–5 would be to discretize it and solve the resulting equations on a computer. Establishing N-point grids in α and β with $\alpha_1 = \beta_1 = 0$ and $\alpha_N = \beta_N = T$, we have

$$K_S(\lambda + \alpha_i) \cong \sum_{j=1}^{N} K_S(\alpha_i - \beta_j)h(\beta_j)\,\Delta\beta_j,$$

which can be put into matrix-vector form,

$$\begin{bmatrix} \vdots \\ K_S(\lambda + \alpha_i) \\ \vdots \end{bmatrix} = \begin{bmatrix} \cdots \\ K_S(\alpha_i - \beta_j)\,\Delta\beta_j \\ \cdots \end{bmatrix} \begin{bmatrix} \vdots \\ h(\beta_j) \\ \vdots \end{bmatrix}.$$

We recognize this equation as a special case of that encountered earlier when estimating a scalar random variable from a random vector using a linear estimator. The solution to this matrix equation then yields an approximate solution to the integral Equation 10.6–5.

A second approach to solving Equation 10.6–5 involves the Karhunen–Loève (K–L) expansion of Section 8.5. Using Mercer's Theorem (Equation 8.5–6) we can write

$$K_S(\alpha - \beta) = \sum_{n=1}^{\infty} \sigma_n^2 \phi_n(\alpha)\phi_n^*(\beta), \qquad 0 \le \alpha, \beta \le T,$$

where σ_n and ϕ_n are respectively the K–L coefficients and eigenfunctions of the covariance K_S.

Inserting this equation into the integral in Equation 10.6–5, we have

$$\begin{aligned} K_S(\lambda + \alpha) &= \sum_{n=1}^{\infty} \sigma_n^2 \phi_n(\alpha) \int_0^T h(\beta)\phi_n^*(\beta)\,d\beta, \\ &= \sum_{n=1}^{\infty} \sigma_n^2 h_n \phi_n(\alpha), \end{aligned}$$

where the h_n are the coefficients in the expansion of $h(t)$ using the orthonormal functions $\{\phi_n(t)\}$. Thus we have that $\sigma_n^2 h_n$ are the coefficients in an orthonormal expansion of $K_S(t + \lambda)$ as a function of t for fixed λ. Since these coefficients may depend on λ, we write them as $k_n(\lambda)$. We thus have

$$k_n(\lambda) = \sigma_n^2 h_n(\lambda),$$

where

$$k_n(\lambda) \triangleq \int_0^T K_S(\lambda + \alpha)\phi_n^*(\alpha)\, d\alpha.$$

Then $h(t)$ is given by the infinite sum,

$$h(t) = \sum_{n=1}^{\infty} h_n(\lambda)\phi_n(t)$$

$$= \sum_{n=1}^{\infty} \frac{k_n(\lambda)}{\sigma_n^2}\phi_n(t). \tag{10.6–6}$$

This sum will exist if the infinite series

$$\sum_{n=1}^{\infty} \frac{|k_n(\lambda)|}{\sigma_n^2}$$

converges to a finite number.

This solution may seem quite abstract and unintuitive, but viewing the prediction operator in K–L coefficient space, it seems quite reasonable. Inserting Equation 10.6–6 into the equation for the optimal prediction,

$$\hat{S}(t + \lambda) = \int_{t-T}^{t} h(t - \xi)S(\xi)\, d\xi,$$

we get

$$\hat{S}(t + \lambda) = \int_{t-T}^{t} \left(\sum_n h_n(\lambda)\phi_n(t - \xi)\right)S(\xi)\, d\xi,$$

$$= \sum_{n=1}^{\infty} h_n(\lambda)\left(\int_0^T S(t - \beta)\phi_n(\beta)\, d\beta\right),$$

with $\beta = t - \xi$. This equation takes each K–L component of S over the interval $[t - T, t]$ and then multiplies this random variable by $h_n(\lambda)$, the ratio of the variances of the components of $S(t + \lambda)$ and $S(t)$. This seems very reasonable because K–L components with different n are orthogonal. Note that in the limit as λ tends to zero the ratio $(k_n(\lambda))/\sigma_n^2$ tends to one, so that the prediction tends to $S(t)$ itself. This, in fact, means that $h(t)$ tends to an impulse and that formally at least we would then have

$$\delta(t) = \sum_{n=1}^{\infty} \phi_n(t).$$

Just as we did in the discrete-time case, we could go on to derive the predictor when the data is contaminated by noise. However, we shall choose the related problem of estimating the noise-free signal itself based on the whole observation interval $[a, b]$. This is thus a smoothing problem, that is, *subproblem* 3 under the general linear estimation problem statement at the beginning of this section.

Example 10.6–2: (Estimating a nonstationary signal in white noise.) Let the observations be given by $Y(t) = S(t) + W(t)$ for $t \in [a, b]$ with $S \perp W$ and the desired signal given as $X(t) = S(t)$ with $\mu_S(t) = 0$. The observation noise $W(t)$ is white with $\mu_W(t) = 0$ and $K_W(t, u) = \sigma_W^2 \delta(t - u)$ and $K_S(t, u)$ is a given covariance function.

The method of solution will be to use the K–L expansion of S in solving Equation 10.6–3a specialized to the smoothing case. In this regard we have

$$K_{XY}(t, u) = K_S(t, u)$$

and

$$\begin{aligned} K_{YY}(t, u) &= K_S(t, u) + K_W(t, u), \\ &= K_S(t, u) + \sigma_W^2 \delta(t - u). \end{aligned}$$

Thus Equation 10.6–3a becomes

$$K_S(t, u) = \int_a^b h(t, v) K_Y(v, u)\, dv, \qquad a \le t, u \le b \tag{10.6–7}$$

and

$$\hat{S}(t) = \int_a^b h(t, v) Y(v)\, dv, \qquad a \le t \le b.$$

Using Mercer's Theorem (Equation 8.5–6), we have

$$K_Y(t, u) = \sum_{n=1}^{\infty} (\lambda_n + \sigma_W^2)\phi_n(t)\phi_n^*(u),$$

since $Y = S + W$. Inserting this in Equation 10.6–7, we get

$$\begin{aligned} \sum_{n=1}^{\infty} \lambda_n \phi_n(t)\phi_n^*(u) &= \int_a^b h(t, v)\left[\sum_{n=1}^{\infty} (\lambda_n + \sigma_W^2)\phi_n(v)\phi_n^*(u)\right] dv \\ &= \sum_{n=1}^{\infty} (\lambda_n + \sigma_W^2)\phi_n^*(u) \int_a^b h(t, v)\phi_n(v)\, dv. \end{aligned}$$

As earlier, we try a solution of the form

$$h(t, v) = \sum_{n=1}^{\infty} h_n \phi_n(t)\phi_n^*(v).$$

Inserting this into the previous equation, we obtain

$$\sum_n \lambda_n \phi_n(t)\phi_n^*(u) = \sum_n (\lambda_n + \sigma_W^2) h_n \phi_n(t)\phi_n^*(u),$$

which is an equality for

$$h_n = \frac{\lambda_n}{\lambda_n + \sigma_W^2}, \qquad n \ge 1.$$

Solving for $\hat{S}(t)$, we get finally

$$\begin{aligned} \hat{S}(t) &= \int_a^b h(t, v) Y(v)\, dv \\ &= \int_a^b \left[\sum_n h_n \phi_n(t)\phi_n^*(v)\right] Y(v)\, dv \\ &= \sum_{n=1}^{\infty} \left(\frac{\lambda_n}{\lambda_n + \sigma_W^2}\right) Y_n \phi_n(t), \end{aligned}$$

which has the system diagram shown in Figure 10.6–1.

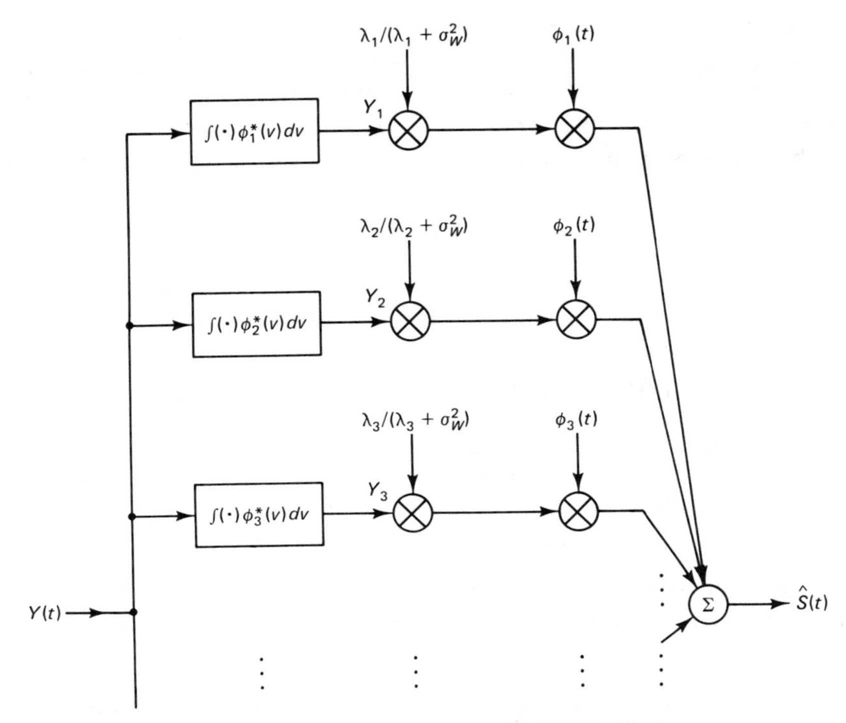

Figure 10.6–1 System diagram for LMMSE estimate.

The equation for the optimal MSE (Equation 10.6–3b) becomes

$$
\begin{aligned}
\varepsilon_{\min}(t) &= K_S(t,t) - \int_a^b h^*(t,s)K_S(t,s)\,ds \\
&= \sum_n \lambda_n \phi_n(t)\phi_n^*(t) - \int_a^b \Big[\sum_n h_n^*\phi_n^*(t)\phi_n(s)\Big] K_S(t,s)\,ds \\
&= \sum_n \lambda_n \phi_n(t)\phi_n^*(t) - \sum_n h_n^*\phi_n^*(t)\Big[\int_a^b K_S(t,s)\phi_n(s)\,ds\Big] \\
&= \sum_n \lambda_n \phi_n(t)\phi_n^*(t) - \sum_n h_n^*\phi_n^*(t)\lambda_n\phi_n(t)
\end{aligned}
$$

by Mercer's Theorem. Combining and simplifying, we get

$$
\begin{aligned}
\varepsilon_{\min}(t) &= \sum_{n=1}^{\infty} \lambda_n(1 - h_n^*)\phi_n(t)\phi_n^*(t) \\
&= \sum_{n=1}^{\infty} \left(\frac{\lambda_n\sigma_W^2}{\lambda_n + \sigma_W^2}\right)\phi_n(t)\phi_n^*(t).
\end{aligned}
$$

While in theory this approach requires an infinite number of parallel sections in the system shown in Figure 10.6–1, a good approximation can often be obtained by implementing only the sections where λ_n is above a selected threshold.

Infinite Time Interval - Stationary Processes. In the finite time-interval case, we have found solutions to the LMMSE estimation problem through the K–L expansion. Now we turn to the infinite time-interval case for WSS processes where we find Fourier transform-based solutions.

Substituting infinite limits into Equations 10.6–2 through 10.6–4, we obtain

$$\hat{X}(t) = \int_{-\infty}^{+\infty} h(t, v) Y(v)\, dv, \tag{10.6–8}$$

$$K_{XY}(t-u) = \int_{-\infty}^{+\infty} h(t, v) K_{YY}(u-v)\, dv, \tag{10.6–9}$$

$$\varepsilon_{\min}(t) = K_{XX}(0) - \int_{-\infty}^{+\infty} h^*(t, v) K_{XY}(t-v)\, dv. \tag{10.6–10}$$

In Equation 10.6–9 we suspect that $h(t, v) = h(t - v)$. Making the substitutions $\tau = t - u$ and $\sigma = t - v$, we obtain

$$K_{XY}(\tau) = \int_{-\infty}^{+\infty} h(\sigma) K_{YY}(\tau - \sigma)\, d\sigma.$$

Thus the optimal filter is linear shift-invariant (LSI). The above equation is just a linear convolution, so taking the Fourier transform we obtain

$$S_{XY}(\omega) = H_u(\omega) S_{YY}(\omega),$$

where the subscript u has been introduced to indicate an unrealizable (that is, uncausal) estimate. The solution, at frequencies for which $S_{YY}(\omega) \neq 0$, is

$$H_u(\omega) = \frac{S_{XY}(\omega)}{S_{YY}(\omega)}. \tag{10.6–11}$$

At other frequencies one may set $H_u = 0$. This solution is called the *unrealizable filter* or *smoother*. The estimate 10.6–8 is

$$\begin{aligned} \hat{X}(t) &= \int_{-\infty}^{+\infty} h_u(t - v) X(v)\, dv \\ &= h_u(t) * X(t). \end{aligned}$$

The MSE (Equation 10.6–10) becomes $\varepsilon_{\min}(t) = \varepsilon_{\min}(0) \triangleq \varepsilon_{\min}$, where

$$\begin{aligned} \varepsilon_{\min} &= K_{XX}(0) - \int_{-\infty}^{+\infty} h_u^*(\tau) K_{XY}(\tau)\, d\tau \\ &= \frac{1}{2\pi}\int_{-\infty}^{+\infty} S_{XX}(\omega)\, d\omega - \frac{1}{2\pi}\int_{-\infty}^{+\infty} H_u^*(\omega) S_{XY}(\omega)\, d\omega, \\ &= \frac{1}{2\pi}\int_{-\infty}^{+\infty} \left[S_{XX}(\omega) - \frac{|S_{XY}(\omega)|^2}{S_{YY}(\omega)} \right] d\omega \end{aligned}$$

by Parseval's Theorem and Equation 10.6–11.

We next present some examples of applications to the various estimation subproblems.

Example 10.6–3: (LMMSE smoother for WSS data.)

$$X(t) = S(t)$$
$$Y(t) = S(t) + N(t) \qquad \text{with } S \perp N$$

then

$$H_u(\omega) = S_S(\omega)/[S_S(\omega) + S_N(\omega)]. \tag{10.6–12}$$

The error expression becomes

$$\begin{aligned}\varepsilon_{\min} &= \frac{1}{2\pi}\int_{-\infty}^{+\infty}\left[S_S(\omega) - \frac{S_S^2(\omega)}{S_S(\omega) + S_N(\omega)}\right]d\omega \\ &= \frac{1}{2\pi}\int_{-\infty}^{+\infty}\frac{S_S(\omega)S_N(\omega)}{S_S(\omega) + S_N(\omega)}\,d\omega.\end{aligned}$$

Example 10.6–4: (Noisy predictor for WSS data.) Let $X(t) = S(t + \lambda)$ with $\lambda > 0$ and take $Y(t) = S(t) + N(t)$ with $S \perp N$. Then

$$\begin{aligned}H_u(\omega) &= S_{XY}(\omega)/S_{YY}(\omega) \\ &= e^{+j\omega\lambda}S_S(\omega)/[S_S(\omega) + S_N(\omega)],\end{aligned}$$

which has the system diagram shown in Figure 10.6–2.

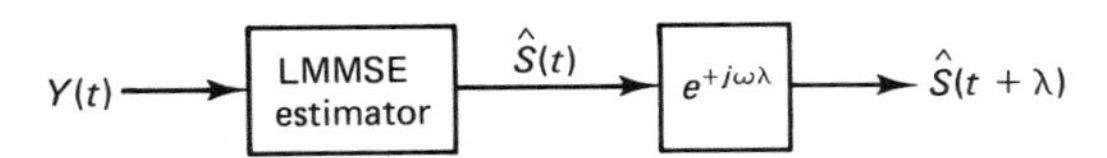

Figure 10.6–2 Optimal linear prediction (uncausal).

Example 10.6–5: (LMMSE estimate of derivative.) Let $X(t) = S'(t)$ and the observations be $Y = S + N$ with $S \perp N$. Then

$$K_{XY}(\tau) = K_{S'S}(\tau) = \frac{d}{d\tau}K_S(\tau) \leftrightarrow j\omega S_S(\omega)$$

so

$$H_u(\omega) = j\omega\frac{S_S(\omega)}{S_S(\omega) + S_N(\omega)}.$$

The structure of the estimator is shown in Figure 10.6–3.

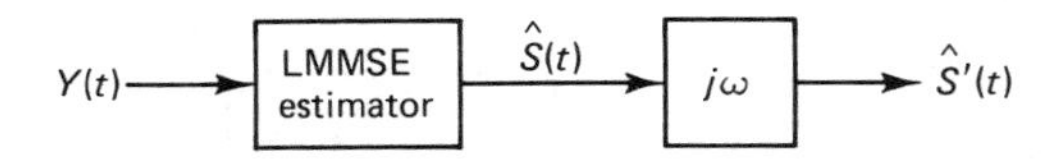

Figure 10.6–3 Optimal LMMSE estimation of derivative.

Causal (Realizable) Wiener Filters. Now we turn to the case of LMMSE optimal filtering with the practical constraint that the impulse response of the optimal filter be one-sided. We simplify the notation by treating only the *real valued* case and also assume that all the random processes considered are at least WSS and possibly stationary. The observation interval will be $I = (-\infty, t]$. The Wiener-Hopf equation for the optimal filter (Equation

10.6–4a) then becomes

$$K_{XY}(\tau) = \int_0^\infty h_c(\tau) K_{YY}(\tau - \sigma)\, d\sigma; \qquad \tau \geq 0, \qquad (10.6\text{–}13)$$

where we have incorporated the facts that the optimal filter must be LSI and that $h_c(\tau) = 0$ for $\tau < 0$ due to causality. Equation 10.6–13 does not have a simple Fourier transform solution because the equality in this equation only holds for $\tau \geq 0$. Thus the solution in the general case is not obvious by classical methods. However, there are special cases where the solution is straightforward.

Example 10.6–6: (White-noise observations.) If we want to estimate the WSS random process $X(t)$ given observations of a white random process $Y(t)$, then in this case the solution of Equation 10.6–13 is easy. Assume

$$K_{YY}(\tau) = \sigma^2\, \delta(\tau),$$

then substituting in Equation 10.6–13, we immediately obtain

$$h_c(t) = \frac{1}{\sigma^2} K_{XY}(t) u(t). \qquad (10.6\text{–}14)$$

The filtering error is then given by

$$\varepsilon_{\min} = K_{XX}(0) - \frac{1}{\sigma^2} \int_0^\infty K_{XY}^2(\tau)\, d\tau,$$

which can be compared to the random sequence counterpart (Equation 10.5–8).

We have seen that the solution is simple if the observation process $Y(t)$ is a white noise. If we do not have a white observation process, we can filter the process with a causal LSI filter called a *whitening filter* to transform the observations into a white noise process. If this whitening filter has a causal inverse† then the realizable Wiener filter estimate based on the white process will still be LMMSE optimal. This must be so since the Wiener filter could undo the whitening transformation by performing the causal inverse. This strategy to solve Equation 10.6–13 in the general case of non-white observations is illustrated in Figure 10.6–4. The existence of such a whitening filter will be discussed below.

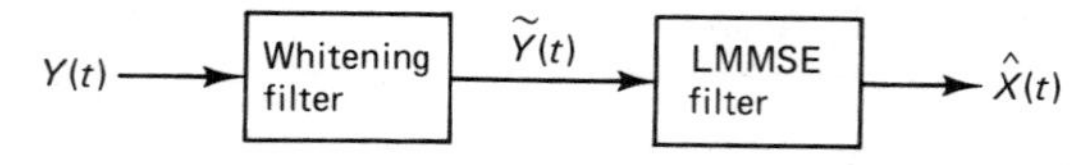

Figure 10.6–4 Wiener filter for general observation process Y.

As indicated in Figure 10.6–4, we call the process at the whitening filter's output $\tilde{Y}(t)$. It is a continuous-time *innovations process* corresponding to the observations $Y(t)$. This is completely analogous to the discrete-time case for

† A causal inverse is an inverse filter which is also causal.

random sequences studied in Sections 10.3 and 10.4. Thus once again we see that the linear, causal estimation problem becomes much easier once we have an innovations representation. The whitening filter is obtained, and thus its existence is demonstrated, by *spectral factorization*, which is discussed next.

Spectral Factorization. The concept of spectral factorization is that of factoring a given power spectral density (psd) into two factors. The first factor is to be causal with a causal inverse. The second factor, necessarily the conjugate of the first factor, is therefore anticausal† with an anticausal inverse. In the application to Wiener filtering, we want to factor the psd of the observation process Y. Then the inverse of the resulting causal factor will become the desired whitening filter.

We assume the psd S_Y is given and is rational and positive for all ω. We can write

$$S_Y(\omega) = P(\omega)/Q(\omega),$$

where P and Q are polynomials with real coefficients. Since we assume the random processes are real valued in this subsection, it follows that the psd S_Y is even in ω. Thus we can rewrite the polynomials P and Q as functions of $-\omega^2$,

$$A(-\omega^2) \triangleq P(\omega) \quad \text{and} \quad B(-\omega^2) \triangleq Q(\omega),$$

where the minus sign is just for convenience. Next we define the functions of a *complex variable*:

$$P(j\omega) \triangleq P(\omega), \qquad Q(j\omega) \triangleq Q(\omega)$$
$$A((j\omega)^2) \triangleq A(-\omega^2), \qquad B((j\omega)^2) \triangleq B(-\omega^2).$$

Then inserting $s = j\omega$, we obtain

$$P(s) = A(s^2), \qquad Q(s) = B(s^2)$$

which yields the psd as a function of the complex variable s,

$$S_Y(s) = \frac{A(s^2)}{B(s^2)},$$

which must have the four-fold symmetry shown in the pole-zero diagram of Figure 10.6–5. This symmetry arises since a pole or zero at $s = s_1$ must also occur at $s = -s_1$, this in addition to the conjugate symmetry, which arises from the assumption that A and B have real coefficients.

Because of this symmetry, we can factor A and B as

$$A(s^2) = C(s)C(-s),$$
$$B(s^2) = D(s)D(-s),$$

† An anticausal impulse response is one that satisfies $h(\tau) = 0$ for $\tau > 0$.

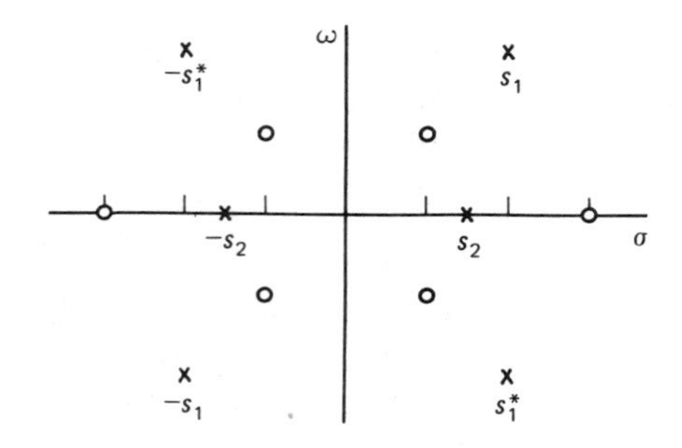

Figure 10.6–5 A symmetrical pole-zero plot of $S_Y(s)$.

where $C(s)$ and $D(s)$ have zeros only in the left half-plane. On defining

$$F^+(s) \triangleq \frac{C(s)}{D(s)},$$

we have the following spectral factorization:

$$S_Y(s) = \sigma^2 F^+(s) F^-(s). \qquad (\sigma^2 = 1).$$

The required whitening filter is then just

$$W(s) = [F^+(s)]^{-1},$$

which is causal with a causal inverse because of the location of the zeros of C and D in the left half-plane.

Example 10.6–7: Let

$$S_Y(\omega) = \frac{2k}{\omega^2 + k^2}, \qquad k > 0.$$

Spectrally factoring, we obtain

$$S_Y(\omega) = \left(\frac{\sqrt{2k}}{j\omega + k}\right)\left(\frac{\sqrt{2k}}{-j\omega + k}\right)$$

Thus

$$F^+(s) = \frac{\sqrt{2k}}{s + k}.$$

Complete Solution for Causal Wiener Filter. We must find $K_{X\tilde{Y}}$ or equivalently $S_{X\tilde{Y}}$, which is

$$S_{X\tilde{Y}}(\omega) = S_{XY}(\omega) W^*(\omega). \quad \dagger$$

The corresponding causal impulse response is then, by Equation 10.6–14,

$$g(t) = \frac{1}{\sigma^2}[K_{XY}(t) * w(-t)]u(t),$$

where $w = \text{IFT}\{W\}$. Note that $g(t)$ is nonzero even though $w(-t)$ is anticausal because K_{XY} is a two-sided function.

† It should be understood that to obtain $W(\omega)$ from $W(s)$ we replace s by $j\omega$ and then write the function of a real variable as $W(\omega) = W(j\omega)$.

The overall impulse response is then the convolution

$$h(t) = w(t) * \frac{1}{\sigma^2}[K_{XY}(t) * w(-t)]u(t),$$

We can write the overall solution in the frequency-domain by introducing the notation

$$[F(\omega)]_+ \triangleq \int_0^\infty f(t)e^{-j\omega t}\,dt,$$

and writing

$$H_c(\omega) = W(\omega)\frac{1}{\sigma^2}[S_{XY}(\omega)W^*(\omega)]_+$$

so that the system function becomes

$$H_c(\omega) = \frac{1}{\sigma^2 F^+(\omega)}\left[\frac{S_{XY}(\omega)}{F^-(\omega)}\right]_+, \tag{10.6–15}$$

which is the general form of the causal Wiener filter.

The general procedure is summarized as follows:

1. Factor the input psd into F^+ and F^-.
2. Form S_{XY}/F^-.
3. Find the realizable part $[S_{XY}/F^-]_+$.
4. Set H_c as in Equation 10.6–15.

Depending on the relationship between X and Y, various solutions can be obtained. As an example we will illustrate the causal estimation of a lowpass signal process in white noise.

Example 10.6–8: (H. L. Van Trees) [10–9] Let the observations be given by

$$Y(t) = \sqrt{p}X(t) + W(t),$$

where the signal psd is

$$S_X(\omega) = \frac{2k}{\omega^2 + k^2}, \qquad k \geq 0,$$

and the noise psd is flat

$$S_W(\omega) = \sigma_W^2,$$

with $X \perp W$. The variable p is the average power in the "transmitted signal" $\sqrt{p}\,X(t)$. We wish to estimate the signal $X(t)$ using a causal Wiener filter. We first evaluate the psd of Y,

$$S_Y(\omega) = \sigma_W^2\frac{\omega^2 + k^2(1+\alpha)}{\omega^2 + k^2},$$

where $\alpha \triangleq (2p/k\sigma_W^2)$. Factoring this psd, we obtain

$$\begin{aligned} S_Y(\omega) &= \sigma_W^2\frac{j\omega + k\sqrt{1+\alpha}}{j\omega + k}\cdot\frac{-j\omega + k\sqrt{1+\alpha}}{-j\omega + k} \\ &= \sigma^2F^+(\omega)\cdot F^-(\omega). \end{aligned}$$

Next we evaluate S_{XY},

$$S_{XY}(\omega) = \sqrt{p}S_{XX}(\omega)$$
$$= \frac{2k\sqrt{p}}{\omega^2 + k^2}.$$

Thus

$$S_{X\tilde{Y}}(\omega) = S_{XY}(\omega)/F^-(\omega)$$
$$= \frac{2k\sqrt{p}}{(j\omega + k)(-j\omega + k\sqrt{1+\alpha})}.$$

We can evaluate the "causal part" of this by making use of the partial fraction expansion,

$$S_{X\tilde{Y}}(\omega) = \frac{c_1}{j\omega + k} + \frac{c_2}{-j\omega + k\sqrt{1+\alpha}}.$$

Thus

$$[S_{X\tilde{Y}}(\omega)]_+ = \frac{c_1}{j\omega + k} = G(\omega),$$

where

$$c_1 = \frac{2\sqrt{p}}{1 + \sqrt{1+\alpha}},$$

as determined by the above partial fraction expansion. The overall system function thus is

$$H_c(\omega) = \frac{2\sqrt{p}}{\sigma_W^2(1 + \sqrt{1+\alpha})} \cdot \frac{1}{j\omega + k\sqrt{1+\alpha}},$$

which is seen to be a first-order lowpass filter with pole at $k\sqrt{1+\alpha}$.

10.7 SUMMARY

This chapter has extended the estimation theory of Chapter 5 for random variables and vectors to the estimation of random sequences and processes. We started by reviewing the fundamental role of the conditional mean. Because of the nonlinearity of the resulting estimates in the general non-Gaussian case, we turned to linear estimation and introduced the orthogonality condition as the necessary and sufficient condition for this constrained optimality. Innovations sequences were then introduced as a data transformation that orthogonalizes the data, thus simplifying linear estimation (cf. to Theorem 10.2–2(b)). The Kalman filter was derived as the optimal MSE estimate for a Gauss-Markov signal in white noise. Since this linear estimator uses only second-order properties, the Kalman filter is also the optimal linear estimator for wide-sense Markov signals. The final two sections were devoted to Wiener filters for random sequences and random processes respectively. The Wiener filters, which predate the Kalman filters by more than ten years, are not restricted to a wide-sense Markov signal model, but do have the restriction to WSS signal and observation models. Also, the Wiener approach leads to closed-form solutions for some common signal models, whereas the Kalman approach generally does not.

PROBLEMS

10.1. Use the orthogonality principle to show that the MMSE

$$\varepsilon \triangleq E[(X - E[X \mid Y])^2],$$

for real-valued r.v.'s can be expressed as

$$\varepsilon = E[X(X - E[X \mid Y])]$$

or as

$$\varepsilon = E[X^2] - E[E[X \mid Y]^2].$$

Generalize to the case where **X** and **Y** are real-valued random vectors, that is, show that the MMSE matrix is

$$\begin{aligned} \boldsymbol{\varepsilon} &= E[(\mathbf{X} - E[\mathbf{X} \mid \mathbf{Y}])(X - E[\mathbf{X} \mid \mathbf{Y}])^T] \\ &= E[\mathbf{X}(\mathbf{X} - E[\mathbf{X} \mid \mathbf{Y}])^T] \\ &= E[\mathbf{X}\mathbf{X}^T] - E[E[\mathbf{X} \mid \mathbf{Y}]E^T[\mathbf{X} \mid \mathbf{Y}]]. \end{aligned}$$

10.2. Conclude that the limit in Equation 10.1–3 exists with probability-1 by invoking the Martingale convergence theorem 6.5–4 applied to the random sequence $G[N]$ with parameter N defined in Equation 10.1–4. Specifically show that $\sigma_G^2[N]$ remains uniformly bounded as $N \to \infty$.

10.3. (Larson & Shubert [10–10]). A Gaussian random sequence $X[n]$, $n = 0, 1, 2, \ldots$ is defined by the equation

$$X[n] = -\sum_{k=1}^{n} \binom{k+2}{2} X[n-k] + W[n] \qquad n = 1, 2, \ldots,$$

where $X[0] = W[0]$, and $W[n]$, $n = 0, 1, 2, \ldots$ is a Gaussian white noise sequence with zero mean and variance of unity.

(a) Show that $W[n]$ is the innovations sequence for $X[n]$.

(b) Show that $X[n] = W[n] - 3W[n-1] + 3W[n-2] - W[n-3]$, for $n = 0, 1, 2, \ldots$ where $W[-3] = W[-2] = W[-1] = 0$.

(c) Use the preceding result to obtain the best two-step predictor of $X[12]$ as a linear combination of $X[0], \ldots, X[10]$. Also calculate the resulting mean-square prediction error.

10.4. Let $W[n]$ be a sequence of independent, identically distributed Gaussian random variables with zero mean and unit variance. Define

$$X[n] \triangleq W[1] + \ldots + W[n] \qquad n = 1, 2, \ldots.$$

(a) Find the innovations sequence for $X[n]$.

(b) Let there be noisy observations available:

$$Y[n] = X[n] + V[n], \qquad n = 1, 2, \ldots,$$

where $V[n]$ is also a white Gaussian random sequence with variance σ_V^2, and $\{V[n]\}$ is orthogonal to $\{W[n]\}$.

Find the recursive filtering structure for computing the MMSE estimate

$$\hat{X}[n \mid n] \triangleq E[X[n] \mid Y[1], \ldots, Y[n]].$$

(c) Find the recursive equations specifying any unknown constants in the filter of (b). Specify the initial conditions.

10.5. A random sequence $Y[n]$, $n = 0, 1, 2, \ldots$, satisfies a second-order linear-difference equation

$$2Y[n+2] + Y[n+1] + Y[n] = 2W[n], \qquad Y[0] = 0,\ Y[1] = 1,$$

with $W[n]$, $n = 0, 1, \ldots$, a standard white Gaussian random sequence (that is, $N(0, 1)$). Transform this equation into the state-space representation and evaluate the mean function $\boldsymbol{\mu}_{\mathbf{X}}[n]$ and the correlation function $\mathbf{R}_{\mathbf{X}}[n_1, n_2]$ at least for the first few values of n.
Hint: Define the state vector $\mathbf{X}[n] = (Y[n+2], Y[n+1])^T$.

10.6. In our derivation of the Kalman filter in Section 10.4, we assumed that the Gauss-Markov signal model (Equation 10.4–1) was zero-mean. Here we modify the Kalman filter to permit the general case of nonzero mean for $\mathbf{X}[n]$. Let the Gauss-Markov signal model be

$$\mathbf{X}[n] = \mathbf{A}\mathbf{X}[n-1] + \mathbf{B}\mathbf{W}[n], \qquad n \geq 0$$

where $\mathbf{X}[-1] = \mathbf{0}$ and the centered noise $\mathbf{W}_c[n] \triangleq \mathbf{W}[n] - \boldsymbol{\mu}_{\mathbf{W}}[n]$ is white Gaussian with variance $\boldsymbol{\sigma}_{\mathbf{W}}^2$, and $\boldsymbol{\mu}_{\mathbf{W}}[n] \neq 0$. The observation equation is still Equation 10.4–2 and $\mathbf{V} \perp \mathbf{W}_c$.

(a) Find an expression for $\boldsymbol{\mu}_{\mathbf{X}}[n]$ and $\boldsymbol{\mu}_{\mathbf{Y}}[n]$.
(b) Show that the MMSE estimate of $\mathbf{X}[n]$ equals the sum of $\boldsymbol{\mu}_{\mathbf{X}}[n]$ and the MMSE estimate of $\mathbf{X}_c[n] \triangleq \mathbf{X}[n] - \boldsymbol{\mu}_{\mathbf{X}}[n]$ based on the centred observations $\mathbf{Y}_c[n] \triangleq \mathbf{Y}[n] - \boldsymbol{\mu}_{\mathbf{Y}}[n]$.
(c) Extended the Kalman filtering Equation 10.4–11 to the nonzero mean case by using the result of (b).
(d) How do the gain and error-covariance equations change?

10.7. (Larson & Shubert [10–10]). Suppose that the observation equation of the Kalman predictor is generalized to

$$\mathbf{Y}[n] = \mathbf{C}_n\mathbf{X}[n] + \mathbf{V}[n],$$

where the $\mathbf{C}_n$, $n = 0, 1, 2, \ldots$ are $(M \times N)$ matrices, $\mathbf{X}[n]$ is a $(N \times 1)$ random vector sequence, and $\mathbf{Y}[n]$ is a $(M \times 1)$ random vector sequence. Let the time varying signal model be given as

$$\mathbf{X}[n] = \mathbf{A}_n\mathbf{X}[n-1] + \mathbf{B}_n\mathbf{W}[n].$$

Repeat the derivation of the Kalman predictor to show that the prediction estimate now becomes

$$\hat{\mathbf{X}}[n] = \mathbf{A}_n[(\mathbf{I} - \mathbf{G}_{n-1}\mathbf{C}_{n-1})\hat{\mathbf{X}}[n-1] + \mathbf{G}_{n-1}\mathbf{Y}[n-1]],$$

with Kalman gain

$$\mathbf{G}_n = \boldsymbol{\varepsilon}[n]\mathbf{C}_n^T[\mathbf{C}_n\boldsymbol{\varepsilon}[n]\mathbf{C}_n^T + \boldsymbol{\sigma}_{\mathbf{V}}^2[n]]^{-1}.$$

What happens to the equation for the prediction MSE matrix $\boldsymbol{\varepsilon}[n]$?

10.8. Here we show how to derive Equation 10.5–2 using property (b) of the $\hat{E}$ operator in Theorem 10.2–2.

(a) Use property (b) in Theorem 10.2–2 iteratively to conclude,

$$\hat{E}[X[n] \mid \tilde{Y}[-N], \ldots, \tilde{Y}[+N]] = \sum_{K=-N}^{+N} g[k]\tilde{Y}[k],$$

with $g[k]$ given by Equation 10.5–4.

(b) Use the result of Problem 10.2 to show that $\lim_{N\to\infty}\left(\sum_{k=-N}^{+N} g[k]\tilde{Y}[k]\right) \triangleq$

$\sum_{k=-\infty}^{+\infty} g[k]\tilde{Y}[k]$ exists with probability-1 when X and Y are jointly Gaussian.

10.9. We observe a noisy signal on $[0, T]$ of the form $Y(t) = S(t) + V(t)$ where the signal $S(t)$ and noise $V(t)$ are orthogonal, that is, $S \perp V$. The signal and noise correlation functions happen to have Karhunen–Loève (K–L) expansions with the *same finite* set of N orthonormal basis functions,

$$\{\phi_n(t)\}_{n=1}^{N}.$$

Letting

$$Y_n \triangleq \int_0^T Y(t)\phi_n^*(t)\,dt,$$

$$S_n \triangleq \int_0^T S(t)\phi_n^*(t)\,dt,$$

and

$$V_n \triangleq \int_0^T V(t)\phi_n^*(t)\,dt,$$

we have

$$Y_n = S_n + V_n \qquad n = 1, \ldots, N.$$

(a) Show that $E[S_n \mid Y_n] = [\lambda_n/(\lambda_n + \nu_n)]Y_n$ where $\lambda_n \triangleq E[|S_n|^2]$ and $\nu_n \triangleq E[|V_n|^2]$

(b) Set $\hat{S}_n \triangleq \hat{E}[S_n \mid Y_n]$ and show that the estimate

$$\hat{S}(t) = \sum_{n=1}^{N} \hat{S}_n\phi_n(t)$$

satisfies the orthogonality condition,

$$E\big[(S(t) - \hat{S}(t))Y^*(\tau)\big] = 0 \quad \text{for} \quad 0 \le t, \tau \le T,$$

that is, the estimation error at time t is orthogonal to the observed data at all times τ in the observation interval.

10.10. Solve part (b) of Problem 10.9 by using the linearity properties of the operator $\hat{E}$ that were established in Theorem 10.2–2.

10.11. Assume that an analog communications channel carries two message waveforms,

$$R(t) = M_1(t) + M_2(t),$$

which can be modeled as mutually orthogonal, WSS random processes with psd's $S_{M_1}(\omega)$ and $S_{M_2}(\omega)$. Show how to construct optimal linear estimates of both M_1 and M_2 using only one Wiener filter:
(a) Assume $E[M_1(t)] = E[M_2(t)] = 0$.
(b) Assume $E[M_1(t)] \neq 0$ and $E[M_2(t)] \neq 0$.
(Hint: Note the symmetry in the MSE expression at the end of Example 10.6–3.)

REFERENCES

10–1. K. S. Miller, *Complex Stochastic Processes.* Reading, Mass.; Addison-Wesley, 1974, pp. 76–80.

10–2. S. A. Tretter, *Introduction to Discrete-Time Signal Processing.* New York: John Wiley, 1976.

10–3. L. R. Rabiner and R. W. Schafer, *Digital Processing of Speech Signals.* Englewood Cliffs, N. J.: Prentice Hall, 1978.

10–4. R. E. Kalman, "A New Approach to Linear Filtering and Prediction Problems," *Journal of Basic Eng.*, Vol. 82, March 1960, pp. 35–45.

10–5. R. E. Kalman and R. S. Bucy, "New Results in Linear Filtering and Prediction Theory," *Journal of Basic Eng.*, Vol. 83, Dec. 1961, pp. 95–107.

10–6. A. N. Kolmogorov, "Interpolation and Extrapolation of Stationary Random Sequences," reprinted in *Linear Least-Squares Estimation* (ed., T. Kailath). Stroudsburg, Penn: Dowden, Hutchinson and Ross, 1977.

10–7. N. Wiener, *Extrapolation, Interpolation and Smoothing of Stationary Time Series.* Cambridge, Mass.: M.I.T. Press, 1964.

10–8. A. V. Oppenheim, A. S. Willsky, and I. T. Young, *Signals and Systems.* Englewood Cliffs, N. J.: Prentice Hall, 1983.

10–9. H. L. Van Trees, *Detection, Estimation and Modulation Theory: Part I.* New York: John Wiley, 1968, pp. 488–489.

10–10. H. J. Larson and B. O. Shubert, *Probabilistic Models in Engineering Sciences: Vol. II*, New York: John Wiley, 1979.

INDEX

D

E

J

K

L

M

Q

R

S